THEORY OF LEGAL SCIENCE

SYNTHESE LIBRARY

STUDIES IN EPISTEMOLOGY,

LOGIC, METHODOLOGY, AND PHILOSOPHY OF SCIENCE

VOLUME 176

THEORY OF LEGAL SCIENCE

Proceedings of the Conference on
Legal Theory and Philosophy of Science,
Lund, Sweden, December 11-14, 1983

Edited by

ALEKSANDER PECZENIK
Faculty of Law, Lund University, Sweden

LARS LINDAHL
Faculty of Law, Lund University, Sweden

and

BERT VAN ROERMUND
Faculty of Law, Tilburg University, The Netherlands

D. REIDEL PUBLISHING COMPANY

A MEMBER OF THE KLUWER ACADEMIC PUBLISHERS GROUP

DORDRECHT / BOSTON / LANCASTER

Library of Congress Cataloging in Publication Data

Conference on Legal Theory and Philosophy of Science
(1983: Lund, Sweden)

Theory of legal science.

(Synthese library; v. 176)
Includes index.
1. Jurisprudence–Congresses. 2. Law–Philosophy–Congresses.
I. Peczenik, Aleksander. II. Lindahl, Lars. III. Roermund, G. van
IV. Title.
K225.C656 1983 340'.1 84-16026
ISBN 90-277-1834-2

Published by D. Reidel Publishing Company,
P.O. Box 17, 3300 AA Dordrecht, Holland.

Sold and distributed in the U.S.A. and Canada
by Kluwer Academic Publishers,
190 Old Derby Street, Hingham, MA 02043, U.S.A.

In all other countries, sold and distributed
by Kluwer Academic Publishers Group,
P.O. Box 322, 3300 AH Dordrecht, Holland.

Printed in The Netherlands

TABLE OF CONTENTS

PREFACE ix
KJELL ÅKE MODÉER, Opening Speech xi

CHAPTER 1: THEORY OF SCIENCE AND THEORY OF LAW 1

BERT C. van ROERMUND, Synopsis 1
BENGT HANSSON, Recent Trends in the Philosophy of Science 5
ENRIQUE ZULETA PUCEIRO, Legal Dogmatics as a Scientific Paradigm 13
AULIS AARNIO, Paradigms in Legal Dogmatics Towards a Theory of Change and Progress in Legal Science 25
MATTI SINTONEN, Pragmatic Metatheory for Legal Science 39
OTTO KAMSTRA/FRANK KUNNEMAN, On Making Implicit Methodologies Explicit 53

CHAPTER 2: ONTOLOGY AND EPISTOMOLOGY IN LEGAL SCIENCE 67

BERT C. van ROERMUND, Synopsis 67
HECTOR-NERI CASTANEDA, Ought, Reasons, Motivation, and the Unity of the Social Sciences: The Meta-theory of the Ought-Is Problem 71
ALEKSANDER PECZENIK, Legal Data. An Essay about the Ontology of Law 97
STIG JØRGENSEN, Pluralis Juris 121
JAN M. BROEKMAN, Changes of Paradigm in the Law 133
BERT C. van ROERMUND, Legal Norms: a Transformational Approach 145
JACQUES LENOBLE, Epistemology and Validity in Law 163
CSABA VARGA, Is Law a System of Enactments? 175
W.H. BALEKJIAN, The Concept of "Fact" in the Physical Sciences and in Law 183

CHAPTER 3: OBJECTIVITY AND RATIONALITY OF LEGAL JUSTIFICATION 189

ALEKSANDER PECZENIK, Synopsis 189
GÖRAN HERMERÉN, Objectivity in the Social Sciences 195
OTA WEINBERGER, Objectivity and Rationality in Lawyer's Reasoning 217
NEIL MACCORMICK, Coherence in Legal Justification 235
JERZY WRÓBLEWSKI, Paradigms of Justifying Legal Decisions 253
SHELDON L. LEADER, Monism, Pluralism, Relativism and Right Answers in the Law 275
MARTIN P. GOLDING, Discovery and Justification in Science and Law 295
GUNNAR BERGHOLTZ, Reasons and Causes in Connection with Judicial Decisions 307

CHAPTER 4: TECHNICAL RATIONALITY IN THE LAW 315

ALEKSANDER PECZENIK, Synopsis 315
MAREK ZIRK-SADOWSKI, Legal Rationality Among Different Types of Rationality 319
AART HENDRIK DE WILD, Paradigms of Legal Research; Empirical Science and Legal Dogmatics 335
ROBERT S. SUMMERS, Goal Reasons in Common Law Cases - Are They Predictive? 345
PER OLOF EKELÖF, Teleological Construction of Statutes 357
ENRICO PATTARO, Reason, Law and History 363
MATS MATTSSON, The Rule of Law in Legal Reasoning 373

CHAPTER 5: SOME SPECIAL TOPICS CONCERNING RATIONALITY AND LEGITIMACY IN THE LAW 383

ALEKSANDER PECZENIK, Synopsis 383
ALEJO DE CERVERA, An Ubiquitous Paralogism in Legal Thinking 385
MATS FLODIN, Power of Tolerance - On the Legitimacy of a Legal System 401
ROLAND SCHÖTT, Sir Edward Coke's Legal Conservatism 413

YEMIMA BEN-MENAHEM/HANINA BEN-MENAHEM, Popper's Criterion of Refutability in the Legal Context 425

CHAPTER 6: CRITICISM AND DEVELOPMENTS IN PARTICULAR AREAS OF THE LAW: PROPERTY, CONTRACTS, AND TORTS 437

LARS LINDAHL, Synopsis 437
MARSHA P. HANEN, Theory Choice and Contract Law 441
JAN HELLNER, Trends in Legal Science Relating to Contracts and Torts 455
GÖRAN SKOGH, The Economics of Trade Laws 469

CHAPTER 7: INTERDISCIPLINARY BRIDGES BETWEEN LEGAL RESEARCH AND OTHER SCIENCES 483

ALEKSANDER PECZENIK, Synopsis 483
JØRGEN DALBERG-LARSEN, On Bridging the So-Called Gap Between Normative Legal Dogmatics and Empirical-Theoretical Social Science 487
FRANÇOIS OST/MICHEL van de KERCHOVE, Towards an Interdisciplinary Theory of Law 497
VITTORIO VILLA, Legal Science and Hermeneutic Point of View 509
HUBERT ROTTLEUTHNER, Legal Theory and Social Science 521
KAZIMIERZ OPAŁEK, Integration Between Legal Research and Social Science 531

CHAPTER 8: ANALYSIS OF LEGAL NORMS AND JURISTIC PROPOSITIONS 551

LARS LINDAHL, Synopsis 551
JAN EVERS, Karl Olivecrona's Theory of Legal Rules as Independent Imperatives 553
TORE STRÖMBERG, Norms of Competence in Scandinavian Jurisprudence 559
RICCARDO GUASTINI, A Tentative Analysis of Two Juristic Sentences 571

CHAPTER 9: LOGICAL AND PREFERENCE-THEORETICAL STRUCTURES IN THE LAW 583

LARS LINDAHL, Synopsis 583

ANTONIO A. MARTINO/CARLO BIAGIOLI/PAOLO MARIANI BIAGINI/GILBERTO DINI/FIORENZA SOCCI NATALI/DANIELA TISCORNIA, Automated Analysis of Legislation 587
B. GHITA E. HOLMSTRÖM, Rights and Practical Possibilities 607
LARS LINDAHL, Requirements, Urgency, and Worth 617
CARL MARTIN ROOS, The Property Right of Sweden Today - Or a Requiem over an Outdated Way of Argueing 639

LIST OF PARTICIPANTS 649

INDEX OF NAMES 653

PREFACE

This book presents papers read at the Conference on Legal Theory and Philosophy of Science, held at Lund, December 11-14, 1983. The main idea of the conference was to articulate, to test, and to apply scientific rationality with regard to the domain of law and legal reasoning. Although very different views were presented, the general impression is that this idea may have some future, as it yielded many surprising and stimulating discussions in Lund.

The editors express their gratitude to Torsten Söderbergs och Ragnar Söderbergs Stiftelser, Marianne och Marcus Wallenbergs Stiftelse, Craffoordska Stiftelsen, Swensonska Fonden, Einar Hansens Forskningsfond, Svenska Institutet and Lunds Universitet for subsidizing the conference. We also express gratitude to Lund University for help of various kinds, including free rooms, assistance with copying materials, and thousands of other necessary things. We thank all people whose work made the conference possible, first of all Mr.Roland Schött who during a long period acted as the secretary of the Organizing Committee, Mr.Mats Mattsson and Mrs.Irena Peczenik who helped with so many things necessary for the meetings, and many students and members of the staff of Lund University Law School.

Finally, we are grateful for support received from Tilburg University, in particular the Faculty of Law, in editing this volume. Several people advised in matters of linguistic accessibility: mrs.Anti Bax, mrs.André Tingloo (and members of the "Werkverband taalvaardigheid Engels, mr.Hans Boerakker, and mr.Stuart Silvers. They are, however, not to be held responsible for what remains of conference-pidgin.

Mr.Wim Voermans and mrs.Ria Boerakker corrected typoscripts: the latter also composed the index of names. Not just gratitude, but admiration is proper for the professional way in which the manuscript was typed camera-ready by mrs.Marianne Kok - who accepted a heavy responsibility - , assisted by mrs.Jacqueline de Meester.

Aleksander Peczenik
Lars Lindahl
Bert van Roermund

OPENING SPEECH

Kjell-Åke Modéer, The Vice-President of Lund University

On behalf of Lund University I have the honour to wish you all, participants in this international conference on Legal Theory and Philosophy of Science, very welcome to Lund and to our university.

Perhaps you know that you have now arrived at Ultima Thule as the Roman authors called this Nordic outpost of Europe. A pale December sun cast its long fading rays over you, members of this meeting, when you strolled through Lundagård, the small but charming university garden, now open to the public, between the white-frozen limestone cathedral and the main building of the university.

When the French philosopher René Descartes, latinized Cartesius, in the early winter of 1649 arrived in Stockholm at the court of the Swedish Queen Christina, he noted: "The thought of man is freezing here in the wintertime, like the water."

The frozen water some years later, in 1658, made it possible for the Swedish military forces of that time, with its heavy artillery and its overweighted King, to walk over the ice of the Danish Sounds and channels towards the Danish capital of Copenhagen. The result was the treaties of peace through which Sweden received the Danish territories east of the Sound - today the southern part of Sweden, including Lund.

It was in the same cold climate as today that Lund University was opened in January 1668. The students who arrived at this new university had to wear warm clothes when they attended lectures in the sidechapels of the cathedral, where the first classrooms of the university were located. The students, however, probably thought it was worth this physical suffering in the wintertimes. Because it was here that the young people, most of them with their roots in the former Danish peasantry, suddenly heard the voices of Europe.

The new Swedish leaders gave the students their new

ethics. In the seal of the newly founded university they could observe the symbol of the Swedish sovereign: the crowned lion, with one of its paws resting on an opened book and another raising a sword. The motto of the university was to be seen together with this picture. It was, and is, "Ad utrumque paratus" that is: Prepared for both of them: The book and the sword.

For the students of that time it meant that they had to prepare themselves so that they could work not only with the book but also with the sword. It also meant that they should be able to protect themselves, their books and their studies when they had to face military threats. In Europe of the 17th Century the people lived in crises, with war as a natural threat. The people had to be prepared to take up their weapons at any time to protect their country, their families and themselves, but also their knowledge and even their honour.

The ethics of the new university in Lund actually encouraged the members of the academic community to protect the book, its contents and ideas. The book became the symbol for the fruits of knowledge, the result of mental activities. The seal symbolized the protection of those values, the laws of the mathematicians, the physicists, the astronomers as well as the jurists, all laws guaranteed by the university.

The new philosophical idea of interest for the jurists that swept all over Europe in the 17th Century was the new secularised Natural Law.

The members of the Swedish government searched among the European intelligentsia and they scored a great success when they managed to attract Samuel Pufendorf, Professor of Natural Law and jus gentium in Heidelberg, to the small and newly opened University in Lund.

Pufendorf represented the modern philosophy of Descartes, which undermined the Christian doctrine of faith with the help of reason, the "ratio". Pufendorf who had received his education in Jena and Leiden had been inspired there by the disciplines and methods of the natural sciences. His best known works, which were completed and first printed here in Lund in the early 1670's were _De jure naturae et gentium_ and _De officio hominis et civis_. They both belong to the classics in legal and philosophical literature. They both are printed in hundreds of editions and are translated into several languages.

Quite rationally, from incontrovertible principles of

reasons, Pufendorf wanted to deduce the laws, that gave the structure of the state and civil social life. He argued that moral philosophy and jurisprudence could not be related to Christian religion or morality. It was a question for the human reason. The abominable detestable atheist Hobbes was his master.

Pufendorf initiated here in Lund with his works a scientific discussion within the university not only with his colleagues in the Faculty of Law but also with the theologians and the philosophers.

The theories of natural law were essential for the lively scientific discussion in Germany and in the Nordic countries of the time for more than a century. With the first appointed professor, the faculty of law in Lund actually also got its perhaps best known professor.

Another renowned researcher and professor in the Lund Law School was Karl Olivecrona. He belongs to our century and it is just a couple of years ago since he died.

Olivecrona shared with Pufendorf a strong passion for truth and an uncompromising scientific moral stand. His first work of jurisprudence was Law as Fact dating from 1939, now available in a new edition of 1971. Like the works of Pufendorf, those of Olivecrona have been widely acclaimed abroad, and Law as Fact has been used as a textbook all over Europe. Olivecrona saw his startingpoint in the theories of Hugo Grotius and Samuel Pufendorf, and continuously returned to them and their theories in his later works.

The organizers of this conference designed an emblem to illuminate the leaflets and posters of the programme, representing Karl Olivecrona's portrait in the middle of a globe above the university building. Perhaps this symbolic picture will get a realistic connotation, thus, that we shall, here and now, sense the Spirit of Saint Charles during the discussions of this conference.

But heavenly inspiration aside, it is also, and perhaps even more realistic to see Olivecrona's face as a symbol of the continuous research for moral and legal rights that has existed at our university from its beginning and up to our day. In his important work Law as Fact Karl Olivecrona is emphasizing "The Concept of a Right in Legal Theory". And there he introduces his theories in this field by saying: "The fundamental legal concepts are not exclusive professional tools for the jurists. They belong to the common stock of concepts that everybody needs for his contacts with his fellowmen. Expressly or implicitly we are using them in everyday life, buying things, taking a room in an hotel, or paying debts. In this connection they are vehicles for attaining practical ends. But they are also essential ingredients in our view of the community in which we are living, and of the world at large. When we speak of governments, of elections of state officials, of new laws and new taxes, of trials and punishment, of peace and war, we are using legal concepts. We could not, for instance, understand a newspaper without being familiar with them. They determine our whole outlook on society. Therefore they are necessary means of conveying information as to what is going on in the world". This fundamental text can also, as I see it, be a motto for this conference.

Today we are living in a world of crises; political, economical, commercial but also scientific crises. The more we are moving into a computerized world or even space, we also are moving into an even more anonymous, complex, technical and for many an inunderstandable society. It is our responsibility as scientists to make sure that jurists are not reduced to playing the part of technicians or social engineers without any context or social frame. With these aspects in mind the jurists can play a more conscious part in society - also as guarantors of human rights and legal security. "In periods of crisis we need a larger vision", professor Harold J.Berman, Harvard Law School, says in his recently published book "Law and revolution". It's an extremely good book which deals with the formation of the western legal tradition.

Another wellknown Harvard Professor of Law, and also Justice of the Supreme Court in the U.S., Oliver Wendell Holmes, once said to one of his classes of law students: "Your business as lawyers is to see the relation between your particular fact and the whole frame of the univers". The Bostonian lawyer Holmes, grievously wounded in the civil war, was the scholar who synthesized the experience of centuries of Anglo-Saxon jurisprudence in his book _The Common Law_, a classic in its field. He knew from life that without a universal context particular facts are wholly precarious. Holmes's words are about 100 years old, but not antiquated at all. Actually they are very relevant also to this conference. During these days you will discuss central, urgent and common problems in an interdisciplinary way. Jurists will meet colleagues from the humanities and the social sciences. But is is also an international conference. Members of the Ultima Thule will meet colleagues from all over the world. You will discuss recent trends in your disciplines. You will try to define the paradigms of your research, the interactions between legal research and the social sciences.

The organizers have done their utmost to make this conference a success, a success for you as participants but also for our university. Your work will be in Holmes's words to explain the relations between your particular fact and the whole frame of the universe. On behalf of the university I wish you all success with your work here in Lund. May the cold climate outside, however, be a stimulating and activating factor for your work! I am convinced that the cartesian metaphor about the frozen water is not applicable to your thoughts here. It is better to find the wet metaphor in another, more revolutionary, natural phenomenon in Ultima Thule, the icelandic geyser, always ad utrumque paratus, prepared to boil, prepared to continuous convulsions and eruptions like a firework!

CHAPTER 1

THEORY OF SCIENCE AND THEORY OF LAW

SYNOPSIS

Bert C.van Roermund

What kind of enterprise is 'legal science'? This question has been the issue of a rather scholastic dispute between the protagonists of (on the one hand) traditional jurisprudence and (on the other) contemporary legal theory. The former, preoccupied with the parochial nature of legal doctrine, is apt to depict the latter as a 'legal theory without law' (Kunz). And the latter, patently impressed by the universal claims of analytical theories of science, is ready to answer this verdict by calling the former a 'methodology without method'. The sheer vanity of this discussion, manifest in the almost ritual exchange of arguments, ever since - about a hundred years ago - Von Savigny's 'scientification' of law deposited the term Rechtswissenschaft (science of law) on the European continent - is in itself a fact crying out for explanation.

But perhaps the explanation will be of mainly historical importance, as the scene seems to change rather rapidly over the last decade. No doubt this is due to developments of each of the colours defended by the protagonists just mentioned. As far as philosophy of science is concerned, Hansson summarizes some recent trends. He describes general methodology today as a continued investigation of the extent to which scientific conceptual frameworks condition scientific results; an effort undertaken in permanent exchange with methodological studies in special domains. The latter show us the relevance (in heuristic as well as justificatory contexts) of certain unfalsifiable principles and

A. Peczenik et al. (eds.), Theory of Legal Science, 1–4.

indefeasible models which specify how the local properties of different theories are interrelated in such a way as to form criteria of 'good science' and, thus, 'scientific results'.

As to legal dogmatics, Zuleta Puceiro's contribution is powerful evidence of recent evolutions within its self-conception. Only recently, contemporary legal theorists have become aware of their being tributary to the Historical School of Law, its presuppositions regarding the dichotomy history/system and the political preoccupations underlying its methodological plea. The author invites us now to consider this dichotomy as a deeply rooted and effective fact of legal scientific practice, rather than as a methodological value. This insight gives a specific flavour to the overwhelming interest of legal theorists in a philosophy of science which gives an account of rationality in its historical, evolutionary and its pragmatic, societal aspects: it is not a mere interest in self-legitimation and a new hypostatic conception of 'legal reality'. Indeed, as Zuleta Puceiro puts it, one acknowledges that scientific language is a political phenomenon, contributing to the configuration of authoritative structure.

Aarnio's essay is remarkably explicit in this very respect, as he conceives of the ever-recurring, positivistically secured, reference to valid law: not as a necessary representation of a "reified" legal reality, but rather as a strategy to create an epistemic environment in which juridical problems can be perceived, negotiated and communicated, in view of upholding basically political values like justice, legality, constitutionality, equity. Methodological interpretation of these values is parasitic on recognising their politico-moral predominance in our legal culture.

The author distinguishes between three different concepts, 1) the disciplinary matrix of legal dogmatics,2) paradigms in legal dogmatics and 3) legal dogmatic theories. The disciplinary matrix consists of four elements: (1) the positivistic assumption that legal regulations have some justificatory force *per se*, greater in legal reasoning than in moral reasoning; (2) the assumption that some texts, for example, statutes, are the sources of law; (3) the assumption that there are some specific methodological rules for legal dogmatics and (4) the assumption that both such values as equity and such as constitutionality and legality play a great role in legal dogmatics. The disciplinary matrix defines legal dogmatics. Paradigms, on the other hand, are interpretations of the matrix, giving, for instance, some details of the doctrine of the sources of the law, and some methodological rules. Each paradigm specifies in this way the criteria of correctness for legal dogmatic theories.

This implies, on the one hand, a certain reluctancy of legal dogmatics towards liberal proliferation of truth claims and unlimited, functional 'plurality of theoretical paradigms'. On the other hand, it implies open participation in the intersubjective discussion about reliable conceptual tools and social policies with regard to these values, which no one (and no institution) can monopolise. Saying that the hard core of legal dogmatics is immune agains societal changes, is, according to Aarnio, equivalent to endorsing the view that only those changes should be captured by legal scientific 'progress' which allow for persistent identifiability of genuine legal problems; which is in itself a political and, indeed, moral option as to the evolution of society.

Within this fundamental approach (though with interestingly differing elements) Sintonen's contribution may be seen as an attempt to articulate more specific conclusions about the structure and dynamics of legal theories. His frame of reference is the so-called 'structuralist' view on scientific theories, proposed by Sneed and Stegmüller (and their disciples). Its basic tenets are (i) the concept of a scientific theory as the ascription of a fundamental predicate, semantically defined by elegant informal set-theoretical instruments, and (ii) the inadequacy of the dichotomy 'theoretical/empirical'. Although Stegmüller has warned his readers against too Kuhnian mis-interpretations of his theoretical concern, he is also deeply convinced of its capacity to give an account of some of the ideas Kuhn had in mind, e.g. the notions of 'a scientific community using a theory', of 'scientific progress' and of'intertheoretical network relations' All of these characteristics seem to get some intuitive, though initial, plausibility in the context of legal dogmatics, when Sintonen sets out to sketch some developments in Finnish theory of ownership.

A piece of sociology of legal science is offered by

Kamstra and Kunneman, referring to legal research in The Netherlands during the last two decades. They attempt to relate their description to some methodological advice, which reminds one of both Hansson's and Aarnio's arguments concerning the relevance of explicitating and comparing methodological assumptions of (legal) specialties, rather than prescribing overall canons of *the* sciences or, for that matter, *the* legal science.

It is difficult, and perhaps unnecessary, to formulate the general upshot of this chapter. Kuhn's much celebrated and despised work has undeniably attracted mutual attention as well as critical evalution from two camps which until recently virtually ignored each other, but for the gratuitous quarrels within legal theory about the legitimacy of extrapolations from logical positivism into legal positivism and *vice versa*. But it is still a long way to regain the awareness of Ancient Greece, where dogmatic disciplines (primarily medicine and law) were sciences *par excellence*, empirically well-founded *through* their incisive explicitation of values in the life of the *polis*.

RECENT TRENDS IN THE PHILOSOPHY OF SCIENCE

Bengt Hansson

Philosophy of science is a subject with two sides. The first side faces science and scholarship in a unified way, is concerned with general scientific methodology, and studies the growth of scientific knowledge and the general rôle of science in society. It is related to and draws from the history of ideas and general philosophical positions, in particular the empiricism/rationalism dimension in epistemology.

The second side faces the special sciences, studies their methods and their conceptual frameworks, and is often carried out in cooperation with specialised scientists and scholars.

These two sides give mutual inspiration to each other, and it is obvious that philosophy of science would be a very barren field if it had to be restricted to one of the sides only. But it is also true that, from time to time, the interest of philosophers of science has been predominantly with one of the sides, at the expense of the other. As an example, it seems safe to say that the focus has been on the side of general methodology to an unusually large extent during the last two decades. So, if I am to say something about recent trends in philosophy of science, it is not enough to relate the trends within each type of study, but it is also necessary to say something about their relative importance. Already now, I take the premise that I think that the second side is again attracting increased interest, chiefly because there is felt to be a need for empirical underpinning of general theories which tend to drift too easily.

No doubt, the side facing general methodology is the one that has attracted most public attention in the last decade. In particular, the name of Thomas Kuhn and the notions of paradigm and scientific revolution have become commonplace in the scientific community. Many of the questions which have come to the fore in the discussions of Kuhn and his followers are suitable for representation in diagrammatic form. As a point of departure in the diagram

A. Peczenik et al. (eds.), Theory of Legal Science, 5–12.

one takes a naive, somewhat caricatured version of classical positivism. This view is one where scientific truths are easily obtained and accumulated, where a basic aim of science is to collect such truths, where theories are essentially generalisations and thus true or false in a straightforward sense, and where truth itself is unproblematically seen as a correspondence between a sentence and reality.

It goes without saying that real positivists subscribe to such theses in much more sophisticated versions, and that any philosophical position will be grossly misrepresented when summarised in such brutal brevity without any indication of the internal relationship between the theses or the reasoning behind them. But nevertheless, a man of straw is often exactly what is needed when one wants to systematize the various attacks on a position, because that is often the way in which the opponents tend to see their object of attack.

In the diagram (many parts of which I have borrowed from Ilkka Niiniluoto of Helsinki), the four boxes represent four theses on classical topics in the philosophy of science. They are arranged from left to right as they become more fundamental. The first line beneath the boxes represents the classical positivist position and the fact that it accepts all four theses. The minimal departure one can make is to give up the idea of true facts as easily attainable and replace it by a picture of science as an ongoing process which approaches truth through some mechanism or other without ever reaching it. Marxism-Leninism, with its doctrine of dialectical progress, and Popperianism, with its interplay between conjectures and refutations, are two variants of this idea, which thus in no way specifies a unique position, but only a common trait. This trait is represented by the second line.

But if the full truth can never be attained, can the pursuit of this truth really be the aim of a rational enterprise like science? To this question, Kuhn's answer differs from Popper's. Truth is not absent in Kuhn's scientific world (and it should be noted that the incommensurability thesis and the breakdown of communicability do not necessitate abandonment of truth values, even for theoretical statements), but it is not a primary aim of science. What exactly Kuhn puts in its place cannot be stated in a few words, but certainly ability to solve problems is an essential part of it. In our diagram, a position like Kuhn's can therefore be represented by the third line.

Kuhn still thinks of theories as essentially linguistic entities. Other philosophers of science have questioned this assumption. The so-called non-statement view, elaborated most notably by Joseph Sneed and Wolfgang Stegmüller, sees a theory as a structure, often cast in mathematical form, which serves as a mould into which some parts of reality can be fitted. For several theoretical reasons this idea is an important alternative to the received linguistic way of attacking problems in the philosophy of science. On such a view, the question of truth or falsity of theories cannot occur, and the Sneed-Stegmüller approach is therefore represented by the fourth line. Another philosopher of science who would be represented similarly, although he sees theories linguistically, is Pierre Duhem.

The problem which Kuhn faced. -viz. how a rational enterprise could have as its aim the pursuit of something unattainable- can also be solved in another way: truth <u>can</u> be a realistic aim if it is no longer a correspondence with reality, but something else, such as coherence or consensus. Typically, such views are most often held by philosophers outside the analytical tradition, such as Jürgen Habermas and Hans-Georg Gadamer, but also by some analytical philosophers, as for example Michael Dummett and Hilary Putnam.

The lesson to learn from the diagram is that there are many ways away from naive positivism, and that analytical philosophers have tried most of these. It is probably safe to say that the picture of Kuhn and his work which has become widely spread is more conditioned by his "revolt" against Popper than by a close acquaintance with the contents of his writings, and that Kuhn is better viewed as one who triggered off a movement towards epistemological relativism without having gone too far himself. (I here use "relativism" as a generic term for views that knowledge is not so much dependent on objective correspondence with reality). Kuhn is in no way a stranger in analytical philosophy -it is worth noting that "The structure of scientific revolutions" appeared in the series "International Encyclopedia of Unified Science", a platform for the logical empiricists after they had fled to the United States in the 1930's- although he is in opposition to one prominent line of thought in that tradition, viz. Popper's theory of the growth of knowledge.

On the contrary, there is a strong touch of nominalism in the analytical tradition, i.e. a tendency to avoid letting abstract or theoretical notions stand for something with an independent existence. This touch manifests itself in many ways: in an antimetaphysical and antireligious attitude; in an interest in the possibility of reducing one science to another, more basic one; in a taste for reducing larger fields of knowledge to small sets of basic principles, i.e. to axiomatise them; in the use of conceptual analysis as a major tool; etc. One should also remember that Rudolf Carnap's major work was entitled "Der logische Aufbau der Welt". Although it was of course essential for the logical positivists that all reductions ultimately ended in observation sentences, whose truth were firmly linked to reality, it cannot be denied that they also, by letting a small observational base carry a huge superstructure of conceptual constructions, paved the way for what I have called epistemological relativism when the objectivity of the base was later questioned by analytical philosophers. So if the recent trend in general methodology can be described as a continued investigation of the extent to which scientific conceptual frameworks condition scientific results, it is a trend which has received inspiration from the Kuhn debate, but which has also an analytical tradition to fall back on.

In the philosophy of the special sciences, the general trend is to explore the viability of the idea that conceptual apparatuses and broad ideas about what constitutes "good" science are predominant determinators of scientific results. Such an exploration need not, and in fact seldom does, use the notions of paradigm, crisis, incommensurability or revolution, but usually proceeds with a more open mind about the various forms in which epistemological relativism may present itself. Perhaps the best way to illustrate this trend is to present a few examples. I will do so, and I believe that some of these may inspire to similar reflections within legal theory and thus be of special interest to my audience today.

In physics, it has often happened that a special theory has been constructed for a certain phenomenon, and that such theories have later been subsumed under a more general theory, or linked in some other way. What is seemingly the same notion is employed in different empirical theories and is given a certain set of "local" properties in each such theory. In a "global" theory it may or may not be possible to reconcile all these properties and it becomes necessary to establish what may be called <u>inter-theoretical links</u> between the partial theories or <u>model elements</u> as they are sometimes called. Such links may say e.g. that the concepts of mass in the law of gravitation and Newton's third law are one and the same, or that temperature in the context of the gas law is to be construed as the average kinetic energy of the molecules as that concept is used in mechanics. In my opinion it is obvious that such partial model elements and intertheoretical links occur in many other sciences besides physics, including legal theory. The interesting thing about the links are that they are extra-theoretical assumptions or assertions, or at least part of a theory at a higher level. Therefore, they can often be viewed as unfalsifiable guiding principles or as parts of a positive heuristics in Lakatos' sense. An interesting precise result about nets of linked model elements is that the global properties of such nets are not uniquely determined by the sum of the local properties of each model element. This gives a theoretical opening to the problem of emergent properties in empirical science.

In biology, much interest is still focused on the status of evolutionary theories, where a threat of unfalsifiability always lurks. What seems arbitrary in many versions of evolutionary theory is the level on which natural selection works. It is not obvious that it is possible to maintain simultaneously selection mechanisms between different species, between different individuals of the same species, and between different genes. Most observed behaviour is readily explained if one is allowed to choose a level of explanation <u>ad hoc</u>, but the simultaneous explanation of <u>all</u> behaviour is another matter.

In the social sciences, interest is perhaps more scattered on many unconnected topics, probably because of the absence of generally accepted comprehensive theories. One such topic is the status of psychological theories, where some claim that purportedly empirically based theories are in fact only reflections of common-sensical explications of the observations. Others discuss the conception of mind, which has long been thought too abstruse to be admissible in psychology or sociology, but which has recently been made respectable by associaton to certain ideas of Wittgenstein's. Yet another example is the discussion whether anthropological notions like "religion", "sacrifice" and "marriage" (and, perhaps, "right" and "duty") are ontologically autonomous or perhaps reducible to psychological terms.

As a conclusion, I would like to mention a way in which the philosophy of biology has contributed a new idea to the general theory of scientific progress. Scientific progress and the replacement of one theory by another is often thought of in analogy with the evolution of biological species. New ideas arise like mutations, and the fittest one, i.e. the one that does best in confrontations with empirical tests, survives. On such a view, theories with the same content will fare equally well, or badly, i.e. the relevant criterion for sameness is similarity of content.

But in this respect the analogy with biology breaks down. In an evolutionary sense, sameness is determined by common history. Evolutionary unrelated phenomena may be remarkably similar. A case in point is the eye of the octopus, which is structurally very similar to the eye of a vertebrate, but has an evolutionary history which is completely independent. On the other hand, as dissimilar structures as bones in the middle ear of some vertebrates and jaw bones of certain reptiles share their evolutionary history and are therefore identical in an evolutionary sense. Since to understand a structure is so closely related to knowing its evolutionary history, this evolutionary sense of sameness cannot be denied relevance.

If this observation is then applied to the case of the evolution of scientific theories, it implies that there is a possibility that scientific progress (evolution) is not necessarily dependent only on the content of the theories (the present structure). There is empirical evidence that this is in fact the case: highly successful theories have had precursors which have led to no progress at all, although they had about the same empirical content, and theories with a common history have developed quite different contents (Kant and German idealism, logical positivism and phenomenology). This observation suggests that the selection mechanism does not work through the matching of theories with empirical observations only, a finding which goes well with the above-mentioned trend towards epistemological relativism.

To sum up: Thomas Kuhn has no doubt set the scene for today's philosophy of science in a way. The discussion has moved away from technical details, concerning e.g. logical properties of justification or verification, towards much broader questions about various aspects of the general rôle and claims of science. But Kuhn has not set the scene in the sense that it is his own theories that are discussed -paradigms are not much heard of today, and genuine scientific revolutions have turned out to be harder to discern the closer one looks for them. Instead, his influence has been transformed into a widespread recognition of and interest in the important rôle that conceptual frameworks and heuristic ideas play in science. Much of the current discussion in the special sciences can be described as an exploration of the viability of this insight, and the results are fed back as seeds of new ideas in the field of general methodology.

One can say that Kuhn has triggered off a shift towards an increased interest in the impact of the scientific tools on the finished product and away from an earlier view which put more emphasis on the raw materials. This, I think, is the lasting effect of the Kuhnian revolution in philosophy of science.

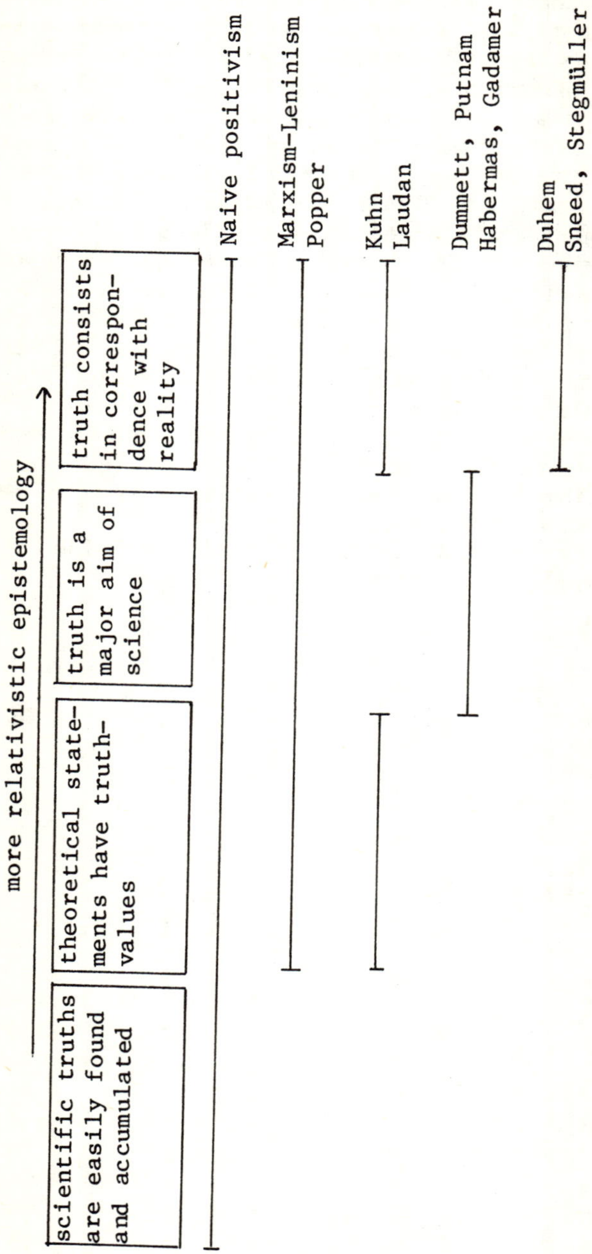
more relativistic epistemology
scientific truths are easily found and accumulated
theoretical statements have truth-values
truth is a major aim of science
truth consists in correspondence with reality
Naive positivism
Marxism-Leninism
Popper
Kuhn
Laudan
Dummett, Putnam
Habermas, Gadamer
Duhem
Sneed, Stegmüller

LEGAL DOGMATICS AS A SCIENTIFIC PARADIGM

Enrique Zuleta Puceiro

1. SCIENCE, TRADITION AND SCIENTIFIC CONSENSUS

One of the most significant traits of the new attitude of legal theory towards the problem of scientific change is, without doubt, the recovery of the dichotomy history/system as a basic focal point. In this sense we can recognize an expressive relationship with the two periods of greatest splendor in the modern theory of legal method: the Elegante jurisprudence of the 16th Century and the Historical School of Law. During both periods, rigorous conceptualism and recognition of the historicity of law constituted complementary and mutually co-relative deep convictions of a shared scientific program, projected towards society with the spirit of an almost prophetical message. In both cases, the ulterior imbalance of scientific practice brought theory to a restauration of formalism, first under the form of the great constructions of rationalist natural law and later under the form of German conceptualist jurisprudence. In turn, practice secluded itself based in middle-range theoretical models and doctrines basically founded in the idea of a jurisprudential law. With this frustration of its initial aims, the recognition of the fact that the balance of a given scientific knowledge depends upon a fruitful co-existence between history and system, has subsisted as shared evidence and as an undeveloped program of contemporary legal theory.

The main task in the present situation is, perhaps, the transition from the consideration of the dichotomy history/system as a value or as a norm to a recognition of the same as a fact, deeply founded and effective in scientific practice. Science is often distinguished from ideology in the sense that it does not recognize leaps in the dark or processes of spontaneous generation conditioned by conjunctural cognitive interests. The current interest of jurists in the recent methodology and theory of paradigms lies precisely in the attraction of a more realistic image

A. Peczenik et al. (eds.), Theory of Legal Science, 13–24.

of science which has shed both faith in indefinite progress as well as in the ingenuous belief in invention as an arbitrary process of isolated wills. Science is, basically, tradition which is communitarily constructed and updated in the solution of problems conventionally recognized as such; it is valorative consensus nourished by implicit presuppositions, among which are possible e.g. creative effort, innovative proposals and inspired openness towards new interpretations and discoveries.

In the case of legal science, this recognition of the historical condition of all scientific knowledge belongs to its original patrimony. In the first lines of his System des heutiges römischen Rechts, von Savigny writes: "When a science like legal science rests on the uninterrupted efforts of many centuries, it constitutes an heritage of incalculable value whose quiet and pacific possession is enjoyed by our generation".
(1). The initial debates of the Historical School were over, and the idea of what, in present-day terms, could be called a "mature science" was fully shaped. Von Savigny himself manifests it in his reconstruction of the scientific discourse of the modern age, in which he adds to the recognizable influences of Doneau, Cujacio and Domat, that of humanism, Elegante jurisprudence, the commentators and the line of evolution which stretches to that of the pandectists before Savigny. His description of science as a cultural reality would surely elicit the enthusiasm of a present-day sociologist of science.

"Besides proven truths -von Savigny wrote- which form the basis of this heritage, we have at our disposal all of the endeavors of the scientific spirit, properly or improperly directed, whose experience shows us the road to follow or to be avoided, thus being able to add somehow to our efforts the work of the past centuries. But: to renounce because of presumptuousness or because of laziness the advantages of our position or to content ourselves with a merely superficial glance at the work of our ancestors, abandoning perhaps that influence which they should exert over our exposition, would be to repudiate this rich heritage, destroy the community of scientific convictions and break the living continuity of progress, without which the community of convictions would degenerate to a situation of complete stagnation".

The reality of turning points and moments of break in these traditions prove that discontinuity has a dialectical relationship with deep continuities in the traditional background of science.

"It is true -he writes- that at times this continuity and the unity of ideas are apparently broken due to the serious disagreements separating the depositaries of science, even within the same period; dissidences which are even more pronounced when two different periods are compared; but for this reason it is even more necessary that from time to time the individual endeavors and results that science offers us be concentrated under a single point of view, not with the intention of accepting or rejecting certain doctrines, but with the intention of resolving opposition at the core of a higher unit, the only road along which science can walk with a surer foot"(2).

The length of the quote is compensated for by the richness of its content and the possibility of a translation to our present-day language. From the perspective cf the founder of modern privatist dogmatics(3), science is a convergent activity, in which "dissidences" (anomalies) are periodically overcome by means of a process of concentration carried out from a unitary point of view (paradigm), not with the intention of rejecting or accepting doctrines (in the name of substantial criteria of truth) but with the intention of resolving them in a "higher unit" (disciplinary matrix) which will make scientific progress possible (operative consensus). This consensus would, in turn, be the basis for a scientific community of convictions (scientific community). There would be a strong relationship between consensus, paradigms and scientific community, just like the relationship which is established between the community of convictions, the recognition of the operative virtuality of tradition, the organic unity between history and system and the very idea of scientific progress. Science is nourished -in the opinion of Von Savigny- "by the periodical revision of the works of our ancestors, the criticism of their errors, the confirmation of their truths and a new takeover which according to our strength makes us take a step towards the definitive goal. The aim of my work is to subject the present era to the application of this procedure".

In Von Savigny's concept, the task of science is multiple in dimensions and perspectives. Construction, historical reconstruction, formal argumentation, typological elaboration, ideological criticism are some of the aspects of a task which can hardly be reduced to the scheme of a science limited to the knowledge and description of legal norms and of the relations they constitute among the facts they determine(5).

The history of the frustration of this originally programatic plan has yet to be made, but the evidence of it forms part of that bundle of conclusions which seems to be accepted today without question, even when the explanations and appraisal of the fact could still be under discussion. A present-day reading of Von Savigny's System reveals both the correctness of the diagnosis of the legal science of his time, as well as the appreciable distances between the program of origin, the effective construction of the conceptualist system and the later evolution of the same throughout the tradition of legal dogmatics, from Ihering and Gerber to Kelsen.

2. PARADIGMS AND SCIENTIFIC PROGRESS

A century and a half after this diagnosis, the question of the reason for this persistence is a main concern for the sociology of legal science. Challenging programs and alternatives for transformation which have accompanied it from the beginnings, legal dogmatics continues conditioning what is essential in legal research. Beyond divergent practices, at times crystalized in programs of reform - as in the cases of Free law school, sociological jurisprudence, jurisprudence of values or legal "principles" -, the dominating factor is that of convergence. Positivist legal dogmatics means thus a genuine disciplinary matrix of schools, theories and debates. The scientific construction of dogmatics by Von Savigny and Von Ihering offers the archetypal experience of a "science made up of paradigms", which explains that "familiar air" that identifies diverse and at times contradictory products, united, however, in the same ascientific and cultural "ethos"(6).

The resistance to change in legal dogmatics as a hegemonic paradigm suggests two main questions. The first refers to the problem of the relations between scientific theories and social change, and includes questions which generally refer to the incidence of changes in the social frameworks of knowledge on knowledge itself. Like any scientific construction, legal dogmatics recognizes a certain set of implicit presuppositions - both of a strictly theoretical nature as well as cultural, ideological and political - , which underlie and condition the community of scientific convictions which Von Savigny speaks of. The problem consists - in elucidating whether that conditioning influence is carried out in a deterministic manner or simply in the rather diffuse way characteristic of ideological influences. In principle, it would seem that in "organic" periods in which the received knowledge seems to lie on more or less fixed presuppositions about society, science or the role of the scientist in society, these types of questions lose the prime importance they assume in rather "inorganic" periods, in which such presuppositions tend to lose their tacit nature. Thus they have primary interest for scientists, who in that way become themselves obliged to transcend the strict horizons of their specific discipline in order to pose questions which are, to some extent, related to the basic form of questioning, characteristic of philosophical thinking.

The second type of questions have more to do with changes inside a theory or with what in present-day terms could be called changes in the theory. Apart from the problem of whether social change is reflected or not by changes in theories somehow derived from the changes in their underlying presuppositions, this second type of question refers to the changes derived from the growth of knowledge, from the processes of replacing one theory by another, more advanced, and even from debates within theories.

The scope of the first type of questions is usually relegated to the field of sociology, psychology or history, disciplines which seem more apt for the analysis of contexts of discovery. The second type belongs to epistemology and to the logic of scientific research, as specific instruments for the analysis of contexts of justification. An important part of the debate on the new philosophy of science arises precisely from this need to delimit and demarcate various disciplinary scopes, supposedly marked off by the nontransferable and specific nature of their respective objects of study. Difficulties arise when, after adopting the inner point of view of these specific disciplines, it is noted, however, that their borderlines are relative, diffuse and debatable, reflecting a community of problems which overrules possible criteria for a priori demarcation.

To what extent do these problems have repercussions for the scope of legal science? From a historical point of view the epistemological concern of legal science is prior or at any rate contemporary to other fields of social sciences. As a result, it is no wonder that in some of its present tendencies, legal theory is especially sensitive and receptive to the evolution of scientific ideas, particularly in those areas of legal theory which have little to do with what could be defined as "academic philosophy of law". It is with scientifically and philosophically concerned jurists where one can note a greater emphasis on dealing with old and always pressing problems linked to the problems of the growth and evolution of legal theories and legal institutions through new conceptual perspectives. In the case of the theory and methodology of scientific paradigms one can hardly speak of a simple "trend" or of a mere assimilation of systems of thought, generated in other fields of knowledge and then "translated" to the realm of legal science. It is, rather, a question of constitutives or "normal" problems of modern legal sciences which are more obvious for the the jurist - than for the philosopher in the academic sense. These problems are strictly connected with the deep meaning of the modern scientific tradition of law, but we can find in the new philosophy of science new motives for updating then for redefinition and perhaps better exposition.

A.Peczenik notes that the degree of popularity attained by the methodology of scientific paradigms results from the fact that it emphasizes two things. First, that science must take something for granted; it must have points of reference as well as starting points. Second, the problem of the incommensurability of scientific theories(7). The transition between alternative scientific theories would take place not in a deductive way but rather in a way basically comparable to the experiences of conversion in which a given scientific community chooses between alternative frames of reference responding to almost ideological motives. Besides that, we can add some additional reasons. First of all, the proposal of the methodology of paradigms goes far beyond the methodological sphere. Basically, it constitutes an alternative view of science and of culture itself, related to the main concerns of the human sciences in a period which could somehow be termed as "postmodern". The rejection of the image of science as a linear or accumulative process must be understood as a reaction against the very notion of progress. This is also a redefinition of the scheme of relations between subject/object proposed by the received theory of scientific knowledge. From this new perspective, scientific progress implies something even broader than a change in the models and patterns of scientific work. It implies an alternation of the images of reality really functioning as implicit presuppositions of scientific constructions. This point, that may be discussed in the realm of natural sciences, takes on great importance in the case of social sciences.

In the case of legal science, a normal paradigm of science e.g., that of legal dogmatics recognizes a set of implicit presuppositions consciously or unconsciously incorporated and "taken for granted" by those who take part of a given scientific community. It is unquestionable that the formulation of the methodological canons of legal science in the work of Von Savigny or Von Ihering presupposes a set of images, not fully explicit, regarding society, the social role of State and legislation, the social function of law and jurists and the relations between institutions and social change. These images fulfil a role of essential mediation between the subject and the object of the scientific work. Between the community of jurists and its main concern -legal reality- there exists a much closer interaction than in the case of those who deal with natural phenomena. Thus, while atoms and cells do not receive any significant influence from the image which physicists and biologists have of themselves as scientists, social phenomena can be immediately and profoundly conditioned by the image that the social scientist has of himself and of the nature of his own activity(8). That is to say, that the paradigms which make up the normative structure of a given science are, in turn, a function of a previous and more basic image, of a previous main conceptual frame of reference related to the image about the jurist's role in society. Thus, historical changes in the image of jurists in society has significantly altered the empirical basis of legal science.

3. THE PROBLEM OF LEGAL DOGMATICS

There is also a second point of interest of the methodology of paradigms in legal research. It derives not so much from the Kuhnian scheme of revolutionary transformations in scientific knowledge as from the theory of scientific communities. The shift of the main point of attention from the traditional image of the isolated scientist to the more complex image of a socio-cultural and ideologically oriented scientific community offers new explanations about unsolved traditional questions. A normal scientific paradigm such as that of legal dogmatics provides not only an example of "maximum relevance" but above all a common point of reference, a frame for basic orientation which nourishes both those who, sharing deep metaphysical views, diverge on specific interpretations, as well as those who diverge on fundamental points coinciding on specific aspects, observations or interpretations. Besides its heuristic functions, a normal paradigm of science has a normative function which enables it to operate as a mechanism of inclusion, and above all, of _exclusion_. A paradigm is, therefore, an institution that stabilizes and consolidates social processes of distribution of intellectual authority and power in society(9). The scientific status of a theory essentially always depends on the emergence of a communitary consensus and on the authority which a normal paradigm of science confers. There is a link of co-implication and circularity between the concepts of "paradigm" and "scientific community". All scientific language is a political phenomenon in the sense that it contributes to the configuration of a certain structure of authority.

In the case of legal science, the implications of these ideas seems to me obvious. The theory of the jurist proposed by Savigny in his _System_ is simply the systematization of the modern experience of a "jurist's law"(10). The result of scientific work is expressed by its contribution to the process of growing rationalization of legal and political structures. This instrumental rationality constitutes thus a frame of authority to the extent it represents a general criterium for judging knowledge as a phenomena of intellectual and political authority in society.

It is obvious that in the present debate on legal dogmatics, there is an inevitable conceptual ambiguity, that explains to a large extent confrontations and resistances which are not wholly justified(11). The starting point is the very vagueness of the expression "dogmatics", which, depending on what point of view is being dealt with, means both a specific level of analysis of the legal phenomenon as well as a particular approach of legal science, an ideological conception of science, a normative model of scientific practice or an attitude towards the object of legal knowledge. All of these conceptual senses are valid and can be justified in their scientific usage, but under the condition of a previous agreement on the length and limits of our words. The traditional discourse on legal dogmatics is not very clear on this point, as can be shown by reviewing the debate on the "object" of legal dogmatics.

The theory of dogmatics is usually not very eloquent with regard to this problem. One would say that the "object" is not clearly defined and explained. On the contrary, it is rather "presupposed". When speaking of "lower jurisprudence", Von Ihering speaks, for example, of "the matter", "the principles", "the originary legal substance"; all these expressions evoke, without mentioning, the norms or the rules which constitute the object of the statal will. In turn, "higher jurisprudence" must be applied exactly to positive law (12). This is the traditional conception later adopted by most present-day legal doctrines. "Legal dogmatics -A. Peczenik, writes- consists on the interpretation and systematization of valid legal norms"(13).

At this turn A. Aarnio explains that legal dogmatics is the doctrinal or analytical study of law, and implies two phases: the interpretation and the systematization of legal norms(14). Legal dogmatics would be the scientific expositon of positive law: positive law is presupposed to be the object of the jurist's work i.e., of interpretation and systematization.

There arises, however, the problem of the verifications of this first and central meaning at the level of scientific experience. Are norms really the object of the activity of legal experts? At the lower phase of interpretation this could -at least provisionally- be accepted. At the higher phase of systematization, the fact that the basic materials are the results of the previous level of the interpretation would also presuppose the norms, but as a sort of mediate object of research. However, such a conclusion would hardly be in accord with the real scope of the scientific theory of legal dogmatics. On this point Von Ihering proposes as object of scientific work the legal institutions and concepts, understood as "legal bodies", "entities" or "logical individualities". The norms are to some extent "presupposed", as mediate reference point of the "concept" e.g., property, usufruct, proprietor, responsibility, etc. This explains the descriptive nature of the language of legal science; the main issues are not the prescriptions - the norms - but the factual referents of the language or norms - the concepts -. Although from the point of view of its emergence the language of norms is prior to that of legal dogmatics, it would be difficult to simply affirm that it is its referent. The referent of both languages is the same. The language of norms delimits and processes the reality of law. The language of legal dogmatics refers to that same reality, although already delimited by the language of norms. There is a substantial referential identity. Analysis, logical concentration and systematization - using Von Ihering's terms - are thus ulterior phases, expounded on the basis of concepts and notions - a basis which the language of norms also refers to.

The preceeding observations introduce a second broader meaning of the expression "Legal dogmatics", which makes it possible to see the first one in rather relative terms. From this traditional perspective of legal dogmatics, the object of legal science would not be acts of legislators, but rather concepts or notions formulated starting from continuous scientific traditions, whose borders surpass the strictly positivist approach. Concepts are not a-historical substantive realities, but rather "points of concentration" (Von Ihering) of broader processes of evolution which can only be envisioned from perspectives of analysis capable of integrating their complexity and globality. The historicity of the object of legal science conditions the historicity of scientific knowledge itself. From this perspective, legal dogmatics would not be what legal theory canonically establishes as a model of legal science, but rather what its effective tradition exhibits as such.

From this new perspective, the nature and tasks of legal science differ significantly from what is commonly accepted among those who share the hegemonic view of legal dogmatics. It surpasses the level of purely analytical study - identification, interpretation, systematization of given positive norms - and faces, on the other hand, a complex strategy in which the still vague and unseizable notion of interdisciplinarity plays a main role. This poses not only conceptual problems but also some important difficulties of implementation. Open to an interdisciplinary approach, legal dogmatics could inevitably clash with obstacles mainly derived from the multiparadigmatic and unstable condition of the rest of the social sciences. The difficulties are obvious, but not sufficient, in my opinion, to excuse us from the task. The interdisciplinary borderlines are loosing their original sharpness and make already possible the alternative of a historically and sociologically oriented legal theory capable of a natural re-reading of the scientific tradition of legal knowledge. Capable, lastly of and organic integration of the perspective of both its implicit presuppositions as well as of the substantial unity existing between scientific systems and the historical framework in which they appear.

NOTES

(1) Savigny, F.K.: System des heutigen römischen Rechts. Aalen, Scientia Verlag, 1981, p. IX.
(2) Savigny, F.K.: Op cit., pp. IX and X.
(3) For a broader justification, see Zuleta Puceiro, E.: Paradigma dogmático y ciencia del derecho Madrid: Ed. Revista de Derecho Privado - Ed. de Derecho Reunidas, 1981.
(4) Savigny, F.C. de: Op cit. p. 42.
(5) In that sense, the very well known definition of Kelsen, H.: Teoría pura del derecho. Transl. R.J. Vernengo. Mexico: UNAM, 1981, p. 89.
(6) Zuleta Puceiro, E.: "Teoría jurídica y crisis de legitimación", in Anuario Argentino de Filosofía Jurídica y Social, 2 (1983).
(7) Peczenik: The basis of legal justification. Lund, 1983, p. 130.
(8) Friedrichs, R.W.: Sociología de la sociología. Transl. M. Horne. Buenos Aires: Amorrortu, 1977, p. 69.
(9) For a broader illustration of this argument, see

Pocok, J.G.A.: Politics language and time, Essays on political thought and history. New York: Atheneum, 1973, pp. 13-41 and 276-291, and Wolin, S.: "Paradigm and political theories" in Gutting, G.: Paradigms and revolutions. Applications and appraisal of Thomas Kuhn's philosophy of science. Notre Dame/London: Univ. of Notre Dame Press, 1980, pp. 160 ss.

(10) See the classic Koschaker, P.: Europa y el derecho romano. Transl. J. Santa Cruz Teijeiro. Madrid: Ed. Rev. de Derecho Privado, 1955, pp. 247 ss.

(11) In this respect, see De Wild, A.H.: Progress in legal science, communication (non-edited) to the XIth IVR World Congress. Helsinki, 1983, p. 2; Aarnio, A.: Legal point of view, six essays on legal philosophy. Helsinki, 1978, p. 149.

(12) Ihering, R.: L'esprit du droit romain. Transl. O. de Meulenaere. Bolonia: Forni Editore, 1969, vol. III, p. 63.

(13) Peczenik, A.: The basis of legal justification, cit., p. 118.

(14) Aarnio, A.: "Paradigm articulation in legal research" in his Philosophical perspectives in jurisprudence. Helsinki: Acta Philosophica Fennica, 36 (1983), p. 210.

PARADIGMS IN LEGAL DOGMATICS
Toward a Theory of Change and Progress in Legal Science(1)

Aulis Aarnio

1. THE POINT OF DEPARTURE: KUHN'S CONCEPT OF A PARADIGM.

It need hardly be reminded that the notion of paradigm, which occupies a central place as an opinion-divider in today's discussions in philosophy and sociology of science, owes most of its present status to the work of Thomas S. Kuhn. As is usual with those scholarly works that more or less drastically challenge the received views and assumptions in a particular intellectual field, Kuhn's work also has occasioned many different sorts of reactions. In the course of the study reported in this paper, we have gained the impression that the most controversial points in Kuhn's theory, for instance the perplexing contrast between periods of "normal science", i.e. paradigm-boundness, and "revolution", i.e. paradigm change, begin to make sense as soon as Kuhn's paradigm concept is taken as an operative principle of examining the development of a <u>particular</u> scientific field or discipline. In fact, Kuhn himself has patiently laid stress on this point by underlining that paradigms or the like can be discerned only amongst certain <u>specialties</u> in sciences and not "all-embracingly" as spanning over whole scientific branches or disciplines. For the analysis of the development of legal science, this observation is extremely important. If it is overlooked, Kuhn's research program remains without its essential historical and sociological dimension. This is also why we have chosen to speak about paradigms <u>in</u> legal dogmatics, not about paradigms <u>of</u> legal research or legal dogmatics. This is an important point to notice.

As is more or less generally known, Kuhn provides us with at least one explicit and extensive, albeit condensed, definition of a paradigm. We shall quote this <u>in extenso</u>: "(Paradigms) I take to be universally recognized scientific achievements that for a time provide model problems and solutions to a community of practitioners"(2).

A. Peczenik et al. (eds.), Theory of Legal Science, 25–38.

If we use the notation X for the notion of paradigm itself, we thus have the following criteria:

(1)X is a universally recognized scientific achievement.

(2)X functions as a research guidance for a time.

(3)X provides model problems and solutions (i.e., the primary example which shows scholars how they are supposed to do their trade).

(4)X is held by a community of practitioners.

X is a paradigm within a given scientific specialty if X meets all these Kuhnian conditions. From all these criteria flows the Kuhnian idea of a normal science which is an activity that uses past achievements as models and guidances for formulating and solving new problems about the subject matter (i.e. about the "world"). Typically, the function of the paradigm lies in its being the central cognitive resource for this normal-scientific work: scholars agree to accept a given paradigm as the basis of future work (which does not mean that they also accept it as a finished product). This communal judgment of the scholars (pointed out in criterion 4) is a central feature of Kuhn's original theory of scientific rationality: having met with puzzling or anomalous problems within their field, scholars frequently strive to accommodate them by constructing analogies between them and the previously known.

2. LEGAL DOGMATICS AS A PARADIGM-BASED FIELD OF INQUIRY

The basic framework described above has set the tone for our analysis of paradigmatic structures in legal dogmatics as well. Our analysis contains one important systematic standpoint about Kuhn's own conceptual apparatus itself. We maintain that the distinction (which Kuhn himself introduced after the first edition of his Structure book) between disciplinary matrixes and scientific exemplars is very important and posits great strategic explanatory potential for the study of legal dogmatics. As is relatively well known, Kuhn holds that disciplinary matrixes typically consist of (a) symbolic generalizations, (b) shared models thematizing the object of study, (c) commitments to certain shared values guiding scientific work, for instance the selection of hypotheses, and (d) common scientific exemplars, i.e. paradigmatic problem solutions(3).

The disciplinary matrix seems to function as a kind of mediating link between the social context and the cognitive orientation of research. For instance in legal dogmatics, it lays down certain "institutional imperatives"(4) which make up the normative structure of how legal inquiry is supposed to be conducted in given societal circumstances. The matrix helps us explain why the normal science of legal interpretation has the form it has - in legal dogmatics, the form of interpreting and systematizing the norms of valid law.

It seems important to notice that Kuhn's original idea of the matrix as explanatory principle was in all probability intended to underline that "normal science" should by no means be seen as an intellectually <u>closed</u> undertaking. Kuhn has frequently been misinterpreted as advocating the idea of a "blind observance to rule" in scientific inquiry. In fact, however, the notion of disciplinary matrix seems to be designed by him to explain the problem-communication and negotiation processes in a given specialty, and also to explain why in the course of these negotiations the problematic of a given scientific field also may develop in new directions, usually by way of an increasing "anomaly consciousness" in the field.

This seems to be one of the points which Kuhn himself sought to incorporate in his analysis by way of talking - we admit, sociologically somewhat inadequately - about scientific "revolutions". What Kuhn obviously intended by talking about revolutions as paradigm shifts was that the <u>interpretations</u> of the basic matrix of a given field may change and in fact do change. This often takes place as a consequence of taking up under study new exemplars whose study has a sort of feedback effect on the paradigmatic interpretatory framework. Recall Kuhn's point that "... a new theory, however special its range of application, is seldom or never just an increment to what is already known. Its assimilation requires the reconstruction of prior theory and the re-evaluation of prior fact, an intrinsically <u>revolutionary</u> process that is seldom completed by a single man and never overnight(5)." What this seems to mean is that the assessment of the rationality of eventual new interpretations of the matrix is not simply a matter of following transcendent rules but depends essentially on the social origins and context of these interpretations(6).

In our study we maintain that the disciplinary matrix of legal dogmatics consists of the following elements: (1) assumptions about the subject matter of legal interpretation, (2) assumptions about the doctrine of the sources of valid law, (3) assumptions about the current methodical principles and rules of legal interpretation and systematization, and (4) assumptions about the values and valuations governing legal dogmatic interpretation and its objectives(7). This basic furniture of the matrix provides an, as it were, quasi-transcendental cluster of prerequisites that are essential for any dogmatic inquiry in order to serve its tasks in the human community(8). It is also essential because it lays down the condition that legal knowledge is only possible as an answer to genuine legal problems (the notion of "genuine legal problem", to which I shall return in Section 3, is a core notion in our reconstruction of the matrix).

What, then, are paradigms in legal dogmatics in their proper sense? In our study we suggest that paradigms are interpretations given to the basic elements of the matrix. That is, paradigms, in various social and historical contexts, fill out with cognitive content the "assumption variable blanks" (1)-(4) in the basic definition of the matrix. As for the assumptions about the subject matter of legal interpretation, it seems to be characteristic of any legal dogmatic thinking that these assumptions show a more or less legal positivistic content: legal regulations are thought of as rules issued by societal sovereign and they need not be backed up by any elements outside the positive law. In addition, interpretations of law are frequently assumed to be substantiable by reference to valid law. It should be pointed out that it is these assumptions that seem to function as some sort of a necessary minimum standpoint of legal positivism, in the overall matrix of legal dogmatics. Werner Krawietz has formulated this by saying that all juristic argumentation and systematization, in addition to its being socially motivated, has also its own, particulary juridical rationality, which above all is manifested in the ever-recurring resort to valid law and legal system(9). We are inclined to say that this assumption of juridical rationality plays a very central role in any possible paradigm of legal dogmatics - as a quasi-transcendental prerequisite. The need to have secured reference to valid law lays down the basic imperative of providing an answer to a genuine juridical problem.

On the other hand, various paradigms may greatly differ from one another as to how they fill out the rest of the blanks in the matrix basic frame - in particular, regarding assumptions about the current doctrine of legal sources and assumptions about particular methodical principles. A greater degree of uniformity can, however, be discerned among the paradigmatic answers to the fourth criterion, the values and valuations governing legal interpretation and its objectives. Values such as legality, constitutionality, and equity seem to be prevailing here: it is hard to imagine that a viable legal dogmatic paradigm would delete these values from its range of activity and objectives. These values also can be conceived or reconstructed as values that govern the actual dogmatic scholarly inquiry, i.e. values that in some significant way guide the selection of interpretations and justifications in legal inquiry(10). Here these values seem to play a role functionally similar to the role of values governing the selection of explanatory hypotheses in science.

Conceptualist legal dogmatics (Begriffsjurisprudenz), the jurisprudence of interests and the analytical approach in legal dogmatics are good historical examples of various interpretations of the basic matrix and thus appear to be legal dogmatic paradigms in the proper sense. They all seem to meet the criteria laid down in Kuhn's original definition of paradigm which we quoted earlier(11). What is more important, their historical emergence, transformation and eventual deterioration/renewal can be conceived of as a process of normal-scientific research in legal dogmatics. In this normal science, the introduction of new exemplars - representative and acute cases of new problemata in law - has frequently resulted in a transformation of previous paradigms into new ones. Sometimes, and not infrequently, this takes place by way of "rethinking" the old paradigm.

Before we shall proceed to discuss this problematic of scientific change, we would like to say something about our study as far as the relation between paradigm and juridical (or legal dogmatic) theory is concerned(12). Legal dogmatic paradigms are interpretations of the basic matrix. These interpretations above all seek to uphold the subject-matter and valuative assumptions of the matrix: the ideal of finding out valid law, the ideal of legality, and so on. Paradigms pursue to do justice to these ideals by way of given doctrines of the sources of law and given methodical principles. In this way they provide the "epistemic environment" for actual juridical problem-solving work which employs legal dogmatic theories. This is to say that these theories are always articulations of a given paradigm of legal dogmatics. It is in the course of this articulation that the actual work with representative exemplars from juridical practice takes place.

We define a legal dogmatic theory in a deliberately broad manner as a cognitive whole which produces substantiated (well-justified) solutions to given juridical problems. The basic elements of such a theory are, first, factual and valuative assumptions about the object of the theory (the subject matter of legal regulation), second, legal principles, third, propositions about the norms regulating the subject-matter problem in question, and fourth, current systematization principles. An example of such a paradigm-embedded theory is the analytical theory of ownership in Finnish civil law of the early 1950's, a theory which actually at that time introduced a new paradigmatic approach - the germ of it - so to speak with its "built-in" intended applications for other branches of law as well(13).

All this should show us that to speak about paradigms and theories in legal dogmatics is to speak about two sides of the same coin. The normal-scientific activity uses theories and exemplary analyses (e.g. the analysis of the concept of transfer of ownership) as its vehicles. It is also at the level of theories that the reproduction and transformation of paradigms - above all, by way of the work with the exemplars - in legal dogmatics occurs. We hope that this shows how misleading it is to talk in a "reificated" manner about paradigms as closed, thing-like entities in legal inquiry. The whole notion of paradigm in fact refers to a whole cluster of intellectual activities: individual and communal perception of significant problems, negotiation of the meanings of juridical terms and concepts, evaluation of received ways of thought in view of juridical problem-solving, devising of fresh arguments for upholding the overarching basic intellectual and social virtues of the matrix: accordance with valid law, legality, constitutionality, and so on.

3. CHANGE AND PROGRESS IN LEGAL DOGMATICS

Our previous discussion suggests that legal dogmatics as a scholarly inquiry has its home, as it were, between two ideals of science. Legal dogmatics, in view of its important societal function of upholding the values of justice and legality, faces the necessity of being deliberately "modest" and theoretically "restricting itself" as far as any alleged overall change and proliferation of ideas in it is concerned. The internal consistency demanded of it in fact prevents it from upholding a substantial "plurality" as regards different intellectual alternatives of its development. Anything will simply not go.

In this respect, legal dogmatics shows an interesting affinity with advanced natural sciences. Significant factors delimiting possible change are locked down by the first and fourth constituents of its matrix: the (more or less) legal positivist basic assumption and the ideal of upholding legality, constitutionally and other values so dear to us, in society. On the other hand, legal dogmatics is a "science of meaning": it is a use of intersubjectively reliable conceptual tools for purposes of human self-evaluation, control, and regulation of social action whose justification is not a matter of science only but also a matter of free and morally responsible human deliberation. On this dichotomy(14), legal dogmatics and the science of law in general seem to have a double-faceted role. In relation to this, we would like to conclude this paper with some viewpoints presenting ideas we have developed in our study.

Our basic intuition is that the legal dogmatic matrix, with its four basic sets of "assumption blanks" which must be filled out with concrete paradigmatic and theoretical content in view of successful juridical problem-solving, serves as a kind of minimum precondition for the subsistence and perception of genuine juridical problems in the legal order. It is characteristic of a genuine juridical problem that the legal order in its "original position", before proceeding to solve the problem, is at the outset uncertain about how to juridically evaluate and assess this problem, its content, or the societal states of affairs related to it. There is a persistent and ever-recurring need to tackle with genuine juridical problems, as a built-in determinant in the matrix.

This means that intellectual inspirations and motivations about law are liable to result in the emergence of a new or fresh dogmatic paradigm only provided that they contain germs of such theories (at this stage, often still implicit ones) as are able to give us fresh answers to pressing genuine juridical problems. In addition to this, these potentially fruitful motivations and intuitions must provide viable interpretations of the dogmatic matrix itself. A much-debated question e.g. is whether change and innovation in legal dogmatic thought can be triggered off by a tendency of thought which is a more or less purely legal-theoretical one. Our answer to this is that there do occur tendencies of thought in legal theory which fail to meet the requirements of the dogmatic matrix, since these tendencies are not capable of providing answers to genuine juridical problems (even though they may, for instance, point out well-argued solutions to societal problems at large). In this way, the matrix functions as a "grid" which lets through only certain paradigm candidates which subsequently will develop and become articulated - via theory building in different branches and institutes of law - into paradigms. This idea can be summed up in the formulation that a paradigm must be a productive interpretation of the matrix(15).

The basic model of scientific change (and, eventually, progress) in legal dogmatics that we submit to discussion is a "contextualist" one(16). Those paradigms and their theories in legal dogmatics that once have established themselves in a successful manner, i.e. have passed through the matrix "grid", function smoothly as long as they (1) provide a solid basis for the perception of genuine legal problems and thus for the communal judgment of legal scholars, (2) provide a legitimating framework for legal dogmatics itself as a discipline, and (3) as long as some legal interpretations produced by them can be effectively chosen in society as the lawful and right ones in view of societal policy. Eventually, developments contributing towards a change emerge: societal forces occasion a need to widen or specify the scope of what counts as a genuine juridical problem, neighbouring sciences (e.g. social sciences) may bring forth essential information for the re-evaluation of the factual side of juridical problems, and so on. These developments may bring about disintegration in legal scholars' community affiliations (the social dimension) and consensus feelings about principles of legal interpretation (the cognitive dimension). Subsequent reproduction and transformation of dogmatic paradigms, then, usually brings about again an integration of those social and cognitive elements in legal science and its community which due to these "external" factors have got disintegrated. By that time, the image of legal inquiry very probably changes.

The idea in this model is to suggest that the matrix-based integration of research ideals and assumptions in legal dogmatics in its social setting contains, so to speak, always the "seeds for its own transformation"(17). However, an essential point in the model is the idea of the immunity of the "hard core" of the matrix itself against societal changes. In fact, progress in legal science can occur only provided that there persist genuine legal problems, identifiable as such in virtue of the matrix' cognitive resources(18), as well as eventual solutions for them.

Accordingly, we are inclined to regard the question of a replacement of the legal dogmatic matrix itself with something else as one that is more or less peripheral to the problem of scientific change in legal dogmatics. Very likely an (abrupt) ceasing of the dogmatic knowledge-interest would occur only in conditions where totally new sorts of institutions occupy the role of law and its scholarly study! Such conditions cannot and (hopefully) need not be so far envisaged in present-day societies. For instance the recently much debated problem of justifying civil disobedience within the framework of Western legal systems signals no dying-off of the dogmatic basic questions; rather it brings new and fresh problems in the picture, problems that perhaps occasion new paradigmatic developments sooner or later(19). Cases such as civil disobedience however show that nobody can set out to define the notion of scientific progress in legal dogmatics without bringing it into some connection with the problem of social progress.

In society, probably in all so-called advanced societies based on exchange and negotiation of interests, there is a genuine "interest of knowledge" in the direction of genuine juridical problems and legal dogmatics. This is the "hard core" of the legal dogmatic matrix. It is perhaps a fruitful paradox to discover that the "purely juridical" but also evolving character of this interest of knowledge only can be realized and subsequently effected by joint attempts of philosophy of science, sociology of science, and the "internal viewpoint of legal dogmatics"(20). We believe that Kuhn was one of the first to point out an equivalent situation as regards the essentially evolutionary rationality of natural scientific inquiry. That is one of the reasons why his ideas have played such a prominent inspiring role in our study as well.

NOTES

(1) This paper was prepared in collaboration with Jyrki Uusitalo. It presents some key arguments and conclusions of an analysis of scientific development in legal dogmatics, undertaken in a team research project. For the complete report, see Aarnio et al. (1984).
(2) Kuhn (1970), viii.
(3) Kuhn (1970), 182 ff.
(4) This term has been suggested by Zuleta Puceiro (1984).
(5) Kuhn (1970), 7. Emphasis added by AA.
(6) This formulation stems form Gary Gutting's excellent editorial introduction to Gutting (ed., 1980).
(7) In relation to the the original Kuhnian conceptual framework, it should be pointed out that in legal dogmatics it may be difficult to distinguish symbolic generalizations and shared models as elements of the dogmatic matrix from each other. However, various sorts of symbolic generalizations in legal dogmatics do seem to serve a function in upholding and transmitting these model construals. In dogmatic methodology and its history, several symbolic "generalizations" (the term may be a bit awkward here), particularly centred on the exact notion of a legal norm, obviously can be discerned.
(8) The cognitive juridical interest manifested in these elements of the matrix is not just a contingent societal fact. Obviously interests like this must be studied by a discipline which is grounded in empirical studies but also involves an irreducibly philosophical element. In legal theory, Harris' studies of the transcendental prerequisites of legal inquiry seem to provide a good example, for instance his ana-

lysis of the basic legal science "fiat" which among other things is designed to bring out the importance of the "legal aspect of society" in relation to the Kelsenianly inspired "pure-norm" conception of legal science and legal system (see Harris 1982).

(9) Krawietz (1981), 315.

(10) Peczenik's juristic operationism seeks incorporate to explicitly the guiding function of values as principles of legal inquiry. See, for instance, Peczenik (1983), in particular chapters 5 to 7.

(11) We won't go here into details about how these traditions and tendencies of thought meet the Kuhnian paradigm criteria. For a detailed account, see Aarnio et al. (1984), in particular chapters 2 and 5. Anyway, it is a fact that these tendencies did grow up amidst given specialties of jurists and acquired their standing via given exemplar studies in juridical theory and practice.

(12) In Aarnio et al. (1984), the concept of a juridical theory is examined especially by Juha Pöyhönen. See also Pöyhönen (1981).

(13) A more detailed analysis of this example is presented in Aarnio (1983), 217-220, as well as in Aarnio et al. (1984), chapter 4. See also Matti Sintonen's contribution in the present volume.

(14) This juxtaposition is inspired by Lorenz Krüger's (1981) discussion of the differences between natural sciences (which in virtue of their claims to truth face theoretical self-restrictions) and human sciences (which posit insights and inspiration for free agency and cultural pluralism). Cf. also Vittorio Villa's contribution in the present volume.

(15) In this connection, additional insights are supplied by Krawietz' recent studies of theory substitution in legal science (Krawietz 1983) as well as by Zuleta Puceiro's studies about the emergence of the dogmatic paradigm (Zuleta Puceiro 1981, 1984).

(16) An elegant manifesto of the contextualist approach in science studies is Whitley (1980), where emphasis is laid on the need to examine the intellectual and institutional contexts of scientific inquiry as well as their interplay.

(17) This model of scientific development draws heavily on the model suggested by Peter Weingart in the sociology of science. The integrative frame of the basic

matrix is seen as essentially open for societal influences. A socially induced disintegration within it may be followed by subsequent reintegration (which means the consolidation of new paradigms as matrix interpretations). This of course implies that scientific change in legal dogmatics is always closely interwoven with societal change. This is not only a commonplace, since the scientific cognitive level of legal dogmatic research also acts as a sort of "intervening variable" which by way of paradigm renewal feeds the social influences back into society and thus plays a role in self-induced change of social system as well. See Weingart (1974, 1976). -We won't discuss here in any detail the question of the explanation of the original emergence of the dogmatic matrix itself. Historically, its emergence is related to the emergence of modern centralized state power and the subsequent consolidation of "genuine juridical knowledge-interest" (cf. Zuleta Puceiro 1981, 1984). An institutional prerequisite for the emergence of the matrix was also the formation of particular and novel specialties of dogmatic research in various developmental stages of European legal thought. These specialities formed by scholars who realized the social and political implications of the budding dogmatic research interest.

(18) That is to say, paradigms as matrix interpretations always must be there as resources for legal scholars' and jurists' perception of genuine juridical problems. By means of these matrix resources, jurists actively make these problems visible. This emphasis is closely related to the similar function of paradigms in natural science in Kuhn's original account, underlined recently e.g. by Barnes (1982, chapter 3). Independently of debates on Kuhn, Knorr (1980a, 1980b) has emphasized these aspects as well, in her studies of natural scientific reasoning.

(19) A characteristic attempt to reproduce the dogmatic matrix by way of partial new interpretations, in face of the problem of civil disobedience, is the insightful essay by Ralf Dreier (1983).

(20) A pioneering work employing such a conjoint perspective is Dalberg-Larsen (1977). However, we are inclined to emphasize the change in the internal paradigmatic structure of legal dogmatics more than

he seems to be. Dalberg-Larsen's main aim is to analyze legal dogmatics and its development as a more or less theoretically uniform tendency of thought (as a quasi legal-theoretical orientation, in fact). For this reason, he has not incorporated the evolution of the matrix itself via its changing interpretations into his picture.

REFERENCES

Aarnio, A.: 1983, 'Paradigm Articulation in Legal Research', in Aarnio, Philosophical Perspectives in Jurisprudence, Acta Philosophica Fennica Vol. 36, Helsinki, pp.209-221.

Aarnio, A., N. Jääskinen, J.Pöyhönen, J.Uusitalo: 1984, Paradigms, Change and Progress in Legal Dogmatics, Gummerus, Helsinki.

Barnes, B.: 1982, T.S.Kuhn and Social Science, Columbia University Press, New York.

Dalberg-Larsen, J.: 1977, Retvidenskaben som samfundsvidenskab, Juristforbundets forlag, Copenhagen.

Dreier, R.: 1983, 'Widerstand und ziviler Ungehorsam im Rechtsstaat', in P.Glotz (ed.), Ziviler Ungehorsam im Rechtsstaat, Suhrkamp Verlag, Frankfurt a.M., pp.54-75.

Gutting, G.: 1980, ed., Paradigms and Revolutions: Applications and Appraisals of T.S.Kuhn's Philosophy of Science, University of Notre Dame Press, Notre Dame and London.

Harris, J.W.: 1982, Law and Legal Science, 2nd ed., Clarendon Press, Oxford.

Knorr, K.D.: 1980a, The Manufacture of Knowledge. Toward a Constructivist and Contextual Theory of Science, Pergamon Press, Oxford.

Knorr, K.D.: 1980b, 'The Scientist as an Analogical Reasoner: A Critique of the Metaphor Theory of Innovation', in K.D.Knorr, R.Krohn, R.D.Whitley (eds.), The Social Process of Scientific Investigation. Sociology of the Sciences. Vol. IV, D.Reidel Publishing Company, Dordrecht-Boston-London, pp.25-52.

Krawietz, W.: 1981, 'Rechtssystem und Rationalität in der juristischen Dogmatik', in A.Aarnio, I.Niiniluoto, J.Uusitalo (eds.), Methodologie und Erkenntnistheorie der juristischen Argumentation, Rechtstheorie Beiheft 2, Verlag Duncker und Humblot, Berlin (W), pp.299-335.

Krawietz, W.: 1983, 'Theoriesubstitution in der Jurisprudenz', in Krawietz, Recht als Regelsystem, Franz Steiner Verlag, Wiesbaden, chapter VI.

Krüger, L.: 1981, 'Unity of Science and Cultural Pluralism', in R.Haller (ed.), Science and Ethics, Grazer philosophische Studien 12/13, pp.167-185.

Kuhn, T.S.: 1970, The Structure of Scientific Revolutions, 2nd ed., University of Chicago Press, Chicago.

Peczenik, A.: 1983, The Basis of Legal Justification, Infotryck AB, Malmö and Lund.

Pöyhönen, J.: 1981, 'The Role of Theories in Legal Dogmatics', in A.Aarnio, I.Niiniluoto, J.Uusitalo (eds.), Methodologie und Erkenntnistheorie der juristischen Argumentation, Rechtstheorie Beiheft 2, Verlag Duncker und Humblot, Berlin (W), pp.127-136.

Weingart, P.: 1974, 'Toward a Sociological Theory of Scientific Change', in R.D.Whitley (ed.), The Social Processes of Scientific Development, Routledge & Kegan Paul, London and Boston, pp.45-68.

Weingart, P.: 1976, Wissensproduktion und soziale Struktur, Suhrkamp Verlag, Frankfurt a.M.

Whitley, R.D.: 1980, 'The Context of Scientific Investigation', in K.D.Knorr, R.Krohn, R.Whitley (eds.), The Social Process of Scientific Investigation. Sociology of the Sciences, Vol. IV, D.Reidel Publishing Company, Dordrecht-Boston-London, pp.297-321.

Zuleta Puceiro, E.: 1981, Paradigma dogmatico y ciencia del derecho, Editoriales de derecho reunidas S.A., Madrid.

Zuleta Puceiro, E.: 1984, 'Scientific Paradigms and Growth of Legal Knowledge', forthcoming in Proceedings of the 11th World Congress on Philosophy of Law and Social Philosophy, Franz Steiner Verlag, Wiesbaden.

PRAGMATIC METATHEORY FOR LEGAL SCIENCE

Matti Sintonen

1. In their classic work Normative Systems Carlos Alchourrón and Eugenio Bulygin applied to legal science methodological tools developed, as they put it, in other fields of knowledge. They had in mind especially the methodology of mathematics and physics, and the notion of a system employed in these fields of exact science(1).

Alchourrón's and Bulygin's primary sources of inspiration were Rudolf Carnap, Alfred Tarski and G.H.von Wright. Their claim was that the logical study of (axiomatic) systems, initiated by these writers, involved unexplored potentialities for legal science - in particular legal dogmatics. In this paper I want to take the proposal seriously and ask how legal science - or legal dogmatics - might profit from being modelled on natural sciences. As it turns out, my conclusions are much the same as those of Alchourrón and Bulygin. While it is certainly true that the mode of being (or subsistence) of a normative order is markedly different from the mode of being of natural order, this truth does not invalidate the thesis known as methodological monism. The admittedly different ontologies of legal dogmatic theories and, say, physical theories, ought not have an undue impact on methodology.

Although my account - which is bound to remain rather sketchy - is in the spirit of Alchourrón and Bulygin, there is a fundamental difference. It should no longer come as a surprise that the tools developed by Carnap and Tarski (and von Wright) involve a shortcoming: they are essentially syntactic and semantic. According to Carnap explication of central concepts in empirical sciences (and mathematics) is carried out within (interpreted) model languages. To attain a required degree of precision and objectivity it was required that all reference to language users is to be excluded. A conceptually hygienic rational reconstruction makes no reference to the people who construct theories, search for explanations, or make legal systematizations(2).

The revolt against non-pragmatic rational reconstruc-

A. Peczenik et al. (eds.), Theory of Legal Science, 39–52.

tions was initiated by Thomas Kuhn who made the notion of a scientific community central(3). At the same time there was among some philosophers of science growing dissatisfaction with Carnapian reconstructions for, it seemed, they treated of theories as "finished products"(4). There was need for a metascientific account which could handle theories in the process of becoming. There is a relatively recent proposal by Joseph Sneed, Wolfgang Stegmüller and Wolfgang Balzer which gives to both these new ideas a new edge. In what follows I propose that this new way of understanding theories contains unexplored potentialities for legal dogmatics. I can only pick out two specific problem areas for closer examination. First, the proposal throws light on what puzzle-solving might amount to in legal dogmatics. Secondly, it can be used to explicate continuity and change in legal dogmatics.

2. Let us first have a look at mature scientific theories as they are understood by Kuhn. According to him mature theories characteristically enable puzzle-solving activity. A mature theory contains a set of *symbolic generalizations* which are taken to be unproblematic and which are used in problem-solving. It also incorporates some shared *models* and metaphysical assumptions which specify, roughly, the types of entities there are and their fundamental mode of operation (e.g. the basic forces, etc.). as well as a set of *values* emphasized by members of a scientific community. These values may include e.g. precision, simplicity, scope or explanatory power. Finally, there are the *exemplars* or paradigm instances of successful applications of the theory's symbolic generalizations to certain problems.

The disciplinary matrix so characterized has both a heuristic (guiding) and a constraining force. It specifies what problems are legitimate and what types of solutions are acceptable. Roughly, only solutions which leave symbolic generalizations intact and which conform to metaphysical assumptions and values are serious candidates.

The proposal by Sneed, Stegmüller and Balzer gives this Kuhnian conception a new and more precise form(6). According to their structuralism a *Kuhn-theory* $\underline{T}$ is a quadruple $\langle \underline{K}_o, \underline{I}_p, \underline{N}_p, \underline{H} \rangle$, understood as follows. $\underline{K}_o$ is the basic theory-core of the Kuhn-theory (roughly, the set of symbolic generalizations or fundamental laws) given through their models. The fundamental laws tell us what entities are to be the models of the theory, i.e. what entities make the theory true.

$\underline{I}_p$ is a set of intended applications, roughly what the theory is introduced to deal with. We shall assume that there are paradigmatic exemplars of applications which are thought by the founding father of the Kuhn-theory to fall squarely with the domain of the theory's applications. The set is an open-ended set of classes of applications generated by similarity relations: any concrete particular similar enough with some such exemplar falls into some subclass of $\underline{I}_p$.

$\underline{N}_p$ is a paradigmatic theory-net. At any given time in the historical development (evolution) of a Kuhn-theory there is not just one theory-element but several theory-elements which form a net-like structure. The founding fathers of the theory T may have derived from the fundamental laws $\underline{K}_o$ certain more specific laws to deal with some classes of application in $\underline{I}_o$. This is to say that the core $\underline{K}_o$ only contains a vague articulation, a basic idea, and that future developments specialize the basic idea to distinct classes in $\underline{I}_p$.

The set $\underline{H}$ introduces a positive and negative heuristic(7). Roughly, it contains the values of the scientific community and an indication of what expansions are "admissible" developments of the existing theory-net.

Now one of the consequences of this view is that it allows us to take a fresh look at scientific growth. Empirical progress is progress in which new special laws are added to the existing net, thus extending the theory to cover so-far unconquered classes of applications. Similarly, there can be normal-scientific progress within already established specialized theory-elements. This consists of clarification, of solutions to apparent anomalies, and of increased empirical support: it is showed that an ever increasing number of structures indeed are models of the specialized theory-elements. Theoretical progress occurs not when the net is extended to new classes of applications but when new more restricted laws are added.

It is then obvious how a Kuhn-theory serves as a disciplinary matrix which both defines problems and sets standards for their solution: $\underline{I}_p$ provides food for thought and $\underline{K}_o$ the cutlery. Similarly, the proposal tells us what normal-scientific progress is versus revolutionary progress. A revolution occurs when it gradually dawns that certain intended applications cannot be turned into models of the core at all - or can be turned into models only by violating some of the hard core-assumptions and values of the Kuhn-theory. What follows is a new basic theory-element, and evolution within a new Kuhn-theory based on that element(8).

3. Applying the notion to legal dogmatics is possible only if we can find, within legal sciences, theories which have enough structure to be amenable to axiomatization. Fortunately, this requirement can be met, though perhaps not with the rigour obtainable in natural scientific theories. For we do have in legal dogmatics systematizations which Alchourrón and Bulygin call normative systems(9).

According to them we can have a (formal) language $\underline{L}$ capable of expressing states of affairs and their (legal) consequences. The language contains propositional variables, logical connectives, and deontic operators which can be used to express normative sentences of the form 'Something $\underline{d}$ ought to be done when $\underline{p}$' etc. A normative system then is a deductively closed set of sentences such that at least some of its deductive consequences are normative ones. Assuming that the valid legal order of a community contains distinct 'fields' or 'suborders' we can say that the legal order consists of a set $D_1, \ldots, D_n$ of classes of the deductive consequences of normative systems $S_1, \ldots, S_n$, as follows:(10)

$$D_1 = Cn(S_1) = \{A \mid S_1 \vdash A\}$$
$$D_2 = Cn(S_2) = \{A \mid S_2 \vdash A\}$$

$$D_n = C_n(S_n) = \{A \mid S_n \vdash A\}$$

And now to the parallel between non-normative theories and normative systems. The first obstacle we encounter is the unavailability in legal dogmatics of symbolic generalizations. There simply do not seem to be mathematical formulas in legal dogmatics. This peculiarity may have to do with the ontology of legal-dogmatic theories, i.e. their not dealing with the causal order of things in rerum natura. (The relevant causal order has to do with judicially relevant causal chain.).

But there certainly are metaphysical commitments, values and exemplars. Here we come to the fundamental theme touched briefly in the beginning of this paper. Legal dogmatic research is conducted within scientific communities the members of which share a disciplinary matrix. The items in such a disciplinary matrix do not necessarily appear in the finished product but belong to the pragmatically circumscribed background. Just as specific non-normative theory-elements for some domains of application are derived from, and seen in the light of, a basic theory-element-core K_o, so specific normative systematizations for different suborders are derived (though in a looser sense) from a legal-dogmatic paradigm.

In a pioneering work on the problem of growth of knowledge in legal dogmatics Aulis Aarnio writes that legal-dogmatic research must necessarily meet the following conditions(11):

(i) it must include conceptions about the origin of law, the validity of legal order, the concepts of norm and normativity, and the objects of legal research;

(ii) it must specify the sources of law and how they are applied;

(iii) it must specify a cluster of methodological rules and principles for interpretation of sources, acceptable inferences, etc.;

(iv) it must incorporate values and evaluations which can be used when 'balanced' judgments are made.

(i)-(iii) relate directly to the ontology of legal dogmatics and (iv) to values. A few remarks are needed. The values in (iv) are not necessarily cognitive values of a legal theorist but also practical values associated with the will of the legislator, so to speak. But we could remedy this by adding to (iv) such cognitive values as simplicity, comprehensiveness, etc., which guide the construction of systematizations.

As to exemplars, there is scarcely any discipline, natural sciences included, which fits better Kuhn's description of how research students are trained. Law students learn by learning paradigm solutions and by attempting to model new cases with them. Obviously the habit stays with legal-dogmatic activity. Moreover, these exemplars fall into distinctly perceived classes: hence the talk of different suborders D_1, D_2, ...,D_n.

4. There are in legal dogmatics no fundamental laws to serve as the core $\underline{K_o}$ of a Kuhn-theory. Consequently, we cannot think that the different systematizations of D_1,

$D_2,\ldots,D_n$ of the legal order are set-theoretic specializations of a core theory. Yet we can talk of more loosely identifiable 'macro-theories' which consist of some fundamental ideas (analogous to K_o of a Kuhn-theory), of paradigmatic exemplars (for the suborders D_1, D_2, ..., D_n), and of paradigmatic nets $\underline{N}_p$ and H which incorporate the positive heuristic.

Now, one of the attractions of the structuralist theory-notion is that it locates the fundamental moving force of normal-scientific growth of knowledge. A scientific community which holds a Kuhn-theory claims that the core $\underline{K}_o$ is applicable to all classes of applications generated by $\underline{I}_p$. However, at any given time this applicability has been demonstrated only for some classes. Since the Kuhn-theory specifies also the unexamined and recalcitrant applications, it locates, so to say, Kuhnian puzzles: it provides both the puzzles and the tooles for solving them.

Assume then that legal-dogmatic traditions such as essentialism and analytical legal theory are loose analogues of Kuhn-theories. We would then have essentialist and analytical systematizations of various legal suborders. They provide specific solutions to specific problems which in fact are generated intensionally (via the similarity relations and the exemplars). To use Aarnio's words, these specific systematizations are articulations of legal-dogmatic paradigms(12).

If this suggestion is followed, we can in fact trace the evolution of the essentialist or analytical Kuhn-theory. It would, I presume, look as follows. First, there is the initial breakthrough of the paradigm, the formation of the disciplinary matrix and the establishment of systematizations in some suborders. Progress could then be of the following kinds.

First, new systematizations within the bounds of the disciplinary matrix are proposed. The fundamental ideas and values which have been succesfully applied elsewhere are extended to new legal sub-orders. Secondly, progress can be made within a sub-order. New better systematizations may solve problems which constitute anomalies for previous ones. Third, there can be increase in understanding which comes, not through solutions to new problems, but through better solutions to old ones. Assume that there are already codified suborders. A new Kuhn-theory may manage to show that a limited number of new assumptions can be applied in the old sub-orders. The result is unification and better understanding of the legal order as a whole.

5. Let me next turn to continuity and change. It is easy to see what these amount to in evolution: although progress of the above three types occurs, the fundamental idea, values, and most exemplars, remain the same. The problem is with revolutions in which a new Kuhn-theory dislodges the old one. For there are characteristically changes in language, in the problems felt to be 'legitimate', and in the values. The result is (alleged) semantic incommensurability and incomparability.

It has recently become clear that the problem of incommensurability for natural-scientific theories has been exaggerated. Growth of knowledge characteristically conforms to what the physicist Niels Bohr called the Correspondence Principle (CP)(13). Physicists characteristically think that even radical theory-changes preserve continuity. A predecessor theory T_1 remains a limiting case of a successor theory T_2. T_2 entails that T_1 is almost right, and therefore T_2 explains, characteristically correctively, the successes and failures of T_1(14).

The structuralist theory-notion, suitably amended, can formally show how continuity is salvaged. The successor theory T_2 is usually formulated within a richer language than T_1. There may then be a translation function from the language of T_1 to the language of T_2. At the same time there is correspondence between the models of T_2 to those of T_1: the latter are (perhaps approximately) 'reduced' to the former(15).

The unavailability of fundamental laws in legal dogmatics makes it impossible to employ here all these formal results. There is also another obstacle. Alchourrón's and Bulygin's notion of a normative system is essentially syntactic in nature, but the idea of explicating continuity and change in normative systems by using tools developed in recent philosophy of science presupposes the use of logico-semantic tools. Now this latter shortcoming is, perhaps, not so serious as it first appears, for also normative systems can be studied by model-theoretic means(16). Though there are (or need not be) 'hard-core' axioms for (legal) normative systems, we can study inter-theory-relationships between them.

It is, then, of some interest to note that CP seems to hold in legal theory too. I can here merely provide an intuitive sketch of a theory-change in legal dogmatics, more specifically in the systematization of property laws. The example I have chosen also illustrates how a legal-dogmatic paradigm in the sense of section 3 works.

Prior to the rise of analytical legal theory (in Finland) ownership was regarded as a person's unrestricted subjective right to a property(17). As a result, there were e.g. tricky problems in exactly when ownership has been transferred. The analysis of Simo Zitting brought about a change, for he split the thus-far unitary concept into a field of related rights, such as the owner's possession, his protective rights against previous owners, creditors etc., and his powers to transfer, pledge and lease the property. Some of the timing problems vanished at once as pseudo-difficulties - Zitting had "systematized in a new way the conceptual apparatus which deals with ownership", as Aarnio writes(18).

How do we analyse this conceptual change? Aarnio has suggested that the old 'essentialist' theory has the structure T_1 and Zitting's analysis that of T_2(19):

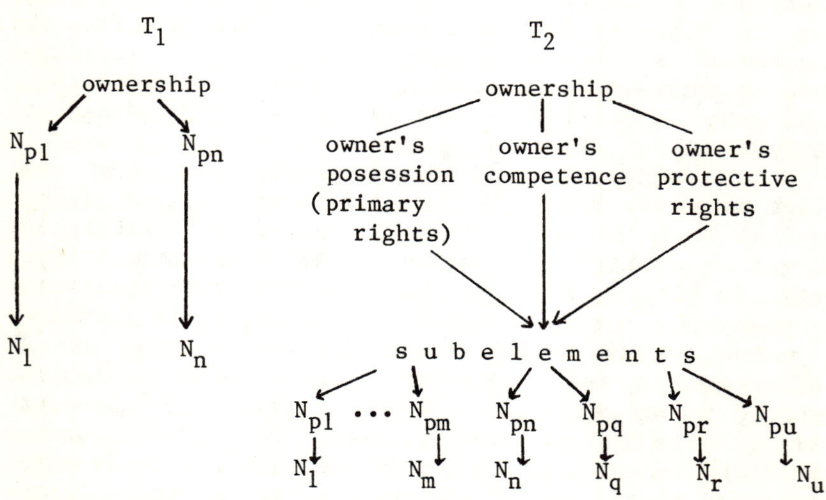

Here $N_1, \ldots, N_u$ are norms and $N_{p1}, \ldots, N_{pu}$ norm propositions, i.e. (linguistic) claims about norms.

Now Aarnio says that systematization T_2 has more discriminatory power than T_1, and therefore allows its holders to identify in a richer way the various possible ownership-issues. But what does it mean to say that transition from T_1 to T_2 involves growth of legal knowledge?(20). First of all, Aarnio says that T_2 may be able to solve more problems than T_1. Secondly, there may be qualititative change if T_2 shows T_1 to be false. T_1 and T_2 are incommensurable in that their interpretations of the norms $N_1, \ldots, N_u$ are different, and in that T_2 is capable of solving problems T_1 is not. Despite this discontinuity Aarnio says that T_2 is 'more developed' than T_1, and that the change is progressive.

My proposal is that we have here a legal-dogmatic change in which an essentialist systematization (an analogue of a natural-scientific theory-element) T_1 is replaced by an analytic successor systematization T_2. There is change in that T_2 gives solutions to (some, at least) anomalies of T_1. T_2 may also show T_1 to be false: T_2 gives a different solution to some problems in their shared domain (e.g. a verdict based on T_1 may have been wrong, from the point of view of T_2).

Now was this change from T_1 to T_2 evolutionary or revolutionary? Can we regard the two systematizations as successive articulations of the same set of fundamental ideas? An unambiguous answer is difficult to give precisely because there are no mathematical theory-element cores. But there are good reasons to think that the introduction of anti-essentialist ideas constituted a revolution, and that property laws were the first suborder in which the new ideas were successfully applied. If this hunch is right, as I think it is, the same anti-essentialist ideas then found their way to other suborders of the legal order. The result is progressive evolution within a new legal paradigm. Needless to say, the introduction of the paradigm may effect regrouping of suborders or increase in depth-understanding in this sense: to the extent the analytical ideas can be extended to other suborders, the result is better understanding of the legal order as a whole. Old problems are seen in an entirely new light. And even a cursory glance at such evolution shows that the disciplinary matrix provides heuristic guidance as well as the puzzles: there are the gaps in (analytic) systematizations, and constraints on what kinds of fillings are acceptable.

It requires of course detailed investigations to show that also revolutionary changes of legal-dogmatic theories (normative systems) obey CP . But it is hardly conceivable that T_2 gives a radically new picture of the legal order. The norm-propositions $N_{p1},\ldots,N_{pn}$ of T_1 are used to deal with concrete legal problems, and if the norm-propositions $N_{p1},\ldots.,N_{pu}$ of T_2 entail radically different problem-solutions, T_2 has no chance of being accepted in the legal community. A realist would of course say, with reference to the parallel employed here that T_2 has no chance of being true. It is intuitively plausible at least that even though T_2 slices the normative order differently, the norm-propositions $N_{p1},\ldots,N_{pn}$ remain approximately true, and T_1 remains a limiting case of T_2.

It is quite natural of course that very different kinds of relations may obtain between T_1 and T_2, such as the translations of the norm-propositions of T_1 into the language of T_2 . The idea of modelling normative systems along 'Kuhnian lines' is a heuristic one. But I can see no reason why precisely the same intertheory relations could not be found in non-normative and normative systems.

6. Let me end up by noting that the task sketched here is mere continuation of the programme started by Alchourrón and Bulygin. One of their basic ideas was that the legal realists attack on attempts to construe legal science as a deductive system was misplaced. For, when criticizing the principles of natural law and proposed deductive systematizations they attacked the very idea of a system. The alternative, they thought, was an empirically founded legal dogmatics.

But, Alchourrón and Bulygin write, a science's having an empirical nature is in no way incompatible with its having a deductive structure(21). As I said in the beginning, Alchourrón's and Bulygin's aim was to apply to legal science the methodological tools developed by Rudolf Carnap and Alfred Tarski and von Wright. The new notion of theory briefly sketched in section 2. is in the same spirit. But it is richer still in potentiality, for it allows a tolerably precise treatment of pragmatic factors.

Normative systems in the sense of Alchourrón and Bulygin have their place within the new setting, but these systems are embedded in pragmatic legal-dogmatic paradigms, the analogues of Kuhn-theories. The reason why my proposal is a continuation of Alchourrón's and Bulygin's ideas can be seen if we look at the basic motivation behind Carnap's explications(22). Carnap wished to confine to purely syntactic and semantic tools because for him (as for all logical positivists) pragmatic equalled subjective or psychological. While such elements may still enter theory constructions (as different weights assigned to different cognitive values), there is nothing subjective or psychological in the pragmatic theory-construct as such.

NOTES

(1) See Alchourrón and Bulygin 1971, p.53.

(2) Carnap later came to realize the need for rational reconstructions of pragmatic aspects. The following quotation from his Intellectual Autography (Carnap 1963, p.56) illustrates this: "A few years later after the publication of the book The Logical Syntax of Language, I recognized that one of its main theses was formulated too narrowly. I had said that the problems of philosophy or of the philosophy of science are merely syntactical problems; I should have said in more general way that these problems are metatheoretical problems. The narrower formulation is historically explained by the fact that the syntactical aspect of language had been the first to be investigated by exact means by Frege, Hilbert, the Polish logicians, and in my book. Later, we saw that the metatheory must include semantics and pragmatics; therefore the realm of philosophy must likewise be conceived as comprising these fields".

(3) Kuhn 1962.

(4) See e.g. Hanson 1958 and 1971, Toulmin 1953, 1961 and especially 1972. A good introduction to these antipositivist ideas is Shapere 1965.

(5) Kuhn 1970, pp.271-272.

(6) See Sneed 1971 and 1976, Balzer and Sneed 1977, and Stegmüller 1976 and 1979. Needless to say I can only provide a non-technical intuitive sketch of the structuralist theory-notion. I should add that I do not fully agree with the entirely set-theoretic underpinnings of the notion. It seems to me that a

more traditional model-theoretic interpretation of these ideas, as given in Pearce and Rantala 1983, is more appropriate. However, these details are peripheral for our present intuitive concerns.

(7) The idea of construing the set of admissible net-expansions as a heuristic is due to Niiniluoto 1980. See also Sintonen 1984, Chapter 5.

(8) For an account of natural-scientific evolution (the Newtonian theory), see Moulines 1979.

(9) Alchourrón and Bulygin, op.cit.

(10) Ibid., IV.4. See also the definitions in the Appendix, pp. 180-196.

(11) See Aarnio 1983b, pp. 215-216.

(12) Aarnio, op.cit.

(13) See Krajewski 1977 and Pearce and Rantala (forthcoming).

(14) For corrective explanation and growth of knowledge, see Sintonen 1984.

(15) See Pearce and Rantala 1983 for discussion of these relationships.

(16) I have in mind, in particular, Hintikka's semantics for normative systems in terms of model sets and model systems. See Hintikka 1970.

(17) Aarnio 1983a and 1983b.

(18) Aarnio 1983a. For further details, see also Aarnio 1977 and 1983b.

(19) Aarnio 1980.

(20) Aarnio 1983b, p.211.

(21) Alchourrón and Bulygin 1971, p.53.

(22) Notice that Alchourrón and Bulygin start the first chapter of their 1971 in Carnap's spirit: "Our aim is to give an explication of the concept of a normative system in order to examine the formal properties of such systems: completeness, consistency and independence. The explication or rational reconstruction of a concept is the method by which an inexact and vague concept - which may belong to ordinary discourse or to a preliminary stage in the development of a scientific language - is transformed into an exact or, at least, a more exact concept.

BIBLIOGRAPHY

Aarnio, A., 1977, On Legal Reasoning, Annales Universitatis Turkuensis, Ser. B - Tom 144,

Turku, 1977.

Aarnio, A., 1980, Tieteellisen edistyksen kysymyksiä oikeustieteessä, (Questions of Scientific Progress in Legal Science, in Finnish), Suomen Akatemian iulkaisuia 6/1979, pp.17-87, Helsinki, 1980.

Aarnio, A., 1983a, 'The Development of Legal Theory and Philosophy of Law in Finland', in Aarnio, 1983c, pp.9-46.

Aarnio, A., 1983b, 'Paradigm Articulation in Legal Research', in Aarnio, 1983c, pp.209-221.

Aarnio, A., 1983c, Philosophical Perspectives in Jurisprudence, Acta Philosophica Fennica, Vol. 36, 1983.

Alchourrõn, C. and Bulygin E., 1971, Normative Systems, Springer-Verlag, Wien and New York, 1971.

Balzer, W., and Sneed J., 1977, 'Generalized Net Structures and Empirical Theories I', Studia Logica, Vol.36, pp.195-211.

Carnap, R., 1963, 'Intellectual Autobiography', in P.Schilpp (ed.), The Philosophy of Rudolf Carnap, Open Court, LaSalle, Illinois, 1963.

Hanson, N.R.,m 1958, Patterns of Discovery, Cambridge University Press, Cambridge 1958.

Hanson, N.R., 1971, Observation and Explanation : A Guide to Philosophy of Science, Harper and Row, New York, 1971.

Hintikka, J., 1970, 'Some Main Problems of Deontic Logic', in R.Hilpinen (ed.) Deontic Logic: Introductory and Systematic Readings, pp.59-104, D.Reidel, Dordrecht-Holland, 1970.

Krajewski, W., Correspondence Principle and the Growth of Science, D.Reidel, Dordrecht,Holland, 1977.

Kuhn, T., 1962, The Structure of Scientific Revolutions, University of Chicago Press, Chicago 1962.

Kuhn, T., 1970, 'Reflections on My Critics', in I.Lakatos and A.Musgrave (eds.) Critisicm and the Growth of Knowledge, Cambridge University Press, Cambridge 1970, pp.231-278.

Niiniluoto, I., 1980, 'The Growth of Theories: Comments on the Structuralist Approach', in J.Hintikka, D.Gründer, and E.Agazzi (eds.), Pisa Conference Proceedings, Vol.I, pp.3-47, D.Reidel, Dordrecht, Holland, 1980.

Moulines, C.U., 1979, 'Theory-Nets and the Evolution of

Theories: The Example of Newtonian Mechanics', Synthese 41, pp.417-439.

Pearce, D. and Rantala, V., (forthcoming), 'Continuity and Scientific Discovery', Communication and Cognition.

Pearce, D. and Rantala, V., 'New Foundations for Metascience', Synthese 56, 1983, 1-26.

Shapere, D., 1965, 'Introduction' to D.Shapere (ed.), Philosophical Problems in Natural Science, MacMillan, New York, 1965.

Sintonen, M., The Pragmatics of Scientific Explanation, Acta Philosophica Fennica, Vol. 37, Helsinki 1984

Sneed, J.1971, The Logical Structure of Mathematical Physics, D.Reidel, Dordrecht, Holland, 1971.

Sneed, J., 1976, 'Philosophical Problems in the Empirical Science of Science: A Formal Approach', Erkenntniss 10, pp.115-176.

Stegmüller, W., 1976, The Structure and Dynamics of Theories, Springer-Verlag, New York, Heidelberg and Berlin, 1976.

Stegmüller, W., 1979, The Structuralist View of Theories. A Possible Analogue of the Bourbaki Programme in Physical Science. Springer-Verlag, Berlin, Heidelberg, and New York, 1979.

Toulmin, S., 1953, The Philosophy of Science, Hutchison, London, 1953.

Toulmin, S., 1961, Foresight and Understanding, Hutchison, London, 1961.

Toulmin, S., 1972, Human Understanding, Vol.I, Princeton University Press, Princeton, 1972.

ON MAKING IMPLICIT METHODOLOGIES EXPLICIT*

Otto Kamstra, Frank Kunneman

This article consists of three interrelated parts: the first describes an impasse in which the development of scientific research in the field of law seems to be caught in the Netherlands. A description of the legal sciences in terms of one science consisting of a number of scientific specialties, each with its own specific object definition and (often implicit) methodology will, we hope, relieve jurists working along traditional lines of any doubts as to their scientific quality, while at the same time creating a climate in which new scientific specialties can flourish. In the second part we discuss and integrate scientific publications concerning subject-matter and method in Dutch scientific research in the field of law within a dynamic scientific framework, defining scientific work in a number of specialist fields in terms of different stages of development. Starting from the concepts "rule" and "stages in thinking", the third part is an attempt to develop a workable strategy by which methodologies in scientific field of law can be made explicit.

1. AN IMPASSE IN THE DEVELOPMENT OF LEGAL RESEARCH

Traditional legal research is exclusively concerned with rules and with the application of legal knowledge; in other words, with legal decision making in concreto. There seems to be little need to reflect explicitly on the methodology of such research. The relatively young social sciences are remarkable for their attempt to claim the label of "scientific knowledge" for obtained research results, through making their methodologies explicit. Pleading the "peculiar nature of law" however, jurists through the years have been persuaded of sufficient reason to be able to ignore the "necessary conditions" for scientific knowledge (eg. falsifiability) formulated in other fields of science(1). At the beginning of the Sixties, the

A. Peczenik et al. (eds.), Theory of Legal Science, 53–66.

universities in our country were faced with a sudden increase in the number of students, partly caused by the sharp rise of the birth rate after the second world war and by the state grant-policy. The Faculties of Law also expanded quickly in a short time(2). This of course led to the necessary expansion of the teaching staff, also in a relatively short period of time. Among other things, these teachers were bound to undertake scientific research, which traditionally consisted of commenting on, and systemizing of, complicated legal questions, topical descriptions of parts of the legal system and elucidating disputable rulings.During these years there was, at first, no explicit reflection on the methodologies involved in such research. The change came when "social consciousness" rather than legal tradition became a determining factor in some of this research(3). Commentators of laws and jurisprudence came to reflect on social needs, thereby coming into contact with other (social)sciences and related methodologies. The number of publications on methodologies of legal research increased somewhat. At the height of the wave of democratisation which marked the end of the Sixties and the beginning of the Seventies, small groups of critical, socially conscious students and members of the scientific staff were to be found at the Faculties of Law, all advocating "different law". Among other things, this resulted in the introduction of the social sciences in the law faculties. Filling up the gap of which jurists had become aware, was left entirely to sociologists, criminologists and others. It has now become clear that, while sociologists have produced useful and fruitful work, they have been unable to fill that gap entirely.

Moreover, there are problems concerning the integration of empirical social scientific knowledge with normative legal knowledge. Introducing sociologists into the Faculties of Law did however afford the opportunity to copy their art of making the rules of theoretising and research explicit. In retrospect it has been less easy than was anticipated to realise the ideal of the Sixties: the integration of social scientific and legal research results. Alas, even fruitful reciprocal criticism is impossible. There is no frame of reference for exchange and criticism, for the subject-matters and methods of both disciplines seem to be of a different order, on which dilemma there was not (and indeed, could not have been) sufficient reflection originally. We wish to look back, and reflect on the question of what to do next. In our view, the development of not strictly dogmatic legal research has been arrested and the situation now resembles something of a stalemate.

Our starting point contents that it is up to jurists themselves to develop a frame of reference for scientific research in the field of law, which encompasses every aspect of law as a subject for research. The different parts of that field we define as scientific specialties. The legal-scientific field is therefore "the collection of subjects of research relevant to law". By using the concept of the scientific field of law we are now able to briefly summarise the previous paragraph. We contend that research at law faculties concentrated for a long time on one part only of the scientific field of law, namely dogmatic legal research. When interest in other parts of the field increased social scientists were invited to try their hand, the implicit premisse being that this would cover the entire field and that it would be possible to integrate different disciplines. It is our explicit view that the scientific field of law encompasses more than the parts mentioned and that, as yet, there is no question of integration. The rest of this article is devoted to an attempt to make large parts of that scientific field both workable and integratable. In order to do this, we draw attention to a smaller part of that field: publications on subject-matter and method of legal research(4).

2. PUBLICATIONS ON SUBJECT-MATTER AND METHOD OF LEGAL RESEARCH IN A DYNAMIC SCIENTIFIC FRAMEWORK

Looking back over the century, it seems to us that the increase in publications on subjectmatter and method of legal research in the Netherlands is characterised by acceleration, the same number of publications being added in an ever shorter space of time(5). This accelerated increase of what we shall call, for the sake of brevity, "methodological publications", can be regarded and interpreted in several different lights. We have chosen an interpretation within a dynamic scientific framework while drawing on the work of De Mey(IV). According to De Mey(6), there is a relationship between qualitative and quantitative changes in methodological orientation of any scientific specialty and the stages in its development (life cycle). He distinguishes between four stages and characterises the related "methodological orientation" within the scope of Radnitzky's(XII) "dominant values" as follows: Pioneer stage: originality, philosophical, programmatic. There is a definite willingness to take risks, an openess towards new points of view and an emphasis on creativity and fruitful ideas. Building stage: verification, non-philosphical, productivity. A tendency to be (too) readily satisfied with the degree in which empirical data support theories. Stage of internal criticism: consistency. Dominating methodological values dictate (too) strict adherence to empirical support for theories, precision and transparency of theories. Stage of external criticism: apologetic, philosophical controversy. Pointed criticism by those who have opted out of the discipline; those who remain are dominated by methodological values such as dogmatisation, immunisation and trivialisation.

Our reference to the accelerated increase of the number of methodological publications is purely quantitative. Our preliminary conclusions with regard to qualitative changes as categorised by De Mey, are as follows: On the one hand we find publications on methodological subjects concerning such areas as criminal, civil and constitutional law, often designated by jurists as the "dogmatic legal specialties". On the other hand we find publications on different attempts to develop new areas alongside of dogmatic specialties by stipulating new objects and/or by importing new methodologies from other sciences.

We shall first elaborate on the distinction between dogmatic and non-dogmatic legal subjects, placing them in the classification derived from De Mey. This will enable us to divide the main body of publications into a number of sub-categories. In conformity with De Mey's classification we place the dogmatic legal subjects in the fourth stage, that of external criticism. Few of those working in these fields concern themselves with methodological problems(7). A minority asks themselves such questions as: "can law be studied scientifically?"(8). Of this minority, some show a tendency towards immunisation(9), often appealing to "the peculiar nature of law"(10). Others think of "scientific" as Erklärend or "hypothetically deductive"(11). Another part of this minority considers "scientific" as Verstehend and describe a legal methodology or way of reasoning(12). Yet again, some of those with a methodological interest attempt to construct a logic adequate to normative reasoning which, as logica minor, could form the nucleus of a legal methodology or logica major(13).

We meet mostly disinterest in methodological problems in the dogmatic subjects and also a tendency to immunisation, as well as attempts to make a legal methodology explicit. The fourth stage in which we placed these subjects can also, perhaps more adequately, be characterised in Böhme's terminology, as the stage of finalisation.

Böhme's classification(II) is especially geared towards the value of "social relevance". Finalisation concerns the application or use of available scientific knowledge for the benefit of goals outside of science. Böhme is not concerned with application in itself, but with influencing the development of any science by external goals set by society or by specific groups outside of science(14).

Traditionally, "dogmatic legal research"(15) is attuned to changes in legislation and extra-scientific influences. Of late some groups have come into existence who may be able to influence the development of object and method of legal science(16) from the outside, such as the feminist, ecological and city-renovation movements: focusing on a social problem, the relevant law is grouped around it(17).

The vast majority of researchers in dogmatic legal areas take little or no part in methodological reflection, a small minority does: either in terms of defensive immunisation, or attempting to make explicit those implicit methodologies that supposedly exist.

Alongside of the dogmatic legal subjects(18) we have found new subjects, at least attempts to stipulate new objects of research and/or to implantate methodologies deriving from other sciences into the field of legal science as a whole. Adhering to the distinction between legal science and the scientific field of law(16), we would designate those working alongside of the dogmatic legal subjects ... etc. as we researchers in the scientific field of law(19). Such researchers appear to have their bread buttered on many sides, borrowing object definitions and/or donor methodologies from the social sciences, philosophy (especially theories of scientific methods), linguistics and mathematics (especially informatica).

Without doubt the researchers in the scientific field of law orientated towards or originating from the social sciences make up the largest group. They form scientific cliques, have their own congresses and journals and are institutionalised at a professoral level(20). Still following De Mey's classification, this group is in the third stage, that of internal criticism(21). Researchers in the scientific field of law orientated towards linguistics(22) and mathematics (especially informatica)(23) are in the first, the pioneer stage(24). We suspect, and empirical research could bear us out, that the last two orientations were initiated by social scientific researchers in the field of law(25).

Those orientated towards philosophy (especially theories of scientific methods) are in the fourth stage, as are the legal sciences. They are prone to philosophying about problems brought up by jurists and they contribute philosophical points of view in legal debate. Their behaviour is paradigmatic, i.e. in accordance with widely accepted conventions regarding object definition and often implicit methodology.

Researchers in the scientific field of law levy external criticism at legal scientific research. Researchers in legal science, as has been said above, are predominantly paradigmatic, while a small group concerns itself with parrying external criticism or attempting to make implicit legal methodologies explicit. The nature of the external criticism is implicit; existing research is not so much branded as implicit or wrong, rather suggestions are made for tackling other than traditional subjects and/or tackling traditional subjects in other ways according to an explicit donor methodology.

In short, we have seen that the number of publications on methodology in law shows an accelarated increase. We have tried to provide qualitative information on those publications (not more than 200 in 80 years), although we must add that these are guesses based on running research (perhaps one could call them educated guesses): We see the scientific field of law as encompassing scientific specialties concerning law and subdivided into different areas. Within this scientific field we find both specialties in the pioneer stage, in that of internal criticism (e.g. sociology of law), and in the stage of external criticism (e.g. legal science).

Our analysis of publications enabled us to distinguish

a number of sub-categories:

- attempts by legal scientists to make a supposedly present but implicit methodology explicit;
- attempts by scientists in the field of law to introduce other object definitions and/or methodologies alongside of legal subjects;
- attempts by legal scientists to thwart attempts by scientists in the field of law.

It is up to empirical research to show whether more distinctions are necessary and the extent of the literature produced within the different areas.

We have attempted to outline the flow of methodological publications within professional literature on law. We discovered that many of these publications were by authors who do not limit themselves to legal science but who have a wider orientation, often derived from other sciences. "Derived from other sciences" presumes demarcations between sciences/disciplines and, for any such derivation to be successful, the possibility of integrating knowledge. The latter forms a problem in research in the scientific field of law where both normative and empirical opinions play a part.

We would be stretching the bounds of this article by exploring problems such as demarcation and the integration of knowledge in the scientific field of law as we have introduced it. We do however propose to develop an aid to make methodologies in that scientific field explicit.

3. ON MAKING A METHODOLOGY EXPLICIT

Proceeding from the concept of "rule" we can elaborate on that of "methodology", for a methodology encompasses:

- rules for judging the <u>validity of reasoning;</u> for most sciences alethic (ordinary) logic is sufficient; jurists also need deontic logic to allow for normative reasoning;
- rules with regard to <u>conceptualisation</u>, which cover the conventions regarding the relationship between conceptual statements and operational statements.Critical-rational social scientific research for example, requires that conceptual statements lend themselves to operationalisation;
- rules with regard to <u>theorisation</u>; such as the cluster

of rules called "testability" and those called "grounded theory".

The relationship between such rules in concreto derives from the postulates (such as those of critical rationalism) from which the scientist concerned proceeds, or can be seen to have proceeded. It is also possible to elaborate on the concept of methodology by taking the concept of "stages in thinking" as a starting point(27). Stages in thinking can be ordered into an entity, as is illustrated by the empirical cycle which De Groot(III) realised. A stage in thinking is a more or less closed unit of information processing to make a distinction between different categories(28). Its "more or less closed" nature can be illustrated quite simple: someone who has to add two large numbers can start immediately and work stolidly on; but he can also stop half way having realised that he has made a mistake, or that he is not adding but subtracting.

Stages in thinking are not the smallest parts of a process of thought in the sense of atoms, i.e. the smallest parts which actually exist. They form the parts of the process of thought which any one researcher chooses to consider the smallest part for the benefit of his goal of research. The following "smallest parts" or stages in thinking are in our view the most appropriate to methodological analysis, because as methodologists we want to know the kinds of solutions to the problems that are grouped together in 'stages'; e.g. what is the logical structure of those problems, which rules and conventions do researchers adhere to in working on those problems, what kind of arguments are considered 'important' etcetera.

- orientation: the processing of information, whether ordered or not into information specifically arranged, e.g. (different) figure-ground segmentation. The leading question, in this stage is What is going on? The researcher formulates the outlines of a hypothesis.
- corroboration: the processing of information of a certain type (e.g. originating "externally") in such a way as to allow for comparison with information of a different type (e.g. originating "internally") in order to determine a degree of similarity. The researcher is looking for instances that are similar to what he is expecting according to the hypothesis. The leading question, in this stage is: What does it mean? The researcher is looking for at least some instances that

verify the hypothesis and (re)formulates it as precise and as general as possible in accordance with those instances.

- concretisation: the processing of distinctions which can at a conceptual level be designated as procedures for separating and combining things in reality. The leading question, in this stage is: How can it be refuted? The researcher gives definitions (e.g. operational ones) of (the main) concepts and points to instances (e.g. groups or events) that are covered by the hypotheses, but not yet observed, and he formulates predictions.
- falsification: resembles verification but aims at determining the degree of difference between two types of information. The researcher is looking for instances that do differ from what he is expecting according to the hypothesis. The leading question, in this stage is: Does it stand testing? Predictions, derived from the hypotheses are tested and, if all conditions for testing are full-filled, are either rejected or accepted.
- evaluation: the processing of information into criteria (e.g. theoretical and social relevance) and into summaries (e.g. of theories or research data) in order to determine the degree of similarity and/or difference between criteria and summaries. The leading question, in this stage is: So what? The researcher indicates whether or not the test-results corroborate his hypotheses and/or other hypotheses and, sometimes, suggests new formulations.

The following example may clarify the previous part. A legal researcher knows a legal theorie (A) that is assumed to cover verdicts in a specific field of law (e.g. contract law). He discovers a lot of anomalies, e.g. verdicts that are not in agreement with (A), and asks himself what is going on? May be, he guesses, theory (B1), with some new concepts in it, explains these verdicts in a more appropriate way. But what does (B1) mean? After studying some more verdicts, he reformulates his original hunch (e.g. (B1)) a few times, and writes down (B5); he is sure that (B5) covers the verdicts he studied and is consistent with legal doctrine.

Instead of publishing (B5) directly he asks himself how can it be refuted? and formulates some predictions. After that, he looks for verdicts that are covered in principle by (B5) and establishes whether it does stand testing. When we assume that he is lucky, he finds out that those new verdicts fit the (B5) predictions better than those deduced from (A). Then he starts working the 'so-what'-question. He gives his arguments pro (B5), preferable by pointing out that anomalies that are inherent to (A) do not appear when using (B5), and by showing that (B5) is consistent with legal doctrine.

These stages in thinking could be used to form a logical reconstruction of scientific research. We propose a research cycle, analogous to De Groot's empirical cycle which distinguishes between stages of observation, induction, deduction, testing and evaluation. Our research cycle encompasses stages of orientation, corroboration, concretisation, falsification and evaluation. As far as research in the field of law is concerned, there are two reasons for deviating from the well known empirical cycle: the emphasis it places on empirical research and the critical-rationalist postulates implicit in it(29). In our view, the empirical cycle forms the logical reconstruction of empirical research based on critical-rationalist postulates, i.e. it forms a specification of the research cycle. The research cycle allows for a meaningful construction of both empirical and discursive research irrespective of the postulates chosen.

Empirical research with a symbolic-interactionist orientation as formulated by e.g. Blumer(I) and Denzin(V), both seem to fit into the research cycle, as do decision-making research(30) and legal research.

In the previous pages we have extended the concept of methodology into an entity of rules and an entity of stages in thinking. Each stage in itself is of interest, but in combination they form a more adequate conceptual framework for making a methodology explicit. In brief, in our view "making methodologies explicit" entails discovering postulates and rules and fitting them into stages of thinking. Both empirical and discursive research can contribute to making legal methodology explicit.

Concluding remark

As we have already mentioned, we hope that making the methodology of legal research explicit, both traditional and less traditional, will relieve the researchers concerned of any doubts as to its scientific quality, while at the same time creating the climate for the fruitful exploration of new scientific specialties in the scientific field of law

NOTES*

(1) See for example Mulder, Klanderman and Van der Velden, NJB 1981, pp.61 and following, especially p.74; also G.C.J.J. van den Bergh in AA 1973, p.334.

(2) In 1983 the Faculty of Law of the University of Amsterdam will enrol more than 1200 first year students (NRC-Handelsblad 5.2.1983).

(3) In 1975 for example, the journal "Recht en Kritiek" (a leftwing critical law review) appeared for the first time.

(4) We are concerned here with research in the Netherlands and Belgium.

(5) We collected our data on this subject both from synoptic articles by Rombach(XIII) and Klanderman (IX) and from relevant dissertations (Polak(X), De Wild (VI) and Van Hoecke(XV)). These authors do not claim to be exhaustive and, of course, neither do we. Taking the fact into consideration that they partly cite the same publications, we have drawn up the following table which in our view lends support to the notion of "accelarated increase":
1900-1939, i.e. in 40 years 40 publications
1940-1961, i.e. in 22 years 30 publications
1962-1973, i.e. in 12 years 32 publications
1974-1979, i.e. in 6 years 37 publications.

(6) De Mey(IV), p.149-154.

(7) They work stolidly and paradigmatically on, unconscious of any methodology. Moreover, we must add that a total of less than 200 publications as a result of said accelerated increase represents but a fraction of the total number of professional publications in the field of law.It is therefore not inconsistent to remark that we see an accelerated increase while "a small minority is responsible for these publications".

(8) Langemeijer, G.E., "Is rechtswetenschap een wetenschap?", Rechtsfilosofie en Rechtstheorie, 1974-1.

(9) See Radnitsky's values in the fourth stage.

(10) See for example Kisch in NJB 1971, p.1178 and following, or G.C.J.J. van den Bergh in AA 1973, p.334.

(11) See for example T.H.Degenkamp:"Anders dan Scholten" in WPNR, nr. 53 14, p.518: "The science of law is

therefore an ordinary empirical science concerned with studying human behaviour in a society".

(12) See Perelman(XI).

(13) See Soeteman(XIV).

(14) De Mey discusses Böhme. We owe this paragraph to De Mey.

(15) For the sake of brevity we have not touched on the position of the more "modern" subjects such as the law of social insurance and administrative law.These subjects belong in our view somewhere between the dogmatic subjects and those mentioned later, such as ecological law.

(16) The terms legal science and scientific field of law are both used by jurists. This terminological dispute masks a philosophical controversy. We define the dogmatic subjects as legal science and reserve the term scientific field of law for all subjects which study law. "Law" of course, has many definitions.

(17) The Faculty of Law at the University of Amsterdam has instituted a part-time professorship for ecological law, while womens studies is now able to claim staff members.

(18) We define the modern subjects mentioned in note 16 here for the sake of brevity as dogmatic subjects; methodologically speaking, they are almost identical.

(19) Those working in dogmatic legal subjects we have termed jurists. Together jurists and scientists in the field of law we term as "those working in the scientific field of law".

(20) Too few people work in psychology of law and anthropology of law (together 10 staff members) to form an independent organisation. Sometimes they join the big brothers criminology and sociology of law, sometimes their mother sciences psychology and cultural anthropology.

(21) "Intellectually the group has obviously conquered a new territory and the major task is now to organize it in a well-ordered fashion". De Mey(IV), p.152.

(22) See for example Van den Bergh, G.C.J.J. and Broekman, J.M., preadvies voor de Koninklijke Notariële Broederschap 1979.

(23) Their "behaviour" is identical to that of psychologists of law and anthropologists of law as mentioned in note 20.

(24) See for example A.H.J.Schmidt, dissertation Leiden,

forthcoming.

(25) Franken(VII) wrote the first publication on jurimetics in the Netherlands. He belongs to the school of Enschedé, who is noted for advocating sociology of law in the Netherlands. Franken derived the necessary research methods from the social sciences. Perhaps the growing interest in "language and law" also originates in an interest in socio- and psycho-linguistics.

(26) There are many definitions of the science of law and many sub-divisions. See for example D.H.M.Meuwissen, AA 1979, 11, p.22 and following, or Gijssels(VIII).

(27) In the language of computer programmers a stage in thinking is comparable to part of a programme in which information is transformed. Cognition psychologists such as Piaget explore the development of stages in thinking in young people growing up.

(28) Someone "writing down what he hears" processes auditive distinctions into visual distinctions; someone transposing tables processes one quantitive distinction into another.

(29) The latter is best expressed by the name testing stage.

(30) Crombag et. al. work with the empirical cycle. Their work can be seen as an attempt to teach law students to solve cases in a methodologically conscious fashion. To us, the problems they encountered are an indication of something not quite right somewhere; they offer their students the empirical cycle which is, in our view a methodology too much geared towards a special type of research.

(31) In NNR 1982 nr.2 we reported just such an attempt. However, we used the empirical cycle and not the research cycle formulated here. It was partly the friction caused by that use that led us to the idea of a new sub-division in stages of thinking.

* A dutch version of this article is published in "<u>The Netherlands Journal for Legal Philosophy and Jurisprudence</u>, vol.13, 1984, no.1, titled "Rechtswetenschappelijk werkterrein; verkenning en verkaveling".

BIBLIOGRAPHY

I Blumer,H., Symbolic Interactionism, New York, 1969.

II Böhme, G. et.al., 'Die Finalisierung der Wissenschaft', Zeitschrift für Soziologie(1973), p.128-144.

III De Groot, A.D., Methodologie, Den Haag, 1961.

IV De Mey, M., The Cognitive Paradigm, Boston, 1982.

V Denzin, K., The Research Act, Chicago, 1970.

VI De Wild, A.H., 'De rationaliteit van het rechterlijk oordeel', Kluwer Deventer, 1979.

VII Franken, H., 'Maat en Regel', Arnhem, 1975.

VIII Gijssels, J., Van Hoecke, M., Wat is rechtstheorie?, Kluwer Antwerpen, 1982.

IX Klanderman, J.H.M., Mulder en Van der Velden, 'Rechtstheorie in Nederland', NJB 1981, blz.61 t/m 75.

X Polak, J.M., Theorie en Praktijk der Rechtsvinding, Zwolle, 1953.

XI Perelman, Ch., 'Juridische Logica der leer van de argumentatie', Antwerpen/Amsterdam, 1979.

XII Radnitzky, G., 'Towards a Theory of Traditions in Science', Communication & Cognition 6 (1973), 15-46.

XIII Rombach, J., Hedendaagse opvattingen over rechtsvinding en rechterlijke funktie, deel I en II, in: WPNR 5249/5250, 1974.

XIV Soeteman, Norm en Logica, Zwolle 1981.

XV Van Hoecke, M., De interpretatievrijheid van de rechter, Kluwer Antwerpen, 1979.

CHAPTER 2

ONTOLOGY AND EPISTEMOLOGY IN LEGAL SCIENCE

SYNOPSIS

Bert C. van Roermund

Although the materials in this chapter are obviously diverse in subject as well as approach, one may try to indicate some common lines of force.

One of the main tasks for legal theory is to give an account of "knowing what the law is". This involves, roughly, both the question of the internal structure of this alleged knowledge and the question of its semantic value. At least implicit answers to these two issues lie at the bottom of many a daily and candid discussion among lawyers, e.g. whether or not the law has changed on a certain point, whether A rather than B is right in claiming that, according to Swedish law, one ought to..., or, indeed, who of the Law School's Students should get an insufficient mark.

We may wish to claim that legal knowledge is, in any case, knowledge of norms and, thus, the content of deontic statements. But it is far from clear how these statements relate to e.g. (valid of invalid) norms themselves, as well as to (true or false) statements of facts. Neither of the following radical strategies seems appropriate: (i) to reduce norms to facts; for as long as we acknowledge - and Balekjian in his article offers some sound reasons to do so - that facts are contents of (true) statements, rather than observed events or things, we would ultimately be forced to admit that there appear to exist two sorts of facts. The question of how to distinguish the two types would almost certainly confront us with the difficulties we wanted to avoid by reducing norms to facts in the first place; (ii) to separate two autonomous realms of knowledge; for that would leave us with the problem how we can tell which realm to enter for which puzzles? Wouldn't that presuppose a third, overriding realm of knowledge,

A Peczenik et al. (eds.), Theory of Legal Science, 67–70.

which challenges the allegedly absolute autonomy of the previous ones?.

Castañeda maintains that it is of great preliminary importance to distinguish between different senses of "autonomy" in this respect. He points out that there is, indeed, conceptual autonomy involved in practical thinking, ultimately based on the specific and irreducible, formal structure of its contents (practitions). His second thesis is that deontic modalities characteristically provide a subtle epistemic bridge between practical and theoretical operations of (one and the same) reason, i.e practitions and propositions. The fact that they present logical relationships of equivalence and/or implication between basically different thought contents does not, however, imply that they are, in the final analysis, reducible to and identical with either sort of these contents. Rather, their complex identity emanates from and reflects our unitary reason, as it deals with one and the same world in which we want to do certain things. This unity is gradually mapped by their belonging to (systems of ...) systems. Different social sciences study different, but necessarily related, segments of this system of bridging implications (e.g. those qualified as "legal"). That is why they, themselves, form a complex unity.

The word "system" is prevalent in Peczenik's paper too, but in a clearly different dimension, forming a matrix with Castañeda's. He claims that "validity" of norms is an _emergent_ property, i.e. presupposes a system of relations. He investigates the ontological commitment involved in this presupposition as far as the law is concerned, aiming at an empirical account of the relational "knots" which can be said to "exist" (in one or more of three different senses) within a network of different components. Peczenik defines a valid legal norm as a complex or tuple of inscription(s) or utterance(s) and their normative meaning (i.e. linguistic units), further related to sets of overt acts performed by certain agents, their mental processes, and other norms; moreover it is the object of reflective, socially oriented and evaluative endorsements by interpreters of the system.

This is, to wit, a (possibly liberal) ontological account of _all_ valid social norms. What makes such a norm specifically legally, valid has nothing to do with a parochial ontological structure, but rather with certain restrictions on its conceptualisation. As far as I can see,

Peczenik introduces here an amended version of the Kelsenian Grundnorm; amended, firstly, because it is more differentiated as to e.g. notions like "sovereignty", competent bodies, sources, hierarchy of norms; secondly, because it refers to interpreters' evaluations, rejecting lex iniustissima, e.g. the extreme immorality of a legal system. This reference is, however, not to be seen as an ultimate appeal to natural law, but as an account of the fact that many persons do not accept such systems and agree on doing so. One might say that such a consensus is a necessary precondition (comparable, but not identical, with Kelsen's Wirkung?) of the existence of valid law. The epistemological upshot of all this is, that "legal validity" is, according to Peczenik, epistemically constructed, via non-deductive steps or transformations, from properties which don't suppose this same conceptualised system. That is to say: legal validity is an "emergent" property.

The concept of a system in a legal context cannot fail to attract attention to what is commonly referred to as the core of legal science: dogmatics. Is this predicate a respectable rudiment of a parochial vocabulary, or rather the source of immunity against any critical approach to law? Broekman (cfr. Lenoble, Van Roermund) and Jørgensen seem to take opposite stands here. The latter assumes a variety of equivalent functional "faces" of the law requiring a variety of possible scientific approaches; dogmatic method would be just one (possible and necessary) approach, interrelated to others. It is representative of "the internal point of view" on law (cfr. Peczenik), in iuxtaposition to the empirical methods characteristic of "the external point of view".

Broekman argues the thesis that prima facie different and changing theoretical approaches towards law, the closer they come to it, turn into strategies of legitimatizing the mutual reference between the dogmatic core of legal thinking and predominant values of our (formal-legal and, thus, cartesian) culture. He denies that there is an immediate link of extrapolation from legal theoretical "paradigms" to dogmatic method. Rather the reverse is true: basically the pressure of dogmatic instrumentalism pervades the various legal theoretical conceptions.

Van Roermund's plea for a "transformational approach" tries to bring the last thesis down to a level of theoretical analysis which permits one to reveal instead of conceal the autoreferential character of law. He sees the conditional forms of legal enactments as semantic surface structures, derived from more general strategies of social problem solving by a set of restrictive operators with specific constraints on their parameters. Castañeda's theory of normative institutions is referred to as a model of this approach. This thesis seems furthermore to be compatible with Varga's, who argues the view that changes of law are not (only) the results of formal enactments, but of its socio-political context.

Lenoble's contribution returns to a question already hit upon at the end of Castañeda's article: the normative vs. descriptive character of legal science itself. Is it merely normative language within quotation marks? Lenoble denies that norm-statements can be a real meta-language in relation to the norms they pretend to describe, because of necessity they have to take into account the performative enactment of a norm which constitutes (part of) its meaning. Now this is a rather provoking view; firstly because it seems to defy consciously the accusation of committing a speech act fallacy, and secondly, because it does so in order to undermine the distinction between the internal and the external point of view on law, which is taken for granted in so many discussions.

This pattern-card of topics in epistemology and ontology related to law and legal science is by no means exhaustive, nor is the content of this chapter to be distinguished sharply from the contents of other chapters. Special reference deserve, in this respect, chapter I (theory of science) and Chapter III (Objectivity and rationality of legal justification)

OUGHT, REASONS, MOTIVATION, AND THE UNITY OF THE SOCIAL SCIENCES: THE META-THEORY OF THE OUGHT-IS PROBLEM (abridged version)

Hector-Neri Castañeda

I. THE OUGHT-IS PROBLEM

1. The traditional formulation of the Ought-Is Problem

As it has been traditionally conceived the Ought-Is Problem consists of a complex of problems about the nature of ought-judgments and 'ought'-sentences, and their kin, like "One ought not to inflict pain on others" and "It is wrong to kill", which is tantamount to, or entails, "One ought not to kill". Of course, the formula "the nature of ought-judgments and 'ought'-sentences" is far from perspicuous. It needs some clarifications, before it can demarcate a unified and worthwhile bundle of issues.

The nucleus of the Ought-Is Problem has been the question whether there is something unique, common and peculiar to all ought-judgments that sets them apart from so-called empirical or factual truths. Sometimes the issue is put in loftier terms: whether the realm of Ought is autonomous or not.

Patently, a great obscurity surrounds both the notions of ought-judgment, empirical or factual truth, and autonomy. As we shall see, 'autonomy' covers a tier of at least five distinct sets of issues. We must elucidate these issues. Now, the Problem -- as it necessarily happens with the major philosophical problems -- is not amenable to simple arguments or observations, but demands a solution deployed in a suitable comprehensive theory of practical thinking and reasoning, that encompasses the two subject-matters being alluded to under the key words 'Ought' and 'Is'.

Here we cannot engage in the development of such a comprehensive theory of Ought and Is(1). On this occasion our chief purpose is much more modest, namely: to contribute to the elucidation of the Ought-Is Problem itself, by

A. Peczenik et al. (eds.), Theory of Legal Science, 71–96.

establishing some crucial surface theses which, having preliminary explanatory power, must fix boundary conditions for any proposed solution (in the way that Boyle's and Charles' laws function in the physical theory of gases), staking at those boundaries some absolutely binding distinctions that no comprehensive theory worth considering can fail to reckon with and illuminate. In brief, here we are dealing with the _metatheory_ of the solution to the Ought-Is Problem.

2. Some basic assumptions

An urgent preliminary clarification is this. Some decades ago Meta-ethical Emotivism ran rampant, and it propagated the so-called non-cognitivist view of ethics, which still lingers on(2). According to this view, there are no genuine ethical judgments or statements, or truths, with their appropriate techniques of verification, there is only ethical language, constituted by ethical words, which by virtue of a battery of variegated roles, are used to vent feelings and express attitudes, but, above all, to influence behavior and subtly guide conduct. For Emotivism ethical discourse was in no way descriptive of any special aspect of external or of mental reality, or of any special domain of thought contents: ethical discourse was, for Emotivism, essentially and characteristically, not so much a means of thinking, but a way of _doing_ and of _making people do_ things. It was peculiar to Emotivism to use the word 'ethical' in a very general way. On the one hand, the word was an umbrella for both the axiological expressions ('good', 'bad', 'valuable', 'worthy', and their ilk) and the deontic vocabulary (which will be illustrated below); on the other hand, under the word 'ethical' Emotivist philosophers gathered _all_ the practical uses of these two types of expressions, regardless of whether they were moral or ethical (in the ordinary senses of these words), thus correctly pitching their unitary discussion at a general level. Obversely, the non-cognitivism of Meta-ethical Emotivism is more radical than what the name suggests: it is a non-cognitivism about all _practical_ non-descriptive expressions. This non-cognitivism is clearly independent of the views pertaining to the problem of universals. Although Nominalism about abstract entities has seemed very congenial to Emotivism.

It will take us to far afield to enter into a discussion of the status of universals or the roles of general terms, or a discussion of the peculiar claims of Emotivism. Yet to carry on we must take a stand. Thus, we shall simply shun the negative theses of non-cognitivism(3). We shall continue to speak of ought-judgments, and even of ought-statements. Besides, we shall assume that ought-judgments have implication relationships, and that the principles governing such implication relationships are the subject-matter of the discipline called deontic logic. We must, therefore, acknowledge that ought-judgments have values analogous to truth and falsity. Specifically, we must presuppose that the truth analogue is not only analogous concerning the logical role of truth in implication relationships, in it being the value preserved in implication. It is also analogous with respect to the psychological role of truth: it is the preferred value of our mental attitudes addressed to them, in particular, it is the value we desire our ought-judgments to have when they function as premises in our deliberations.

We may use the words 'true' to predicate and 'truth' to refer to such a value. But we shall not insist on this terminology: we must insist only on the existence of the logical values of deontic judgments and on the varied roles they play in our experience, in our practical experience of acting intentionally, adopting plans, and deliberating about what to do. The words 'valid' and 'invalid' can be used to attribute them. Our terminological nonchalance does not merely aim at placating those philosophers who refuse to say that deontic judgments are true or false, it is partly generated by a deep respect for such philosophers' well-founded desire to highlight the significance of the difference between the grounds, on which we accept an empirical truth or postulate a scientific hypothesis or theory, and the reasons for which we adopt or come to believe a deontic judgment. To be sure, we can muster good theoretical reasons for classifying ought-judgments as statements or propositions; nevertheless, here we shall not go into this topic either. (See Thinking and Doing, Ch.7).

Palpably, regardless of whether an 'ought'-sentence expresses a truth or not, most of us agree that what 'ought'-sentences express may be said to be correct, or valid, and that this correctness, or validity, is grounded on certain facts. It is, e.g., because pain is such a nasty thing that we ought not to inflict pain except in special circumstances; and obviously the ought-judgment "It is right to inflict pain whenever one fancies one may enjoy doing so"(= "there is no ought about not inflicting pain"), even if one is not allowed to call it false, or not true, is certainly incorrect, or invalid. This connection between an ought-claim and its grounds is definitely within the scope of the Ought-Is Problem. One thing that has to be clarified is whether, or not, that connection is of such a nature as to allow of a fundamental difference in type between ought-to-do judgments and their grounds.

Obviously, the 'ought'-sentences relevant to our discussion are not so much those sentences that merely contain the word 'ought', but those that in given contexts of utterance formulate or express deontic ought-judgments. Not to be entirely dogmatic we register here that a necessary condition for an 'ought'-sentence to express an ought-judgment is that the external negation 'ought-not to do' (= it's not the case that ... ought...) be non-equivalent to the internal negation 'ought not-to-do'. Another necessary condition is that the 'ought'-sentence either formulates a claim upon our motivation that could conflict with other ought claims in a deliberation, or formulates the result of deliberation after the weighing of such claims. Palpably, words can be used ambiguously. Therefore, we shall for convenience speak alternatively of judgments and concepts or of sentences and expressions.

3. Deontic structures and the Ought-Is Problem

Let us clarify the intended span of the word 'ought' as it appears in the name of the Ought-Is Problem. It has been generally understood to include all deontic expressions, not only of the <u>ought</u> or <u>obligation</u> type (like 'ought', 'obligation', 'duty', 'must', 'is required', 'has to', 'the correct [right,appropriate] thing to do', and 'should'), but also of the <u>wrong</u> or <u>prohibition</u> type (like 'prohibition', 'wrong', 'interdiction', 'unlawful', 'illegal', and 'forbidden'), the <u>permission</u> type ('permitted', 'permissible', 'may', 'right', 'allowed', etc.) and the <u>liberty</u> type ('being at liberty, free, to do [or not to do]', 'it being allright, or optional, to do [or not to do]', etc.).

Thus, the Ought-Is Problem is the problem of the nature of <u>all</u> deontic judgments, regardless of which of the above types they belong to.

4. A grave error in the justification of morality

Morality is undoubtedly very pervasive. It is fundamental to other systems of rules, including the legal rules. Morality molds the character of many institutions; it is the main fountain of both the criticism of institutions and the establishment of revised institutional obligations. Hence, it has been easy for some philosophers to identify the conceptual autonomy of morality with the conceptual autonomy of practical thinking in general. And the Ought-Is Problem has been narrowly understood as the *problem of the autonomy of morality*. Yet this is a grave error. Obviously a truly amoral (or immoralist) agent can still make his mind about what to do in his circumstances, and he may engage in attempts to act rationally, though amorally, so that he engages in deliberations to determine what he *ought* everything being considered to do. In particular, nothing in logic or psychology need prevent a person set in carrying out immoral behavior from earnestly fretting about what he *ought* to do, everything considered, in order to attain his immoral purposes. This *motivational ought*, which enters in deliberation and the solution of conflicts of duties and desires, is the fundamental *ought* to which Kant adjudicated what he called respect for the law, but confused with the moral ought:

> "...respect, which means merely the consciousness of the submission of my will to a law [=a motivational and overriding ought-principle] without the intervention of other influences on my mind. The direct determination of the will by the law and the consciousness of this determination is respect." (*Grundlegung*, Akk. edition, vol.IV,p.401n.).

The error of equating morality with practical thinking, or the moral ought with the motivational ought, has typically been committed in attempts to provide a justification of morality. Patently, if morality is equated with practical thinking, then no agent can escape morality: the mere fact that an agent has adopted the simplest intention would *ipso facto* put him under the constraints of morality. Of course, it would be nice if by the very nature of agency nobody could get out of morality. Unfortunately, however, things are not that simple: since the distinction between the moral ought and the motivational ought is of the greatest conceptual importance, nature just *couldn't* have been so accommodating as to produce an empirical identification of the two: the distinction must definitely be thinkable. Indeed, our pedagogical practices aim only at producing their partial convergence. Here lurk very exciting issues.

At any rate, here we understand the Ought-Is Problem to be the general (external) problem of the nature of practical thinking. The problems of the conceptual and alethic autonomies of practical thinking are central aspects of the Ought-Is Problem. A fortiori, we regard the problems of the conceptual and alethic autonomies of morality to be merely special cases of major problems within the Ought-Is Problem.

5. The systemic nature of deontic judgments and institution schemata

As the preceding discussion hints at, there are many deontic structures. The motivational deontic structure occupies, naturally, the central place in the context of rational and reasoned agency, the agency that is not only rational in what it accomplishes, but, furthermore, has reasons, and ensues from the agent's endorsing consideration of such reasons. Moral deontic structures -- and there are excellent reasons to think that morality includes a normative system containing more than one deontic structure(4) -- are pivotal to the deliberations typical of human agency. Legal deontic structures are crucial to the well-being of most persons. Yet these are only the most general and pervasive deontic structures. Every institution we belong to, every casual episode of co-operation, as in the case of a most trivial promise, is constituted by a deontic structure or a network of such structures. Thus, a most significant and important fact about deontic concepts is their systemic nature. They belong into tightly organized systems: the obligatoriness, wrongness, or permittedness of an action is relative to a deontic system. We shall, thus, add subscripts to the deontic expressions we are considering or using, e.g., $ought_i$, $wrong_j$, where the subscript signals the particular normative structure involved. These subscripts are like the adverbial phrases we use in daily life for the same purpose, e.g., we say "One ought morally to honour one's parents" "Everybody is in the United States legally required to file an income tax return yearly", and "Marion has the duty, according to the rules of her job, to record all the calls she receives".

An immediate and important corollary of the systemic nature of deontic concepts is that the equivalences among them, which we have accepted above (in Section 3), must be understood as limited to uniformed deontic judgments, i.e., as valid only within each deontic family. Thus, "For every action A, A $ought_i$ to be performed, if and only if not doing A is $wrong_i$ (or forbidden)" and "John is at liberty to travel by car, if and only if John is neither obligated to travel by car nor obligated not to travel by car" are valid equivalences.

An institution is constituted by systems of rules that determine the duties and rights of its members, the DO's and DONT's of each office within the institution. We may regard an institution in abstract, or an institution schema to be a system of networks of rules or ought-judgments, each network constituted by one or more deontic structures. Thus, an institution schema is a system of systems of systems of ...of deontic structures arranged in a hierarchy, and related to one another in diverse ways. The study of an institution is precisely the study of both the particular rules characteristic of each deontic system and the particular interconnections between the constitutive deontic systems.

6. The comprehensive Ought-Is Problem

Consider an institution I whose normative schema is normative system N. Then N is a system of judgments (or sentences) that stipulate the obligations, permissions, and interdictions characteristic of I. We call all such judgments (or sentences) deontic. We assume that the deontic concepts of obligation, permission, interdiction, and deontic liberty are either literally interdefinable, or at least equivalently related, so that it is sufficient to consider just one of them in discussing the logical structure of a deontic system. Let normative system N consist of a hierarchical network of deontic systems D1, D2, ..., Dn. Hence, in treating the logical structure of N, or in dealing with the Ought-Is Problem for N, we can focus on a sequence of deontic concepts, one for each deontic system, D_i composing N. Thus, if we have chosen obligation as primitive in the logical regimentation (or reconstruction) of N, we confront the sequence of obligatorinesses O_1, O_2,

..., O_n, where each obligatoriness O_i is characteristic of, and can, therefore, represent the corresponding deontic structure D_i. Of course, it may turn out that the most efficient treatment of the mentioned problems for a normative system N involves the confrontation of a deontically non-uniform representative sequence in which obligatorinesses alternate with forbiddennesses or permittednesses or libertynesses. At any rate, the Ought-Is Problem for normative system N is the manifold of problems pertaining to how each of the different deontic concept $O_i (i = 1, 2, ..., n)$ relate to one another, how they differently relate to the empirical facts we establish by ordinary experience or by scientific research, how they connect to the relevant agents' motivational structures, and how they constitute the schema N of institution I.

Now let M be the motivational hierarchy of an agent A. Then M has a deontic structure D* as its underlying normative schema. We can also pose the Ought-Is Problem for D* (and for M). We ask, thus, how the deontic judgments (or sentences) constitutive of D* relate to the empirical facts and the factual truths of the world, including the psychological and sociological facts about agent A. We ask also about the general connections between D* and the institutions to which A belongs, and between D* and <u>all</u> the norms that A recognizes.

To sum up, as a consequence of the preceding assumptions and distinctions, we must understand the <u>Ought-Is Problem</u> even more generally yet, namely: both as the manifold of (external) questions about how normative systems and motivational <u>ought's</u> connect to the so-called factual or empirical truths, and as the manifold of (semi-external) questions about how normative structures within a comprehensive institution schema relate to each other, and how the different normative structures relate to the motivational deontic structure, first, of a rational agent and, second, of a non-rational one, say, because he suffers from weakness of the will.

II. THE MANY FACES OF AUTONOMY

As noted above, the Ought-Is Problem is often equated with the question whether morality or practical thinking is autonomous. But this equation does not, however, represent a unified topic. There are so many different issues blended together under the autonomy heading, that the next item on our agenda must be to separate the main concepts of autonomy and their strands. We proceed from the most fundamental to the most peripheral species of autonomy.

(A) Conceptual Autonomy:

a) Fundamental Conceptual Autonomy of Practical Thinking: In practical thinking an agent thinks thought contents, among which there is at least one peculiar concept that is wholly irreducible to the concepts composing what is thought of in contemplative or descriptive thinking, which concept is characteristic and constitutive of practical thought contents.

a') Fundamental Conceptual Autonomy of Practical Discourse: The sentences of practical discourse contain at least one expression that is wholly irreducible to, or analyzable into, non-practical terms.

b) Weak Deontic Conceptual Autonomy: Deontic thinking is autonomous in that at least one deontic modality, perhaps ought, is not analyzable into, or reducible to, purely empirical or factual terms.

c) Strong Deontic Conceptual Autonomy: No deontic modality is reducible to non-deontic terms.

(B) Alethic Autonomy:

a) Fundamental Alethic Autonomy of Practical Thinking (and Discourse): The [non-logically] valid or truth-like thought and speech contents of practical thinking and discourse - which are truth-like both in being preserved in implication and in being psychologically preferred -, whatever they may be called (valid, or correct, or legitimate, or even true, for that matter) are truth-like by conditions that cannot be exhaustively formulated in non-practical terms.

b) Weak Deontic Alethic Autonomy: The truth- or validity-conditions of no [non-logically valid or truth-like] deontic judgment are exhaustively formulated by empirical and scientific propositions.

c) Strong Deontic Alethic Autonomy: The truth- or validity-conditions of a [non-logically valid or truth-like] deontic judgment cannot be exhaustively formulated in non-deontic terms.

(C) Logical Isolation:

a) Logical Isolation of Practical Thinking (and Discourse): No [non-logically valid, non-logical truth-like] practical thought content (or sentence) is logically equivalent to a non-practical thought content (or sentence).

b) <u>Weak Logical Isolation of Deontic Thinking (and Discourse)</u>: No [non-logically valid, non-logically truth-like] deontic judgment is logically equivalent to an empirical or scientific statement or proposition.

c) <u>Strong Logical Isolation of Deontic Thinking (and Discourse)</u>: No [non-logically valid, or truth-like] deontic judgment (or sentence) is logically equivalent to a non-deontic thought content (or sentence).

(D) <u>(Strictly) Logical Autonomy</u>

a) <u>Strictly Logical Autonomy of Practical Thinking (and Discourse)</u>: No set of non-practical (specially empirical and scientific) premises implies a [non-logically valid, or non-logically truth-like] practical thought content (or sentence).

b) <u>Weak Deontic Logical Autonomy:</u> No set of empirical and scientific premises implies a [non-logically or non-logically truth-like] deontic judgment.

c) <u>Strong Deontic Logical Autonomy</u>: No set of non-deontic premises implies a [non-logically valid or truth-like] deontic judgment.

(E) <u>Logical Sterility</u>:

a) <u>Logical Sterility of Practical Thinking (and Discourse):</u> No set of practical thought contents (or sentences) implies a [non-logically true] non-practical (in particular, an empirical) thought content (or sentence).

b) <u>Weak Deontic Logical Sterility</u>: No set of deontic judgments implies a [non-logically true] empirical or scientific proposition (or sentence).

(F) <u>Moral Autonomy</u>

As for deontic judgments in general, morality has also suffered strong and weak theses proclaiming the conceptual, the alethic, the logical isolation, the strictly logical autonomy, or the logical sterility of morality, depending on whether moral deontic judgments are contrasted with merely empirical and scientific truths or with all non-deontic judgments. There is, however, a further axis along which new even weaker theses can be formulated, namely, the contrast between moral and non-moral deontic judgments. Sometimes, besides, a new dimension is brought in, to wit: the contrast between moral deontic judgments and moral non-deontic judgments. Moreover, there is, of course, the adducible contrast between the moral Ought-to-Be and the moral Ought-to-Do.

(G) Normative Suffusiveness

The normative or prescriptive aspects of expressions cannot be distilled from the descriptive ones. The word 'mother', for instance, does not merely have a biological meaning; its very meaning includes its moral implications.

REMARKS. 1. The contrast between (A)a) and (A)a') is illustrative. It is supposed to be applicable to all the remaining theses. 2. The connections between these two rows of theses depends to a large extent on one's views about the ways in which thinking, language, and the use of language relate among themselves. 3. The words within square brackets in the above statements of theses are sometimes left out in the disputes about the autonomy of practical thinking or about the autonomy of morality. Such omission yields the immoderate or ambitious versions of the theses in question. On the other hand, if those expressions are kept in and only the brackets are deleted, we have the formulations of the moderate versions of said theses.

The above theses are the chief issues around which turn the perennial dialectics of the Ought-Is Problem.

III. THE CONCEPTUAL AUTONOMY OF PRACTICAL REASON, THE UNITY OF EXPERIENCE AND OF THE WORLD, AND BRIDGING IMPLICATIONS

1. The fundamental question about conceptual autonomy

Human living is not merely biological, but rather a historical phenomenon. To live as we humans do is not merely to be the container of chemical reactions, but to be an agent constructing a biography from inside. To live as a human is to adopt intentions and plans of life, to establish effective and best means for such plans, and to attempt to execute them, often with the help of others whom we must, and sometimes want, to enlist in cooperation. And to live through all of that is to display rational agency, i.e. to cause changes in the world that issue either from special episodes of thinking or from events that obtain in an appropriate frame of mind. Practical thinking is the special thinking directly involved either in the causation of such changes, which count as human actions, or in the formation of the appropriate frame of mind. Thus, the conceptual autonomy of practical thinking is tantamount to the irreducibility of some thought content, or part or component thereof, thought of in practical thinking.

What in a content thought of practical thinking can be the ultimate feature or element of such content that bestows upon practical thinking its special autonomous status and character?

This is the fundamental question at the very heart of the Ought-Is Problem. We cannot tackle it here, but we must be clear both about its centrality and about the main issues within its scope.

2. A moderately rich example: Sister Theresa's deliberation

Obviously, to appreciate how the different issues are packed in the Ought-Is Problem we must consider situations in which they are likely to appear. We must not, therefore, examine simple ought-judgments, but must try something more complex, at least a piece of deliberation, where we can see the main pattern of practical reason. The following example does comply with this principle of sound methodology:

(D) Sister Theresa is facing a conflict between a promisory duty and a legal duty; in the process of finding a way out she deliberates, and part of her reasoning (out loud for our philosophical convenience) proceeds thus:

1. I ought, inasmuch as I promised John, to go after 9 p.m. to keep Martha company tonight, since she's very sick.
2. I ought, inasmuch as the city curfew is still valid, to stay home after 9.
3. I must choose: to go or not to go --that's the question!
4. What shall I do? What ought I to do?
5. Now, I am in circumstances...
6. In cirumstances ... one ought to keep one's promise rather than obey the city ordinance.
7. Therefore, I ought, everything relevant being considered, to go out after 9 p.m. to keep Martha company tonight.

To be sure, the language of Sister Theresa's is a bit

stilted. She uses the word 'ought' uniformly, except for her use of 'must' to express her motivational <u>ought</u> at step 3, thus manifesting her awareness of the deontic unity of her deliberation. A native speaker of English might use other deontic expressions, e.g., perhaps substituting 'must' or 'have to' for 'ought to' at step 7, or even at 2, perhaps putting 'should' instead of 'ought to' at step 1. She is careful to signal the adverbial modifiers of her duties in conflict, and monotonously but rigorously does so by using the locution 'inasmuch as', she is somewhat verbose in expressing the toti-resultant in-balance character of her motivational <u>ought</u> at step 7. Yet she allows the context to provide the indication of such a character of this <u>ought</u> at step 5. Now, in spite of the unstylish, perhaps even unidiomatic aspects of Sister Theresa's speech, such aspects neither palliate her predicament nor touch adversely on the point, the structure, or the logic of her deliberation.

3. Deontic vs practical conceptual autonomy

Let's return to the exegesis of data. As is abundantly clear, there is no loss of generality if we focus our examination of deontic judgments on ought-judgments. Let us, therefore, return to Sister Theresa's deliberation. In our canonical notation, the crucially practical ought-judgments in her deliberation are of the canonical form:

(f) I ought Rly to A == ought-Rly(I to A)
Deontic modality: <u>ought-Rly</u>
Argument: <u>I to A</u>

Undoubtedly, rational action has to do with reasons, and, as we shall see in some detail (Part IV), deontic modalities have as chief role to point to reasons for action. This suggests that the irreducibility of practical thinking, in which we establish to find what is rational and even reasonable to to, must consist of the irreducibility or unanalyzability of at least one deontic modality. This suggestion has to be pursued to the hilt, and one must pursue it with a mind open to important alternatives. For one thing, (a) the conceptual autonomy of practical thinking may turn out to be a chimera. For another thing, (b) it may turn out that deontic modalities are irreducible, in which case they provide practical thinking with a measure of conceptual autonomy. But autonomy may <u>ultimately</u> lie deeply grounded on the irreducibility of some element in the suitable arguments for deontic modalities, i.e., something, a constituent, a formal aspect, of a deontic argument, e.g., Sister Theresa's I to A, deeply seated in her deliberation (D) above.

Of course, it seems natural, given ordinary assumptions about logic and thinking, to jump to the conclusion that Sister Theresa's thought content I to A is a proposition. Thus, the deontic modalities would have propositions as arguments, and either (true-or-false) propositions or other (valid-or-non-valid) contents as values. However, this natural theoretical leap must be taken only after very careful and wide-ranging scrutiny of practical thinking and discourse. Without pressing the point, I want to register at this juncture that my view, which has proven very fruitful(5), has been, for over thirty years, that deontic modalities are mixed modalities, and, thus, provide a subtle bridge between the contemplative and the practical sides of reason: they have special and fundamentally practical contents (practitions) as arguments, and have, on the other hand, propositions as values: this mixed character of deontic operators is a syntactico-semantical reflection of their pragmatic role of connecting reasons for doing with inclinations to do.

4. Intending and intended content as the ultimate targets of deliberation

To reinforce the methodological requirement of scrutinizing practical thinking and discourse fully and widely, let's introduce as an illustration a very significant datum. Evidently, the ultimate goal of a deliberation is NOT merely to reach an ought-everything-considered conclusion. A conclusion of such a type is just the penultimate conclusion the deliberator is aiming at. Such an ought-conclusion establishes the convergence of the balance of the agent's reasons with the balance of his inclinations, as he sees them. But he must go on, not necessarily in articulated speech, but at least in mental attitude: the deliberator must go on to acquire the appropriate state of intending. Thus, in Sister Theresa's stilted deliberation (D), she continues (D) to include the following step:

> (8) Therefore, I shall (am going to) keep Martha company after 9 p.m.

What sort of conclusion is this? Again, it is natural to

leap to the view that Sister Theresa's *I shall (am going to) keep Martha company after 9 p.m.* is a proposition. Indeed, it is true or false that she will keep Martha company after 9 p.m. But it is far from clear that what Sister Theresa is assenting to when she makes the inferential step (8) is the proposition that she will keep Martha company after 9 p.m. She is *not* predicting her action. She may even disbelieve that she will in fact carry out her intention, because she believes the curfew is very tightly maintained by the police force. It seems --but, again, I do not want here, dealing with the Meta-theory of the Ought-Is Problem, to press my view-- that Sister Theresa's conclusion (8) is not a mere contemplative, but an intrinsically practical content. If so, then *in it* we should find the feature, element, or structure that ultimately establishes the conceptual autonomy of practical thinking and of morality.

5. An issue of unity within practical thinking

Now we have a sub-problem: What is the connection between Sister Theresa's deontic argument [A] *I to keep Martha company tonight after 9 p.m.*, embedded in her deontic judgments in (D) of the form *I ought to keep Martha company tonight after 9 p.m.*, and her intended content [B] *I shall (am going to) keep Martha company tonight after 9 p.m.* at step (8)?

This sub-problem opens new theoretical forks. (On my view the connection is identity of thought content: asserted at (8), not asserted at the other steps in (D). This identity is a central piece in a unitary and comprehensive view of practical thinking).

6. Unanalyzability vs logical equivalence

Some philosophers defend the conceptual autonomy of practical thinking, or of morality, by attacking that deontic (or moral) judgments are equivalent to non-deontic (or non-moral) propositions, statements, or whatever. Since equivalence is co-implication, such defenders are forced to support the logical autonomy, or the logical sterility, of deontic judgments (or morality); often such defenders travel both ways, and end up proclaiming the logical isolation of deontic judgments (or of morality). Here we have an avoidable chain of errors.

Let's focus on logical equivalence. Reflection shows that identity, which is what the reducibility or analyzability of deontic judgments requires, is not identical with logical equivalence. Indeed, not even the strongest possible logical equivalence amounts to identity, namely, the equivalence that depends on the fewest possible principles of logic. Generally speaking, the more principles govern a domain, the more convergencies, or equivalences, are bound to develop.

A proposition P is most strongly equivalent to its corresponding selfdisjunction P or P. Nevertheless, they are palpably different: they have different truth-conditions, even though the truth-conditions of P are included in those for P or P. Clearly, another devastating simple example is the logical equivalence between P and P v (P & Q).

Likewise, two properties can be logically equivalent and yet be non-identical. A powerful example is this: the properties expressed by Fx and Fx v (Fx & Gx) are both distinct and as strongly logically equivalent as is feasible. Their mutual equivalence hinges only on the minimal set of rules governing quantificational logic; but they have different extensions.

In short, there is no need or reason to be frightened by logical equivalences. The mere fact that two thought contents are logically equivalent is a long way from threatening with the insertion of a reduction schema or analysis.

7. Bridging principles of implication undergird the unity of the world

What we have just espied about logical equivalence holds, *a fortiori*, both for weaker forms of equivalence and for implication relationships. Yet some naturalists as well as some of their anti-naturalist opponents have not realized that the conceptual heteronomy of practical thinking, of deontic discourse, or of morality is neither identical to nor implied by the existence of implication relationships between practical thought contents (or deontic judgments, or moral statements) and non-practical contents of any sort one cares to seize upon.

It might be thought that it would be nice if some sharp and shiny principles as the following were true:

(PT) *Pointcaré's Thesis*: Imperatives cannot be

derived from premises none of which is an imperative(6)

(HG) Hume's Guillotine: Ought-judgments are not implied by premises none of which is an ought-judgment(7).

Yet such alleged principles are not nice at all. They disturb the unity of the world and of reason required for rational and reasonable action. Nor are converse proposals any better, like:

(HT) Hare's Thesis: No indicative can be derived from premises containing imperatives unless it can be derived from the indicative premises alone(8).

Theses like the above are dangerous, because they foment a segregationist point of view. In their favor one can say that they are compatible with, and even suggest, a useful view of logic according to which practical contents in general, deontic judgments in particular, and moral judgments have logical relationships among themselves in parallel with those of propositional logic. This is certainly progress over the Meta-ethical Emotivists. But parallelism is not enough. We must have a full unity between what we think practically and what we think contemplatively.

Reason is a mechanism of survival. And this means, chiefly, the agent's survival by acting succesfully in accordance with his beliefs and plans in a world at least partially hospitable. But it is one and the same world that the agent cognizes and changes to suit his plans. To be effective reason must, in both cognition and purposing, have ways of thinking that oneness and sameness of the world. Reason must have as its disposal principles that constitute that oneness and sameness; such principles are principles of thought content, and, thus, also constitute the oneness and the sameness of practical and contemplative reason. As Kant well knew and insisted upon: There is only one reason, but two uses of it - theoretical and practical. Those principles of unity of world and of reason are principles of bridging implications that connect the different sectors of the world to one another. Thus, all segregationist principles, like the ones quoted above, commit the great sin of running against the unitary structural "grain" of reason. But they are dangerous because they foment a certain procrastination toward the most urgent need of intensive research to discover the whole network of principles of implication that thoroughly bridge the realm of contemplative thought contents and the realm of practical thought contents, with the natural inclusion of the bridging principles that connect the different sub-domains, e.g., that of deontic judgments, and that of moral statements. This is the sense in which segregationism is obscurantist: it impedes progress.

Some segregationist philosophers have argued for their segregationist views on the basis of an old slogan. Patently, to the extent that the slogan furnishes some support to segregationism, the slogan also runs against the unity of world and reason. Thus, in exorcising this slogan that as wrought out so much trouble in the disputes about autonomy, we are reinforcing our hold on Kant's insight. Here is the evil slogan:

> (S) Nothing is contained in the conclusion of a valid inference that is not contained in the premises.

Obviously, direct applications of (S) yield: Hume's Guillotine, and Poincaré's Thesis: (No <u>ought</u> [imperative] in the premises, no <u>ought</u> [imperative] in the conclusion), as well as many other similarly segregationist results. If the equivalences between deontic modalities we have assumed are not promoted to the status of analyses, then (S) yields drastic results like: No <u>may [wrong, liberty] in the premises</u>, no <u>may [wrong,liberty]</u> in the conclusion. This is already a serious hurdle for those who endorse slogan (S). They can accept the intra-deontic equivalences we have referred to <u>only</u> at the cost of taking them as reduction principles.

But even the ploy of taking logical equivalences as reduction-schemata fails miserably in the face of equivalences like the ones introduced in the preceding section. Consider the following, where '<==>' means logical equivalence.

(1) P <==> P v P; (2) P <==> P v (P & Q);
(3) Fx <==> Fx & (Fx & Q);

In (1)-(3) we have something that although it is not contained in the premise on the left side of '<==>' occurs on the right side. But now the ploy of taking logical equivalences to be analyses or reduction schemata is inoperative. The reason is straight-forward: as analyses(1)-(3) would be circular: they contain the analysans in the analysandum -- or the other way around if you wish.

It is worth-emphasizing that in schemata (1)-(3) the

values of the propositional or predicate variables need not be logically valid or true. Hence, (1)-(3) also cause havock for weak logical isolation and autonomy.

The preceding examples show that there is an unbridgeable gap between reduction or identity and logical equivalence.

Now, against the segregationist policy fostered by (S) and spoused by (PT), (HG), and (HT), we must proclaim that the converging unity of the knowable world and the knowing reason is so profound and crucial, that it requires pervasive bridging principles of implication to bind together all thought contents. More specifically, we endorse the following super-meta-principle:

> (TH*) All families of concepts, or properties, or entities, are, by virtue of belonging to one and the same world, connected by general structural principles of implication, which establish that unity, in such a way that:(i) the more intimately related two families of concepts or properties or entities are, the more specific and less formal their bridging implications must be; (ii) the more distant and different two families of concepts, properties, or entities are, the more general and hybrid their bridging implications must be.

Equivalences (1)-(3) above illustrate they (TH*): they link different types of logical compounds to their components. Furthermore, equivalence (3) provides also a bridge between properties (or propositional functions) and propositions.

Of course, not all bridging principles of implication are subsumed under equivalences. Thus they establish even more forcefully the falsity of slogan (S). Two trivial, yet fundamental bridges of experiential unity are the following, where '==>' means logical implication:

(4) P == Possibly(P); (5) not(P) ==>(P & Q).

Principle (4) is a bridge linking non-modal alethic logic to modal alethic logic. Principle (5) is even more pervasive and trivial: it bridges monadic to dyadic propositional connectives.

8. Some practical non-trivial bridging deontic implications

The whole point of practical thinking is to connect

beliefs and thoughts, on one side, inclinations to action, on another side, to action. As we shall see in more detail in Part IV, deontic concepts have been especially devised to connect reasons for doing to inclinations to do things. Since inclinations to action underlie our passional nature and our capacity for action, we should expect to find crucial bridging implications relating desires or wants and powers to do things with reasons for doing. That's precisely what so-called (by Kant) hypothetical imperatives do. To make a long story short, here is a fundamental schema of deontic-psychological bridging implication:

> (H1) X has a want *i* that *p* be the case
> It is causally necessary that:
> if that *p* obtains, then X performs A at *t*.
> At *t* it is within X's power to perform A at *t*.
> *Therefore*, X ought$_i$ to (or should$_i$) perform A at *t*.

Clearly, different purposes, plans, objectives, desires, and wants give rise to different *should* or *prima facie* ought-structures. The novelty here is, apparently, the suggestion that each psychological structure worthy of being called a purpose (or a having of a purpose) is constituted or characterized by a unitary deontic structure all by itself. But this is precisely what happens when an agent considers the demands arising from his passional nature as rational claims for him to weigh, with all other claims he recognizes impinging upon him, in a deliberation in search of what to do. This crucial point must be connected to the other aspects of the deontic connection between reason for doing and inclinations to do.

IV. OUGHT-JUDGMENTS AND THE LINK BETWEEN REASONS AND MOTIVATION

1. Prima facie ought's and their deliberational roles

Let us focus on the 'ought'-sentences in Sister Theresa's deliberation (D), in which the adverbial modality Rly is signaled by the rather complex adverb heralded by the locution 'inasmuch as'.Let's call them *inasmuch* ought-judgments. They play *in* Sister Theresa's deliberation a variety of roles, among which we must stress the following:

(a) The *inasmuch* deontic judgments represent claims, based on reasons, upon Sister Theresa's motivational hierarchy, i.e., her inclinations to action that - given an appropriate mental framework in her, e.g., an intentional framework, or insofar as she can be moved by reasons - can be activated by her conception of the action involved in the judgment;

(b) They signal, through the adverbial *inasmuch* modality, the nature of the reasons in question;

(c) They represent *in* the context of her deliberation Sister Theresa's recognition of the validity of the reasons alluded to in the adverbial modality;

(d) Furthermore, in the context of deliberation the thinking or the asserting of such qualified ought-judgments, which are premises accepted by the deliberator, reveal an internalization of both the reasons and the motivational claim in the deliberator's motivational hierarchy; hence, they manifest in that context Sister Theresa's awareness of such an internalization, i.e., of the fact that those reasons, though not overridingly, do incline her to act in the direction of performing the action she acknowledges that she ought, inasmuch as ..., to do.

(e) The qualification signaled by the locution 'inasmuch as', which may be underscored by a companion use of the deontic expression 'should' in place of 'ought to' or 'must', is, on the one hand, a partial rational qualification in the sense that it signals that *only* a proper subset of *all* the relevant reasons is operative in determining the motivational claim of the deontic judgment in question;

(f) On the other hand, that qualification signaled by 'inasmuch' is also *only* a partial inclination to action in the deliberating agent.

(g) The partiality of both reasons and inclinations expressed in the *inasmuch* ought-judgments is hints that a conflict may be present and that a consideration of a larger, perhaps the total network of relevant reasons should take place, or is in fact already taking place.

This partiality of the motivation and of the reasons involved is what Sir William David Ross saw and pointed to when he contrasted *prima facie* duties with duties *sans phrase*(9). His topic of study was a fully rational and wholly moral agent. Thus, he meant to be making a distinction between two types of duty within the moral ought. He saw clearly and fully that for a moral agent - whose motivational ought is imbued with morality - there can be conflicts of moral claims and that the solution, which is perforce supposed to be moral, has to be both motivational and toti-resultant (he himself used this latter term), and is wrought out by the consideration of all the relevant reasons that conflict. Ross was excellent on the general motivational aspects of conflicts of duties, except that he did not see that the _moral_ aspects of the duties in conflict and of the totiresultant duty, hence, also of the deliberation itself as a whole, are detachable. (See Section 1.7 above). Well, he didn't notice the systemic character of deontic judgments either; thus, he did not appreciate the systemic separateness of each of the _prima facie_ duties in conflict.

2. The motivational ought and the linking of reasons for doing to inclinations to act

The _motivational_ ought (must, have to) is also variously involved in deliberation (D). Like the _prima face_ ought's, it connects reasons for doing and motivation or inclination to do. But it permeates the whole deliberation and furnishes _the unity of deliberation_ by appearing both in the premises and in the conclusion. More explicitly, the motivational ought-judgments have the following crucial functions:

(a) An ought-deliberation as a whole is a practical reasoning in which the agent-deliberator aims at _evolving_ (or discovering, if you wish) a motivational-ought conclusion he can rationally endorse, from some motivational-ought presuppositions, some _prima facie_ ought-premises with systemic adverbial modalities R_1ly,...,$R_n$ly, and some factual assumptions Fl,...,Fn that function as deontic grounds or reasons for doing actions A_i (for i=1,...,n), i.e., actions A_i which he ought $R_i$ly to execute.

(b) The motivational-ought premises of a deliberation are either structural premises that present formal aspects of the agent's motivational hierarchy, or second-order premises establishing a ranking among the _prima facie_ ought-premises in conflict.

(c) The ranking wrought out by the exercise of deliberation is meant by the agent to be a motivational ranking of the impact of the grounds or reasons alluded to by his adverbial or inasmuch modalities, or that he explicitly considers, which supportively lie behind the prima facie ought premises in conflict.

(d) This ranking is, a fortiori, a motivational ranking of the conflictive ought-premises themselves.

(e) The motivational ought is toti-resultant with respect to reasons. It represents to the consciousness of the agent the total balance of the reasons impinging upon his doxastic and motivational system.

(f) The motivational ought is also inclinationally toti-resultant. The impact of reasons on his motivational hierarchy, which impact the agent expects to bring forth through his deliberating, is simply the re-arrangement of all his inclinations to action, thus creating in his organism the appropriate paths for the causation of his doing what he ought, everything considered, to do. But this does not necessarily involve mobilizing the requisite energy or causation along such paths. On the one hand, that such a mobilization not obtain is indeed required for the succesful accomplishment of future plans, which must include conditional intentions. In conditional intentions either further rehearsals of intention or thinking episodes, especially perceptual ones, may furnish the needed energy. But a deliberation process may fail to supply energy to the causal paths it creates in the deliberator's body, even if the ultimate conclusion, like Sister Theresa's step 8 of (D), is of the form Therefore, I shall (will,'m going to) do A now, because, e.g. the agent-deliberator is akratic.

This is all we can say here about the major issues alluded to under the headings "The Ought-Is Problem" and "The Autonomy of Morality". Let us turn now to a very large issue routed in the conceptual autonomy of practical thinking.

V. THE DEEP UNDERGIRDING UNITY AND THE AUTONOMY OF THE SOCIAL SCIENCE

Naturally, the networks of bridging implications provide at once the overall unity of the world and the fundamental unity of all the sciences. Such a unity is a hierarchical unity from above, from the perspective of rational agency, bestowed upon the whole of the thinkable by the pervasive unifying principles of practical reason. This is, not, however, the reductionistic unity from below that has been advocated by many philosophers of science. The two conceptions of unity need not collide, but here we cannot go into the way to steer the two so as to avoid a collision.

Assuming that practical thinking is conceptually, but of course not logically, autonomous, then there is no reduction of practical to non-practical sentences (or thought contents). Then we can find fissures in the unity from above of all the sciences. These are descriptive-normative fissures. We can distinguish the natural or descriptive sciences from the normative sciences modulo a chosen fissure.

Apparently all the disciplines customarily called social sciences have to do with norms or deontic statements. Let's use the term in this way.

Some writers hold that social scientists do not make deontic claims, that they simply study the claims made by others, and that they are, thus, mentioning, not using normative or deontic statements. This may be true. But this is immaterial for sound methodology.Social scientists may mention deontic statements, but they concern themselves with the systems to which the statement they cite belong. Thus, the logical arrangement of such statements with their implications is what counts. We can for convenience jump into the system and live it even if vicariously, and then, when we have finished, put quotation marks at the beginning and the end. The real research work will proceed in between the quotes and the unquotes.

Social scientists study, and thus think, thought contents built upon the basic contents of the following canonical form, distilled above:

(f1) X ought Rly to A.

This formula represents the autonomy and the overall unity of the social sciences. To the extent that --as we have intimated above-- the deontic concepts and the practical structuring of the fundamental practical contents are conceptually irreducible, in (f1) we have the double basis of the conceptual autonomy of the social sciences.

The underlying fundamental logical unity of all the social sciences lies there, in the logical structure of both the fundamental practical contents and the deontic modalities.

To the extent that there is no alethic autonomy for practical thought contents, except inasmuch as practical psychological states constitute relevant truth(-like) conditions, there is a further overall unity of _all_ the sciences.

The differences between the diverse social sciences all lie compacted in the institutional or motivational adverbial modality (variable Rly). Each social science, conceived as the study of behavior that conforms to rules, studies institutions, i.e., networks of rules. Thus, _each social science is just the study of a hierarchical system of adverbial modalities of the form_ Rly _in (f1)_, i.e., the study of the truth conditions of deontic judgments (deontic sentences) that arise from the adverbial institutional modalities in question. The unity of the system of adverbial modalities is the unity of the social science determined by such a system.

NOTES

(1) One comprehensive theory of practical thinking is formulated in Hector-Neri Castañeda, _Thinking and Doing: The Philosophical Foundations of Institutions_ (Dordrecht:D.Reidel Publishing Company, 1975).

(2) Perhaps the most recent manifestation of non-cognitivist emotivism is Bruce Aune's "Castañeda on Believing and Intending", in James E.Tomberlin, ed., _Agent, Language, and the Structure of the World_ (Indianapolis: Hackett Publishing Company, 1983).

(3) For a discussion of the best Emotivism see Hector-Neri Castañeda, "Logic and Ethics: Stevensonian Emotivism Revisited", _The Journal of Philosophy,_ 64(1967):671-683, and 774.

(4) For an argument for, and a detailed exposition of, the view that the institution of morality is a system involving _at least_ three deontic structures --called energetic, ethic, and meta-ethic--, see Hector-Neri Castañeda, _The Structure of Morality_(Springfield, Illinois: Charles Thomas Publisher, 1974), Chs. 1 and 8.

(5) The fruitfulness of the thesis of the mixed nature of deontic modalities spreads in several directions. The basic thesis of the proposition/practition distinction

plays a crucial role in the explanation of volition, as is discussed in Thinking and Doing, Ch.7, Section 3. It allows of a deep and rich account of weakness of the will (akrasia); indeed, it looks as if the mystery of akratic intentional choices is caused simply by the neglect of the proposition/practition distinction. See the Tomberlin volume cited above, pp.405-407. A noteworthy joint application of this distinction and the thesis of the mixed nature of deontic operators lies in a unified solution to all the known so-called paradoxes of deontic logic. For this see, "The Paradoxes of Deontic Logic: The Simplest Solution to All of Them in a Fell Swoop", in Risto Hilpinen, ed., New Studies in Deontic Logic (Dordrecht: D.Reidel Publishing Company, 1981).

(6) PT was explicitly formulated by Henri Poincaré, Dernières pensées (Paris: 1912), p.225. Richard M. Hare hails it in The Language of Morals (Oxford: Clarendon Press, 1952).

(7) Hume's Guillotine is suggested by David Hume in his celebrated passage of his Treatise of Human Nature, at I, I, i. On Hare's view ought-judgments turn out to be a special type of imperatives, thus on that view HG becomes a special case of Poincaré's Thesis. See The Language of Morals, pp.164, 168-169, among others.

(8) This thesis of logical sterility proposed by Hare in The Language Morals, p.28, is an attempt at weak symmetry with Poincaré's Thesis.

(9) See William David Ross, The Right and The Good (Oxford: The Clarendon Press, 1930), pp.19ff.

LEGAL DATA. AN ESSAY ABOUT THE ONTOLOGY OF LAW *

Aleksander Peczenik

1. REALITY

1.1. The problem of Legal Data

The central part of European "legal science" (Rechtswissenschaft) consists of "legal dogmatics" (Rechtsdogmatik), interpreting and systematizing legal norms. Legal dogmatics is analytical-evaluative in principle. Its method is similar to judicial method, although legal scholars have no power to decide cases and think in a more general manner than judges.

The purpose of the present paper is to develop some ontological assumptions a legal dogmatician must make in order to ensure consistency and coherence of his science. I leave out the question whether these assumptions are true in some non-dogmatic contexts.

According to the simplest empiricist idea, the characteristics of a science are as follows: The basic material of science consists of the observed data. One can test scientific laws and theories *via* a comparison with propositions describing data. But *what* are the data of legal dogmatics? If they merely consist in legal norms, expressed in statutes, precedents and other legal sources, then why does a legal dogmatician interpret these sources in accordance with his evaluative attitudes? On the other hand, if the data of legal dogmatics also consist of evaluations generated by the legal scholar, then why does he ascribe his results to valid law, that is, says "according to valid law, one should do X"? May one use evaluations to discover what the law actually says? May one do so to discover any part of reality? Is the law a part of reality discoverable by means of evaluations? *What* is law?

A. Peczenik et al. (eds.), Theory of Legal Science, 97–120.

1.2. Metaphysical Relativity

Such questions, "answered by serious thinkers in ...many diverse, strange, and even paradoxical ways"(1) are metaphysical. The validity of any metaphysics is relative. A metaphysics presupposes a background theory that defines the concept "real"(2) and states precisely what to regard as individual objects, their parts, their kinds, etc.(3). There may be many metaphysical systems, "all such systems being wholly comprehensive and mutually incompatible, but all equally valid descriptions of one's reality."(4) One has no way of showing conclusively which metaphysics is right. Very different metaphysical theses may appear right at different levels of reflection. One cannot easily tell whether they should be interpreted as mutually inconsistent, and whether a choice must be made between them. For example, which of the following five metaphysical theories is the best one?

Hägerström's theory. Reality includes material objects and mental processes. Material objects exist in time and space. What cannot be placed in time and space does not exist. Propositions about objective values, rights, validity and binding force of the law, not referring to anything extant, have no truth value(5).

Quine's theory. In spite of ontological (metaphysical) relativity, one can say generally that to exist is to be included in the range of values of variables of our language. Included are, thus, not only physical bodies but also numbers, classes, etc. On the other hand "propositions and attributes are disfavored by some irregular behaviour in connection with identity and substitution".(6)

Castañeda's theory. The world consists of properties and operators. An operator operates on properties and forms individual objects. An individual object is thus a bundle of properties. Another operator operates on entities and forms sets(7).

Bunge's theory. Neither properties nor individuals are independently real. Only things are independently real; a thing is an individual together with its properties. An individual is real (although not independently real) if, and only if, it is material. Every real object is either a system or a component of a system. A property is real if, and only if, at least one existing individual possesses it. Conceptual existence of sets, relations, numbers, etc., is not real but fictitious(8).

Cossio's theory. There are four kinds of objects: the ideal, natural, cultural and metaphysical. Cultural objects are real, given in experience, and positively or negatively valuable. The law is a cultural object of a special type, namely an egological object, consisting of human behaviour (in its intersubjective interference) and meaning, related to value. One cognizes the law through a cognition of values, from the point of view of a "protagonist", not an impartial observer(9).

1.3. Levels and Concepts of Existence

To avoid controversies, some philosophers single out various levels of existence or different "worlds". Popper's theory constitutes the most famous example:

> "We can call the physical world 'world 1', the world of our conscious experiences 'world 2', and the world of the logical contents of books, libraries, computer memories, and suchlike 'world 3'."(10)

One may doubt whether the three 'worlds' exist in the same sense. Let me thus distinguish three senses of "existence"(11).

Material objects $exist_1$, in time and space.

Mental processes $exist_2$, in time and in a person's consciousness.

Concepts, properties, sets, numbers, contents, etc., $exist_3$. They have no time dimension. Neither can one locate them in a dimension analogous to space.

Both mental processes - $existing_2$ - and contents, concepts, etc. - $existing_3$ - can be either descriptive (theoretical) or normative *vel* evaluative (practical).

Some entities exist in a mixed sense. For example, human action is a complex of an organism ($existing_1$) and intention ($existing_2$). Truth is a property, $existing_3$), if there is a correspondence relation between a proposition ($existing_3$) and "reality", $existing_1$, or $existing_2$, or $existing_3$, depending on the content of the proposition.

Although concepts, properties, numbers, etc., certainly do not $exist_1$ (materially), they "exist"$_3$ (conceptually, perhaps fictitiously), since one uses them as values of existential variables, to organize one's knowledge of "world 1" and "world 2". If "fictitious existence" is a contradiction, one may use another word, for example "subsistence".

2. NORMS

2.1. Valid Social Norms

One cognizes a valid social norm as an indivisible unit, a Gestalt, yet one may philosophically reconstruct it, singling out its different components. A norm is thus a complex (a so-called tuple) of the following components:

1) "World 1" entities, $existing_1$; an inscription or an utterance in a physical sense, that is, printed characters, voice; or many such inscriptions or utterances; or more intricate entities, e.g., action showing attitudes.

2) "World 3" entities, $existing_3$; the normative meaning (content) of this inscription or utterance.

The normative meaning involves the fact that the norm qualifies a human action or another event as prescribed (obligatory), permitted, prohibited(forbidden), etc.(12). It is related to the idea, $existing_3$, of events; the events may themselves $exist_1$. Its understanding is a result of a complex network of causal relations, $existing_1$, between the innate learning capacity and culture bound learning process; analysis of this process exceeds the limits of this paper(13). Understanding itself belongs to "world 2" of mental processes, $existing_2$. Viewed deeply, a norm is thus a complex of interrelated components (a tuple) of many entities of all three "worlds". The above analysis gives only a preliminary insight into a complexity of this "tuple".

In order to be a social norm, a norm must be a component of a further complex (another tuple) of the following entities:

3) The act of norm-creation. This is an intricate complex (tuple) including one of many human beings, $existing_1$, (the norm-givers), the intention to create a norm ($existing_2$) and the norm itself ($existing_3$).

Or:

4) The acts of obeying, endorsing, interpreting, etc., the norm. These are even more intricate complexes (tuples) of entities of all three "worlds".

These acts are further related to the following components:

5) "World 2" entities, $existing_2$. For example a per-

son's memory of an inscription or utterance with a normative meaning; or his intention to create, obey, endorse, interpret, etc. this norm; or many such memories, intentions, etc. of many persons.

In order to be a valid norm, a norm must be related to the following things:

6) An explicit or implicit meta-norm, $existing_3$, saying that the first norm ought to be observed(14).

Or:

7) Another, explicit or implicit, power-conferring norm, $existing_3$, saying that the creators of the first norm, for example a legislator or persons exercising some social practice, are to be regarded as competent to create a valid norm(15).

Many of these entities are interrelated in a strong sense: They are incapable of independent existence outside the complex. For example, an intention ($existing_2$) to create a norm is, in fact, an intention to create a content, $existing_3$, and thus presupposes some understanding of this content.

This analysis of the existence, social character and validity of a norm is certainly not complete. It shows, however, two things: 1) no norm $exists_1$, in the sense in which cars and cows do; 2) norms exist in a more complex sense.

2.2. Evaluative and Rational Cognition of Social Norms

One gains knowledge of valid social norms from an external and from an internal point of view. For example, an empirical sociologist can state, from an external perspective, that people regularly act in a given way; he can conclude that they follow some norms. One may ask whether such a research from an external point of view can be value-free. The question is controversial but the affirmative answer to it is thinkable.

On the other hand, one may gain knowledge of valid social norms from an internal point of view, that is, in order to learn what to do. Knowledge of valid social norms, viewed in internal perspective, is practical and value laden. From an internal point of view, one utters genuine norms, not merely propositions about norms. One also utters interpretative statements, having a complex meaning. To explain this fact, let me single out three ideal types of practical statements.

a) Emotional statements, for example "This is a fine day!", or "I would like to drive a Ferrari!", or "Do not touch this flower!", etc., expressing a spontaneous attitude of an individual towards a real or imagined situation.

b) Rationally oriented practical statements(16), for example statements of intentions, imperatives, etc., expressing a rationally oriented attitude of an individual towards a real or imagined situation. Such statements result from an attempt to rework emotional statements rationally.

Let me remind you here of a theory of rationality(17). A necessary condition of rationality is a practice of having or giving reasons, supporting a conclusion. It seems that all kinds of reasons consist of the following "atoms": facts, values and logical inferences. "Facts" are entities of any of the three "worlds". Values are related to weighting(18) reasons, that is, evaluating whether one reason in a given situation is more important than another, _inter alia_ from the moral point of view. Logical inferences are correct in all or at least some reasonable interpretations of vague words. Reasons can be more or less good. Good reasons fulfil, _inter alia_, the demand of logical consistency and are oriented towards two regulative ideas, the idea that one should be able to express both description and evaluation generally and the idea that one should be able to produce a coherent totality of reasons, in which many statements support many others.

c) Interpretative statement(19) expressing cognition and endorsement of valid social norms. They express a social and rational attitude of an individual towards a real or imagined situation. Interpretative statements result from an attempt to rework emotional statements not only from the point of view of consistency, coherence and generality but also from the point of view of one's expectations of acceptability, that is, consensus from other rational individuals. On the other hand, interpretative statements express norms themselves, _cf. infra_.

2.3. Subjective Component of Valid Social Norms

We have seen (_see_ section 2.1. _supra_) that a valid social norm is an intricate complex (_tuple_) of an utterance or inscription and its meaning, further related to some acts, such as norm-creation, interpretation etc.; to mental processes, such as memories and intentions; and to other norms, deciding about its validity. Among other things, the norm is related to

4) the acts of endorsing, interpreting and obeying the norm.

We have also seen (<u>see</u> section 2.2. <u>supra</u>) that one can view these acts in external perspective, sociologically stating how others endorse, interpret and obey the norm, or in internal perspective, involving one's own interpretation, endorsement of and obedience to the norm. From an external point of view, the norm is thus related to

4e) the fact that other people endorse, interpret and obey it.

From an internal point of view, the situation is more complex. Let me start from en epistemological thesis, formulated from an external point of view but concerning the question how a given interpreter gains knowledge from an internal point of view.

ETe) An interpreter gains knowledge of the norm from the internal point of view through evaluative interpretation, paying attention to socially established validity criteria and involving rationally and socially oriented endorsement. Other persons may test this interpretation.

The corresponding ontological thesis is this:

OTe) Viewed in internal perspective adopted by an interpreter, a valid social norm has precisely the content that, at the same time, fulfils socially established validity criteria, corresponds to the socially and rationally oriented interpretation performed by this interpreter, and is acceptable to many other people.

The interpreter himself can formulate this ontological thesis from the internal point of view, in the following manner:

OTi) A valid social norm has precisely the content that simultaneously fulfils socially established validity criteria, should be accepted from a rational and social point of view, and is acceptable to many other people.

A <u>valid</u> norm thus does not exist for a given interpreter unless he can endorse it in a rationally and socially oriented manner. One's rationally and socially oriented endorsement is conceptually necessary for the norm's validity.

According to the socially established validity criteria, the norm is valid only if related to many other things (see section 2.1 item 1-7 supra). Consequently:

OTec) Viewed in internal perspective, adopted by an interpreter, a valid social norm has precisely the content that simultaneously (1) fulfils socially established validity criteria, that is, corresponds to the system of relations between the utterance or inscription in question, its meaning, and some acts, mental processes and other norms; and (2) can be endorsed in a rationally and socially oriented manner by the interpreter in question; and (3) is acceptable to many other people, interpreting it in a socially and rationally oriented manner.

If a social norm is valid, the corresponding interpretative statement is correct. This correctness is related to the truth of propositional meaning components of this statement, see infra. I define truth as correspondence with facts, and thus not as a mere consensus, but we will see that the facts with which the interpretative statement corresponds include some consensus.

TC) Viewed in internal perspective, adopted by an interpreter, an interpretative statement concerning a norm is correct if (1) it corresponds to the socially established validity criteria, that is, to the system of relations between the utterance or inscription in question, its meaning, and some acts, mental processes and other norms; and (2) it can be endorsed in a rationally and socially oriented manner by the interpreter in question; and (3) it is acceptable to many other people, interpreting it in a socially and rationally oriented manner.

These three conditions are mutually irreducible. Condition 1 thus differs from condition 3. Socially established validity criteria thus normally follow social acceptability, yet in some cases may lead to socially unacceptable results. Condition 2 also differs from condition 3. When Interpreter 1 rationally and socially endorses the norm (condition 2), he adapts its content to the good of others, by his own standards of rationality (e.g., of optimalization of generality and coherence) and goodness. When he states that the norm is rationally and socially acceptable to others (condition 3), he applies _their_ standards of rationality and goodness.

Let me now assume that an interpretation has already passed the evaluative test of rational and social endorsement (condition 2) and the factual comparison with the established criteria (condition 1). One can, then, consider the following fragmentary _truth_ condition:

FTC) Interpretation performed by Interpreter 1, in brief Interpretation 1, is true only if it is acceptable to many other interpreters belonging to the set including Interpreter 2, Interpreter 3, Interpreter 4, etc.(20).

One can ask, moreover, for some preconditions of acceptability. For example, one can consider the following precondition:

PCA1) Interpretation 1 is acceptable to many interpreters if, and only if, it is identical with the interpretation they would present in a rationally and socially oriented manner.

This precondition is, nevertheless, too strong. Various interpreters often appear to agree with each other, although their interpretations are merely similar, not identical. Hence, the following precondition is better:

PCA2) Interpretation 1 is acceptable to many interpreters if, and only if, is it relevantly similar to the interpretations they would present in a rationally and socially oriented manner.

One may now ask: why is acceptability possible? Because of the similarity of interpretations which various rationally and socially oriented persons. But why is this similarity of interpretations itself possible? Two factors explain this possibility. First, all the interpreters must fulfil the same demands of consistency, generality and expected acceptability, and correspondence with the same factual and evaluative data; _see_ section 2.2. _supra_. Second, all of them resemble each other to some extent, since they have the same innate and culture-bound dispositions, _i.e._, share a form of life(21).

In the light of this analysis, interpretative meaning may be analytically divided into four components: (1) pure emotive meaning, which one may express in a norm, neither true nor false; (2) rational emotive meaning, which one may express in a true or false proposition about what is the optimally rational reworking of one's emotions; (3) social interpretative meaning, which one may express in a true or false proposition about the expected rational acceptability by many other interpreters; (4) descriptive meaning concerning criteria, to be expressed in a proposition saying whether the interpretative statement conforms to the socially established validity criteria. Correctness of interpretative statements is an optimal approximation of truth of all three propositions indicated in items 2)-4). Assume now that two interpreters, 1 and 2, interpret the same norm, N. Their interpretations, N_1 and N_2, can be similar to each other and acceptable, yet incompatible, e.g. when they solve 99% cases in the same way but 1% in incompatible manners. Yet N_1 and N_2 can be simultaneously correct, each adapted optimally to social validity citeria, social expectations of acceptability and rationally reworked emotions of two different interpreters.

3. THE LAW

3.1. Valid Legal Norms

Valid legal norms constitute a subclass of valid social norms. The need of many such subclasses results from the need of division of labour; the law has one "job" to do, morality another, etc. We thus need the law when we need a relatively centralized, sanctioned, bureaucratic, predictable and just decision procedure.

Similarly to other social norms, a valid legal norm is a complex (tuple) of inscription(s) and a normative meaning, further related to some acts, mental processes and other norms. It has a content that fulfils established criteria, corresponds to one's rationally and socially oriented interpretation and is acceptable to others. Differences between legal and non-legal norms do not concern their ontological structure, but merely the following questions:

1) The content of legal norms. Legal norms are components of a system that contains some norms claiming sovereignty of the law, a sole right to authorize exertion of physical force, a right to regulate all situations, etc.(22).

2) The structure of the system of legal norms. Legal norms constitute a hierarchy in which higher norms decide about the proper way to create lower norms(23).

3) The structure of legal reasoning. One can doubt everything but not everything at once. For that reason, the totality of human reasoning is divided into sub-classes, each taking some premises for granted. Even physics takes something for granted, as the famous discussion about paradigms amply shows. In legal reasoning, one takes for granted some norms of reasoning(24), that is, source norms and reasoning norms. The former say that statutes, precedents, legislative history, juristic literature, etc. are legal sources and assign a weight to each source. A source norm says, e.g., that all courts and authorities must use statutes and may use juristic literature in the justification of decisions. The latter are oriented towards requirements of rationality. They are thus oriented towards consistency: Many well-known "collision-norms" thus demand that the lawyers arrange incompatible legal norms into priority-orders, for example according to the principle that a higher norm (*lex superior*) has priority before a lower one. Many reasoning norms are also oriented towards generality: Legal judgments must thus be supported by universal premises; one should interpret similar statutory provisions similarly; in some cases, one should rely upon reasoning by analogy, etc. Legal reasoning norms are also oriented towards coherent support, from both the sources of the law and moral considerations. The former is expressed in all reasoning norms demanding respect for statutes, precedents, etc. The latter shows itself, *inter alia*, in the principles demanding that an individual's freedom should be protected from judicial arbitrariness, and that only a cautious reasoning by analogy is admissible in criminal law. Many reasoning norms are also oriented towards justice. The norms of legal reasoning constitute a set that, taken as a whole, is peculiarly related to many concepts, including the concepts of law, legal reasoning and legal ought. One who is simultaneously refuting a significant number of these norms and still trying to establish the content of the law or to perform legal reasoning, would have to think incoherently or to change many concepts he uses. Yet, one can disregard some norms of legal reasoning and formulate new ones , if important reasons for doing so exist.

4) Some relations between persons involved in creation and interpretation of legal norms. In order to be legally valid, a norm must be somehow related to some of the following acts. The act of norm-giving, performed by legis-

lators and precedent-creating courts. The acts of persons, often lawyers, who endorse and interpret the law. The acts of persons who observe the law. The acts of persons who recognize such legal acts as purchase, nomination of an official, etc. All such acts are interrelated complexes ("tuples") of a given human being ($existing_1$), his intention ($existing_2$), for instance the intention to create or to interpret a norm, and the norm itself, $existing_3$. The mode of existence of such complexes as regards legal norms is similar to the mode of existence of corresponding complexes in the case of other social norms. Yet, the law is more complex than some other social norms. For example, the acts of legislation are seldom performed by a singe individual. On the contrary, they are ascribed to a "legislator", further analyzable as a relation between several persons, norms, etc.

5) Content of some thoughts, $existing_2$, associated with norms. Some mental processes of persons engaged in law-making and legal reasoning often have a content referring to mental processes of other persons. For example, a legal researcher understands a norm _via_ ascribing it to an actual or hypothetical intention of a physical person or, more often, of a construct called "legislator"(25). Let me leave open the question whether such an ascription is necessary to legal validity.

6) The difference between legal and non-legal ideas of "the ought". Legal validity is explicable as a conceptual relation, $existing_3$, of a legal norm to a meta-norm, saying that the former ought to be observed, or saying that its creator has a power to create valid law. To many interpreters, this "ought" exists only if the legal system as a whole is weakly acceptable, that is, it is not extremely immoral (lex _iniustissima_)(26). This condition of weak acceptability is _fulfilled_ if (1) the legal system is not _lex iniustissima_ from the point of view of rationally and socially oriented evaluations of a given interpreter (that is, legal decider or researcher); and (2) many other interpreters would not consider it _lex iniustissima_, if they had performed its rationally and socially oriented interpretation. A valid legal norm is thus related to the interpreters' evaluations. Yet, the relative importance of evaluations is different at different stages of the process of justification(27) of legal knowledge. It is not great as regards establishing the fact that a given normative system as a whole, for example "the

Swedish law", is a legal system. It is greater as regards establishing legal validity of the sources of the law in this system, for example, as regards establishing the fact that a given Swedish statute is valid law. Finally, it is greatest, as regards establishing the interpreted content of legal norms, expressed in some legal sources or unwritten.

7) Consequently, one may distinguish two levels of the law, the legal sources (the "input law") and the interpreted ("output") law. Because of the role of evaluations in legal interpretation, these levels are not identical. Yet legal interpretation is supposed to justify knowledge of the law, often (especially in clear cases) vaguely identified with knowledge of legal sources. Courts and authorities have a legal duty to base their reasoning on legal sources, even if they are obscure.

3.2. The Concept of Law

These characteristics of legal norms enables one to define the law. The law is thus a system of interrelated components. One component consists in legal norms. Other components consist of various entities to which legal norms are related.

In other words, the law is a complex (tuple) of legal norms (themselves being complexes of inscriptions or utterances and their meaning), some acts and mental processes of various persons involved in legislation, interpretation, etc., some other norms, and finally, each interpreter's weak evaluative endorsement. This applies also to fragments of the law, such as the law of torts etc.

3.3 The Problem of Intersubjectivity

Such a definition of law leads, however, to the problem of intersubjectivity. I have thus expressed three inter-related theses:

Viewed in internal perspective, adopted by an interpreter, a valid social norm has precisely the content that fulfils socially established validity criteria, is acceptable from this interpreter's evaluative point of view, and is acceptable to many other interpreters.

Viewed in internal perspective, adopted by an interpreter, a valid legal norm has precisely the content that fulfils socially established criteria of legal validity, is weakly acceptable from this interpreter's evaluative point of view (<u>see</u> section 3.1) and is weakly acceptable to many other interpreters.

Viewed in internal perspective, adopted by an interpreter, valid law is a complex (tuple) of some facts, such as the existence of some norms, acts and mental processes of others, and this interpreter's weak evaluative endorsement.

One may regard such theses as queer. Valid social norms, <u>inter alia</u> legal and valid law, are said to exist, let it be in an intricate manner; we all assume that there is only one Swedish law, the same for all interpreters. On the other hand, the law, legal norms and social norms in general, appear to be complexes of existing facts, identical to all observers, and evaluations, subjective and, thus, <u>not</u> identical to all interpreters. My law is, thus, not identical with your law, though they are similar enough to enable our common discussion.

One cannot easily eliminate this apparent contradiction. Let me merely point out that this difference between inter-subjective facts and subjective values is less sharp than one might expect.

Facts. In my mind, there $exists_2$ a perception of some data. I arrange it rationally and conclude that some facts $exist_1$, or $exist_2$ in other minds, or $exist_3$. In your mind, there $exists_2$ a perception of some data. You arrange it rationally and conclude that some facts exist. We then agree that the facts you think about are the same as the facts I think about.

Values. In my mind, there $exists_2$ an evaluation, that is, an evaluative thought. I arrange it rationally and conclude that something has such and such a value. In your mind, there $exists_2$ an evaluation, that is, an evaluative thought. You arrange it rationally and conclude that something has such and such a value. Then we agree that we have been thinking about the same "something" and the same value.

Where lies the difference? It seems that the factual claim to objectivity, "We all perceive the same facts", is stronger than the evaluative claim to objectivity, "We all perceive the same values". If one explicates a value as a tuple of some facts and one's emotions, this difference becomes clear: Only facts are the same to all observers, emotions are "private"; the tuple as a whole is partly common to all, partly "private".

One problem concerning this weak inter-subjectivity is that one may grade it, whereas "objectivity" is not so graded. Yet already this weak claim to objectivity justifies the theses that (1) the law exists, though it has an evaluative component, and (2) one uses evaluations to gain knowledge of the law.

3.4. Legal Data and Scientific Data

These characteristics of the law enables one to discuss the "data of legal science". Science is based on data. The data of "legal science" must consist of information about all components essential to law, their relations, and for each interpreter, of his own rationally and socially oriented evaluations.

It is useful to compare such legal data with scientific data. One way to do so starts from Imre Lakatos's reconstruction of science. According to Lakatos, a given research programme, that is, a series of scientific theories, contains a "hard core", protected by auxiliary hypotheses; one should direct objections and counter-examples against the auxiliary hypotheses, never against the "core". The research programme is no longer progressive but rather degenerative, and often gives way to another one with another "core", when it no longer produces theories with greater and greater empirical content(28). Aart de Wild applied these ideas to the non-evaluative component of legal reasoning(29). However, an interpreter of the law not only describes enacted legal norms but also adapts them to the "new" evaluations he continually expresses. To include this evaluative side of legal interpretation, one must freely modify Lakatos' theory.

To achieve this adaptation, let me assume that evaluations expressed by the interpreter constitute a part of legal data. The sources of the law, such as statutes, precedents, legislative history, juristic professional literature etc., constitute another part of legal data. The interpreter thinks about both kinds of data, and these thoughts $exist_2$. The data exist themselves in different senses. The evaluative data exist as tuples of emotions ($existing_2$), contents and other facts. The content of legal sources $exists_3$ (conceptually) in relation to some other data, $existing_1$ (materially) or $existing_2$ (mentally). On the other hand, the legal sources are products of rationally and socially oriented evaluative deliberation of their authors, for example the legislators and judges. Because of this rationality factor, they are analogous to theories and their "cores", albeit evaluative. Remembering the relation of legal norms to human behaviour etc., one can thus classify the legal data, as follows:

1) Data in the descriptive sense, $existing_1$, $existing_2$ or $existing_3$;
   a) data of empirical social science and common sense knowledge about the behaviour of legislators, judges, ordinary people, criminals, etc., creating, interpreting, obeying and violating the law; data about their mental processes;
   b) data about norms expressed in statutory enactments, precedents and other legal sources.
2) Data in the normative sense, that is, evaluations expressed by an interpreter in the process of rationally and socially oriented deliberation.

These data correspond to some theory "cores" and to some theories. One can single out the following theory "cores":

1) "Cores" of descriptive theories, of social science, linguistics, common sense, etc., if any;
2) "Cores" of evaluative, rationally and socially oriented, deliberation: the socially established products of deliberation performed by others, that is, statutes, precedents, and perhaps other legal sources; the interpreter must interpret them and may not totally refute them.

Finally, one can single out the following types of legal theories:

1) Ordinary descriptive theories of social science, linguistics, common sense, etc.
2) Theories of practical (evaluative) legal dogmatics about the "expert meaning" of the legal sources, and, in general, about the "real law" whose cognition involves evaluative endorsement.

This philosophical reconstruction assumes that common sense data belong to a more primitive level than the law.Yet a holistic insight reveals the fact that data, theory cores and theories of common sense, empirical science and legal dogmatics are intertwined, thus giving rise to such well-known phenomena as causal feed-back and hermeneutical circle(30). To understand legal dogmatics, one must somehow join the reconstructivist and the holistic point of view.

3.5. Legal Validity As an Emergent Property of a System of Relations

This characterization of the law enables one, finally,to conclude that the properties of "being legal" and "being legally valid" are emergent. Let P be a property of a system, X, or of its components. Then P is a resultant propterty if, and only if, P is a property that some components of X possess independently from their membership in X. Otherwise P is an emergent property of X(31).

I have said above that valid law is a complex of many components, that is, norms (themselves being complexes of inscriptions or utterances and their meaning); acts; mental processes; and, finally, each interpreter's weak evaluative endorsement. This complex has the property of "being legal" and the property of "being legally valid". One component of this complex, that is, the set of norms, also has these properties; so do members of this sets individually. The norms in question are thus legal and legally valid. Yet, they have these properties only because of being components of the complex. In isolation from the acts of legislation, interpretation, etc., and in isolation from intentions and other thoughts accompanying these acts, one cannot assign the property of legal validity either to individual norms or to their system. We know that a norm is legal or legally valid only if we know something about the circumstances of its creation or interpretation. This conclusion is obvious as regards rules of enacted law; these are legal owing to the act of legislation. This conclusion however, is valid also as regards unwritten principles of law. These are not merely moral principles but also legal, owing to the fact that an interpreter of law justifies them in relation not only to moral appropriateness but also to such materials as statutes, precedents, legislative history, etc., themselves being products of some acts.

Whereas both the whole complex of norms, acts, mental processes, etc., and the norms constituting a component of this complex, have the properties of being legal and being legally valid, other components of this complex do so in a slightly different sense. An act of legislation can be both legal and legally valid. An act of interpretation of a legal norm can be legal, although not in exactly the same sense as the norm itself; on the other hand not called "legally valid". Legal norms are sometimes called "valid law" but one does not call the acts of legislation or interpretation "valid law". One may note more such subtle differences. The important point is, however, that even if some components of the discussed complex possess both the property of being legal and the property of being legally valid, they do so only because of their being components of this complex not in isolation from other components(32).

4. "JUMPS"

4.1. Non-deductive Steps in Legal Reasoning

Let me now connect these ontological reflections with an epistemological theory I have developed in some other works.

Legal reasoning involves many non-deductive steps. One may need such a step to derive a conclusion about valid law from premises that do not mention valid law; for example, to derive the conclusion that the Swedish Constitution of 1974 is valid law from the description of the Swedish society etc. One may also need such a step to derive a conclusion about some legal sources from other sources; for example, to derive the conclusion that one should follow, in Sweden, preparatory legislative materials from the premises included in Swedish statutes and precedents. Finally, one may need such a step to interpret the law, inter alia to derive a conclusion about the "expert meaning" of a legal norm from the legal sources.

A justified non-deductive step ("jump") is performed from p to q if, and only if, the following conditions are fulfilled: (a) Truth (or rightness, etc.) of p is a sufficient reason for truth (or rightness) of q; and (b) p does not deductively entail q.(33)

One can try to present as extensive a part of legal reasoning as possible in terms of logical deduction. One can thus convert a "jump" into deductive reasoning through adding or changing some premises or inference rules. However, these "new" premises or inference rules require justification. By definition, they cannot be analytic (logically necessary). They can be certain (evident) from the point of view of legal justification, contextually sufficient when one litigates, writes legal treaties, and is otherwise engaged in a legal discourse. On the other hand, one cannot convert a "jump" into deductive reasoning through adding a finite number of premises or inference rules that are certain from the point of view of deep (not merely legal) justification, that is, when one gives to one's conclusion as much support as one meaningfully can give at all (not only in the legal discourse but, for instance, also in a moral or political debate). For example, the premise "one ought to observe the constitution", indubitable for a lawyer, is open to a philosophical or political doubt, for example in an Anarchist context. In particular, no evaluative premises and inference rules, regardless how convincing for lawyers, are beyond all doubt(34). In many cases, a conversion of a "jump" into deductive reasoning requires bridging premises or principles, connecting different domains, for example, law and morality(35).

4.2. "Jumps" and Ontology of Law

I have said above that the properties of "being legal" and "being legally valid" are emergent properties dependent upon a relation (or a system of relations) between norms, acts, mental processes and, finally, each interpreter's weak endorsement. Moreover, the whole legal system is understandable only in relation to these properties. Now, one cannot derive emergent properties from resultant properties only. It is thus logically impossible to deduce the properties Y has only in relation to X from properties Y has in isolation from X. Especially, one cannot deduce the properties of "being legal" and "being legally valid" from non-legal properties of a norm, such as its content.

One reason in favour of this thesis is the fact that legal validity of a norm, viewed in the internal perspective of a particular interpreter, involves his weak endorsement of the norm. One cannot deduce an endorsement of a norm from its content. This deduction is impossible because it would assume additional premises, justifiable by evaluations only. These would never be beyond all doubt.

One can deduce some propositions about emergent properties from a set of premises containing some propositions about resultant properties and some additional premises. For example, one can deduce some propositions about legal validity from some propositions about the content of some norms together with such additional premises as "*lex iniusta non est lex*", expressing some factual and moral standards of legal validity. These additional premises are, however, not certain (not evident), *inter alia* because some of them, e.g., the moral standards, express evaluations.

Consequently, one cannot convert the reasoning step from propositions about such resultant properties of legal norms as their content to propositions about the emergent property of legal validity through adding a finite number of certain (evident) premises or inference rules.

The role of evaluations in legal cognition is thus a sufficient ground for an ontological conclusion and for an epistemological conclusion. The ontological conclusion is that the law has emergent properties. The epistemological one is that the law is cognizable *via* "jumps". The role of evaluations is sufficient ground for these conclusions, provided that one endorses the weak claim of value-statements to inter-subjectivity, *see* section 3.3. *supra*.If one refutes this claim, one must endorse unreasonable theses, such as "there is no law", or "evaluative interpretation gives no knowledge of the law".

This is the point of the present theory. One needs the idea of "jumps", emergent properties, conceptual existence, practical rationality, etc., in order to save the legal discourse from devastating but over-simplified philosophical criticism, expressed by various schools of Legal Realism, value nihilism, etc.(36)

NOTES

(*) I express my gratitude to Lars Lindahl for a very important discussion.

(1) H.L.A.Hart, *The Concept of Law*, Oxford 1961, p.1.

(2) *Cf.* W.V.O.Quine, *Ontological Relativity and Other Essays,* New York 1969, pp.53 ff.

(3) *Cf.* N.Goodman, *Ways of Worldmaking*, Sussex 1978, pp.7 ff.

(4) H.N.Castañeda, *On Philosophical Method,* Bloomington, Indiana 1980, p.19.

(5) *Cf.* A.Hägerström, 'Sellbstdarstellung', in: R.Schmidt

(ed.), Die Philosophie der Gegenwart in Selfbstdarstellungen, vol.7, Leipzig 1929, pp.111 ff. Cf. K.Marc-Wogau, Studier till Axel Hägerströms filosofi, Falköping 1968, p.113.

(6) Quine, Ontological (n. 2 supra), p.98. Cf. pp.91 ff.

(7) Cf. Castañeda, 'Thinking and the Structure of the World. Discours d'ontologie', Philosophia. Philosophical Quarterly of Israel 4 1974, pp.10 ff.

(8) Cf. M.Bunge, Scientific Materialism, Dordrecht-Boston-London 1981, pp.20 ff. and 165 ff.; and Bunge, Treatise on Basic Philosophy, vol.3, Dordrecht-Boston 1977, pp.99 ff.

(9) Cf. C.Cossio, La teora egologica del derecho y el concepto juridico de libertad, 2 ed. Buenos Aires 1964, pp.54 ff.Cf. J.Wróblewski, 'Law and Liberty in the Egological Theory of Law', Österreichische Zeitschrift für öffentliches Recht 16 1966, pp.4 ff.

(10) K.R.Popper, Objective Knowledge. An Evolutionary Approach, Oxford 1972, p.74.

(11) Quine, Ontological (n.2 supra), pp.99-100, grants that there are different senses of existence, but there is no confusion.

(12) A.Peczenik, Podstawy semantyczne logiki norm, Kraków 1963, unpublished doctoral thesis. Cf. Peczenik, 'Norms and Reality', Theoria 34 1968, pp.117 ff. (=Essays in Legal Theory, Copenhagen 1969, pp.9 ff.).Cf. Peczenik, The Basis of Legal Justification, Lund 1983, pp.7 and 137.

(13) Cf. Peczenik, 'Empirical Foundations of Legal Dogmatics', Logique et Analyse 12, 45 1969, pp. 35 ff. (=Essays, n.12 supra, pp. 45 ff.)

(14) Cf. Peczenik, Basis (n.12 supra), pp.6, 17 and 137; references to W.Lang, G.H.v.Wright, K.Olivecrona and H.Ofstad.

(15) Cf. , e.g., T.Strömberg, Inledning til den allmänna rättsläran, 8 ed. Lund 1980, p.89.

(16) For example, deliberative statements, cf. Castañeda, Thinking and Doing, Dordrecht-Boston 1975, pp.25 ff.

(17) Re an older version of this theory of rationality, cf. Peczenik, Basis (n.12 supra), pp.84 ff. The present version is a result of my discussion with Robert Alexy in August 1983. Alexy elaborated a stronger theory, cf. A.Aarnio, R.Alexy and A.Peczenik, 'The Foundation of Legal Reasoning', Rechtstheorie 12 1981 pp. 257 ff.; R.Alexy, Theorie der juristischen Argu-

mentation, Frankfurt/M 1978, pp.221 ff.

(18) *Cf.*, e.g., T.Eckhoff, 'Guiding Standards in Legal Reasoning', *29 Current Legal Problems* 1976, pp.205 ff.

(19) *Cf.* Aarnio in Aarnio, Alexy and Peczenik (n.17 *supra*), pp.427.

(20) Aarnio formulated a stronger thesis: "The interpretation for which one can achieve the rational acceptance of the widest audience has the greatest societal relevance", *cf. id.* p.444. The discussed formulation, weaker than Aarnio's is, nevertheless, consistent with it. I leave out the complex problem to what extent I endorse Aarnio's thesis.

(21) *Cf. id.* and Peczenik, *Basis* (n. 12 *supra*), pp.101 ff. and 176 ff., referring to many works of Wittgenstein and Aarnio.

(22) *Cf. id.* p.14. *Cf.* J.Raz, *The Authority of Law,* Oxford 1979, pp.118 ff.

(23) *Cf.* Peczenik, *Basis* (n.12 *supra*), p.14. *Cf.* H.Kelsen, *Reine Rechtslehre,* 2 ed. Wien 1960, pp. 228 ff.

(24) *Cf.* Peczenik, *Basis* (n.12 *supra*), pp. 33-76; extensive references.

(25) *Cf.* K.Olivecrona, *Law as Fact*, 2 ed. London 1971, pp.128 ff. on "independent imperatives".

(26) *Cf.*, e.g., J.Lucas, *On Justice,* Oxford 1980, p.123. *Cf.* Peczenik, *Basis* (n.12 *supra*), pp. 15 and 143.

(27) *Cf.* Peczenik, 'Taking Laws Seriously', *Cornell Law Review* 68 1983, pp.663 ff.

(28) *Cf.* I.Lakatos, 'Falsification and the Methodology of Scientific Research Programmes', in: I.Lakatos and A.Musgrave (edd.), *Criticism and the Growth of Knowledge,* Cambridge 1970, pp. 132 ff.

(29) Cf. A.H.de Wild, *De Rationaliteit van het rechterlijk Oordeel*, Deventer 1980, pp.55 ff.

(30) I am indebted to Jan Broekman for this observation.

(31) *Cf.* Bunge, *Treatise* (n.8 *supra*), pp.97 ff. and *Bunge, Scientific* (n.8 *supra*), pp. 26 ff. *Cf.* P.Oppenheim and H.Putnam, 'Unity of Science as a Working Hypothesis', in: H.Feigl, M.Scriven and G.Maxwell (edd.), *Minnesota Studies in the Philosophy of Science,* vol.2, Minneapolis 1958, p.15: "there appear certain new phenomena which are 'emergent'...in the sense of being forever irreducible to laws governing the phenomena on the level of the parts".

(32) In this manner, the relation, not its components, is

the primary object of legal philosophy. Such a shift of emphasis from things to relations is characteristic of many lines of development of modern culture, *cf.* J.M.Broekman, *Strukturalismus*, Freiburg-München 1971, pp.155 ff.

(33) *Re* older versions of this definition, *cf.*, *inter alia*, Peczenik, *Basis* (n.12 *supra*), pp. 3 ff. and 70 ff., and Aarnio, Alexy and Peczenik (n.17 *supra*), pp.136 ff.

(34) *Cf.* Peczenik, *Basis* (n.12 *supra*), pp.70 ff. Robert S.Summers has suggested the words "contextually sufficient" and "deep" (*cf.* id. pp.1 and 135. The distinction between the contextually sufficient and deep legal justification corresponds to the difference between the "legal ought" (cf. section 3.1. item 6) supra) and the "overriding ought" (*cf.* Castañeda, *Thinking*, n.16 *supra*, pp.303 ff.). Justification of the legal ought through a derivation back to the presupposed fundamental norms is characteristic of H.Kelsen (*Reine*, n.23 *supra*, pp.196 ff.). For him, it was a consequence of a combination of moral relativism and the objectivity claim of the legal ought. "Although one thus cannot answer the question of rightness of fundamental evaluation in a generally valid manner, the situation is different when one postulates validity of such an evaluation and asks what consequences it has for judging a concrete case", J.M.Jansson, *Hans Kelsens teori mot bakgrunden av hans rättsfilosofiska askadning*, Helsinki 1950, p.33 (my translation). In my opinion, the contextually sufficient legal reasoning is only *relatively* more inter-subjective than moral reasoning, but already this difference of degree justifies autonomy of the former and the special character of the legal ought.

(35) *Cf.* Castañeda, *Thinking* (n. 16 *supra*), p.333.

(36) *Cf.*,e.g., Olivecrona (n. 25 *supra*) *passim*.

ABSTRACT

The law is a system of interrelated components. The sytem includes: (1) linguistic units, *inter alia* norms, that is, relations between inscriptions or utterances and their meaning; (2) behaviour of 'law-givers', courts, legal researchers, persons who perform such acts as purchase, etc.; (3) mental processes of all these persons; and (4) rationally reworked, socially oriented and evaluative endorsement by each interpreter, _inter alia_ a judge or a legal researcher. In internal perspective, a valid legal norm does not exist for a given interpreter unless he can weakly endorse it. He can do so, if the total system of alleged legal norms is not _lex iniustissima_. His evaluative interpretation of the law is true only if it is essentially similar to rationally and socially oriented interpretations that many persons would present in his situation. Legal norms are valid. This property is emergent, that is, presupposing a system of relations; put without the system, no component has this property, yet the system is understandable only in relation to this property.

Some emergent properties, e.g., validity, are cognizable _via_ non-deductive steps ("jumps"). A justified "jump" from p to q takes place if, and only if, the following conditions are fulfilled: (a) truth (or rightness) of p is a sufficient reason for truth (or rightness) of q; and (b) p does not deductively entail q. One converts a "jump" into deductive reasoning through adding some new premises or inference rules and changing some old ones. Such a change or addition makes the passage from p to q deductive, only if the added premise or inference rule is not deductively justifiable on the basis of a finite number of evident (certain) premises.

PLURALIS JURIS

Stig Jørgensen

PLURALIS JURIS(1)

It is a substantial idea of my book that law like any other phenomenon in reality cannot be defined in one and only one way. Any definition which is a mental construct isolating particular elements of the phenomenon and putting them together according to criteria in a pre-existing mental system is in the end governed by some general or special cognition interest. Definitions are tools of the mental activity called science, the purpose of which is to increase the knowledge of man. Definitions or concepts are abstractions of some part of reality, abstractions which make it possible for man to communicate in a symbolic way, information from one individual to another.

It is common knowledge for everybody today that science is not and cannot be entirely objective. Science must be dependent on a meta-science or philosophy of science which lies behind science as a guiding principle. What is known as meta-science does not constitute a subject of its own but is the science of science. Like philosophy it deals with the basis of science, but it does not, however, treat the question of cognition as such as an essential problem but rather focuses on the methods applied by science. The ideal of the so-called logico-empiristic or positivistic meta-science of the Vienna circle, which dominated from the 1920s to the 1950s, was that science be valuefree. But as we have already mentioned, this is an ideal which has not the same credibility after the progress of linguistic analyses of the last decades. We have always known that there has been so to speak built in a moral demand to science in the demand for truth. But truth is not only a moral demand but also a condition of scientific communication and collaboration. The above mentioned positivism of the Vienna circle had the superior purpose of keeping science free from political, ideological and religious usurpation in the growing

A. Peczenik et al. (eds.), Theory of Legal Science, 121–132.

totalitarian tendencies in Europe. In the centre of the theory was the linguistic expression. Only statements, i.e. assertions, which can be empirically verified or which are analytically, i.e. logically, self-evident can be accepted as scientific statements. Metaphysical assertions, which cannot be verified, and statements containing evaluations are of no scientific relevance. These demands on science have been applied with great success in mathematics and the natural sciences since the days of Descartes and Galileo. The new practice is to make the same demands for objectivity on other sciences: the social sciences and humanities.

In opposition to positivism various movements among the cultural sciences have framed a hermeneutic-dialectical or critical-scientific ideal. The point of departure in recent times has to be looked for in the French and German phenomenology of the 1930s and the 1940s. To the functional descriptions of its object adhered to by positivism Heidegger raised the objection that any human activity is based on a conscious or unconscious intentionality. Human conduct is purposive. If therefore science confines itself to a description of human activity that description will at most be a very imperfect one. It will be able only to answer the question "how" and not the question "why". Science cannot deal scientifically with the interpretation of the meaning of the activity. Moreover, linguistic theory acknowledges that language is in itself filled with a hidden meaning and interpretation of reality. It is impossible to apply a linguistic expression without at the same time speaking intentionally. You cannot for instance mention the word "table" in a sentence without having also said something about the use of this piece of furniture. When analysing other words, concepts and sentences one must end up with a theoretical assumption that it is impossible to pronounce on anything whatever without having a preconceived opinion of it. This is the so-called hermeneutic circle. Another branch of phenomenology, so-called structuralism or semiology, seems to regard reality as a reflection of linguistic structures. A more relevant observation is perhaps the underlining by the ideology-critical Frankfurt School of the fact that the basis of cognition is interest. The problem to be solved is that of finding out what factors determine our choice of value-concepts in the formation of theories as well as ideologies. It seems natural to stress on anthropology or human biology, the fact that man is created in such a way that he is able to feel certain needs and consequently to have certain interests.

As a matter of fact Immanuel Kant's theory of cognition underlined in principle the limits, which are set to cognition by the human apparatus of cognition. To this must be added that biological assumption of the influence of our needs on the direction of our interests and so on our interest in the direction of cognition. It has already been mentioned that cognition is an activity, which has an intellectual and therefore a linguistic character. Our cognition is so to speak limited by our linguistic capacity. What cannot be formulated in language cannot easily be made the object of cognition. Analytical philosophy of language has been occupied especially with analysing everyday language on the assumption that it contains the culture or heritage of mankind, which is held to have settled into the language through the process of culture and consequently of cognition. This is exactly what the so-called hermeneutic science claims: that there is so to speak no possibility of cognition and science without regard to the cultural system and tradition lying behind the actual culture and the purposes and goals of man according to his cultural background.

1. MYTH

Seven blind sages were given the task of defining an elephant

> One found that the elephant was a wall, another that it was a pillar, the third called it a snake, the fourth a spear, while number five and six felt certain that it was a whip and a big leaf respectively. The seventh was convinced of its being a thunderstorm.

The parable points to the fact that our cognition is fragmentary delimited by our ability to grasp reality by the instruments we are using in the cognitive process and by the interests, which are lying behind the questions we put. It is obvious that we do not get an answer to questions that we have not yet asked, and we get the answers that our instruments are capable of providing. That does not mean, however, that the outcome of the cognition process is incorrect. It just means that one gets only a variety of true answers and that truth has many faces.

A theory cannot be true as such but it can be of use as a description of certain phenomena. At least the theory cannot be true unless truth is defined as the fact that the theory is not at variance with reality. One cannot exclude the possibility that reality may be described just as truly in another way. One of the classic examples is the description of light as either waves or particles. Both are equally true according to the purpose of the description to be made and the methods used to verify the theory.

This pluralistic or relativistic view of science lies behind the anecdote, which illustrates the dependence of science upon purposes and possibilities. Because of such a dependence scientific descriptions and methods must be in accordance with the varying purposes of science, and a theory, therefore, that fails when confronted with a special problem cannot claim to be true, i.e. to contain a sufficiently comprehensive description. On the other hand, the critic needs no other reason for his criticism than the fact that the theory is unable to cover a single isolated case.

History knows examples enough of science in general or special sciences being forced to change their course and take another direction than just forward. That is why the history of ideas or civilization is as popular with some as it is heretical with others. Critical or macro-oriented scientists are for it, while steady and micro-oriented researchers are against it. The distaste for the comparative perspective laid upon the sciences by the history of ideas is understandable considering the amount of dilettantism and sciolism displayed in the latest twenty years under the name of science-criticism, expecially by the so-called Marxists. Nevertheless the comparative view is an important corrective to any kind of dogmatism.

The science of any period in history of course is part of the culture of that period and must be understood as such.

I cannot go through the entire history of science, particularly legal science, in this short article but have to refer to my earlier writings(2). Like Thomas Kuhn, however, I find quite naturally substantial shifts in the general attitude in science in general, dependent on the world picture of its time and correspondingly changes in the direction, terminology and methods of legal science, which illustrates the dependence of science upon the horizon of understanding and the cognition interest of its time. A reflection on the various purposes or functions of law makes it equally clear that it will not be possible to maintain a monistic conception of law. Like Kuhn one might speak of various paradigms, being the methodical traits, generally accepted by the pursuers of the specific sciences at a given time and in doing so indicate the method to be used according to the purposes of its time. One might also prefer not to choose a specific paradigm for the legal science. In order to avoid the fallacies invariably accompanying any dogmatics that magnifies one out of more scientific aims into the aim, and so to speak takes the object of science to be monistic, one might try to adopt a pluralistic point of view on legal science.

The point of what I have said about science in general is that law and legal science are phenomena of the history of culture, which makes it natural to look upon legal science as part of the comprehensive scientific universe being at the service of the interests in power at the time in question.

The law has always had different functions, some of which are original while others have been added as society grew more and more complex.

It is necessary to look upon law from a functional point of view to avoid the risk of defining away important functions from legal science. A comparative method in the widest sense, that means vertical as well as horizontal, historical as well as international or intersystematic, must necessarily be functional. It does not make sense to compare institutions at different times or places which do not have the same function, just as it would be meaningless to separate different institutions, which have the same function totally or partly. For cultures which are closely related in time or contents, there will be no great difficulties, but the difficulties will increase according to the distance in these respects.

As regards legal science, it is an obvious absurdity to define law in general in relation to the modern political institutions and afterwards to maintain that past and present so-called primitive societies lacking these institutions have (or have had) no legal rules either. For example, it has been a common practice in connection with the neo-positivist theory of science from the 1920s up till today to identify law and state, which has led to the result that many ethnologists dealing with pre-state societies have found themselves obliged to deny the existence of law in such societies. Likewise Karl Marx and the Marxists, or some of them, have prophesied the non-existence of law in the future Communist society where the state will wither away.

2. FUNCTIONS

I can only mention in headlines the different functions of law and the corresponding theories of law built upon them.

2.1 Peace and Order: External Point of View

Whenever people have formed a society there has been a need for means to secure peace with the outside world and order at home.

The function of settling conflicts, however, is not the only function of law and not the most important one either, although it must be presumed that the actual legal rules have developed from this function.

To this external or political aspect of a society-preserving function corresponds the general concept of public utility as the basic value or criterion for balancing the conflicting social interests.

Corresponding to the above mentioned external viewpoint of law one can choose another viewpoint that deals with the internal functions or micro-function of law, connected with internal considerations as the point of departure of individual evaluations and dispositions. Considerations of justice in the widest sense play an important part in this area.

2.2 Justice (Obligation): Internal Point of View

The core of formal justice is the expectation that equal cases will be treated equally. In this sense the concept of equality is socially valuable and related to the rule of law.

Material justice means that there must be equality between the tenders that individuals have to exchange between them;elves and between the evil and the sanction which is laid upon the wrongdoer. Distributive justice means that society distributes goods and resources to the citizens according to their value for society.

Of course, material justice means a lot more than that. In most legislations we find expressions like reasonable, justifiable,common honesty, Treu und Glauben, and so forth, which are expressions of general maxims of material justice. But also the different provinces of law find guiding principles which have developed during the history of culture, creating and signaling the framework of a general value-system reflecting the general conception of man and society. These general values are lying behind our legal systems and will be taken as the starting point for legal argumentation in any particular case. They are so to speak the hermeneutic world picture which silently decides our way of arguing in social and political matters.

3. LEGAL SCIENCE

To these different functions of law correspond different concepts of law and different methods of legal science. Corresponding to the first aspect, law is regarded, politically or socially, as an empirical fact, which shall be dealt with by descriptive methods. Corresponding to the other aspect law is regarded as valid obligations, rights and duties, wich must be interpreted and brought into action as guiding devices for the behaviour of people or as norms for the settlement or deciding of social conflicts. In this respect the legal science must use dogmatic methods, i.e. methods which make interpretations of authoritative texts in order to work out what rule or obligation to apply in a given situation.

Legal science may hypostatize the aspects of law and take one of the functions of law as the function of law and so to speak choose one model or one analogy.

In my book I mention a variety of legal theories which isolate certain aspects of law and certain functions of law and make these isolated items the very core of the concept of law and legal theory. Here, I give these variaties very briefly.

3.1 Realistic Theories

The general feature of this type of legal theories is that they are conceived of as empirical facts which more or less must be dealt with as such by empirical or descriptive methods.

Political theories, having their starting point in the concept of state, regard law as the governing device of society. Cultural theories, on the other hand, regard law more or less as a reflection of the material structure and functioning of society, a conception, which is generally accepted in ethnology and in history dealing with primitive societies. Sociological theories regard the legal system as a social system in cooperation with, and coordinated with, other social systems. Especially the Anglo-Saxon countries concentrate on the behaviour of judges, whereas the so-called system-theories of different kinds are dealing with the social behaviour or legal behaviour in different groups and the interaction between different groups in society.

Psychological theories have been found to compete with social-logical theories especially in the U.S.A., the so-called behaviouristic theories, which regard the actions of man, like the actions of animals, as a response on an external or internal stimulus. Especially the Scandinavian realism has operated with social psychology as a model of law.

3.2 Ethical Theories

After World War II we have experienced a revival of natural law theories. In Germany Gustav Radbruch, in the U.S.A. Lon Fuller, in Scandinavia Frede Castberg have acted against legal positivism, which they accused of being partly responsible for the functioning of the totalitarian system during the 1930s.

3.3 Analytical Theories

Particularly Herbert Hart has elaborated a positivistic legal theorie and argued in a discussion with his successor Ronald Dworkin that you should rather distinguish sharply between law and morality, and speak about immoral law instead of illegal law or "Unrecht" as Gustav Radbruch puts it.

Hermeneutic theories in particular refer to the fact that description, also the legal description, cannot be fully objective as the language itself has built in the valuations and purposes that cause the intentional character of the language. In addition comes the recognition that the description also depends on the cultural background and the ideology behind our political and social systems. Hermeneutic theories therefore, in opposition to the analytical theories, which are mainly occupied with the analysis of the validity of law and legal system, are occupied with the problem of creating harmony between legal rule and legal fact in the process of interpretation and description in the legal decision.

4. PLURALIS JURIS

After this survey of the different functions of law in past and present, and the legal theories built upon them, it must be time to return to the introductory myth about the elephant. The sages define the elephant differently because they caught hold of different parts of it and mistook that part for the whole object.

The mistake of the elephant-men, who were unable to see that they had seized only a fragment of a whole, is easy to understand if we assume that they knew nothing in advance of the whole, which was an animal-organism of a certain kind. It is an experience of the Gestalt--psychology, and of the hermeneutic philosophy of language as well, that parts of reality are perceived and understood as parts of a whole. The hermeneutic and critical meta-science developed from these schools has emphasized the interest of cognition as constituting such fragmentary one-dimensional parts of the total perception. Finally, the general relativity theory has taught us that perception both depends on and influences our measuring instruments. What has been called the hermeneutic circle is just an expression of the fact that one cannot understand the whole without knowledge of the single parts, which in their turn are understandable only as constituent parts of an organized whole. The myth implies both an understanding and a criticism of science or philosophy. Nowadays science should be expected to be familiar with this theory of cognition, which must render it suspicious of or downright immune to the tendency, found in primitive thinking to hypostatize a single aspect of the function of the whole as the essence of things. The fundamental line of thought behind the above description and analysis of a number of different legal theories is just the realization that law has different functions and that legal science must consequently have different departments.

When legal philosophers have attempted to comprise these elements under a common concept of law, this has sometimes been done with reference only to dogmatic legal science, and sometimes they have disengaged themselves from that too and treated legal science as a social science. This latter variant has not often been lastingly successful, because it has been of no outstanding interest to the customers, the legal profession. For the same reason a strictly analytical theory (pure legal theory) has been of no great success either. The debate between legal science and legal philosophy will probably go on until legal philosophy realizes and respects the fact that law has different aspects according to its different functions, and that legal science must therefore apply different methods.

The content of the legal concept in a given situation will depend on the kind of question asked, which in turn depends on the specific role or function performed by the legal actor. The consulting lawyer is most interested in the probability of a certain decision of a prospective lawsuit, whereas the judge is in no need of such an external definition, but must look for an internal information of what rule he is obliged to apply. Dogmatic legal science wants to see law as a system of legal rules validly following from certain sources indicated by the principles of legal sources. Like the judge, legal science is interested in the internal aspect of law, not only in connection with the solution of actual problems but in order to be able to present the individual rules or solutions as parts of a systematic consistent whole. The legal sociology and the social psychology are interested in the current law, i.e. rules or behaviour which are actually complied with or which are felt to be obliging. The cognition interest of politicians is the governing effect or the oppressive character of law. Especially the latter effect, the aspect of power and sanction, must be considered to be important, while other democratic government-models attach greater importance to the public acceptance of the legal system and the individual rules. History of law, ethnology and international law are, for different reasons, uninterested in the aspect of sanction; they want to see law as part of national and international patterns of culture. The church and ideological movements in the widest sense conceive law as the realization of a superior order: the course of nature, the nature of man, the will of God, reason or justice (natural law), and they are unable therefore to respect positive law unconditionally, because to them its content may be invalid, in which case they have a right and a duty to resist it (right of resistance, civil disobedience).

The definition chosen is of a programmatic or ideological nature, connected with the underlying general conception of man and society. A command theory corresponds with an authoritarian state-concept and a collective concept of man. A sociological-functional theory of law may be the expression of a collective concept of man too, since social utility and distributive justice are regarded as the supreme values of society. A theory departing from the individual rights of man looks upon man as an individual and nothing else, and assigns to society the minimum task of securing the individual rights. A pluralistic legal philosophy, regarding man as an individual and a social being as well, recognizes that social utility and individual justice are equal, competing values and that individual justice cannot be reduced to form part of distributive justice. A manifestation of this fact is the increasing tendency of legal philosophy of the last decade to be interested in the reflective (cultural) and individual oriented aspects of law, at the expense of its governing and distributive functions (indeed distributive justice) which have been the main interests of the previous half century.

NOTE

1) Pluralis Juris, Towards a Relativistic Theory of Law, Acta Jutlandica LVI. Social Sciences Series 14, Arhus 1982.

ABSTRACT

A theory cannot be true as such, but it can be of use as a description of certain phenomena. At least the theory cannot be true unless truth is defined as the fact that the theory is not at variance with reality. One cannot exclude the possibility that reality may be described just as truly in another way. One of the classic examples is the description of light as either waves or particles. Both are equally true according to the purpose of the description to be made and the methods used to verify the theory. The science of any period in history of course is part of the culture of that period and must be understood as such.

A reflection on the various purposes or functions of law makes it clear that it will not be possible to maintain a monistic conception of law. Like Kuhn one might speak of various paradigms being the methodical traits generally accepted by the pursuers of the specific sciences at a given time and in doing so indicate the method to be used according to the purposes of its time. One might also prefer not to choose a specific paradigm for the legal science. In order to avoid the fallacies invariably accompanying any dogmatics that magnifies one out of more scientific aims into _the_ aim, and so to speak takes the object of science to be monistic, one might try to adopt a pluralistic point of view on legal science.

The point of what I have said about science in general is that law and legal science are phenomena of the history of culture, which makes it natural to look upon legal science as part of the comprehensive scientific universe being at the service of the interests in power at the time in question.

It is necessary to look upon law from a functional point of view to avoid the risk of defining away important functions from the legal science. A comparative method in the widest sense, that means vertical as well as horizontal, historical as well as international or intersystematic, must necessarily be functional. It does not make sense to compare institutions at different times or places, which do not have the same function, just as it would be meaningless to separate different institutions, which have the same function totally or partly. For cultures, which are closely related in time or contents, there will be no great difficulties, but the difficulties will increase according to the distance in these respects.

CHANGES OF PARADIGM IN THE LAW

Jan M.Broekman

1. A 'STATUS PRAESENS'

The paradigm theory of methodology, developed by authors such as Kuhn and Toulmin, has been adopted by legal scholars as an instrument of legal science(1). However, in its extrapolation of the concept of 'paradigm', legal theory has assumed a rather uncritical attitude towards the concept itself. The concept is used not only to articulate problems of legal science and legal dogmatics, both analytically and descriptively, but also with regard to the more encompassing question of legal change. Can the law change at all, and if so, is this change only part of a general social change, or does it imply a change of the legal paradigm?

In his recent study concerning The basis of legal justification, Aleksander Peczenik(2) stresses, that the popularity of the paradigm methodology arises from two factors. The first is that the paradigm theory demontrates to what extent any science has to take something for granted, i.e. must have its framework and its starting points. According to this viewpoint, one could say that the paradigm theory prevents us from openly admitting that the law is autoreferential. This is a great advantage especially to the legal practitioner because his practice can remain autoreferential even though his reflection upon it tends to indicate the opposite. Furthermore, this suggests that the lawyer has his specifically given starting points as well as his social and moral framework. Specifically given legal starting points legitimate the practice of autoreferentiality.

As a second reason for this popularity, Peczenik mentions the incommensurability problem. The hallmark of the problem is the mutual exclusion of paradigms. The exclusion comes about because in the process of transition from one theory to the other, semantic differentiations occur. These in turn delineate and demarcate the range of appli-

A. Peczenik et al. (eds.), Theory of Legal Science, 133–144.

cation with regard to the respective theories concerned.

On the basis of the foregoing, it may well be inferred that there are two arguments relating to the application and the use of the paradigm theory in the field of legal theory. Both arguments can, therefore, be considered in terms of their epistemology and ideology. Paradigm theory covers the autoreferential character of the law and the theory suggests incommensurability with legalistic features. Both aspects have their consequences for legal practice and legal theory.

The general idea of paradigm theory, notably, that the starting points as well as the framework of a particular science must be taken for granted, is transformed in the law into the judge's duty to motivate his decision. As such, it is at the same time an overall obligatory device for facilitating legitimation through the law. The incommensurability thesis is transformed, on the level of legal dogmatics, into the prohibition to think and speak about the basic assumptions and presuppositions of legal practice. Rather, the injunction is to follow the rules and to interpret the principles of law in accordance with the prescribed methodology.

These two transformations should be regarded as basic elements of legal rationality. It is clear, therefore, that in the field of legal theory the paradigm theory cannot break through the cocoon of legal rationality as defined and determined by legal dogmatics. Accordingly, it does not lead to a reflection upon the law from the standpoint of the philosophy of science. For this reason the paradigm theory in legal science becomes transformed (according to the dominant needs of the legal practitioner) into a process of legitimation. In doing so, the paradigm theory legitimizes an evolutionary interpretation of the concept of law, which is advertized in the contemporary literature as a 'reflexive' or 'responsive' concept(3). The basic attitude underlying this concept requires not only a development and evolution of the law within the framework of culture and society, but also an encompassing rationality on the part of the legal decisionmaker. The rationality of choices and of the general finding of the law becomes the main postulate within legal argumentation theory, within the heuristic process and even within posteriori legitimation processes.

This development has its consequences with regard to another element of paradigm theory, namely the distinction

between the judge's context of discovery (the application of the law according to his set of methodological rules and/or principles) and his context of justification.
If legal rationality is taken as the main principle that governs the behaviour of the decision maker, then one has to come to the conclusion that in the field of legal theory there is no sense in the distinction between the two contexts! They belong to the same continuum of rationality. Both serve the general need for legitimation of the decision. In that case one has to conclude, furthermore, that there is always the closest commensurability between the two contexts. This proximal commensurability is created by the claims of legal dogmatics, being a set of methodological rules and principles that necessarily shape the behaviour of the legal practitioner.

A second issue is closely connected to this observation. Does legal rationality form a continuum in which a diversity of legal paradigms can exist, or is there one and only one legal paradigm possible? This is often suggested. Peczenik doubts indeed whether the diversity of standpoints of the schools of legal thought, such as positivism, functionalism, pragmatism and others, should be called paradigms. "Perhaps there is only one standing paradigm of legal dogmatics, more or less closely connected with 'formalism'(4)". Werner Krawietz holds the same position. He introduces the distinction at three levels, namely, that of dogmatic legal science and its praxis; the level of legal methodology of dogmatics and the level of legal theory and legal philosophy, the latter being the level of presuppositions of the dogmatic legal science(5). The common paradigm of these levels is that of legal dogmatics. Descriptions of paradigms of legal action and legal decision can be given on the level of legal theory. Even an analysis of legal dogmatics can take place, according to Krawietz, on the level of legal theory. But the hidden paradigm, which is ultimately the only paradigm in the law, remains the paradigm of the judge's finding of the law. This paradigm forms a specific unity of practical action and research.

The consequence of this standpoint is, that in legal science the heuristic activity of the decision maker is functionally dependent upon the legitimation of the decision. On this perspective the rationality of the judge is his "juridical programme", or his "programme to develop the law", which is always connected to a positive law as an element of general jurisprudence(6). This programme, however, is not only a programme of future legal action, but it is at the same time a programme of legitimation. Such a programme can be developed only on the basis of legal rationality. The latter is itself a representation of the rationale of legal dogmatics. Thus, the choices to be made as well as the methodological steps that are to be followed in a given juridical programme become the manifestations and the articulations of the basic legal rationality: the rationale of the paradigm, which was called legalism (Shklar) or formalism (Peczenik). Both the rationality and the methodology are, according to Aulis Aarnio, elements of the legal dogmatical paradigm(7). So the juridical programme of the legal decision maker can never escape from the functionalized interconnection between the context of discovery - the heuristic activity of the judge - and the legitimating activity of the judge, the context of justification.

It is apparent that the proximal interconnection between these two contexts may indeed be described as inseparable, at least fundamentally, as both arise from and are expressive of the basic rationality of legal dogmatics. On these premises the conclusion seems to be that there is only one paradigm in the law. The question of change becomes important again. What is legal change? Is it a simple result of a more general social change? Can there be a more encompassing change of the legal paradigm as such and is that type of change an extrapolation of the methodological components of the legal dogmatic paradigm, or of legal rationality in general? Is such change an example of 'change of paradigm' in the sense of a theory of science?

2. A CONCEPTUAL ANALYSIS

It could be clear from this description of the status praesens of the problem, that the above questions cannot be answered without a preceding analysis of the meanings of the concept 'paradigm'. Many meanings are given to that concept. The meanings derive from and differ according to the standpoints and starting points of the various schools in legal theory as well as the dogmatic interests in legal science. It would appear that, given a typology of paradigms, some would and others would not be susceptible to change. In this connection there is an interesting remark by Allott, who suggests an almost natural division of the legal paradigm(8). This author refers to the differentiation between LAW, Law and law, which is traditional in legal thinking. The LAW is thought of in terms of general culture and philosophy. Schools with a diversity of standpoints and perspectives deal generally with the Law, and the law is the sphere of the practitioner and the dogmatician.The concept of paradigm has different connotations according to those three levels.

Our analysis has to take into account three distinct levels of meaning. The first semantic level of the concept of paradigm is that of general culture, the general normative system, and the level of philosophical understanding of the paradigm (paradigm p 1). This is the level of the Marxist understanding of change in the legal paradigm. According to this understanding, the law and the state will be changed as a result of the alteration of the capitalist structure of society.

The second semantic level of the concept of paradigm is that of legal theory (paradigm p 2). Legal theory should be understood here as the theory of legal science. This legal theory is not identical with legal philosophy, since legal philosophy is the philosophy of the theory of legal science. But legal theory is also not identical with legal dogmatics. On this semantic level, the paradigm of the 'Law' functions with regard to questions concerning the theory of legal science as an element of a general theory of science. The theoreticity of the legal science is debated on this level. Three questions assume singular importance here. The first concerns the various issues pertaining to the type and character of theory-formation. The second relates to the specific role of the dogmatician and, by extension, the practitioner, precisely because the dogmatic sphere is the dominant point of reference in legal theory-formation. The third question touches upon the relationship of harmony or disharmony between law, conceived of as an independent unit, persisting as a closed system of thought, and the interdisciplinary character of legal theory. These issues are judged differently according to different starting points, cultural and philosophical presuppositions and views on the ends and means of the legal system in a society(9).

At this level of paradigm, one can find many schools formulating sometimes contrasting ideas. In that regard, one could indeed say, that the Pure Theory of Law has provoked a change of the legal paradigm. The most complex problem is, however, on the third level, notably, the level of legal dogmatics as a theory (Rechtslehre) involving adherence to a set of rules as well as methodological principles that determine the behaviour of the practitioner. This is the level of paradigm p 3 , and in contrast to the preceding levels, the question is hightly debatable if at this level there exists an independent paradigm.

Aulis Aarnio, among others, formulates p 3 as the 'legal dogmatic paradigm'. The components of this paradigm are according to this author: (1) philosophical background assumptions, (2) assumptions concerning the sources of law, (3) methodological rules and principles, and (4) values common to the dogmaticians within the legal system. Nrs. 2 and 3 are components of the dogmatic discourse within the legal discourse, and they are therefore constitutive elements of the autoreferential legal system. Nr.4 looks as if it is external to that system, but this is not the case. It belongs to the system itself. The values shared by the dogmaticians support, in the first place, the legal system, and that only via the juridical articulations of reality, the cultural system with its p 1. A discussion should be initiated concerning the paradigmatic status of Nr.1. The philosophical background assumptions of the judge and the general practitioner are already highly juridical notions. Our culture is, according to the main analyses of Max Weber, a formal-legal culture and as such an easy background for dogmatic behaviour. The question is, whether or not non-formal and non-legal philosophical assumptions are playing a role as constitutive elements of legal dogmatics. It is very doubtful if this could be the case.

In the case of p 1, change of the legal paradigm implies cultural change on a large scale. It implies the reorientation of a society towards changing fundamental norms and values, including economic and legal organization. This is a macrosociological event on a timescale that transcends the individual human lifetime. The answer to the question of change is for p 1 only possible in terms of interpretations of history. Some historians, like recently Arnold Toynbee(10) have grafted this change into the methodological cornerstone of their philosophy of history. Toynbee's 'challenge-and-response' structure is, not unlike Spengler's ideas, the basic element of an evolutionary vision on history. The same has to be said for the LAW.

Problems are different for p 2. The legal paradigms at this second level are more strictly defineable, even in terms of incommensurability, although this can never be the case in absolute terms(11). The changes of the legal paradigms are expressed in terms of change of perspective on legal theoretical questions. Such changes relate mainly to the epistemological aspects of the legal paradigm. Hence the importance of the debate on argumentation theory in contemporary legal theory, on methodology in general and on certain theoretical principles which are elements of p 2.

There seems to be no way to determine exactly, to what degree p 3 has really a comparable paradigmatic character, nor what in this instance the idea of paradigmatic change implies at all. At the level of legal dogmatics (understood as a set of rules and principles governing the practitioners' action) a completely independent and distinct meaning of p 3 does not seem to exist. Practitioners do not regard their methodological behaviour, which is a question of rule-following and value orientated behaviour, as paradigmatic behaviour. The practitioner does not orient himself towards the totality of the legal order. Nor does he actively take problems into account that belong to p 2, such as consistency of the legal system or the problem of completeness(12). He has in mind the actual norm, not the normative system. The rules and principles of his methodological framework, their interrelation and their application are his main concern. If he is attracted by theoretical problems, then the paradigmatic character of legal science in the sense of p 2 is his nearest connotation.

One should conclude from this briefly sketched observation concerning the law (not: the LAW or: the Law) as a paradigm, that the only paradigm of the law is the Law. The 'legal dogmatic paradigm' is hardly a paradigm itself, but a set of methodological rules and principles that has to be followed in any case. Change of the legal paradigm in the sense of p 3 is change of this dogmatic methodology. This is the real issue in the debate on 'responsive' or 'reflexive' law. But that debate suggests that the real problem is on the level of p 2, where the problem of change is at the end only the level of p 3. If changes are said to have been effected, these are performed by practitioners of the law and the changes are in the light of _p 2_ or even _p 1_ _extrapolations_ from existing, dominating dogmatic structures.

A consequence of all this should be formulated as clearly as possible. Changes at the level of _p 1_ are changes of the cultural paradigm that is more or less adequately embodied in the LAW. This paradigm could philosophically be called _cartesianism_. In terms of legal phi--osophy the name would be _legalism._

Changes at the level of _p 2_ do take place, especially in the legal theories of Western Europe. It appears that they do take place less frequently in the USA(13). Peczenik suggests, however, that these changes are less radical and thoroughgoing than one would judge at first glance. According to his view, there is in the realm of _p 2_ only one incommensurability, namely, between formalist and non-formalist movements. The formalists are legalistic theoreticians and their theory is the ideological continuation of traditional legal dogmatics. It is very difficult to describe the standpoint of antiformalism in a comparable positive way.

Peczenik refers to Zuleta(14) and states that antiformalist movements have hitherto been unable to create a new practically operational paradigm. From what is said until now, it is obviously the case that the formulation of such a new paradigm is quite impossible, since there is no _p 1_ that enables and justifies (the ensemble of the contexts of discovery and justification) the edification of such a _p 2_. Our ideas about the latter paradigm are nevertheless constantly at work(15). They have led to the formulation of functionalism, topological jurisprudence or rhetoric jurisprudence and other aspects of a non-formalist paradigm. The metaphysical superstructure, particularly, the _noumenal_ superstructure of contemporary Western culture allows the development of non-formalisms (on the level of _p 1_ to be called non-legalist or post-cartesian) of the paradigm. Yet it failed to provide the complete and autonomous formulation of a realisable paradigm. A real incommensurability between the two does not come into life, since that situation would destabilize Law and order, in other words, both the _LAW_ and the _law._

The _noumenal_ pressure of _p 1_ towards a uniform legal paradigm is mirrored by the _dogmatic_ pressure of _p 3_. The answer to the question of paradigmatic change remains complex. Change is possible as far as p 1 is concerned only in terms of historical interpretation of change; for p 2 in terms of schools and theoretical perspectives, but only under the pressure of p 1 and p 3. For p 3 change is a technical problem of extrapolation from dogmatic structures and the consistency of an evolutionary model of the law.

3. PHILOSOPHICAL CONSIDERATIONS

This insight leads to a diversity of philosophical considerations, which reach far beyond the scope of the paradigm problem. The answer that can be given to the question concerning changes of paradigm in the law is in itself closely connected with and even dependent upon p 1 and all complex elements of the level concerned. Is it possible to deal with the problem of paradigmatic change in another way than from the standpoints of p 1? Is this problem the root of the autoreferentiality of p 3? And, as always, is this question an attempt to formulate an anti-formalist paradigm? Let us go back to the activity of the legal decision maker.His context of discovery is highly dependent upon methodological choices. Even though the choices remain within the dogmatic framework, it may well be observed that these choices do not only possess the decisional character of a choice according to the *voluntaristic* meaning of the word but they have also a *textual* character. This observation should be the cornerstone of a philosophical discussion about the legal paradigm.

It is clear that the *textual* character of decisions is produced and governed by legal dogmatics. The methodological rules and principles, their background values, the commitments and consensus of the practitioners are directed towards and bound by legal *texts*. Reality is said to be present in the files on the desk of the judge. Reality is said to be mirrored in those pages but it is not expressed in how far the legal reality is produced throughout the textualizing processes. The rules concerning the application of norms to situations are rules referring to texts. The social effect of the law can be understood in the same context: any legal activity evaluates social relations in terms of the legal text. The law creates relations when it expresses itself in texts. These texts form a topology of the legal paradigm(16). This textuality is not a linear phenomenon and the textual character of legal discourse is not characterized by linearity. Nevertheless the influence of p 1 makes us read the textuality of p 3 in a linear way! The grammar of legal dogmatics is linear. This holds also for the grammar of the texts, announcements and rules of dogmatics as well as the behaviour that is supposed to be resulting from this mode of reading. p 3 tends to reduce the language of the 'dogmatic legal paradigm' into a sole surface-structure. Linearity is the outcome of these processes of reduction and transformation, which are so characteristic of the dogmatic paradigm. This linearity is presented as if it is the most natural and universal property of language. This move of inversion belongs to what Jean Pierre Faye once called the 'narrative economy' of p 1, which is reinforced in the specific case of p 3. The legal dogmatic paradigm is characterized by the fact that it transforms narrative structures into linear products. Processes of the deep-structure of the language, such as the selection of parameters, negation, suppression and/or reduction of structures of relevance, semantic differentiations and compositions of intertextuality, are not overtly debated as the proper history of p 3. These processes do, however, produce the legal act, the legal decision and its consequences.

Every legal norm, that is every point of orientation of the practitioners' legal interpretation of reality is therefore taken as an element of the surface structure that covers the narrative structure and the proper history of the norm-in-its-relation-to-reality. The 'if p then q' structure, which is dominating the theory of the legal paradigm is a surface-phenomenon, forms a legal reality that is often in contradiction to reality. That 'if p then q' structure is always a tranformation of reality. The relation between p 1 and p 3 appears in a new light. The surface structure of p 3 is legitimized with the aid of concepts of the level of p 1, such as a divine truth, an everlasting value, the command of a sovereign, or rules of the Authority. The relation between p 1 and p 3 appears as a functional relation: the presentation of the law is presented as a representation(17). This fact makes a theory of the legal paradigm ambiguous. The ambiguity is not only in the variety of meanings of the concept itself. There is, as was said before and noticed by other legal theorists, presumably only one meaning of that concept. This meaning is not straightforwardly formulated. A theory of the legal paradigm has to uncover its proper network of connotations, interests and transformational activities. This is especially important with regard to the problem of change. The linear surface-structures of the law suggest change and evolutionary concepts wherever extrapolations from the given dogmatic structures are produced. The fundamental option and the basic norm for a scientific legal theory, in which the problem of change can be developed, has to make visible its own linearity, in other words: its restricted rationality. The linear reading of text, norm and action in the law suggests a universality of legal rationality. This universality can never be the foundation for a theory of paradigm change in the LAW, the Law and the law.

NOTES

(1) Kuhn,R.: 1970,The structure of scientific revolutions Chicago, 2nd ed. St.Toulmin: 1972, Human Understanding, Oxford, Vol.1.

(2) Peczenik, A.: 1983, The basis of legal justification, Lund,p.130

(3) Teubner, G.: 1982, 'Reflexives Recht', ARSP, 'Substantive and Reflexive Elements in Modern Law', Law and Society Review, 1982/83, Nr.17, p.401

(4) Peczenik, A.: op.cit., p.131.

(5) Krawietz, A.: 1979, 'Zum Paradigmenwechsel im Juristischen Methodenstreit', in: Rechtstheorie, Beiheft I, Berlin; - id -, 1978, Juristische Entscheidung und wissenschaftliche Erkenntnis, Wien/New York, p.200-242.

(6) de Wild, A.H.: 1979, De rationaliteit van het rechterlijk oordeel, p.50,52.

(7) Aarnio,A.: 1981, 'On the paradigm articulation in legal research', in Rechtstheorie, Beiheft III, Berlin.
Also in: 1983 , 'Philosophical perspectives in Jurisprudence', Acta Philosophica Fennica, vol. 36, Helsinki, p.209-222.

(3) Allott, A.: 1980, The limits of Law, London, p.2.

(9) Broekman, J.M.: 1982, 'The inner form of legal thinking', Primer Concreso International de Filosofia del

Derecho, La Plata.

(10) Toynbee, A.J.: 1934, 1962, *A study of history*, Oxford U.P.,Vol.*1*, p.271-339, Vol.*2*, p.1-395.

(11) Peczenik, A.: *op.cit.*, p.130.

(12) Alchourron, C.E. and Bulygin, E.: 1971, *Normative Systems*, New York, p.4.

(13) Summers, R.S.: 1982, *Instrumentalism and American Legal Theory*, Cornell, Ithaca & London, p.268-283.

(14) Puceiro, E.Z.: 1981, *Paradigma dogmatico y ciencia del derecho*, Madrid.

(15) Puceiro, E.Z.: 1981, 'System and function in legal dogmatics', in: *Rechtstheorie*, Beiheft II, Berlin, p.349-357.

(16) Faye, J.P.: 1972, *Théorie du récit*, Paris, part I, ch.2.

(17) Broekman, J.M.: 1981, 'Juristischer Diskurs und Rechtstheorie', in *Rechtstheorie*, Vol.XI.

LEGAL NORMS: A TRANSFORMATIONAL APPROACH*

Bert C. van Roermund

1. CONCERNING A PROGRAMME

A dogmatic account of positive law is, in my view, never a theory of law, though many a putative 'theory' of law is an extender, an appendix, or a fashionable appearance of legal dogmatics. As a decisive criterion for an account's being dogmatic I would regard, neither its use of somewhat archaic, parochial, and rigidly defined terms, nor its reference to a more or less open, evolutionary system of norm statements to which it confesses itself. Instead, I propose this one: whether or not the account entails an instrumental notion of truth(1) in the ontological interpretation of its theorems; i.e. whether or not it hypostatizes the dominant interests, ends, and stategies of a certain culture or community into the structure of the world. Thus, we can speak of legal dogmatics in exactly this sense: as a device to develop truth-claims about equivalences ('sameness') in the world, defined on the meaning of a token of that specific paradigm of normativity: a rule.

Not only are, as Wittgenstein says, the use of 'rule' and the use of 'same' interwoven, as are the use of 'proposition' and the use of 'true'(2); interwoven are also, in a dogmatic account of e.g. law, the use of the complex 'rule'/'same' and the use of the complex 'proposition'/-'true'. This texture is, indeed, the very target of the multi-coloured arrows in the Philosophical Investigations. Within this network, truth-claims are assessed or fixed, rather than ascertained or corroborated. I take this criterion to be decisive, because it goes against the grain of what theories hold about themselves when they claim to be, truly, theories, whatever else they claim and whatever they are in fact.

A theory of law then, should resist, the continuous pressure and temptation to fulfill a dogmatic function within its domain of research, e.g. under the guise of 'methodo-

A. Peczenik et al. (eds.), Theory of Legal Science, 145–161.

logy'. Rather, it should give an account of the interrelations and operations constituting the alleged texture of rules and equivalences. It should not present this account as a 'presupposed' philosophical background of the concept of law in general. World views are commonly either too harmless or too dangerous to be accepted as theories, especially by lawyers. Rather, it should provide an incisive analytical model that allows to explicitate the relations and operations mentioned as the very generative structure of every instance that counts as 'a norm of valid --law'. It is the purpose of my argument to develop some prolegomena for such a model(3).

2. LINEAR VERSUS TOPOLOGICAL ACCOUNT OF LEGAL NORMS

Kelsen proposed a general scheme of (legal) norms, that can be framed as

(1) If p is the case, then q ought to be done (A)

where p ranges over circumstances, q over actions, and A stands for a signature of a constitutional authority(4). Many a legal theorist is inclined to believe that legal claims are justified, if they can be inferred from a scheme like (1). If we have both (1) and (2)

(2) p is the case

established, we are justified in claiming in virtue of (1) and (2):

(3) q ought to be done.

The same is assumed to be right, if the claim itself is of the form (1): it is justified by premises of a higher order. The only thing one has to do (quite a job, though), is to find the appropriate values for p and q (and A, if one wants to be 'pure'), or - as one commonly prefers to put it - to find the right premises. These premises, however, are believed to have the same 'if...then.....'structure- as (1) has.

I would call this account of legal norms linear: it presupposes and suggests a one-dimensional order of language in which there are seamless junctions between no matter how many 'meta-levels' of sense and reference. Missing or disfunctioning connections on one level can always be repaired and embedded by 'if-then'-phrases on another level, be they higher order premisses or (what boils down to the same effect) operation rules. This strategy, in my view, is bound to beg an important question: the question concerning the nature, especially the constraints, of the 'if-then'-structure of legal norms itself. It is also known as the problem of 'commitment'. Kelsen used to call it the problem of Zurechnung; and although he was probably right in searching for an account analogous to the notion of causality, I doubt whether he offered an adequate solution by tracing this relation of commitment right back to the Setzungsakt of legal authority. But at least he made it clear that it is not a logical problem either(5).

In conditional deontic logic the problem of commitment is hidden to a large extent by the famous stroke '/' introduced by G.H.von Wright. It seems to me, that Castañeda(6) has raised a crucial objection against this manoeuvre: it does not distinguish between relativity of duties to circumstances and relativity of duties to acts or processes of endorsement, and, hence, to the reason for such endorsements. The latter relativity, he says,

> '...is an entirely different matter. Here we do not have conditional duties, but the adverbial qualifications I have for over two decades been representing by means of the subscripts attached to deontic words.'(7)

Indeed, as Lewis noticed in a different, though related context, we should be aware of the fact

> '...that the if of our restrictive if-clauses should not be regarded as a sentential connective. It has no meaning apart from the adverb it restricts.'(8)

This adverb may engage us in a number of epistemic strategies that generate (our understanding of) the relation of commitment, at the same time, while there is a number of constraints on our description of 'circumstances' ('cases'). To catch just a glimpse of the complex machinery concealed in the bond of antecedent and consequent of (1), compare:

(4) *p* is the case; therefore *q* ought to be done
(5) As *p* is the case, *q* ought to be done
(6) Always when *p* is the case, *q* ought to be done
(7) Because *p* is the case, *q* ought to be done

Note that 'if...then...' in (1) indicates, for all we know, the very *same* relation between the constituents as 'therefore' in (4). Though it is logically interesting, it is epistemologically uninformative to say that one is allowed to state (3) *in virtue of* (1) and (2). Indeed, why not the other way round? It is impossible to analyse 'if...then...' in (1) without distinguishing (the relations between) the connectives for which it can be a stand-in (cfr. (5)-(7) as examples).

To deal with the problem, it is necessary to see phrases like (1) as abbreviations of, probably very complex, different epistemic layers involved by an 'if .. then ...'-structure (in law). I admit that this proposal is only partially new. It is embedded in a long tradition of logical and linguistic analysis. The heart of this historical matter is the old debate on the nature of so-called *consequentiae*(9). From at least as early as the days of Boethins but especially during the Middle Ages, logicians have struggled with the problem of *consecutiveness*. It seems to be involved in conditionals and inferences, though not in the same sense. The debate turns around such questions as: should we refer to the value (truth/false) of a *consequentia* (in case it is a statement) or to its validity (in case it is an inference)? If the latter, aren't at least some *consequentia* 'sound', not in virtue of the form of their constituting expressions or their analytical content, but in virtue of their accidental *positio terminorum* (the so-called 'consequentiae materiales bonae ut nunc')?(10) If we want to take up these suggestions, we'll have to embark on a theoretical account of the topology of terms, which allegedly determines the structure of *consequentiae*. We'll have to go into the cohesion between two orders: *both* the linear order of subjects and predicates, antecedent and consequent in sentences of schema (1) *and* the order of the process that generates the related positions at which such terms can appear. I will call this theoretical effort the search for a *topological* account of legal norms, in contrast to a linear account. It aims at describing and explaining *how* certain terms are moulded in such a way that they can be put into a transparant, linear, 'if ... then ...'-order.

3. THREE DISCURSIVE LEVELS

Legal norms are, in their ordinary linguistic form, not particularly 'iffy'. This can be learned from an analysis of the very phenomena that we meet in dealing with such norms: legal transactions or acts-in-law, like e.g. statutes, contracts, judgments, decisions, wills, summons, requests, etc. The predicate 'act' is, however, misleading here, as are all the predicates just mentioned. To avoid (reproaches of) committing a speech act fallacy, we have to adopt on this phenomenological level a distinction between:

(i) the speech act performed by an agent at some particular time and place, e.g. enacting, issuing, rejecting or endorsing a command;

(ii) the specific, form of what is enacted, issued, etc., viz. the structure of a command, with its charcteristic coherence between its constituents (e.g. someone lower in rank to do something)

(iii) the practical content of this form, e.g. the cat to leave the mat.

It's especially important to recognise phenomena of type (ii), for acts-in-law are a subset of this set. They are not acts in any concrete spatio-temporal sense, but rather a topological pattern, which can be realised also by acts which are not speech acts. For instance: whatever the particular arrangements of a contract are, and whatever the ways in which parties enter into it, they are bound to a specific network of positions or places, e.g. places for agents with a certain ability, objects with certain properties, relations of demand and offer etc. Note that in ordinary (legal) language, the term 'contract' is used to refer to phenomena (i), (ii), and (iii). If we bear these distinctions in mind, it seems a solid strategy to analyse the standard forms of acts-in-law in order to discover the putative generative texture of legal norms. For the sake of brevity, I report a few results from recent research, omitting adequate linguistic argumentation(11).

When we take a sample of such acts-in-law, e.g. in a domain like contract-law, we may discover two things: (a) that there are several levels of discourse in both antecedent and consequent, more or less deeply embedded in one another, and (b) that there are certain variables, which play a role on every level of embeddedment but not the same role on each level.

Level III. The level nearest to the surface of the text, embracing all other textual elements, contains devices to refer to a first person, supported frequently by reference to his title or quality, as well as an operative which specifies the act-in-law according to this quality. This quality (or competence) authorizes the operative. We'll never find the adverb 'legally' added to this operative. The first person doesn't, indeed, refer to the law in the operative, he poses it (in actu).

Level II. A second layer is what can be called the 'ought' level, the level of deontic properties and institutions. On this level the antecedent shows all sorts of normative notions, both in considering what is the case and in deliberating what ought to be done relative to a lot institutional contexts. The consequent shows a normative sentence which formulates what ought to be done everything relevant being considered (or 'all relevant institutional contexts integrated'). The 'ought' of the Considerans (antecedent) and the 'ought' of the Dictum (consequent) have a different logic. E.g. (8) is not tautological:

(8) As/If everyone $ought_C$ to pay his debts, everyone $ought_D$ to pay his debts.

Rather, it seems ambiguous between for instance:

(8a) As/If everyone ought to pay his debts according according to Dutch law, everyone ought to pay his debts according to Dutch law,

which is, of course, tautological, and

(8b) As/If everyone ought to pay his debts according to Dutch law, pay, everyone, your debts!

The antecedent of (8) is some sort of proposition, which can be true or false, depending on the institution referred to. The consequent of (8), however, is in sense (8b) not a proposition. It does not ascribe a property to an individual - something which it is or has - but something to be done by this individual. It cannot be true or false depending on the 'world' of an institution, for this institution is itself at stake as the subject referring (to 'everyone' and their debts) in order to frame a noema which is not a proposition(12).

Level I. Embedded within this normative level, there is another structure, which is, so to speak, the really practical core of the matter in antecedent and consequent.

Most legal claims refer in their Considerans not only to existing legal norms, but also to states of affairs beset by axiological values: they should or should not be. To this antecedent consideration of 'the problem', as a demand for something to be done (rather than something to happen), corresponds the consequent response of the Dictum, selecting something to be done. A statute law, for instance, is not simply to be analysed as a series of articles in 'if...then...'-form. Primarily, the articles belong to the consequent of a larger topological structure: they constitute the response of actions to be done (under certain circumstances), towards the demand to cope with some situation considered as problematic in the (sometimes rudimentary) antecedent.

How are the relations of embeddedness to be understood? If something-to-be-done is presented as a responsive solution to a problem, it is to be evaluated in terms of 'adequacy'. Whatever criteria of adequacy are accepted, a solution is never an adequate solution only because it is a legal solution. That would make the very idea of a legal problem arising inconceivable! On the other hand: a problem (and its solution) arises only within an established framework or world that is prestructured along certain lines of relevance. Such a framework is the institution of law referred to and referring in Considerans and Dictum. The relation of adequacy between demand and response is restrictively qualified by and embedded within the normative modality of (!) this institutional framework. Again, if something-to-be done is presented as an institutional Dictum on a Considerans, it can be evaluated in terms of 'right and wrong'. Whatever criteria of rightness are accepted, a Dictum is never a right Dictum because it is said to be one by legal authority. That would make the very idea of law as an institution inconceivable (reduce it to 'umpire's discretion'). On the other hand: an institution 'referred to and referring' exists only within a framework of procedural distribution of competence to set norms. Such a framework is (forcefully sanctioned) legal jurisdiction. Again, this distribution is not identical with the relevant structures of the institutional framework, but it qualifies the instances which establish it. A specific, admittedly very intuitive pattern emerges concerning the structure of legal claims. If we attend to the hierarchy of restrictive qualifications, we get a pattern of embedded layers of elements in the following positions:

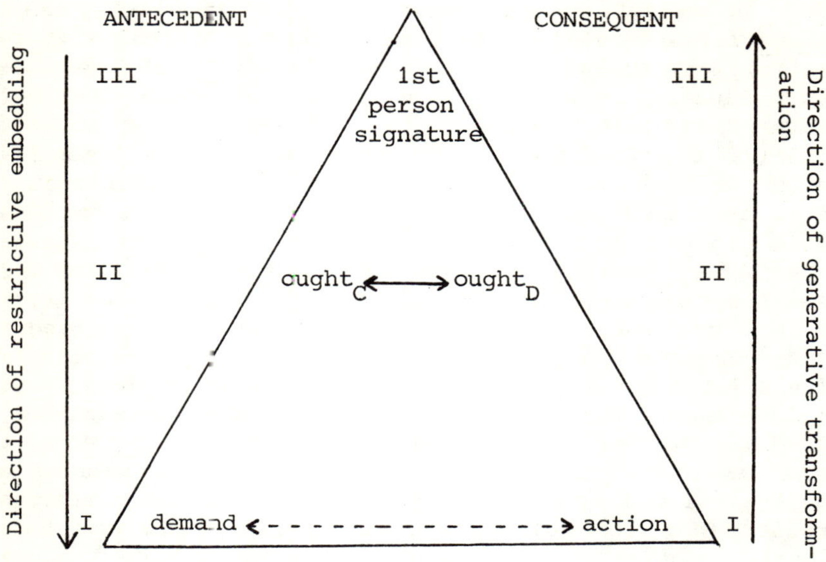

The diagram intends to indicate, that the discursive levels I, II, and III are various, but connected stages in the deep structure of legal norms, which by their very connections (cumulating in the embracing position of a competent first person) generate the tie of commitment between antecedent and consequent. One may represent these connections by operators of various kinds. Pending section 5, which will point to more analytical possibilities, I propose to be satisfied with a triad of barbarian 'philosophemes' to remember the following operations.

- Level I is generated by <u>conceptualisation</u> of an action 'to be done' as a <u>practical</u> response to problematic facts of our (social) lives.
- Level II is generated by <u>modalisation</u> of this demand and response in terms of <u>deontic</u> properties and institutions creating a <u>normative world.</u>
- Level III is generated by <u>distribution</u> of <u>competence</u> to attribute or establish these properties, among those who endorse the institution (the 'inhabitants' of the normative world).

4. PARAMETERS OF DISCOURSE LEVEL I, II, III TRANSFORMATIONAL SHIFTS

As far as I can see, levels I, II, and III are not inferentially connected, at least not in the straightforward sense that a prior level would contain noemata as premisses for a valid conclusion on the next level, or that the output of a prior operator would be automatically the input of the next (a simple and seamless re-ordering of 'the same' elements into different relations). In fact, it seems to me that these elements are not identical on all levels; they are equivalent under certain sets of constraints, relative to a tertium comparationis. Each operation between discursive levels requires a revision of the constraints under which elements can be treated and transferred to the next level as 'equivalents'. Therefore, the connections between the levels are not seamless junctions, but shifts or transformations(13).

Let me explain and illustrate this. Transformations in general are, basically, mapping relations between certain structures, e.g. tree structures in the derivation of sentences or stages in the task environment of an Information Processing System. They are commonly formulated in rules, which are brief accounts of specific rewrite- or 'move'-strategies. In our case, transformations relate levels of discourse. More particularly, they are conceived of as shifts in the parametrization of the elements which certain variables in the domain of the respective operators can take as their values. We may describe these changes by articulating intensionally the alterations in the concepts functioning as variables.

How are we to discover those parameters and their drifts in relation to distinct levels of discourse? We'll have to investigate some particular field or doctrine, case or rule of law. The following paragraphs yield a generalising summary of results I accounted for elsewhere. The reseach was guided by a twofold presumption: (a) that if transformations are involved in the generation of legal norms, they will not be visible at the linear surface of the norm, but for some traces or marks in the argumentative arrangements of textual elements; and (b) that if there are such traces, they will represent characteristically the dominant shade of the set of constraints that goes with each operation. Traces of the first operation (practical conceptualisation, constituting level I) are textual elements referring mainly to 'the nature of things' and having, anyway, a factual connotation. In connection with level II the flavour of argumentation changes over to normative or institutional concepts (e.g. 'the reasonable man', 'justified expectations'). Level III, finally, is specifically characterised by references to procedural orderings.

If we take a sample of legal claims (be they verdicts, decisions, requests, agreements, etc.) pertaining to problems of Dutch contract law (let's say: misunderstanding between parties), we may establish that these traces point to particular conceptual focuses or notional cores. Though it is almost certainly oversimplified, I give a picture of three of such cores, the parametrization of which is slightly altered at the various levels.

Firstly, there is a set A of actors which are conceptually qualified as subjects. If this is full-fledged legal subjectivity, we shall find textual elements that permit us to construct the dogmatic notion of legal subjectivity, i.e. being factually a bearer of rights and duties. In other cases we only find some embryonic or rudimentary indications of these qualifications.

Secondly, there is a set E of ends, interests or wants, qualified as objects possessed by the members of A. Again, if these are full-blooded legal interests, we shall discover fragments of argumentation that exploit the dogmatic definition of a patrimonium. This notion, as the notion of 'possession' in Dutch law, has indeed a factual ring.

Thirdly, we collect elements in the set S of assertions characterising the situation of the foregoing subjects and their patrimony in terms (predicates) of a balance of debts, which is often called the cause of the contract (do ut des; quid pro quo). It is essentially related to the way the subjects themselves conceive of this 'casual pattern'. At level II, we find the same sets A, E and S, but in the specific shade of the transformations to normative standards. On this level, members of A are considered to be not merely subjects, but rational, intentional agents, who can handle their own affairs in a well-ordered frame of mind. A fully dogmatic notion here is 'legal capacity'. Members of the set E are not just interests and ends, but are trimmed as subjective

rights (property, for instance, as (legally) granted possession) in an hierarchical order. The situational traits S are no longer accounted for in terms of (a-)symmetrical relations of debts (a market of demand and supply debts due to and due by...), but in terms of public order and good faith.
As to level III, here we meet parameters of procedural distribution referring to the circumstances at hand. The members of the same sets return once again, to be qualified respectively by ascription of (legal) competence or jurisdiction (A), tactics of suing (E), and the administration of certificates, documents, records (S).
To sum up: the operations on levels (I), (II) and (III) require related contexts or domains of the form C (A, E, S) in this specific sector of law (contract law).

5. PRACTITIONS, DEONTIC JUDGMENTS, AND AUTHORISATIONS

To get a reliable model for analysis, we'll probably have to complicate the composition of these contexts by further empirical research. Moreover, we'll have to articulate the matrix of operators and contexts in a much more regimented way. As to the latter, I'd like to indicate a few, necessarily sketchy, lines.

Castañeda's theory of practical thinking and normative institutions(14) is, in my view, an illuminating framework to comprehend the operations generating discursive levels I and II. He defines a practitional operator P, which produces practical counterparts of propositions: prescriptions and intentions, distinguished by Castañeda as first and second person practitions. They do not show the ordinary subject-predicate form dominated by the copula 'is', but they focus on some causal openings in our conceptions of reality as 'to do' in relation to some agent(15). Practitions cannot be true or false, but they are Legitimate or Illegitimate with respect to a certain context.

The deontic operator 0_i operates on practitions, yielding (deontic) propositions or judgments. These can be True or False, depending on their reference to an institutional 'family' ('i'), which qualifies every deontic operator. Their values, however, are reducible (under certain constraints) to the Legitimacy-values of the embedded practitions. But their discursive function is not reducible to practitions: they allow us to think and speak of actions-to-do (i) in a contemplating way, without the link to immediate action, (ii) but also without abandoning the level of practical thinking and stepping into a second order (meta-)level of propositions about the values of practitions. To this end, the deontic operator generates 'a hypostatic framework' or quasi-ontology of actions and their properties (obliged, permitted, etc.). Only an 'overriding' or 'dominant' deontic judgment, viz. what ought to be done 'everything being considered', leads up to endorsing a (Legitimate) practition.

Castañeda doesn't consider a separate operator for the third level, which would produce distributions of competence (authorisations) out of (overriding) ought-judgments. He seems to include 'procedures' in the set S, which contains situational characteristics. I wonder if it wouldn't be more adequate to define procedures on institutions by means of an authority-operator. This would perhaps also prevent the well-known confusion between the deontic property 'being permitted' and the concept of (legal) competence or 'power' (rechtliches Können). Lindahl has developed an elegant theory as to this point, trying to offer a new interpretation of what has been emphasized by authors like Burtham, Hohfeld, and Kanger(16).

Leaving this question aside, we may ask why Castañeda's theory has to be considered as a transformational account of the relation between levels I and II. The answer is this. The theory presents a powerful model for a quantificational logic of practitions and deontic judgments on a semantic basis, formed by contexts C (A, E, S, v, t), in which A gives a set of agents, E their ends and interests, S the specification of the situation they live in, v the range of practitions open to them and t a time. The logical account itself bears on the way these contexts are moulded by various sets of constraints. For example: attributing Legitimacy-values presupposes, according to Castañeda the 'total description' C^+ (A, E, S, v, t), which we have to construct by:

(i) eliminating all contradictions from a context C (A, E, S, v, t);

(ii) forming the context C'(A, E, S, v, t) as the union of this 'clean' context and the set R of propositions formulating the realisation of the ends in E;

(iii) closing C' under implication (Castañeda, 1975:137).

The same sort of moulding process is going on when Castañeda describes the absolute context K* (A), which is based on the harmonisation of motivational hierarchies of (rational and intentional) agents, considered as 'co-persons'. These are conceptual strategies, required by the transition from Legitimacy-relative-to-a-context to deontic Truth. Obviously, such sets of constraints are not accounted for by (the model of) practical thinking itself; for they are the very strategies that constitute this model. Now then, these constraints define, in my view, the transformational shifts between different levels of (legal) discourse. Only on condition that these shifts are made, we may 'discover' logical inference in legal language. A good deal of this language, however, performs in fact the function of establishing and reïnforcing constraints of the kind mentioned. But these are never expressed, only marked, by the linear order of a claim's text. I would think, that Castañeda's most perspicuous account (which is necessarily underexposed here) of some deep discursive structures in normative language makes transparant the philosophical (or ideological) choices by which it is restricted. It also makes comprehensible where we have to search for the characteristic role assigned to 'dogmatics' in this language: this role has to do with transferring the force of the constraints onto parameters like A, E, and S, i.e. the ways in which values of these variables are 'measured'.

7. SOME CONCLUDING REMARKS

I can imagine the objection that this transformational approach does not give a definition of law or 'legal norm'. I think, however, that it does, though not in the sense of a demarcation. The reader has undoubtedly noticed that in section 4 I suggested parameters which may, each of them, be closer to or farther from rigid legal definitions (e.g. 'subject'-'legal subject'), while the operators remain the same. This may be explained in the following way. The three operations together determine the content and topological form of certain performative acts (say: appointing someone as a representative, agreeing on some contract, etc.), which realise what they say by saying it. These effects are not restricted to a specifically legal world, for they have to do with someone's financial position, his freedom or bare life, his social relations. In other words, via a performative act, the conceptual parameters of contexts C (A,E,S,v,t) get their denotations in social reality. A next instantiation of the model can, therefore, rely on a firmer basis of evidence from social reality to construct these contexts. The circuit of the three operators determining a performative act, is, thus, self-reinforcing. It succeeds increasingly in picking out elements from social reality according to the preferred parameters, while it succeeds increasingly in effectuate (performatively) a (co-)domain according to these parameters, etc. We cannot decide at which point of this continuum we start off to use the word 'law' or 'legal'. And it would, by the same token, also be a fallacy to trace law back to some point zero from where it all emerges (e.g. social ('non-legal') reality). Law-like ways of posing problems and giving answers are part and parcel of our whole cultural life. Dominant social facts and values (on the one hand) and law (on the other) are the two focusses of the same ellipse(17). It is the task of legal theory to show that this ellipse is the product of discursive construction, rather than a hypostatic framework.

NOTES

* As this paper summarizes a part of my dissertation on legal theory and philosophy of science, I express my gratitude to my promotors, Jan M.Broekman (Louvain) and Stuart Silvers (Tilburg). The revision of the conference version of the paper was stimulated (though not sufficiently, perhaps) by discussions with Broekman, Castañeda and MacCormick.

1) Cfr. Wieacker (1970:319-320): "Die Sätze (sc. der Rechtsdogmatik) (...) sind nicht Aussagen freier Forschung über Wahr erklärte Prämissen geknüpft; sie haben (...) 'stipulatorischen' Charakter".

2) Cfr. Wittgenstein, Philos. Invest.par.225.

3) For more detailed argumentation of these points, cfr. Van Roermund (1983), esp. ch.1.

4) This is an interpretative scheme, especially as far as the additions 'to be done' and '(A)' are concerned.

They are, however, based on Kelsen (1960, ch.1 and 3) and (1979).

5) Cfr. Kelsen (1960:79 ff.).

6) (1976:121-124).

7) Ibid. 123.

8) Lewis (1975:11); cfr. Kratzer (1978:232; vv.).

9) For an interesting analysis cfr. Kneale/Kneale (1970:177-197); 274-297).

10) The term 'positio terminorum' is Boëthius' (cfr. Kneale/Kneale: 1978:193). For the medieval classification of consequentiae, cfr. ibid., 279.

11) Cfr. van Roermund (1983:ch.2).

12) I deviate from Kelsen's opinion, who states that references to Rechtsnormen are made in descriptive Rechtssätze, these being the paragon of Rechtswissenschaft. Cfr. Castañeda (1975:193): "...we must note very carefully that expressions like 'by law L' (...) can be naturally used to make a purely quotational statement, as well as a genuine deontic statement".

13) The term is both fashionable and fuzzy in legal theory, as can be readily learned from the discussions in Krawietz-Alexy (Hrsg.) (1983). Cfr. in particular Weinberger's questions (161ft.) to Peczenik's theory. In my view, the first part of Peczenik's definition (ibid:14) would run: "p must/can be replaced by q". I'd leave out the word 'Grund', as well as the idea that transformations can be justified in terms of (contextually determined) rationality. They are part of the definition of (e.g. legal) rationality analogous to the way they define 'well-formedness' of language for linguistic theory. That is why they are the boundaries of a 'context of justification'.

14) Cfr. Castañeda (1975) for a systematic and detailed exposition of his theory.

15) According to Castañeda (1975:281), P is (...) "a focussing mechanism, that spots the relevant parts of a proposition when the agent's thinking causality can connect the proposition whose truth he is to bring about with the mechanisms that carry out the attempt.

16) Vgl. Lindahl (1976:194vv.).

17) Thus, I intend to construct a bridge from an analytical account of legal norms to some very central ideas about law presented by J.M.Broekman (cfr. e.g. his contribution to Krawietz/Alexy (Hrsg.)(1983). Cfr. Van Roermund (1983), esp. the end of ch.3.

BIBLIOGRAPHY

BROEKMAN, J.M., 1983, Die Rationalität des juristischen Diskurses, in Krawietz/Alexy (Hrsg.) (1983), pp.89-116.

CASTANEDA, H.N., 1975, Thinking and doing, the philosophical foundations of institutions, Dordrecht-Boston, 1975 (PSSP,7).

CASTANEDA, H.N., 1976, The twofold structure and the unity of practical thinking, in M.BRAND/D.WALTON (eds.), Action theory, Dordrecht, 1976 (Synthese Library, 97) pp.105-132.

KELSEN,H., 1960, Reine Rechtslehre, 2., vollst. neubearb. und erw. Aufl., Wien, 1960 (1934[1]).

KELSEN,H., 1960, Allgemeine Theorie der Normen, im Auftrag des Hans Kelsen Instituts aus dem Nachlass herausgegeben von K.Ringhofer und R.Walter, Wien, 1979.

KNEALE, W.M.KNEALE, 1978, The development of logic, Oxford, 1978 (1962[1]).

KRAWIETZ, W./R.ALEXY (Hrsg.), 1983, Metatheory der juristischen Argumentation, Berlin, 1983, (Schriften zur Rechtstheorie, 108).

KRATZER,A., 1978, Semantik der Rede, Kontexttheorie, Modalwörter, Konditionalsätze, Kronberg (Ts), 1978 (Monographien Linguistik und Kommunikationswissenschaft).

LINDAHL,L., 1976, Position and change, a study in law and logic, Dordrecht-Boston, (Synthese Library, 112).

LEWIS, D.K.,1975, Adverbs of quantification, in E.KEENAN(eds.) Formal semantics of natural language, Cambridge, 1975, 3-15.

ROERMUND,G.C.G.J.van, 1983, Wetten en Weten, 'Theorie van het recht': een wijsgerige kritiek, ('Legal and illegal knowledge, 'Legal Theory': a philosophical critique'), Leuven-Zwolle, 1983 (diss. Tilburg 1983).

WIEACKER, J., 1970, Zur praktischen Leistung der Rechtsdogmatik, in R.Bubner c.s. (Hrsg.), Hermeneutik und Dialektik, Bd.II, Tübingen (Festschrift H.G.Gadamer).

ABSTRACT.

The object of legal theory is: norms of positive law, not as 'sources' or 'premisses' in legal justification, but as surface structures of such justifications. To give an account of the underlying deep structure(s), a transformational model is constructed in its basic form, by confronting (a) an analysis of the conditional form of legal norms and its topology, and (b) parameters of contract law. Castañeda's theory of practical thinking and normative institution (and Lindahl's theory of 'changing positions') are referred to as an analytical framework for the transformational model.

EPISTEMOLOGY AND VALIDITY IN LAW

Jacques Lenoble

INTRODUCTION

The contemporary reflection on the problem of the validity of juridical norms conducts to a radical questioning of the frontier drawn by the positivist theories of law between an approach which intends to be merely descriptive and a process that tends to integrate the axiological dimension into the study of normative systems. Recent developments in both deontic logic (Bulygin,1982) and the analysis of juridical reasoning (A.Aarnio, R.Alexy and A.Peczenik, 1981) reveal clearly the necessity to question the principle of axiological neutrality which underlies the positivist explanations of the validity in law. Our reflection tries to bring out the aporetic structure which the thesis of the principal positivist theoreticians offers with regard to the question of the validity of norms.

Kelsen's works (Section I), those of Alf Ross and H.L.Hart (Section II) as well those of A.Aarnio and I.Niiniluoto (Section III) will be successively analyzed. In the fourth section, I will discuss the relation between the findings in the preceding section and the theories of J.Habermas (Section IV).

SECTION I

In order to perceive how contemporary philosophy of law is confronted with a recurrent apory - although not qualified as such - but perhaps bound to the dilemma of the prevailing epistemological model, the first author to take into consideration is certainly Hans Kelsen. Our reasoning will be divided into three parts. We will briefly describe the Kelsenian position on the nature of the law proposition; then we will analyze the different and possible interpre-

A. Peczenik et al. (eds.), Theory of Legal Science, 163–173.

tations of this position and distill the doctrine that arises therefrom. As a sequal to this we will attempt to provide answers to the problem related to Kelsen's teaching.

We will recapitulate briefly the essential outline of the kelsenian project. The "pure" theory of law, insofar as its object is built upon the "a priori" category of the ought (Sollen), is in that much outside or beyond any sociological approach to law. The epistemological field of the "a priori" category leads Kelsen to characterise legal propositions as being simultaneously susceptible to the normative and the descriptive. Kelsen does, however, distinguish between the norms (Sollnormen or Rechtsnormen) and the propositions by which they are described in legal science (Rechtssätze or Sollsätze). He observes that the latter may be subjected to the criterion of truth or untruth in spite of their normative character. But the former cannot be brought within the range of this criterion as they must be considered only from the vantage point of validity.

The distinction Sollsatz-Sollnorm, bound to the kelsenian conception of normative science, is far from being clear. What interpretation should be attached to Kelsen's passage in which at the end he qualifies the Sollsatz as descriptive and normative? Various interpretations have been advanced. It is important to mention them briefly because their consideration provides the possibility to understand the issues involved in Kelsen's scientific enterprise. Such understanding is particularly pertinent to the questions we intend to deal with.

A first possible interpretation is to construe the use of the descriptive Sollen worked out by the law propositions of Kelsen as the "inverted commas use" described by Hare (1952). The normative character of the language would only be indirect, "entre guillemets". In a way there would be a reference to Sollen in the law proposition and not a utilization. This thesis was defended by G.Kalinowski (1968) and Hart (1963) both of whom attribute the difficulties arising from that problem to an absence of a clear distinction between the levels of language.

Nevertheless the interpretation of the use of the "ought to be" (Sollen) in the Rechtssatz as "inverted commas use" reduces the complexity of the kelsenian analysis, in the same way that many interpreters, including some of the authoritative ones, have also reduced the kelsenian notion of validity to the criterion of membership of a legal determined system. The researches of J.Raz (1971) and C.S.Nino (1978) prove, on the contrary, that the "ought statements describing the law" refer to a concept of validity bound to that of justification and whose normative nature has very often been overlooked. It is clear that the "justified normativity", in reference to the *Grundnorm*, cannot be taken in an absolute sense but is related to a peculiar point of view: that of the jurist. This lays down the basis for a proper understanding of the famous assertion of Kelsen that "even an anarchist, if he were professor of law, could describe positive law as a system of valid norms without having to approve of this law".

Consequently, we can see that it seems reductionist and even incorrect to construe normativity of the *Rechtssatz* when one neglects the fact that for Kelsen the science of law is defined as normative and that the normativity of law is a "justified" normativity in its own right even though it is also justifiable in terms of social normativity. This observation underlines the main difficulty of Kelsen's enterprise, namely, an attempt to give an analytical description of the juridical phenomenon understood as a normative phenomenon. So, the willingness of H.Kelsen to take into account, on a theoretical level, the symbolic dimension of the juridical phenomenon, (as this does not present itself simply as mere factuality) leads him to an implicit denial of the logical character of normative propositions. Kelsen's *Sollsatz* is ambiguous. The ambiguity is an expression and simultaneously a manifestation of the aporetic structure of legal science. In the science of law this structure establishes itself in an epistemological project which tends to articulate the symbolic analysis of law as a descriptive presentation. Nevertheless it is possible to question if such a deadlock does not arise from the refusal to apprehend juridical normativity in terms of social normativity. Such is precisely the way that will be explored by the authors of Scandinavian realism and the Oxford School.

SECTION II

The evolution of Alf Ross's thesis, one of the principal

representatives of Scandinavian realism, represents a clear interest: his initial attempt to explain juridical validity in exclusively empirical terms will progressively change so as to integrate normativity in the shape of the formal requirements of deontic logic. In opposition to the naturalistic traditions of Bentham and Austin, A.Ross's realism emphasizes that law should not be reduced to imperatives imposed by a power. Rather, what distinguishes it from other human experiences is the "awareness of universal validity", the experience of universal validity. But in opposition to Kelsen, Ross strives to give an account of the symbolic dimension of law in simple empirical terms: descriptive requirements in this respect are brought about by the systematic reduction of the normativity of juridical systems to a block of phenomena of the psychological and sociological order. But is this not to deny every efficacy proper to this symbolic dimension which we have, in other respects, qualified as constitutive of law? In this connection it is nevertheless interesting to note the evolution of A.Ross's theory (cf. Directives and Norms).

Indeed the final evolution of the Scandinavian realist movement represented here by Alf Ross denotes the possible reinsertion, by way of an eminently contemporary form - the logical form - of a normativity in a reading, which intends to be purely descriptive of the validity of norms, including the juridical norms.

H.L.Hart faces the same apories. Hart's position can be defined in comparison with that of A.Ross: on the one hand, he rejects, following the Scandinavian realists, the imperativists' theories and states that the analysis of the juridical phenomenon implies reference to the idea of obligation. On the other hand, he considers that the validity of a juridical norm is based on its conformity to a recognized rule whose existence is a simple question of fact. But, for Hart, there are many possible viewpoints about the juridical system. Indeed, we must distinguish the external viewpoint which is that of an observer who does not accept for himself the rules whose existence he discovers, from an internal viewpoint which is that of a member of the group who accepts them and uses them as guiding rules. Moreover, Hart distinguishes a third viewpoint, that of an observer who, not accepting the rules himself though acknowledging their existence, confirms nevertheless that the members of the group accept them and so refers to the exterior in the way that the members of the group consider them from an internal point of view. What then could signify that such an observer does take into account from the vantage point of the exterior the internal viewpoint of the group? Could external assertions, insofar as they are "expressive" of the internal point of view, be deemed to be real propositions in the logical sense of the word, and as such consequently susceptible to truth or untruth? Do they have a normative dimension? Is the reference to the internal point of view in Hart to be understood as a translation of the classical semiotic distinction between language and metalanguage? Such is, in any case, the interpretation that E.Bulygin gives us (1982). But such an interpretation seems to be problematic. Indeed, could the reduplication of obligation be analysed in terms of a metalinguistic name? How could one make law adhering strictly to the notion of normativity and translate that into factuality without relapsing into the predictionist reduction of Scandinavian realism? It would seem that to express the obligatoriness of a norm within the limits of a juridically given order would imply the recognition of normativity even under a peculiar modality. These questions, if subjected to thorough scrutiny, would reflect how, through the use of a subtler analysis, two authors, namely, R.Hill (1970) and J.Raz (1971), defended Hart's interpretation. Thus, according to J.Raz, the third kind of external judgment defined by Hart - qualified by Raz as "detached legal statement" - is normative but it "does not carry the full normative force of an ordinary normative statement "because" its utterance does not commit the speaker to the normative view it expresses... this third kind of statement is characteristic of the lawyer and the law teacher..." (op. cit. pp.346, 348). In this way, we come back to the Kelsenian position, to which Raz refers explicitly.

This reference to the Kelsenian position helps us to understand how Hart's and Raz's positions reveal again the aporetic structure of the epistemological current underlying the theories of validity defended by the two authors. The authors' claim to bring juridical normativity back to social normativity distinguishes them from Kelsen. However, their willingness to subscribe to the symbolic dimension of the juridical phenomenon, even though it is predicated by a non-reductionistic tendency, and, on the other hand, their readiness to anchor the legal science project upon a descriptivist epistemological model lead to what seems to be a dead-end. Indeed, we must follow E.Bulygin by recognising that to ascribe a normative dimension to the legal science proposition prevents it from being a proposition in the strict logical sense, that is, to be subjected to the rigours of truth and falsity.

SECTION III

The discussion between A.Aarnio and I.Niiniluoto provides once again the occasion to situate the aporetic structure of the empiricist science of law. At the same time this discussion contains elements likely to lead to the solution of this aporetic structure of knowledge of the law. These elements appear quite clearly in Aarnio's position which implicitly recognizes the hermeneutic and axiological dimensions of any theory of law. However, the elements in Aarnio's position are not fully developed in their principle and in their theoretical and practical implications.

For I.Niiniluoto, propositions such as "legal norm n is valid in Finland" presuppose susceptibility to the truth scale in logic. Thus if we accept the empiricist conception of the correspondence theory of truth, then the statement that a valid rule exists is reduced to the discernment of a corresponding block of social facts.

A.Aarnio's thesis, which generates difficult questions of interpretation, seems first of all to be based upon the fundamental premise of the realist position with regard to validity, notably, that the validity of a legal norm cannot be defined merely in empiricist terms, that is, as a social state of affairs (A.Aarnio, Rechtstheorie, Beiheft 2, 1981). Aarnio seems to suggest the subsumption of a concept of validity connected to that of rational acceptability. This must be bound to the ideology of law that asks for "the avoidance of arbitrariness". As part of the argument of the theory, we have to consider that the norm n is valid not only if "it is part of the system authorized by the fundamental norms", and if it is efficacious, but moreover if it "would be generally accepted on the basis of the (dominant) system of values, if matters were to be considered rationally" (Aarnio, 1983).

The non-empiricity of the constructive elements of

validity and the normative character of the retained criterion indicate, I think, that the statement of validity has in itself a normative meaning. The idea that what must be retained as valid is the normative content corresponding to what is established according to an argumentative procedure and following the criterion of rational acceptability, means that the established rule signifies an obligatory rule in the juridical sense and not simply in the sense of a socially accepted practice. It follows that, for us, again the willingness of the philosopher of law to take into account the symbolic dimension of the juridical phenomenon impinges upon the constitutive elements of the "proposition" which accounts for it. The matter is even the more representative because the specifically judged element is qualified as the ideology of the law and forms an axiological principle. Immediately a simple descriptivist point is put aside. We can go even further. It is precisely Aarnio's consideration as well as his inclusion of the too reductionist character of an empirical reading and of the hermeneutic element of legal discourse which lead him to produce a directly normative reading of the validity of juridical norms.This recently acquired position after a long evolution of the theory of juridical validity, allows us to put in more accurate words the epistemological problematic whose elements we try to define: in fact we have the feeling that the circle closes itself at the very moment when the question of validity rediscovers the conditions of a critical break out.

SECTION IV

We could be tempted, as E.Bulygin was in his already mentioned article, to consider that the normative contamination of the law proposition is nothing but perversion in relation to the scientific requirements that the descriptivist epistemological scheme expresses.And moreover, this perversion would be considered as the result of disregard for the ambiguity of deontic sentences. Logical research, on the contrary, would certainly have removed the ambiguity and in consequence even shaped the field of normative propositions in the logical sense of the term.

We will not immediately subscribe to such an interpretation. The perversion of the legal philosopher's positions appears to be rich with insights that certainly affect the claims of the logician concerning the analysis of normative propositions on the basis of the theory of the levels of language. The latter's claims refer particularly to the propositional character of the normative proposition. The problem arises at the starting point of a descriptive project designed to underplay the symbolic dimension of the norm whose proper signification is its normativity.Does the suppression of normativity therefore reveal an inadequacy in the descriptive project? In fact, it would be difficult to give full meaning to the idea of "describing" a rule as valid in the sense that it is obligatory. Does such a description inevitably involve a reduplication of the deontic functor which syntactically makes the norm? We can of course describe the content of a norm. What meaning could be attached to an already established condition of obligation if the conveyor of this meaning has not already assumed this state of obligatoriness? Moreover, our reflection does not intend to deny that where the law propositions denote an external point of view of the first two types described by Hart, such propositions are, insofar as they refer to verifiable facts, simply descriptive and susceptible of truth or untruth. But perhaps it is precisely at the cost of a "desymbolization" of the juridical phenomenon, by ignoring that the studied fact is a normative fact, that this fact is non-reducible to a simple conglomerate of natural or social facts.

We may even wonder, if "this" impossibility to describe juridical norms - except by devitalizing their normativity - is not related to the performative character of these norms. Is the proper effect of this kind of linguistic act which the performative act constitutes, susceptible to metalinguistic expression: could we have some doubts about this? What characterizes the performative act is "its auto-referentiality", it is the circumstance that "l'énoncé exprime la proposition que ce que l'énoncé prétend produire est l'énoncé lui-même (J.de Greef, 1980). But this characterisation would appear to render inappropriate the idea of accounting for the normative effect - if the latter were conceived of in non-factual terms - through the distinction of levels of language. This may well be compared to De Greef's assertion that the performative dimension of a statement renders unnecessary any recourse to the metalanguage suggested by Russell who proposed it for the resolution of the logico-semantic paradox of auto-referentiality, which recoils upon the metalinguistic level and so and infinitum.

That perspective rejoins the critique of positivism developed by J.Habermas with regard to the indissociably hermeneutic and ethical dimensions of the human language. An awareness of the hermeneutic dimension of the reading of any normative phenomenon in terms which claim to apprehend this normativity does not legitimately imply that this signification never lets itself be captured. It is always marked by a precomprehension that it is the sisyphic duty of the hermeneutic to uncode it always over and over again. Consequently one also understands why J.Habermas accords to ordinary language the privilege of a place where the interactions mediate themselves and in that mediative core the hermeneutic sciences are inserted and fulfilled. It is in itself its own metalanguage for it has the possibility to interpret itself (auto-interpretation). At this point of the argument we find again the reflexivity of performative statements. From the same viewpoint it is necessary to take into account that the truth of the normativity as signification can never be elucidated. One could paraphrase here J.Lacan's famous sentence, namely "il n'y a pas de métalangage (affirmation faite pour situer tout le logico-positivisme), que nul langage ne saurait dire le vrai sur le vrai puisque la vérité se fonde de ce qu'elle parle".

But, if so, then consequences ensue on the level of the theory of the juridical norm. Any theoretical consideration of a norm which attempts to integrate the symbolic dimension implies a discourse based on an epistemological project irreducible to the project advanced by the legal positivist theoreticians. That new requirement, to which the emphasis of the aporetic structure of the positivist juridical epistemology leads, may however be met in different ways. We shall only "mention" them here.

The first one immediately tries to pose the problem in terms of legitimacy. This view is principally that of J.Habermas, more particularly retaken by A.Aarnio, A.Peczenik and A.Alexy in the outlines of the theory of law.The question is to re-formulate the issue of legitimacy in the frame of the problematic of the hermeneutic character of the social sciences and so of the normative phenomenon's sciences. The validity criterion, the truth of norms will be found in the conditions of "communication without impediment", that is neither in a classical iusnaturalistic perspective, nor on a normative contents level; but rather on the level of the adoption procedures norms, which may always be reinterrogated.

There is an other way, even more contemporary, which tries to elaborate the elements of thinking about the normative phenomena overdoing any (allegedly "imaginary") belief in the rationality of a meaning in the process of "deciphering". The theory of norms has to break off from such a presupposed rationality. Such a paradigm is the basis of the researches of M.Foucault, J.Lacan, J.Derrida, J.F.Lyotard.

As we have already said, it was important here to highlight at one and the same time the dilemmas of the prevailing epistemological model and the perspectives that contemporary philosophy presents in order to pass beyond the positivist juridical epistemology, further exposition of which is reserved to a work already under way.

BIBLIOGRAPHY

Aarnio Aulis, On truth and the acceptability of interpretative propositions in Legal Dogmatics, in Methodology und Erkenntnistheorie der juristischen Argumentation, Rechtstheorie, Beiheft 2, 1981, pp.33 et ss. (A.Aarnio, I.Niiniluoto, J.Uusitalo eds.).

Aarnio Aulis, Alexy Robert, Peczenik Aleksander, The foundation of legal reasoning, Rechtstheorie, 1981, pp.133-158; 257-279;423-448.

Aarnio Aulis, On the validity, efficacy and acceptability of legal norms, in Philosophical perspectives in Jurisprudence, Acta Philosophica Fennica, 1983, pp.152 et ss.

Bulygin Eugenio, Norms, normative propositions, and legal statements, in Contemporary Philosophy. A new survey, vol.3, Philosophy of Action, edited by G.Fløistad, Martinus Nijhof Publishers, The Hague, Boston, London, 1982, pp.127 et ss.

De Greef Jan, Peut-on fonder des conventions, Revue Interdisciplinaire d'Etudes juridiques, 1980-4, pp.97 et ss.

Gadamer Hans Georg, L'art de comprendre, traduction française par Marianna Simon, Paris, Aubier-Montaigne, 1982.

Habermas Jürgen, Connaissance et intérêt, traduction française par Gérard Clemencon et Jean-Marie Brohm, Paris, Gallimard, coll. N.R.F., 1976.

Habermas Jürgen, Raison et Légitimité, traduction française par Jean Lacoste, Paris, Payot, 1978.

Hare R.M., The language of morals, Oxford, 1952.

Hart L.A.Herbert, Scandinavian Realism,Cambridge Law Journal, 1959, pp.233 et ss.

Hart L.A.Herbert, Le concept de droit, traduction française par Michel van de Kerchove, Bruxelles, Publications des Facultés universitaires St.Louis, vol.6, 1076.

Kalinowski Georges, Querelle de la science normative, Paris, L.G.D.J., Bibliothèque de philosophie du droit, vol.X, 1969.

Kelsen Hans, Theorie pure du droit, traduction française de la 1ère édition de la "Reine Rechtslehre" par Henri Thevenaz, Neuchatel, Editions de la Baconnière, 1953.

Kelsen Hans, Théorie pure du droit, traduction française de la 2e édition de la "Reine Rechtslehre" par Ch.Eisenmann, Paris, Dalloz, collection Philosophie du droit", vol.7, 1962.

Niiniluoto Ilkka, Truth and Legal norms, communication in the 11th World Congress on Philosophy of Law and Social Philosophy, Helsinki, 1983.

Nino Carlos Santiago, Some confusions around Kelsen's concept of validity, A.R.S.P., Bd LXIV/3, 1978, pp.357 et ss.

Raz Joseph, Legal validity, A.R.S.P., 1977, pp.339 et ss.

Raz Joseph, Kelsen's theory of the basic norm, The American Journal of Jurisprudence, 1974, pp.94 et ss.

Ross Alf, Directives and Norms, London, Routledge and Kegan Paul, 1969.

Ross Alf, Tû-Tû, Scandinavian Studies in Law, 1957, pp.138 et ss.

Ross Alf, On Law and Justice, Berkeley and Los Angeles, University of California Press, 1974 reprinted.

IS LAW A SYSTEM OF ENACTMENTS ?

Csaba Varga

Law has two models: an ideological one and an actual one. The former represents the professional ideology characteristic of modern formal law. Notwithstanding, according to the latter, formal enactment can only get social meaning through its social context in a semantic sense (termed as system of law) and social existence through its social context in a socio-ontological sense (termed as legal system). Dworkin's theoretical field is characterized by a jurisprudential approach. The socio-ontological approach offers the promise of deeper theoretical perspectives. Notably, it suggests that law is 1) a historical continuum, 2) an open system, 3) a complex phenomenon shapable by either its description or its contexts alternatively, and 4) an irreversible process. All in all, law is more than a set of rules and even more than a set of enactments: by its very definition it is just at the borderline where legal research and social science are expected to meet.

1. WORKING MODELS OF LAW

According to its ideological model, law is the product of competent state institutions observing established rules of procedure. Consequently, only and exclusively what is enacted by given institutions in given ways is to be considered law, independently of its actual contents, formal coherence, foreseeable realizability, factual realization,etc. This model has some definite shortcomings. First, both the qualification of any institution as competent and the establishment of any rule of procedure are the function of legal enactment. It means that law gets defined by the law itself or, in other words, that the very definiton of the preconditions of the existence of any law is offered by its own self-qualification. Secondly, the model in question has solely a view of the law in books, without having any regard to

A. Peczenik et al. (eds.), Theory of Legal Science, 175–182.

its conceivable or completed action(1). Thirdly, this model is not the image of any reality. It is the outcome of a definite type of wishful thinking: it is a projection of normative requirements. Notwithstanding, in the legal cultures of the western world in general and in their professional practice in particular(2) the ideologically working model has a primary role in identifying what ought to be regarded in theory and treated in practice as distinctively legal(3), the model as such holds as reasserted even if it proves to be imbued with clearly utopian elements.

According to the actual model of law, the meeting of these normative requirements will result, in practice, in an outcome which will necessarily be either more or less than what the ideological model has suggested, and which will differ in one or another feature from it. As outlined elsewhere(4), it is the self-qualification of law as the law that seems in any case to remain the final criterium even if no rigid distinction between the spheres "within the law" and "outside the law" is accepted any longer, even if law is taken and treated as a continuum in an unbroken motion, or even if the self-qualification of law will only be understood within the boundaries of its actual social acceptance. Consequently, it is formal enactment (FE) that will serve as an abutment stone and also as a series of corner-stones in delimiting what is considered and what is to be considered law. In the same way, it is formal enactment that offers itself to be used and commands its own solely decisive use as a starting point and also as a series of turning points in any argumentation within a genuinely legal process. However, formal enactment is not sufficient in itself at all for delimiting the realm of law or for channelling legal argumentation process. In other words, formal enactment is not in a position to circumscribe the province of what is considered and what is to be considered distinctively legal or to characterize the law's general trend, goal-orientation, value-commitment, etc., i.e. to define in an operative manner - sharply and unambiguously - the formal and doctrinal qualities (e.g. conceptual system, logical coherence) of the law or its social potentialities and realities without taking into account its social context.

2. SENSES OF CONTEXTUALITY IN LAW

In connection with law, contextuality has two main fields of action.

According to contextuality in a semantic sense, a system of law(SL) conceived of as a definite set of norms(5) is formed by that part of formal enactments which can be socially relevant and by the social context(SCI) in which the formal enactments are embedded. It is the social context in question that defines the paradigms of the formal enactments by giving their text a truly social existence. By social context in a semantic sense I mean here the sumtotal of presuppositions, preconceptions, value choices, non-formal propositions and enactments, etc. which have a major influence in a given society and, consequently, are instrumental in making formal enactments meaningful in a socially concrete way and, thereby, interpretable and applicable in practice. Hence,

SL = FE + SCI.

According to contextuality in a socio-ontological sense, no bare factuality or a pure declaration or verbal reassertion thereof can stand for the totality (i.e. the inner potentialities, henceforth, the real significance) of any phenomenon. For instance, in the last resort both power and law are backed by coercion, by the potentiality of making use of force, by the guaranteed alternative to resort to "men with peaked helmets" as Max Weber specified it symbolically(6). However, it is neither the factual presence nor the concrete actuality of coercion that is needed in ordinary cases. It is rather the continuously renewed threat of coercion that matters. It is the immanent possibility of resorting to it (a possibility which is imminent at any time as backed by a whole normative, institutional and ideological apparatus) that will be good enough to achieve a general law-observance in average cases. Or, to come back again to law: the question of whether there is any legal possibility to turn anything ideal into real or not, is a function of the system of law concerned, i.e. of the socially accepted meaning of the law's formal enactments. Yet the question of what and how much of this possibility is to be and gets eventually transformed into and implemented in social reality, and also the question of with what result and by-effect and in what manner will all this be done, are a function of the law's social context(SC2) turning the system of law into a socially functioning complex mediating among social complexes. By social context in a socio-ontological sense I mean here the total sum of social considerations, political forces, power conditions, etc. which have a major influence in a given society and, consequently, are instrumental in making the expediency(inevitability of the practical implementation of formal enactments socially evident) tolerated. These, then, act as the driving force of and the responsible agent for the actual functioning of the system of law in question. It is this actual functioning that is termed as legal system(LS) conceived of as the sumtotal of activities carried out in name and socially accepted/tolerated as being within the boundaries of what is called distinctively legal. Hence,

LS = SL + SC2.

3. JURISPRUDENTIAL APPROACH AND SOCIO-ONTOLOGICAL APPROACH

It is very well known how much the seemingly self-evident question Professor Dworkin raised two decades ago, 'Is Law a System of Rules?'(7), has been of a liberating effect to the western legal thinking in general and how much it has become one of the most challenging and promising theoretical reformulations of our time. Though the assertion it has formulated is seemingly just a simple one: law is more than a set of rules (other linguistic expressions of normative contents are to be considered, too), in order to fill the gap between the rules as enacted and the judicial decisions as reached. My question is not a reconsideration of Dworkin's which, in the wave of the repeatedly re-started discussions, seems to gain more and more strength and a reassertion of its foundations. I think that my question is rather a continuation thereof, a continuation, but with a change of underlying assumptions. For Dworkin's question is fairly a jurisprudential one conforming to the basic assumptions of the ideal type of modern formal law as shaped in various ways by Common Law and Continental Law development. By jurisprudential approach I mean here the one which starts from/arrives at the following assumptions: (1) law is something identical with itself, consequently it is something defined/definable in and by itself; hence (2) it has some definite bordering lines making a clearly-cut distinction between the spheres "within the law" and "outside the law"; therefore (3) the only question to be answered is how to enlarge the very notion and/or the texts of the law in order to be able to bridge the gap between the officially fed in-put and the practically realized out-put of a legal process. As added to this and also confronted, but at another level, I argue for a <u>socio-ontological approach</u> which reveals the purely postulated nature of the basic assumptions of the ideal type of modern formal law and, while treating them as components of the jurist's professional ideology having a real function in this law's proper existence, tries at the same time to reconstruct the whole societal context they are embedded in and which makes them function. Should some issues and tendencies of my earlier investigations(8) be confirmed, a socio-ontological approach has the following statements as conclusions: 1) law is but a historical continuum defined through its actual social practice; hence 2) in want of any <u>a priori</u> given demarcation line only an <u>a posteriori</u> description of interactions (e.g. "differentiation" and "unification", "core" and "marginalia") among the diverse homogeneous and heterogeneous impulses is possible; therefore 3) the claim of any control through logical demonstration is replaced by a socio-ontological reconstruction of all the factors which have had a role to play, not negating but constraining the one played by logical control.

Having in mind the many-sided social conditionedness of law, i.e. the very fact that both its objectification (system of law) and its use, as a practical means, of (legal system) can acquire social meaning only in their concrete social context in a way that you can state with transforming the former definitions that

$$LS = (FE + SC1) + SC2,$$

it goes without the need of further explanations that only and exclusively a socio-ontological approach can be instrumental in coping with dilemmas that have remained unsoluble, untouched, or simply ungrounded within the framework of a jurisprudential analysis. To take but a few instances: how to explain <u>la jurisprudence</u> with its well implemented structures and institutions if it has been made up by a consequential judicial practice on the sole basis of some general clauses and principles of the law? (ad 4.1.) how to reconstruct the past development in the formative era of the European law if fairly diverging national systems of laws have grown up through the alleged reception of the same body of classical and post-classical Roman law? (ad 4.2.) how to assess the contradictory claims and trends of the human rights movement today? Notably, how to qualify cases when legal regulation seems to be perfected but meets no appreciable fulfilment, and how to qualify cases when no declaration is made in the law but practice proves to be as fulfilling as possible? (ad 4.3.) or, how to draw the limits of a change in the law imposing even on the power and arbitrariness of the law-maker with the effect that the formal change he enacts cannot touch upon past achievements? (ad 4.4.)

4. CONCLUSIONS

In the final analysis from all that has been said at least four conclusions are to be drawn.

4.1. Law as historical continuum

Law is a historical continuum in an unbroken process of formation. Because it has no social existence of its own without the context making it interpretable (SC1) and setting it in function (SC2), it changes - or may change - its social contents and impact ceaselessly even if there is no change in its formal enactment.

4.2. Law as open system

Consequently, law is an open system. It can only be treated as closed for the sake of its historical reconstruction. For the aim of such a reconstruction is just to reveal which sort of laws "within the books" and "outside the books" may have been in play so as to make the reality of that law in action "deducible" therefrom or "reduced" thereto in the most consistent way.

4.3. Law as complex phenomenon with alternative strategy

Law as a bipartite phenomenon organized together from two distinct sources raises the question of the character and composite nature of its instrumentality. Namely, if law as a working system is composed of formal enactment and its social contexts making it interpretable (SCI) and setting it in function (SC2) and if a change of any of its components may cause a change of the law as a working whole, there is offered a perspective for an alternative strategy. I mean thereby that a struggle for the law can be fought through a struggle for confirming/reforming/revoking its formal enactment and through a struggle for strengthening/reshaping/loosening its social contexts as well, and that any of these alternatives can eventually lead to the same goal as set.

4.4. Law as an irreversible process

The social existence of law is to be seen as an irreversibly progressing process. It is irreversible because any enactment may be revoked with ease, but formal enactment is so thoroughly combined with and filtered through its social contexts that something from the latter will be left irresistibly. Or, to put it another way, law cannot be manipulated in all its components to the same depth.

4.5. The genuinely societal character of law

All these conclusions seem to suggest that law is something more than a set of rules and it is even more than a set of enactments. By its very definition law is just at the borderline where legal research and social science are expected to meet.

NOTES

(1) For the distinction between 'law in books' and 'law in action', see Pound, R.: 'Law in Books and Law in Action', American Law Review, 44(1910)1.
(2) For the theoretical analysis of some common characteristics from a historico-comparative point of view, cf. Varga, Cs.: 'Moderne Staatlichkeit und modernes formales Recht', Acta Juridica Academiae Scientiarum Hungaricae, 25 (1983) in press and Varga, Cs.: 'Logic of Law and Judicial Activity: A Gap between Ideals, Reality, and Future Perspectives', in Z.Péteri and V.Lamm (ed.), Legal Development and Comparative Law - Evolution du droit et droit comparé, Akadémiai

Kiadó, Budapest, 1982.

(3) For the first use of the term, see Selznick, Ph.: 'The Sociology of Law', in D.L.Sills (ed.), _International Encyclopedia of the Social Sciences_, vol.9, MacMillan and The Free Press, New York etc., 1968, pp. 51 et seq.

(4) Varga, Cs.: _Macrosociological Theories of Law - From the "Lawyer's World Concept" to a Social Science Conception of Law_, IVR Helsinki 1983 World Congress Working Group VI Session 4 opening speech, in press, ch.1, or, in more details, Varga, Cs.: _Domaine "externe" et domaine "interne" en droit_, paper presented at the Serbian Academy of Sciences and Arts in Belgrade, October 1983, in press.

(5) For the differentiation between 'system of law' and 'legal system', see Kulcsár, K.: 'Historical Development of the Law-Applying Functions - Social Conditions and Legal Evolution', in Z.Péteri (ed.), _Droit hongrois - Droit comparé_, Akadémiai Kiadó, Budapest, 1970.

(6) Weber, M.: _Gesammelte Aufsätze zur Wissenschaftslehre_, Mohr, Tübingen, 1922, p.403, etc.

(7) Dworkin, R.: 'The Model of Rules', 35, _U.Chi.L.R._ 14 (1967), as reprinted in R.M.Dworkin (ed.), _The Philosophy of Law_, University Press, Oxford, 1977. For the whole range of problems concerned see Dworkin, R.: _Taking Rights Seriously_, Duckworth, London, 1977.

(8) Varga, Cs.: _The Place of Law in Georg Lukács' World Concept_, Akadémiai Kiadó, Budapest, 1984, part 11, and, for the first concise formulations, cf. Varga, Cs.: 'The Concept of Law in Lukács' Ontology', _Rechtstheorie_, 10 (1979) 3; Varga, Cs.: 'Towards a Sociological Concept of Law - An Analysis of Lukács' Ontology', _International Journal of the Sociology of Law_, 9 (1981) 3; Varga, Cs.: 'Towards the Ontological Foundation of Law - Some Theses on the Basis of Lukács' Ontology', _Rivista Internazionale di Filosofia del Diritto_, 60 (1983) 1.

THE CONCEPT OF 'FACT' IN THE PHYSICAL SCIENCES AND IN LAW

W.H.Balekjian

A comparative examination of the concept of 'fact' as used in legal theory as a discipline lying in the sphere of the social sciences, on the one hand, and philosophy of science on the other, suggests that a 'fact' in the realm of science is not 'harder', more real or absolute for the purpose of epistemological enquiry. It cannot be asserted that essential differences exist between 'facts' in the two spheres, because the epistemological components for ascertaining what a fact is, that is, how it is known to be what it is, are fundamentally the same.

In scientific discourse, reference to a 'fact' may be found pointing explicitly or implicitly to the existence of something, process or phenomenon which is assumed to be inherently self-evident, objectively undeniable, that is, characterised by attributes which the given fact is supposed to have. As such, a fact is supposed to be perceivable in a process of objective cognition, and the human mind is supposed to have a correspondingly objective ability to effect such cognition.

When the question is raised, however, as to what objective perception or cognition is, (apart from the more fundamental question as to whether the human mind is endowed with such objective (and finite) perception or cognition,) and what the objectively (and universally) cognisable attributes of a fact are, should be or can be, the enquiring mind is confronted with a number of subjective elements and valuative criteria. While helping the mind to qualify something as fact, these subjective elements and valuative criteria bar the mind from ultimate access to a generally valid definition of 'fact' going beyond the dimensions of formal components.

A formal definition, of general applicability, is, with respect to 'fact', not difficult to formulate, but its suitability as a concept for critical control is exposed to doubt. "Any occurrence taking place in space and time and falling, or potentially capable of falling, into our experience; an experienced event" is a fact; an

A. Peczenik et al. (eds.), Theory of Legal Science, 183–188.

event in turn is a "singular or general occurrence taking a definite place in space and time, which is amenable to direct or indirect observation"(1).

The above definitions imply, rightly, a number of conclusions:

1. A fact or event, while occurring in space and time, may or may not be experienced by the human mind;
2. Experience of facts and observation of events by the human mind are dependent on their respective potential capacity to be experienced or observed by the human mind;
3. While being potentially capable to be experienced or observed, or as a corollary to such potentiality, facts and events may have existed and may exist in space and time, independently from human cognition;
4. As the history of ideas and science shows that the experience of facts and the observation of events is not least dependent on the stage of development or evolution of the human mind, and the tools too it has progressively devised for experiencing facts and observing events have a historical dimension, no finite statements as to the nature and scope of facts can be made as validly binding with respect to the future or infinity.

Thus, in the realm of facts, be it in the sphere of the natural sciences or the social sciences, an element of inevitable relativity has to be confronted and accepted. This is due to the existential limitation(s) of the enquiring mind of the human being, irrespective of the level of its development, training or scientific, philosophical or intuitive virtuosity. The human mind has no means of immediate and direct access to the absolute (or naked) existence of facts without effecting the cognition and understanding of a 'fact' in terms of an interpretation or qualification. The nature or condition of the enquiring human mind being the same in the spheres of legal theory and philosophy of science, it has to be concluded that only gradual and no essential differences can exist between 'facts' in both spheres. The argument of reference to the rigour of and results achieved by the method(s) of the natural sciences is not acceptable for a conclusion in favour of these sciences. The progressive, evolutionary nature of the method(s) and the results achieved therewith in no way support the claim that they prove the existence of absolute (eternally valid) facts in the sphere of science in contradistinction to the sphere of the social sciences or law. For the purpose of scientific, intellectual and philosophical enquiry, discourse and communication, there is no 'fact' without a qualification which the human intellect attaches to it. This statement does not concern the existence or validity of facts, ultimate truths etc. in the sphere of religion or metaphysics. It is limited and rigidly applies to reference to a 'fact' at the level of intellectual endeavour and communication; at such a level an inevitable dependence on a subject/object relationship between the human mind as qualifying agent and the 'fact' as qualified object exists. Hence, for the purpose of philosophical discourse, it may also be claimed that there can be no *per se* existing facts endowed with semantic content independently from the human mind defining or accepting that content.

The definition of the content of a 'fact' is not autonomously dependent on what the experienced fact or observed event yields when a respective method is applied. The level of development of a method and the scientific, intellectual or philosophical curiosity or motivation behind it are dependent on historical and other factors. At the level of method applied to the elicitation of 'facts', the physical sciences and the social sciences, including law, are not as different as the laudable successes of the physical sciences may seem to indicate. Natural sciences too cannot begin or continue without ideas or hypotheses when looking for or checking the veracity of experienced or observed, or hypothetically assumed facts. This may involve the neglect of other facts which may later turn out to be more important or relevant in the light of the experience of later generations of scientists(2).

If for the positivist, "scientific theories consist of sets of highly general universal statements, whose truth or falsity can be assessed by means of systematic observation and experiment", and if the "results of these observations and experiments can be known either with total certainty, or at least with a far greater degree of certainty than anything else, including the theories which are evaluated by reference to them"(3), it is to be added that the concepts of 'truth' and 'falsity', 'systematic observation and experiment' do not carry an absolute meaning; they are subject to the stage of development or history of science. The same applies to 'total certainty'. Human knowledge about a 'fact', for example, the existence of atoms and the forces that affect them, is ultimately historically relative. It may be assumed that atoms existed before their discovery, but such existence can be only inferred and not proved as a fact. Applying the same reasoning to the dimension into the future, their is no scientific guaranty whatsoever that our present knowledge about atoms will be valid for all times to come; that is, the fact 'atom' may as a concept undergo change; it is not an absolute concept, a naked fact unto eternity. This is due to the existential fact that the human mind is unable to design ultimate rules and tools of research capable of unveiling the ultimate essence of reality, of ultimate truth, which the human scientific intellect wants to explore and discover, describe and finalise in a definitive manner. The reality of a 'fact' is not limited to its measurement(s) in terms of size, volume, properties etc. An entity described as fact may be part of something which transcends it and as such is different from it; it may be part of a continuous process which may be a non-repeatable continuity.

The human mind, in spite of its best endeavours (and illusions) has thus far been unable to grasp or define a total, ultimate and finite reality deploying itself through space and (past, present and future) time. May be this reality is so immense and so complex that the human intellect, at least at its present level of development and evolution, is unable to grasp, study it with adequate instrumentation, interpret and subsume it under the notion of an ultimate reality. The situation may be visualised as a sea with unknown shores beyond the horizon, and contents of unfathomable quantity or diversity. What is extracted out of it, for example, fish, with the application of a method, depends on the mesh of the nets used, and the availability or use of the given mesh may be a matter of existential condition or deliberate choice. Whatever the results, they yield no indication, philosophically or scientifically, as to the existence or non-existence of further fish or resources beyond certain experiential parameters. A 'veiled' total reality or total fact may exist; but at a given point on the axes of time or space, only partial knowledge of it may be claimed. When values, views and methods are shared only relativistically will the claim hold that "facts are something we can all observe, identify, and hold in common"(4). "Even for science a fact is not a stubborn entity existing in the outside world, which of itself suggests the hypothesis which will describe it"(5). If in the sphere of science, the given period of development or history, general consensus and/or scientific authority define the contents of posited facts, the situation is no different in the sphere of law: "In the realm of law there is no fact 'in itself', no immediate fact; there are only facts ascertained by the competent authorities in a procedure determined by law"(6). It is therefore necessary to indicate that any reference to an 'ultimate fact' is a qualification of the subject accepting it as 'ultimate' at a given point in time and not of the object referred to as an 'ultimate fact'. A discourse on science, its progress and endeavour to find out the realities (facts) of nature also have not least to do with the sociology of science(7). In such a sociological and historical framework, similarities between 'fact' in the physical sciences and the social sciences, including law, may be best emphasised by referring to the words of Gilles-Gaston Granger: "the technology of the 'human sciences' can stimulate discussions and disagreements of detail, but it is past doubting that every sufficiently well informed observer recognizes that the same spirit of control and rigor that reign in the natural sciences, is at work here, thwarted, it is true, by incomparable difficulties"(8).

A relatively integrated approach to the concept of 'fact' helps to cast a sharper critical light on the unity of problems and limits which human endeavour faces when trying to unveil the dimensions of a universe assumed to be a reality; it also casts a critical light on the question as to how allegedly 'value-free' scientific enquiry by the human mind may or can be: sundry historical, sociological, ideological and other factors condition the direction of scientific curiosity, effort and research undertaken at a given period in human history. A critical evaluation of these points may contribute to reduce the alleged rift between the natural and social sciences, for seeing them more holistically in the context of human existence and action, with possible repercussions on existing axiological spectra. Both spheres of scientific enquiry ultimately deal, at the level of philosophy, with the question of relation and interaction between reason and sense-perception. Both spheres are confronted with the same fundamental perceptions or convictions as to what the global reality and its components surrounding human existence are, and how the unity of subject-method-object approach may be promoted.

NOTES

(1) Yildirim, C.: 1971, Science: Its Meaning and Method Middle, East Technical University, p.71.
(2) "The view that science can begin without ideas, hypotheses, or anticipations of nature and proceed first of all to gather the facts, is nothing less than silly./.../to find out what are the facts is the very purpose of scientific investigation". Cohen, M.R. 1982: Law and the Social Order: Essays in Legal Philosophy, Transaction Books, p.190.
(3) Keat, R. and J.Urry: 2nd edn., 1982, Social Theory as Science, Routledge and Kegan Paul, p.13.
(4) Langer, S.K.: 1942, Philosophy in a New Key, Harvard University Press, p.10; see also A.L.Corbin: 1960, 'Legal Analysis and Terminology' in R.D.Henson (ed.), Landmarks of Law: Highlights of Legal Opinion, Harper & Bros., p.197.
(5) Paton, G.W.: 1964, (3rd edn.) D.P.Derham (ed.) A Textbook of Jurisprudence, Oxford University Press, p.175.
(6) Kelsen, H.: 1952, Principles of International Law, Rinehart, p 265; ibid, (2nd edn.) 1962, Holt, Rinehart and Winston, pp. 388 and 393, and footnote 92, on the relative existence of facts.
(7) Cone, R.A. et al.: 'Scientific Creativity', Johns Hopkins Magazine, 31 (1980) pp.20-25.
(8) Granger, G.-G.: 1983, Formal Thought and the Sciences of Man, Reidel, p.176.

CHAPTER 3

OBJECTIVITY AND RATIONALITY OF LEGAL JUSTIFICATION

SYNOPSIS

Aleksander Peczenik

This chapter deals with several basic problems concerning objectivity, rationality and legal justification. Moreover, it constitutes a useful background for the next two chapters, on technical rationality and on special topics concerning rationality in the law.

The basic idea of Göran Hermerén's paper can be summarized as follows: The content of a research report is lacking objectivity, if it is the case that

(A) it has certain sorts of shortcomings, e.g. is misleading or biased because it omits important facts or aspects, uses unreliable or invalid indicators, and so forth;

(B) it favours the interests of some of the parties involved (i.e. mentioned, affected by the consequences, or financing the investigation) at the expense of some others;

(C) the fact that it favours these parties depends on the fact that it has these shortcomings.

The author argues that there is a vast agreement, *inter alia* between legal scholars of different traditions, that some things are shortcomings and mistakes. This fact makes the proposed definition of objectivity useful. Objectivity is highly desirable in the social sciences, but there are also other ideals, e.g. relevance and readability, and they can occasionally come into conflict with each other. In the conclusion of his paper, the author takes position against some relativistic trends of the modern philosophy of science.

Ota Weinberger discusses objectivity in the following fields: 1) legal decisions, e.g. the demand of impartia-

A. Peczenik et al. (eds.), Theory of Legal Science, 189–193.

lity; 2) justification of legal and social institutions and 3) justification of legal research.

Only logical relations and logical inference are strictly and objectively valid, albeit only relatively to the accepted calculus. But rationality in legal reasoning is not restricted to legal relations. Although the author holds the non-cognivist view, the thesis of non-derivability of "Ought" from "Is" and relativistic views about extra-logical rationality, he also discusses 1) rationality based on social experience, e.g. on the way of life; 2) procedural rationality, e.g. based on coherence. In the last context, he deals with universalization and concludes that our mental constitution is such that only structured fields of objects or ideas are viewed as rational. Concerning these questions, the author seems to come close to the ideas expressed by other authors represented in this volume, namely, Neil MacCormick and Aleksander Peczenik.

Neil MacCormick discusses coherence as a test for soundness of arguments, and distinguishes normative coherence and narrative coherence, the former concerning norms, the latter questions of fact. Coherence of norms is a matter of their "making sense" by being rationally related to the realisation of some common value or values, or to the fulfilment of some common principle or principles. Narrative coherence is a test of truth or probability in questions of fact and evidence upon which direct proof by immediate observation is unavailable. Narrative coherence is, thus, the explicability of the tested proposition in the light of explanatory principles ("laws") of a causal and probabilistic kind. Normative coherence justifies norms because it is agreeable to a certain conception of rationality of practical life, that which requires both universality and also the greatest possible degree of generality in practical principles. Narrative coherence justifies beliefs because (a) it is a necessary condition of the intelligibility of the phenomenal world; and because (b) rationality requires us to make the phenomenal world intelligible. The author concludes that there is a genuine parallelism between normative and narrative coherence.

Jerzy Wróblewski approaches similar problems, although his starting point is quite different, that is, consists of a theoretical model of judicial decision making. He adopts a decisional model, according to which there are the following types of fractional judicial decisions:

a) The decision of validity states that the applied legal norm is valid and applicable.
b) The decision of interpretation states that the norm to be applied has a determined meaning.
c) The decision of evidence states that some facts of the case are admitted as existing.
d) The decision of the choice of consequences states that, according to the applied norms and proved facts, the case has determined legal consequences.

The applied norm is valid in accordance with some rules of recognition and some evaluations. The interpretation is correct in accordance with some first-level and second-level interpretative directives and evaluations. The facts are proved according to some rules of evidence and evaluations. The decision of the choice of consequences is made, finally, according to some directives for the choice of consequences, and according to some evaluations. All the fractional judicial decisions refer thus to some norms or directives . The author calls these norms sources of the law or sources of the decision. The sources are included in the basis of a justifiable final judicial decision.

In the author's opinion, the paradigm of the deep structure of judicial decision is thus the following: The facts of the case have the consequences C_1-C_n according to the rule of decision RD. This paradigm expresses a theory of judicial decision, a theory of justification and the concept of rationality inherent in our legal culture.

Wróblewski's paper can be read as perfectly coherent with MacCormick's article. The fractional judicial decisions are justified if they are supported by some directives. These directives, however, have perhaps something to do with values and principles that, according to MacCormick, decide about the normative coherence. In this manner, the division of labour between the authors is as follows. MacCormick analyses the nature of normative coherence, Wróblewski shows the "fractional" fields to which the notion of coherence is applicable.

Sheldon Leader discusses Ronald Dworkin's question whether there is a single right answer to "hard" legal questions. Leader argues for contingent pluralism, that is, for the thesis that there is usually no single right answer to questions of law, however complex the legal system may be. He starts his argument from a critical examination of Dworkin's position. According to Dworkin, one justification may be better than another in two different dimensions, it may prove a better fit with the body of settled law and it may prove a morally more compelling justification. Leader argues that the second, moral, dimension plays a decisive role in Dworkin's theory. Then, he criticises Dworkin's view that there is at bottom a single truth of the matter. Although he recognizes the possibility of a global moral theory, such as utilitarianism or a rights-based theory, which among rivals emerges as the single best theory, he denies, nevertheless, that such a global theory can provide a unique set of right answers in any system. To find the right answer, we need a local theory, that gives an account of some particular realm of activity such as punishment or promise keeping. He concludes, however, that we will usually have more than one correct set of weightings and elaborations of local theories and principles in satisfaction of the demands of a global theory.

Martin Golding gives a response to a critical account of judicial decision-making, widely held by American Legal Realists, according to which judicial decisions are a function, in part, of the biases and value-predispositions of the judge, and the judicial opinion gives a false picture of the process of decision. The response assumes the well-known distinction between the context of discovery and the context of justification. The crucial issue is whether the reasons given are adequate to establish the conclusion, and not whether the conclusions were products of a hunch, bias or value-predisposition. This response is satisfactory even if Norwood Russell Hanson is right that logic plays some role not only in the context of justification but also in the process of formulation of scientific hypotheses. In such a case, one must follow Carl R.Kordig and make three distinctions: 1) initial thinking, 2) plausibility, and 3) acceptability. Logic is not essential to the first but it is to the second and third. This theory fits as well with science as with the judicial situation. In the conclusion of his paper, the author states that a judge's particular justifying reasons are good legal reasons, if they are acceptable to the losing party, to other persons who may be affected by the result and to the community at large. When good reasons are of each side of a question of law, non judge's conclusion is uniquely correct, but some conclusions are reasonable. This conclusion is quite similar to Sheldon Leader's. A

Scandinavian reader must also notice a striking similarity to the ideas previously expressed by Aulis Aarnio.

Gunnar Bergholtz also departs from the distinction between the context of discovery and the context of justification. His main thesis is, however, that the proffered reasons for legal decisions not only are relevant in the context of justification but also constitute causes of these decisions. He argues that judges have a common disposition to act according to what they think is right or in accordance with valid law. The judge chooses solution R in spite of his disliking it because it is a legal solution. This hypothesis is coherent with the fact that, most of the time, judges are in total agreement on both reasonings and outcomes of cases while laymen disagree wildly. On the other hand, if you take for granted that the judges are influenced by several legally irrelevant dispositions to act, you will have difficulties to explain the prevailing consensus among judges.

OBJECTIVITY IN THE SOCIAL SCIENCES

Göran Hermerén

1. INTRODUCTION

In the philosophical literature, several arguments have been advanced against the claim that objectivity is (a) possible and (b) desirable. Some of these arguments are based on certain assumptions about the object of research in the social sciences - that it contains subjective or non-reducible evaluative (or normative) elements. Others are based on certain assumptions about the relation between the norms and values of the researcher and the social phenomena which are being studied: these norms and values allegedly make it impossible for the researcher to be objective.

Thus one important and wide-spread line of thought can be condensed in the following way:

(1) Research in the social sciences presupposes norms and values

(2) To be objective is to be value-free

(3) Hence no research report in the social sciences can be objective

(4) Hence no researcher should try to be objective

(5) Hence every researcher can write as he or she pleases; "anything goes".

The last conclusion certainly does not follow from the four previous statements. Nor does conclusion (4) follow from the three previous statements. But what about the second premise? Is it tenable?

A basic question here is, of course, how objectivity is defined, or what criteria of objectivity are taken for granted in arguing for conclusions like (3), (4) or (5). The main purpose of the present paper is to propose a definition, or rather a set of criteria, of objectivity which, I shall argue, (i) satisfies certain intuitively reasonable standards of adequacy and which (ii) deprives arguments in favor of conclusions like (3), (4) and (5) of whatever force they may have had, if other definitions or

A. Peczenik et al. (eds.), Theory of Legal Science, 195–215.

criteria had been used.

2. TWO TRADITIONS

As a starting point I would like to call attention to the existence of two influential traditions in philosophy of science. In these traditions opposing or conflicting views concerning objectivity are held.

According to one of these traditions, the social sciences could, and should, be value-free; and hence objectivity is an ideal social scientists should strive for. Let us call this "the positivistic tradition", since the fact-value distinction obviously plays a central role here. According to another tradition, which may be called the "critical or emancipatory tradition", the social sciences are not, and cannot be, value-free; and hence social scientists should not strive for objectivity - they should take sides and make their political sympathies explicit.

In this paper I shall try to show that each of these traditions contains some important insights and some mistakes. If the mistaken ideas are abandoned, it should be possible to combine what is tenable in both traditions in an interesting and fruitful way. The critical point is the underlying assumption in both traditions that objectivity in one way or another presupposes value-neutrality or value-free research. This crucial assumption in untenable in my view.

What is tenable, however, in the first tradition is the view that objectivity is an ideal worth striving for, provided it is (re)defined in a certain way. And what is tenable in the second tradition is the recognition of the important role of norms and values in the social sciences: norms and values may influence the choice of problems, materials, methods, working hypotheses, and forms of presentation. But in saying this, it is obviously important to distinguish between different kinds of norms and values(1).

What is needed, then, is obviously a new set of criteria of objectivity. But before I go any further, I would like to take a look at the kind of examples that give rise to controversies concerning objectivity in research.

3. SOME EXAMPLES

None of the examples below are fictional. They are drawn mainly from the social sciences, but I'll present them here very briefly and schematically. To avoid misunderstanding I would like to stress that I do not wish to pass any judgement here on the works I have used as a model - what I want to say is only that they illustrate cases and contexts in which the issue of objectivity could come up. (What the verdict would be, or should be, in these cases is then another matter).

Suppose that in the history of a totalitarian country a certain strike proved to be ideologically important as precursor of the great revolution. Let us also suppose that two parties were competing for power at that time, and let us for the sake of simplicity call them bolsheviks and mensheviks. Let us say that the bolsheviks won. In the official party history it is claimed that the bolsheviks played a leading role in this strike that anticipated, and served as a source of inspiration for, the great revolution. But much later, more or less by chance, a historian happens to come across leaflets that were handed out by the bolsheviks during that strike. They show that far from playing a leading role, the bolsheviks advised against the strike. Accordingly, the official party history seems to be based on false, forged or non-existing sources, and it does not account for these leaflets.

Let us now look at a perhaps somewhat more complex example. Suppose that there is a great deal of discussion in a particular country about how the parliament works. Different suggestions are made as to how it might be reformed to be more efficient and responsible. Then a sociologist comes along and makes a study of the inner life of the parliament which favors one of these reform plans. The sociologist claims to have used a "hermeneutical method". But is is soon disclosed that the book is to a large extent based on gossip, that the sources have not been analyzed, that his own presence at important meetings leaves a great deal to be desired, and so on. As a result, politicians claim that he gives a misleading and unfair picture of that parliament.

Here is another example, drawing on a criticism of certain theories, which once were used to support the economic policy of a conservative British government. It was later argued that the author of these theories had manipulated his figures in order to prove his thesis that inflation always depends on the fact that too much money is available, and that the rate of circulation (of money) is determined by the amount of money that is available. For example, it was argued that the author more or less arbitrarily had decreased the amount of money available from 1921 to 1955 with roughly 20%, and also made other changes in the official price statistics during the post war period. If these accusations are well founded, questions may be raised as to the objectivity of these economic theories.

Investigations of objectivity can themselves be lacking in objectivity. Suppose an investigation is made of the objectivity of radio, television and other mass media in a particular country. The state has a monopoly on radio and television, and the radio and TV-company asks a political scientist to make an investigation of the alleged lack of objectivity in the radio and television programs preceding a certain important election. This is done by checking how much time is devoted to the various political parties, by counting and classifying positive and negative "perspectives" (value-loaded descriptions) in programs about these political parties, by checking the extent to which these programs deal with issues the political parties themselves want to focus on, and by comparing what is said about these issues to the way they are presented in various newspapers (of different political persuasions).

The conclusions of these investigations are that the accusations of lacking objectivity on the whole are without any foundation. The political parties get roughly the same amount of attention in radio and TV, etc. But soon criticisms are made of the reliability and the validity of some of the indicators used: some of them measure fairness or justice rather than objectivity, the classification of perspectives as "positive" and "negative" depends to some extent on the values of the investigator, and so forth. Besides, even if the radio and television programs focus on what the parties themselves want radio and television to focus on, this does not necessarily mean that the programs are objective. Perhaps some political parties want to cover up certain aspects of their politics, and then the journalists ought to disclose this.

4. THE OBJECT OF OBJECTIVITY

What is "objective" predicated of in the social sciences? What is said, or denied, to be objective? Here we must distinguish between objectivity as a property 1) of researchers, 2) of the methods they use, 3) of particular research results, and 4) of the knowledge in a particular field. Thus we have a series of different but related explicanda, which may be referred to as, respectively, personal objectivity, methodological objectivity, objectivity of particular results and objectivity of knowledge.

These explicanda differ from each other in important respects. Hence the following brief comments may be called for:

(a) it would be dangerous to assume that "objective" and "objectivity" are used in the same sense in all these cases;
(b) the enumeration above makes no claim of completeness - one might add "objective phenomena", "objective judgements", etc.;
(c) no claim is made as to what usage is primary and secondary or original and parasitic.

Thus it may well be the case that "objective" is primarily used about persons as some sort of dispositional concept, referring to certain traits of character. It may be applied subsequently, by metaphorical extension or some other way, to methods and to scholarly results.

Of the four explicanda mentioned above I shall here concentrate on objectivity as a property of research reports. The reason for focussing on results rather than on methods is simply that I regard methods as having instrumental value; they are tools we use when we want to arrive at good or acceptable theories. Like Popper, I find the results more important than the process. But unlike Popper I will concentrate on objectivity as applied to particular results or reports. Thus I will not take his evolutionary approach to the problem(2). It is, moreover, concerning specific results that the discussions (accusations of not being objective) have been most heated over the years.

At this point it may be asked whether objectivity of particular accounts can be discussed without, in the end, bringing in personal and methodological objectivity. It is, however, no objection to the analysis I am going to suggest if it would turn out to be the case that an explication of these concepts of objectivity would involve, in the last analysis, also references to some of the other explicanda. But there are certain prima facie reasons why it is not necessary to bring in the author, and why, at any rate, one should not begin at this side, provided that we are interested in the objectivity of particular results or reports.

For example, even if a person is objective on the whole, he may sometimes make mistakes; thus we cannot conclude that everything he has written is objective. Moreover, we may come across an anonymous treatise. Without knowing who the author is, or what his motives are, we may be justified in accusing it for not being objective, if the treatise is biased in a very obvious way. Third, even if we know the identity of the author, it may sometimes be hard to find out enough about his motives (even with the help of psychoanalysis) to be able to say whether he is objective or not. Finally, suppose the author is well known, as well as his motives. If he was passing on (or using) unreliable or faked data, and was acting in good faith, this does not eliminate the shortcomings of his text.

5. STANDARDS OF ADEQUACY

Before stating or suggesting criteria of objectivity, it may be a good idea to reflect on what standards such criteria or explications should satisfy - and to make them explicit. Here are some proposals:

1) They should be roughly compatible with the way (or at least with some of the ways) in which "objective" and "objectivity" are used in the social sciences.

Since the usage of the social scientists in this respect is not uniform, 1) leaves us with a certain amount of freedom. A particularly important source of information is, of course, debates between scholars from different methodological or theoretical traditions concerning the objectivity of particular research reports.

Fortunately, there are various other standards which may be combined with 1), for example:

2) It should be possible to operationalize the criteria in such a way that it becomes possible in principle to decide whether a particular research report is lacking objectivity.

However, this requirement must not be interpreted too strictly; "in principle" should be stressed. It would be unreasonable to demand that it should be possible to decide whether accusations of lack of objectivity are without foundation in every particular case. Sometimes there is not enough evidence, sometimes the evidence is ambiguous and can be interpreted in several ways, and sometimes this is true also of the conclusions of the report. Thus we simply have to learn to live with the fact that we sometimes must reserve our judgment.

Moreover, for reasons to be given shortly the two standards above should be supplemented by the following one:

3) If the explication or criteria results in a dichotomy (objective vs. non-objective reports), then one of these classes must not by definition be empty.

The reason for this standard is simply that otherwise the criteria would be of little use to those who wonder whether a particular report is lacking in objectivity or not. Of course, it may well turn out to be the case that - as a matter of fact - all dissertations in sociology from a certain university during a particular period of time have had serious shortcomings and have all been lacking objectivity. But this has to be discovered by empirical studies.

Furthermore, an important standard of adequacy - though hard to state precisely - is obviously the following one:

4) The classification that results should be fruitful; investigations grouped together as subjective (or objective) should have at least one further property in common that is of interest from a theoretical or practical point of view.

For example, such further properties may be that investigations lacking in objectivity cannot be relied upon, other scholars cannot build upon their results, they cannot be used as a basis for prediction, planning and rational decision-making, and so forth.

Finally, the last standard to be discussed here specifies what sort of concept (comparative or classificatory) we are dealing with:

5) The classification that results should admit of

degrees - thus it should be possible to say that an investigation is more or less objective rather than having to say that it is either objective or not objective.

The investigations I have made of current usage suggest that if 1) is satisfied, then 5) is automatically also satisfied. But since space does not permit me to go into details here, and 5) raises important issues about the nature of the concept of objectivity, I have preferred to make this standard of adequacy explicit.

6. THE CRITERIA

The basic idea of the criteria to be proposed here is simple. It can be summarized as follows: (the contents of) a research report is lacking in objectivity, if it is the case that

(A) it has certain sorts of shortcomings, e.g. is misleading or biased, because it omits important facts or aspects, uses unreliable or invalid indicators, and so forth;

(B) it favors the interests of some of the parties involved (i.e. mentioned, affected by the consequences, or financing the investigation) at the expense of some others;

(C) the fact that it favors these parties depends on the fact that it has these shortcomings.

In other words, the investigation would not have favored these parties - at least not to the extent that it does - if it had not had these shortcomings, and this is something that in principle could be checked(3).

To avoid misunderstandings I want to stress immediately that all the three clauses above have to be satisfied. Thus it is not enough that only the second one is satisfied - it could very well be the case that one of the parties involved is right. An investigation which shows that this is the case, obviously favors that party, if he or she is involved in an ideological or scientific controversy. But the investigation that shows that he or she is right can, of course, itself be a model of clarity on objectivity. Thus (B) alone is clearly not sufficient for an investigation to be lacking in objectivity. Nor is (A). If an investigation has certain specifiable shortcomings, they can be specified without using the word "objective" or "objectivity"; it is the particular combination of the clauses that is interesting and important.

This schematic formula can be made more precise, concrete and specific in a number of ways, for example, by being more explicit about the nature of the shortcomings, about the kind of parties that are relevant, and concerning the dependency-relation in (C). Since there is little time to go into all this here, I shall have to refer to other writings of mine on this point(4). For example, to forestall certain obvious objections, it is necessary to specify the parties involved in the way suggested within parentheses in (B).

The need for making such restrictions explicit can be seen from cases like the following, which Ingmar Persson has brought to my attention. Suppose a scholar writes (not on a particular historical or social problem but) on the way other colleagues B, C, D, have described and explained this problem. In this survey A makes some mistakes and happens to misquote his colleague B. As a result A's description becomes somewhat misleading; it contextually implies the statement _p_ which is false. If A and B later get involved in an academic quarrel (both apply for the same position, for example), then B can point out this shortcoming in A's work - and since B is referred to in that work, he is one of the parties involved. To take care of examples like this, one would have to add restrictions on who is to count as one of the parties involved, or on the nature of the relevant interests of these parties - or distinguish clearly, as Ingmar Persson has suggested to me, between the contextually implied (false) statement _p_ and the fact that A's text implies p; it should be _p_ rather that this fact which favors the interests of some of the parties involved. I shall have to save a more detailed discussion of these problems for another occasion.

Let me just add that as far as the "if"-problem is concerned, I would be prepared to stick out my neck and replace the italicized "if" preceding the criteria (A), (B), (C) with the stronger "if and only if" - and to regard this as an explication, or rather an explicatory schema. If one would like to condense this schema into an explicative definition, as brief as possible, the result could be stated as follows: to say that an investigation in the social sciences is lacking in objectivity is to say that it favors certain parties involved (as specified above), because it has certain shortcomings(5).

This is a negative definition. Also the criteria above are stated in negative terms. Some people will probably object to that. But there are two reasons for this way of putting things. First, problems of objectivity of the kind relevant here always come up in negative form according to my experience: we are confronted with accusations of lacking in objectivity, rather than with proofs of alleged objectivity. Second, by applying the so-called law of contraposition to these criteria, it is easy to obtain a positive formula: an investigation _X_ is objective, if (and only if) it is _not_ the case that

(A) _X_ has certain shortcomings of the kind indicated above,

(B) _X_ favors some of the parties involved (mentioned, etc.),

(C) (B) because of (A), i.e. _X_ favors some of these parties because it has these shortcomings.

What method could be used to decide whether (C) is satisfied?

The strategy can be outlined very briefly as follows. Identify and describe the interests of the various parties. Summarize each description in a thesis. Check whether the account in question, with its various shortcomings, can be used as an argument for one of these theses. Make up for the shortcomings - delete the false statements, add the facts and aspects overlooked or "forgotten", use representative samples and valid and reliable indicators, and check whether this makes any difference. Is the relevance and the tenability of the account, without its various shortcomings, affected in any way - as an argument for this thesis? If it is not affected in any way, then (C) is not satisfied. If, however, the investigation now is a much weaker argument for the thesis, then (C) is indeed satisfied. (There are variations, depending on whether the interests of the parties can be summarized in theses like _p_ and non-_p_, or in statements like _p_, _q_, _r_..... but I hope this will be enough to give a rough idea of the method).

Let us now for the sake of illustration apply these criteria to examples of the type mentioned above in section 3. I think it is fairly easy to see that the criteria can be applied to such examples:

(A) they have all certain obvious shortcomings: the conclusions are based on false, forged or non-existing evidence, no critical analysis has been made of the sources, the material is incomplete, arbitrary changes have been made in the official statistics, the indicators used are not valid and reliable, and so forth;

(B) the investigations, with these shortcomings, state or imply (logically or contextually) certain things which favor certain of the parties mentioned, affected by the consequences of the investigation, or financing it, at the expense of some of the other parties involved: the bolsheviks, those who want to change parliament in a certain way, the conservative party, the radio and TV-company;

(C) and, finally, if the investigations had not had these shortcomings, they would not favor the parties and interests which they - in their present shape - do favor. This can be seen if new evidence is used, if the evidence is analyzed properly, if the methods handled incompetently are used correctly, and so forth.

7. APPLICATIONS TO LEGAL CONTEXTS

In view of the main theme of this volume, it may be tempting to ask whether criteria of the kind I have advocated can be applied to the area of jurisprudence, legal dogmatics or legal practice.

This is probably a question I should leave open for the discussion or to professor Weinberger's contribution. But it seems to me at least quite plain that there are such applications, and that this is an additional merit of these criteria. A judge or a law scholar can - like a historian or a social scientist - make all sorts of mistakes. He or she (_X_) can also be politically prejudiced in one way or another. If _X_'s conclusions favor one of the parties involved, for instance, the political views of the plaintiff in a court case, and does so because of the shortcomings of what _X_ has written, then we have a case which satisfies the criteria outlined above.

What sort of shortcomings could _X_'s text have? This depends on whether _X_ is a judge or a law scholar. But let _X_ be a law scholar, and let us for a moment return to the

disciplinary matrix that professor Aarnio discusses (cfr. his article in this volume). Such a matrix is made up by the following elements:
(1) legal positivistic background
(2) sources of law
(3) directives of interpretation (methodological rules)
(4) certain values
Mistakes can obviously be made in several of these elements, particularly as far as (2) and (3) are concerned, which could very well be relevant for discussions of accusations for not being objective.

But it may be more difficult to see that there are mistakes of this kind which are not related to a particular research tradition or "paradigm" (a word I personally think should be abolished - it has been, and still is, used in too many senses to be very helpful).

8. SHORTCOMINGS AND RESEARCH TRADITIONS

In other words, I shall argue that there are shortcomings and mistakes which are not just mistakes in a given research tradition (and hence not mistakes and shortcomings in other research traditions). This brings up certain aspects of the issue of relativism; other aspects will be discussed briefly later on in this paper.

If total relativism is to be avoided, it is of course essential that there is a fairly widespread agreement in the scholarly community
- including groups of scholars belonging to different research traditions
- concerning what constitutes at least some basic mistakes and shortcomings of research reports in various fields. It is also essential that this list of basic mistakes can be illustrated by a variety of examples (and not just the ones I have chosen in the beginning of this paper). If the criteria advocated were not independent of the particular examples chosen to illustrate them, then the claim that these particular examples were lacking in objectivity according to these criteria would not amount to very much.

But fortunately there seems to be a vast agreement also between legal scholars of different theoretical and normative persuasions that it is wrong to make false statements, contradictions, circular arguments, mathematical mistakes, statistical errors, as well as to fake evidence, to use non-representative samples, unreliable and invalid indicators, to fail to state and test alternative explanations, to disregard facts that are incompatible with the hypothesis stated, and to present views one happens to like in a simple and clear way but those one happens to dislike in an obscure and abstruse manner. The list could be made longer, but I trust that this is enough for my purposes.

There are, of course, more difficult and dubious cases as well, i.e. cases where scholars with different normative and theoretical frameworks (for example, in favor of conflicting ideals of science or supporting opposing political views) may disagree as to whether (i) a certain alleged shortcoming is a shortcoming or not, or - if they agree that it is indeed a shortcoming - whether (ii) it is a serious shortcoming or not. Is it, or is it not, outweighed by the merits of the work? Are there mitigating circumstances? Which ones? Here opinions may differ.

The late J.L.Austin once wrote: "I dreamt a line that would make a motto for a sober philosophy: Neither a be-all nor an end-all be". I must say that I find it rather frustrating to see how questionally some people try to cope with border-line cases. They tend either to regard every shortcoming as unproblematic and obvious (like 2+2=5), or to say that no alleged shortcomings are uncontroversial, and that there is no real difference between science and witchcraft. The simple truth of the matter, it seems to me, is that there are several classes of shortcomings, none of which is empty, that the boundaries between them are somewhat hazy and can be disputed - but that there are clear cases of shortcomings which are uncontroversial and relevant to the issue of objectivity.

9. MARXISM AND OBJECTIVITY

The criteria outlined in section 6 do not, and are not intended to, distinguish between Marxists and non-Marxists. As is obvious from the examples given in section 3, they cut across this distinction, since it is possible to find examples of both Marxist and non-Marxist investigations which are lacking in objectivity according to my criteria. Thus the situation can be illustrated by a diagram of the following type:

investigations	Marxist	non-Marxist
objective		
not objective		

The important thing is then that none of the four boxes is empty by definition, though, of course, the result of an empirical investigation might show that at a certain time and place some of them are.

But suppose someone were to say: if you are not a Marxist, then you do not know what objectivity is. Since claims of this kind are made occasionally, it might be instructive to contrast the criteria in section 6 above with a criterion of the type: a research report is objective, if and only if what it states and implies is compatible with what is said in Marxist writings. Which are the arguments for and against such a proposal? If the standards of adequacy outlined in section 5 above are accepted, it seems that there are a number of objections to proposals of this type.

For example, these objections could include the following arguments:

1) Such proposals are not compatible with ordinary language, nor with the language of at least a great number of researchers.

2) If Marx's writings, as some people argue, like the Bible contain statements which are false, if interpreted literally, then the consequence seems to be that there are cases in which a statement to be objective has to be false - which certainly is counterintuitive.

3) There is some uncertainty as to what precisely the standard is supposed to be. Which writings are Marxist writings? For example, some Marxists regard Fischer and Marcuse as Marxists, but other Marxists disagree(6).

4) Marxist writings contradict each other on several points, or at least so it appears. There are certainly differences of opinion within the Marxist camp concerning e.g. determinism, the distinction between science and ideology and the relation between theory and praxis. It would obviously be counterintuitive, when both a statement and its negation could be objective.
5) Unless Marxist writings are very vague and contain no precise theses, and unless Marxists are prepared to reinterpret them in every period, such a criterion is likely to stifle the growth of science.
6) Marxist writings cover a lot but not everything. In _Das Kapital_ there is, for example, very little about the rising bureaucracy, the automatization of working plants, or the threats of computers toward the integrity of the workers - for obvious reasons. For writings in areas not covered by Marxists there is no standard; hence disputes concerning the objectivity of research reports in such areas cannot be settled.
7) Consider the statement "certain parts of Marxist writings are lacking in objectivity". This statement may be false but it is certainly not a logical contradiction. However, according to the criterion proposed in this section, such statements are not only false but untenable on logical grounds, since Marxist writings constitute the very yardstick that is used to settle problems of objectivity.
8) Similar problems will arise if - as has been suggested - "Marxist writings" in this criterion above is replaced by "the Bible", "the Koran", or any similar book.

The conclusion one can draw from all this is, of course, that it will not do to suggest criteria of objectivity in terms of compatibility or incompatibility with a limited number of writings in any area. Such a proposal would be contrary to the very idea of a science. To understand this idea, we would need to grasp better concepts like _critical rationality_.

10. THE DESIRABILITY OF OBJECTIVITY

Is objectivity desirable in the social sciences? My answer is _yes_. Is objectivity the only thing that is desirable in the social sciences? My answer is _no_. Both answers are controversial and may be disputed for a number of reasons which cannot be discussed fully here.

Underlying my first affirmative answer is an ideal of science, in which objectivity as a regulative principle is integrated. For those who do not want to distinguish between researchers and political activists, this idea and ideal of science is probably not very appealing. But this ideal of science can be supported by several different arguments (historical, theoretical, utilitarian, democratic, and so forth), some of which I have discussed elsewhere.

Moreover, I hasten to add that I advocate the ideal of objectivity, only if it is interpreted in the way outlined here (in section 6 above), _and_ that I do not think that this is the _only_ ideal scientists should strive for. They should also strive for readability and relevance, as well as a number of other things, which unfortunately sometimes may come into conflict. Without going into these conflicts, I would like to make explicit - to prevent misunderstandings - that scientists should strive for many things, and being objective is just one of these goal.

Thus I do not want to suggest that research has just one goal such as to explain or to produce true theories, and the criteria of objectivity I have suggested do not presuppose this. On the contrary, it seems to me that the disciplines I have some familiarity with have rather different goals which cannot easily be condensed into a single, short and precise formula.

11. THE ISSUE OF RELATIVISM

Is it possible to achieve objectivity? Radical relativists would, of course, deny this. The issue of relativism has been touched upon several times during this conference, and it is hard to avoid when the theme is objectivity. But the issue of relativism is not fatal, at least not to my analysis, provided one is prepared to make some distinctions.

The first distinction I have in mind can be introduced as follows. Consider the following two statements:

1) The position of _X_ can only be determined in relation to something else, for instance, by saying that _X_ is next to _Y_, and

2) The statement that *X* is next to *Y* is not true absolutely but only true in relation to some other statement *Z*.

The relativism expressed by 2) does not follow from 1). While 1) is tenable, 2) easily leads to an infinite regress, if it is generalized to hold for all statements. (From the fact that we need a thermometer of a certain kind to measure temperature it does not follow that the statement "today it is minus 2 degrees centigrade" is only relatively, or relatively relatively, true).

Social scientists, like other scholars, need intellectual standards when they propose and criticize the research reports of their colleagues as well as the assumptions or information on which such reports are based. But these intellectual standards are not ends in themselves. The same goes for methodological rules. They should rather be regarded as rules of thumb, which are abandoned, modified or supplemented if it is discovered that they do not help us to reveal propaganda, myths, false or improbable statements, to make better predictions and to get a better basis for decisions in theoretical or practical matters (including court cases).

Some of these standards are controversial, but it is - as I have already argued - easy to overstate the extent and importance of such controversies. The fact that we now know more in a given area is indicated by the fact that we are now able to solve more problems in that field, that we can make more exact and reliable predictions in that area, that it is possible to act on our theories without being disappointed or frustrated, and so forth.

If the first distinction concerns two kinds of relativism, the second one is about two aspects of meaning. Relativistic theses are sometimes supported by the claim that concepts in different paradigms are incommensurable; hence no rational discussion can take place between those who work within different paradigms. The main argument for claims of this sort seems to be that theoretical concepts like mass and energy have at least partly different meanings in different theories (e.g. those of Newton and Einstein). Clearly "anxiety" in Freudian theories has a somewhat different meaning than "anxiety" has in common usage.

But if the references of these terms are checked, it is often possible to check whether the claims made by those who work within different paradigms are compatible or incompatible; hence the second important distinction is the one between

1) The meaning of the key terms and sentences in a given theoretical context, and
2) what these terms and sentences refer to, given that they are interpreted in a certain way.

This distinction is, of course, well-known in the semantical literature, but its implications have perhaps not always been fully grasped.

Thus let us for the sake of argument grant that the meaning of certain key terms may vary in the sense that the same term has different meanings in different theoretical contexts. However, from the fact - if it is a fact - that the concepts in different paradigms are incommensurable, it does not follow that the claims made by those who work within different theoretical frameworks or paradigms (for example, functionalists, Marxists, positivists) are necessarily incommensurable, and hence that rational discussion between scholars from different theoretical camps is impossible. It is often very difficult, and they usually talk at cross purposes, etc. - but to say this is obviously something different and much weaker than to say that it is *impossible*.

I come to the third and last distinction. Being an ontological realist, I think it is very important to distinguish between the two questions:

1) Does the world, as a matter of fact, have the particular feature *F*?
2) Can we - at the present time - know or decide if the world has the feature *F*?

The answer to the first question could very well be "yes", though the answer to the second might be "no". The distinction is, I think, a simple one. But it has important consequences and it can also be applied to the problem of objectivity. What I want to suggest is simply that it is one thing to say that a certain research report is objective, and another thing to say that we - at the present time - can decide that it is objective.

If this distinction is granted, we get a certain antidote against dogmatism. And it also becomes clear where a certain element of relativism enters into discussions concerning objectivity. The idea is obviously that when we have to pass a judgment on a particular research report as to whether it is lacking in objectivity or not, we have to base our judgment on the evidence, methods and theories presently available. And anyone who knows something about the history of science will know that the evidence, methods and theories available at any particular moment of time are likely to be refined, modified and perhaps abandoned later on. Thus we must be prepared to live with the fact that the judgments we pass are based on what we know now, and that these judgments may have to be revised, as new evidence and methods are discovered or invented.

But in my view this does not diminish the force or the appeal of the principle that objectivity is a regulative idea, an ideal that one should strive for - just as one should strive for veracity, honesty and justice - however difficult it is to attain, and however difficult it is to ascertain that we have reached these ideals in a particular case.

12. CONCLUDING REMARKS

In concluding this paper I would like to end on a perhaps somewhat provocative note by commenting on some recent developments in the philosophy of science which were discussed during the first day of this conference.

Personally, I am not altogether happy with these recent developments, but I also think that there are signs indicating that the tide may soon turn. If a philosophy of science implies that today we do not know more of the world than in the days of Plato, that anything goes, or that the scientist's basis for his claim is nil, or that science is nothing but guesswork, and that there is no distinction between guided or informed guesses and others, or that there is no real difference between science and witch-craft(6) - then I for one would take this as a reductio ad absurdum of that philosophy of science.

The reason for this is not only that I think that these conclusions are patently false, but that nobody - not even the alleged adherents of these doctrines - would be prepared to live according to these conclusions. Suppose somebody is facing a heart operation or is about to depart for a trip around the world with a space craft. I suppose neither Feyerabend nor Popper nor indeed anyone else in his right mind - if they were to enter the operation theatre or the space craft - would say that anything goes or that it does not matter who is the pilot or whether the specialist or the janitor performs the operation.

I would like to add that in concentrating on general and abstract claims of the sort indicated above, philosophers of science have to my view neglected a number of important tasks. In every discipline - also in the law school - a number of concepts, arguments and methods is used. An example of the latter are the various comparative methods which are used in many disciplines. The use of these concepts, arguments and methods are governed by, and based on, a number of tacit rules and assumptions which need to be made explicit, and to be examined critically - by philosophers of science working in close cooperation with the scholars from the field in question.

In this area important work remains to be done, and I think we may look upon this conference as a recognition of the need for further cooperation between scholars in fields like jurisprudence, legal dogmatics and philosophers of science.

NOTES

(1) See e.g. Göran Hermerén, Värdering och objektivitet, Lund: Studentlitteratur, 1972, pp.16 ff, 39 ff,, 64 ff.
(2) See Karl R.Popper, Objective knowledge. An Evolutionary Approach, Oxford: At the Clarendon Press, 1972.
(3) A method which could be used for this purpose is briefly outlined in the next page.
(4) See e.g. my earlier "Criteria of Objectivity in History", Danish Yearbook of Philosophy, vol.14, 1977, pp.13-35 and "Relevans och objektivitet", Statsvetenskaplig Tidskrift, 1979:1 pp. 21-25.
(5) Cf Tore Nilstun, "Utvärdering och objektivitet", in Utvärderingsforskning. Modellkonstruktion, objektivitet och tvärvetenskap, by T.Nilstun et al, Lund: Filosofiska institutionen 1982 (MURA-rapport 3), p.101.
(6) See e.g. chapter 5 of my Aspects of Aesthetics (Lund: Gleerup, 1983) for examples of various Marxist traditions and the controversy about the works by Marcuse and Fischer.
(7) Statements of this kind are made in Paul Feyerabend,

Against Method, London: NLB, 1975. Cf. also Karl Popper, *Realism and the Aim of Science*, 1982.

OBJECTIVITY AND RATIONALITY IN LAWYER'S REASONING

Ota Weinberger

1. SOME PRELIMINARY REMARKS

Objectivity and rationality belong to the most important and universally postulated virtues of lawyers' reasoning. But what does it mean that legal practice and jurisprudence aim at objectivity and rationality? Of course - everybody has some vague idea about objectivity and reasonableness, but it is one of the philosopher's tasks to clarify these terms and to analyse the methodological implications of these postulates.

In jurisprudence the claim for objectivity is mainly concerned with the following issues:
(i) legal decisions, which should take into account all relevant facts and judge impartially and with equal concern the interests and the rôles of all parties concerned;
(ii) the justification of legal institutions, legal regulations, and the conception of justice;
(iii) the recognition of jurisprudence in that respect that the science of the law is directed towards the provision of objective knowledge of the law as it is actually valid in a certain society.

1.1 The Objectivity of the Judge

The judge or any other official who is empowered to lead legal proceedings and to promote legal decisions is supposed to be wise and impartial. He should be wise in the sense of having at his disposal sufficient knowledge of society and human affairs, which enables him to take over the rôle of a secondary legislator within the realm of his discretionary power. Within the field at his discretion the claim for objective reasoning is in the main of the same character as in legislative deliberation.

Impartiality means not to favour one party or the other. But this is not enough: impartiality also embraces the demand that legal valuations should be based on a

A. Peczenik et al. (eds.), Theory of Legal Science, 217–234.

generalizing deliberation. An actual case must be analysed in such a way that it becomes a type defined only by its legally relevant features, and this type is judged from the viewpoint of law and social value standards as well as in the light of policies (to use Ronald Dworkin's term). The decision of the case under consideration should be met as a specific type in taking account of all possible cases of the type. Not only the individual consequences of the decision concerning the case at hand should be weighed, but the consequences of all possible cases of the given type.

There are, of course, plenty of occasions for rational analysis in the deliberation of the judge which may be objective in so far as it can be critical, empirically based and logically sound. But all this cannot work without subjective valuations and thus cannot be objective throughout.

1.2 The Objectivity of Justifying Institutions

Some theories presuppose that there are objective criteria for judging social institutions and legal regulations - the most important of them may be Rawl's Theory of Justice(1) -, but I believe that arguments of this kind are not convincing. They are a persuasive plea for a certain political conception rather than an objectively valid argument.

1.3 Objectivity in Legal Science

The objectivity of science is a common thesis, but in the social sciences it is more of an ideal than a simple fact. Legal knowledge, in particular, is governed by the ideal of objectivity, but - as every lawyer knows - there does not exist any possibility for arriving at complete objectivity. Neither is there any advice of how to verify interpretative theses objectively; there does not exist an exact method in the field of juristic hermeneutics which could provide univocal and objectively valid results.

In contemporary legal philosophy and methodology the problems of rationality are issues of great importance, even though the problem situation itself does not seem clear at all. In most of the discussions 'rationality' is understood in a sense exceeding logical relations and requirements. The philosophy of hermeneutics, Perelman's theory of argumentation, Viehwegian topical jurisprudence as well as Habermas's and Alexy's conceptions or Toulmin's-Peczenik's extralogical transformation theory, they all conceive rationality as transcending logic and logical analysis, and each of these authors uses the term 'rational' in a somewhat different sense.

In a later section I shall express my own views on extralogical rationality and its rôle in jurisprudence and in political argumentation. But as the main problem of this paper I shall consider the question whether there is a method for getting hold of an objectively justified 'Ought'.

2. HUMAN FREEDOM AND THE CLAIM FOR AN OBJECTIVE JUSTIFICATION OF NORMATIVE RULES, DECISIONS AND ACTIONS

Human behaviour is based on and controlled by information processes. It is not only a system of biologically programmed reactions or predetermined behavioural structures. The impulse for activity is anchored in the structure of man as a living being(2); his choices, decisions and programmes for action are directed by an information process which is based on informations about environmental facts, on knowledge of causal laws and on technical knowledge as well as on practical informations expressing attitudes of the agent(3). Practical informations are expressible by sentences which state goals, norms, valuations, or preferences.

The fact that human beings are capable of acting on the basis of information and of realizing acts of choice is a necessary precondition for the idea of responsibility. Judgments on responsibility refer to the relation between our actions and moral, legal, or other social norms. On the first level of valuation, behaviour is judged only as to its correspondence or conflict with some presupposed system of norms. But in modern society, mainly under the influence of ideals of democracy, people do not only feel responsible for their actions with respect to valid social norm systems, but, as members of a democratic society, they also feel responsible for the content of the according normative system itself. Thus, in our cultural setting, as well as in any democratic society, there is a need for a theory of argumentation in political matters, that is, for a theory of de lege ferenda argumentation. Furthermore, we claim that this kind of argumentation should be objectively valid as much as possible. If science could provide us with objective criteria for judging social institutions and with objectively valid argumentation in political and legal matters, then our responsibility for the content of the legal system could be taken for justified and our conscience would be freed from the heavy burden of difficult decisions of such grave moral and social impact. But I believe indeed that we cannot do all this by an exclusive reference to objectively valid justifications. And thus our responsibility cannot be relieved by the recognition of what should be fair law. There are, of course, many attempts based on religion, philosophy, or science to find principles which could objectively justify legal conceptions and institutions, but in my opinion there are no ultimate and objectively valid bases allowing for a cognitive justification of the 'Ought'.

3. ATTEMPTS TO GRASP THE SO-CALLED 'OBJECTIVE OUGHT'

All the anthropological facts mentioned so far, namely our relative freedom to shape our way of life by establishing rules of behaviour and by building up legal and social institutions, as well as the fact of our taking over the responsibility for the content of social rules, compel the philosopher to look out for objectively valid methods of argumentation in the field of social and legal politics.

Some legal philosophers - I would designate them as 'extreme positivists' - start from the fact that legal argumentation de lege ferenda depends in principle on valuations. From this they derive that, in legal practice, justice is concerned only with the correct application of the law, while the measure for the determination of legal decisions is provided only by institutionalized normative standards of the law. I take this opinion for unsatisfactory, because it does not meet the fact that people in modern democratic societies are very much aware of the possibility of shaping institutions and their own ways of life, and that people do feel responsible for the results of the organization of their social activities. If we conceive ourselves as responsible for the content of the legal system and of the existing institutions, then we are in need of rational and objective arguments which could justify our de lege ferenda decisions.

Philosophers have, I believe, submitted three main types of argument for an objectively just Ought:
(i) religious faith,
(ii) anthropological knowledge,
(iii) so-called 'practical reason'.

3.1 Religious Faith as a Source of the Objectivity of Practical Valuations

Faith in religious sources, Holy Books or prophets, as well as so-called 'mystical recognition' are sometimes conceived as providing objective information about which rights and duties people should have. Strong ideological systems - e.g. political marxism(4) - take over a rôle similar to that of religious systems.

To presuppose that holy sources provide us with an objective measure for good and bad and that they therefore furnish the basic criteria for justifying human social duties and institutions seems to be a very questionable foundation for legal and juridical argumentation. It is not necessary to have religious convictions, and, of course, not every reasonable man has religious faith. The normative contents of religious systems diverge in some important respects, and there is no strong evidence that especially our faith is morally superior to the others. Indeed, to accept a religious system as a basis of social valuation reduces the scope of possibilities for shaping our social institutions. Religious beliefs and their institutionalization by church organisations usually tend towards a relatively unflexible conservatism, so that it becomes rather difficult to accommodate institutions and normative regulations to the actual state of social needs. (Cf. the problem of birth-control in view of the explosion of population and the standpoint of the churches). The plurality of religious systems in the world - and often within one and the same state - leaves little chance for mutual understanding and peaceful coexistence, if these religious standpoints are not connected with genuine tolerance. But tolerance cannot be expected where religion is conceived as a dogmatic ready-made answer to all problems of life rather than as a matter of search for a humanistic way of life. And even more so if the church and its hierarchy take over control of the state, of public activities and of the sciences.

3.2 Anthropology and Practical Recognition

I do not exclude the possibility that there are some basic necessary features of all moral or legal systems owing to our anthropological constitution. But I hold that we are unable to draw a sharp division between unchangeable anthropologically given necessities on the one hand and our conceptions which seem to us to be necessary ones because of our education, our historical situation and other sociological determinants on the other. Therefore we should not contend - as some legal positivists do - that anything could be stated as valid law by an empowered legislator. But it is one of the essential features of man that he is in disposition of a range for action and of the essential freedom to shape his institutions, normative regulatives and value standards. There is an important degree of plasticity for moulding human life so that anthropological constants by themselves can provide us with no more than a rather broad frame for shaping our ways of life and our value standards. But anthropology cannot decide objectively about the adequacy of these different social rules and institutions which are possible within that anthropologically given frame.

3.3 Reason as a Source of Practical Knowledge

Attempts, to base the objectively justified 'Ought' on reason alone have been made by reference to one or the other of the two following fields of philosophical analysis:

(i) the field of logic proper or logical analysis, and
(ii) the realm of a material 'a-priori', not only conceived as a formal and operational structure, but as a structure determining the character of the content of this realm itself - the structure of practical knowledge in our case.

A more detailed analysis of both kinds of argumentation seems necessary. I shall bring it forward in the next two sections.

4. LOGIC AND THE OBJECTIVIZATION OF PRACTICAL ARGUMENTATION

Logic can be characterized as the science of formal analysis of thought and of formal relations and operations within the field of thought. In this context 'thought' does not mean a mental act, nor does it mean the content of a concrete mental act, but rather a piece of idealized content of possible mental acts. Thought - in this sense - is linked with language in such a way that a thought is the meaning of a linguistic sign series formed in accordance with the constitutive rules of the respective language. Language is to be conceived as a system of signs and as a system of pieces of meaning coordinated to well-formed sign series.

It is useful to conceive logic as a rational reconstruction of linguistic and argumentative practice, because this conception takes into consideration the modern methods of logic, namely, the use of special languages, the rôle of stipulations and the strictly formal method of inference, and it yields an adequate explanation of the rôle of logic in philosophical analysis.

Two well-known sayings are essential for logical studies: the proverb 'Logique oblige' and Carnaps thesis "In der Logik gibt es keine Moral"(6) which is often referred to as the 'principle of tolerance'.

'Logique oblige' postulates exactness and rigidity of logical forms and operations which are strictly bound by rules of language constitution and by rules of inference; the principle of tolerance opens the door for logical theory into all fields of thought and to every kind of formal operation rules. From this point of view, different kinds of logical systems are equally valid: classical two-valued logic, intuitionistic calculi, many-valued logic, and - I would add - systems of genuine logic of norms, etc.

The character of logical inquiry allows for the following methodologically and philosophically important consequences:

(i) Logical forms, relations and operations, especially logical inference, are objectively and strictly valid. This is a consequence of the construction of such systems on the basis of definitions and of the use of explicit formal rules of inference.

(ii) The adequacy of a proposed system is a mere matter of fact, not a matter of a priori (i.e. strict and necessary) knowledge. The proposed system must be tested in respect to its adequacy for the intended task in the realm of application of formalism.
(iii) The following holds for all empirical fields and for practical philosophy:
(a) there is no knowledge of facts based only on logic, and
(b) there cannot be any practical knowledge justified merely by logical analysis without further premises carrying additional non-analytic content.

I may conclude this section: Logic provides for relative objectivity, namely strictness and objectivity of logical forms and of logical operations. But there is no strictly objective evidence for the adequacy of a proposed logical system, nor can we prove what is good and just by logical analysis alone - that is, without accepted practical premises.

5. MATERIAL 'A-PRIORI' AND OBJECTIVE PRACTICAL KNOWLEDGE

Some philosophical doctrines take 'reason' not only for a contentless formalism or a capacity for formal operations. Instead, they conceive it as a given structure per se (in a platonic sense), so that rational analysis is capable of and has the task of exposing an essential structure, that is, the 'material a-priori' of the field under consideration. There are, of course, some differences between various authors in their methods of how to grasp this material a-priori: some characterize this method as a special kind of analysis, others as a kind of intuitive recognition, or 'Wesensschau',etc. But, by their shared presupposition of the platonic existence of the material a-priori, all these teachings - we could designate them as the 'phenomenological approach' (in a broad sense of the term) - lead to the view that there is, in principle, a method of how to justify objectively what is good and what is bad or what is just and what is unjust, or, at least, that there are objectively valid criteria for such value judgments. I do not share these views, and I am convinced that the phenomenological line of argumentation is deceptive by mistaking the results of an analysis of our ideas for an objectively valid recognition of immanent material features of our mind or of our moral sense.

The 'Ought' as it should be, is not determined by values and principles of justice existing per se, but practical argumentation is based on material value standards and on standpoints which are not given by "reason" itself.

A very interesting author who argues for a phenomenological analysis in legal theory is the late Fritz Schreier. His well-known theory of the "Possible Law" (das "Mögliche Recht") aims at transcending the difference between juristic analyses de lege lata and de lege ferenda(7). He illustrates his phenomenological deliberation by the following example. If we see that the basis of a three-dimensional figure has the form of a square, then we can say about the whole figure that it might be a cube, or a pillar, or a pyramid, etc., but that it cannot be e.g. a sphere.

Analogously we can speak, says Schreier, of possible and impossible legal regulations.

The example of the square basis of the figure shows clearly that phenomenological analysis, and the analysis of all possibilities in a certain field of inquiry, is bound to a presupposed frame of data. And the same goes for practical valuations: there must exist some frame of facts, values or/and regulations, and only then can we rationally develop a field of possibilities of how to complete the body of normative rules(8).

My thesis concerning the justification of 'Ought' by phenomenological analysis reads: Just like our knowledge, so is the justification of legal rules and institutions based on a framework of empirical data, of accepted or presupposed rules, and of value standards and principles. Only within the frame of these material elements can we ascertain the field of possibilities of the phenomena under consideration.

Even such principles like the postulate of universalizability do not provide materially relevant criteria of justice if they are not supported by material decisions, that is, by decisions concerning the criteria for the respective type of cases. Only if based on suitable material criteria, universalization can take place and have a rôle in practical argumentation. Not only the universe of discourse for universalization (whether all citizens, all men, or all living beings) is stated by a material decision, but every universalization of a given case is determined if and only if we have decided what the relevant features of the case are, i.e. the features subjected to universalization. To make the problem clear consider the following case. N has promised M that he will marry her next month, and M has consented. We may say that N is morally obliged to marry M. This individual normative relation can be universalized in different ways according to how we determine the relevant features of the case. If engagement is conceived as essential, we deduce a quite different universalized rule than if promise is taken as the basis for universalization. If we universalize over 'men' becoming engaged we receive a different rule than if we universalize over 'men or women'; etc.

6. EXTRALOGICAL RATIONALITY IN LEGAL REASONING

Practical philosophy and legal theory use the term 'rationality' in a much broader sense than in the sense of correctness of logical form, consistency or valid inference. Rationality in this broader sense has to do with methods of reasoning or of action, with the effectiveness of procedures used in argumentation, in the search for true and useful information and for justified practical convictions. In this sense 'non-rational' does not mean logically false, inconsistent, or incorrectly inferred; rather, it means the choice of an ineffective procedure, unsuitable for the attainment of the given purpose.

There are, of course, some important relations between logical and, so to say, extra-logical rationality. At least, there is the following relation: logical soundness of deliberation is a necessary, though not sufficient, precondition of any kind of rationality in argumentation, valuation or action. I shall plead for the opinion that logical processes have an important rôle to play even in processes of extra-logical rationality as well.

It seems that philosophers have proposed different foundations of rationality in the broader sense. I propose the following tentative systematization of this attempts:
(i) rationality as based on social experience of a certain society in a historically and culturally determined situation (6.1.);
(ii) rationality subsisting in a kind of interpersonal

interaction following certain rules (6.2.);
(iii) rationality as a system of postulates concerning the structure and/or the content of a system of theses, of practical standpoints, aims, or institutions and their normative regulations (6.3.).

Actual teachings to be found in the relevant literature often contain combinations of various of these elements of rationality.

6.1 Rationality Based on Social Experience

The theory of Perelman, 'la nouvelle rhétorique', which is based on pragmatical considerations concerning the process of making plausible some theoretical or practical thesis or of attaining assent from an audience, refers to the state of the respective culture as the decisive basis for rational acceptance of arguments. Cf. e.g. "Ainsi chaque culture, chaque individu a sa propre conception de l'auditoire universel, et l'étude de ces variations serait fort instructive, car elle nous ferait connaître ce que les hommes ont considéré, au cours de l'histoire, comme réel, vrai et objectivement valable"(9). The idea of a universal audience as the essential condition for distinguishing rational plausibility from merely factual subjective belief, does not take account of the quantitative view, nor does it refer to an interpersonal process. Instead, it intends to free deliberation from personal preferences or prejudices.

Aarnio's conception, elaborated in a Wittgensteinian vein, according to which the respective way of life is the very last basis of rationality, leads into the same direction(10). All this reminds me of Kotarbiński's idea, put forward in his praxiology, that principles of effective (good) work are products of experience and cultural development in the long run of the ages(11). Even from the standpoint of evolution in the sense of ethology (e.g. Konrad Lorenz, Campbell etc.) such conceptions are believed to be obvious.

6.2 The Procedural Conception of Rationality

Habermas, Alexy and others(12) - Perelman only in a very restricted sense, namely in so far as he conceives argumentation as an interpersonal process between the speaker and his audience - plead for a procedural conception of rationality. Free and unrestricted discourse between people is conceived as the very essence of rationality, and processes of discourse are presupposed to lead - under expressly ruled conditions - to truth and justified practical knowledge.

Free discussion is a very good means to further philosophical analysis. But do the presuppositions of the adduced authors guarantee the best approach towards truth and justified Ought? Is there always a chance of reaching a consensus? Is a consensus, reached in free discourse, a criterion for discerning truth from falsehood and the valuable from the bad? I believe, that this is not the case. It may be a valid criterion for democratic decisions - but reasonable only on the basis of a good previous argumentation ("Democracy means discussion", says Masaryk) - in order to discern how to act or which fact should be taken for established. But a rationally valid proof or a sound plausibility argumentation should be concerned only with evidence from valid (good) arguments.

6.3 Extralogical Rationality of Systems of Propositions, Norms or Actions

It is important, of course, to test systems of propositions, of norms, or of actions from points of view which transcend mere logical requirements. D.N.MacCormick and others have postulated coherence as an according requirement in respect to narrative or normative texts; beside this, MacCormick has postulated consistency in practical discourse over time and universalizability over cases.

In respect to coherence, I agree with him in so far as, besides logical consistency, there are, indeed, additional requirements for a reasonable text or for reasonable human behaviour. However, there are serious difficulties in defining 'coherence' adequately. I even doubt that there is a unique kind of relation which could be taken for the basis of coherence. There is, e.g., a principle of the ordering of sentences constituting a narration. A series of sentences drawn by change would probably yield an unreasonable text. There may be different possibilities with such an ordering. E.g., the items of a list of publications can be ranked by their date of appearance, by the name of their authors, by their topics, etc. If we are unable to discern a ranking principle in the field of our analysis, we shall have the impression of disorder, or indeed of non-rationality. In the example adduced by MacCormick, where the allowed speed of cars was dependent on their colours(13), we invite a reproach of unreasonableness, because such a rule seems unjustified by the goals which traffic norms should serve. As there are different sources of coherence or different reasons for our reproach of incoherence, the notion of coherence is to be considered as a kind of family notion, which cannot be defined by one exhaustive definition, and it seems preferable to go on in the analysis and disclose the structure of such a reproach of unreasonableness. If we continue our analysis then consistency of behaviour over time may be nothing other than the constancy of our system of aims and preferences. (If this system changes in time, there is no reason for the reproach of incoherency to the agent). Universalization over cases, again, should be thoroughly analysed. Then we may say: universalization depends on establishing types defined by relevant features. If and only if the respective type is established correctly, namely by all its relevant features, then and only then universalization is justified.

It seems that our mental constitution is such that only structured fields of objects or ideas are viewed as rational. Rational ordering is partly an automatic and intuitive process, and partly it is the result of a systematic endeavour. Only systematized knowledge is of practical use.

7. MY VIEWS ON LOGICAL AND EXTRALOGICAL RATIONALITY

Let me summarize my conception of rationality in legal reasoning by the following points:

7.1 The language of Practical Discours

Discourse in practical philosophy and in jurisprudence requires a language which allows for drawing a strict distinction between descriptive and practical sentences. The same distinction has to be made in respect to the concepts used in practical discourse. A correct analysis of problems of action is possible only on the basis of the type of semantics I have called 'gnoseologically differentiated semantics'(14).

7.2 Non-Cognitivism

I hold a non-cognitivist view and the thesis of non--derivability of 'Ought' from 'Is' and of 'Is' from 'Ought'. The very essence of information processing and action control, as well as of our dealing with norms, not only presupposes the semantical differentiation of informations mentioned in section 7.1., but also the non-derivability theses:
(i) informative ought-sentences do not follow from descriptive sentences alone;
(ii) informative descriptive sentences do not follow from practical sentences alone. Moreover, it presupposes a non-cognitivist conception of practical argumentation. Any practical argumentation depends - at least at some points - on decisions, choices, acts of valuation or preference, and the like. A reduction of these elements of action - or of discourse leading to action - to mere recognition is impossible. It is, in my opinion, one of the main tasks of analysis to make those points explicit where elements of voluntary decision have to take place in practical discourse. I am convinced that this conception is not in conflict with postulating that argumentation in jurisprudence can, and should be, rational.

7.3 The Relative Objectivity of Logical Recognition.

Logical relations and logical inference are strictly and objectively valid in the sense explained above, namely within the frame of the accepted system (a logical calculus of a certain kind). As logical inference is only one of the conditions for valid proof, and as the other elements - especially valuations - are not objectively valid, only relative objectivity can exist in practical justifications.

7.4 The Value Thesis

The basis and the starting points of our analyses aiming at justifying practical standpoints, goals, or norms, are some convictions of what is morally good or bad, what is desirable, useful etc. Every person and every community holds some value standards and is able to evaluate objects or situations. This psychological and sociological fact provides the basis for practical discourse justifying norms in a perspective de lege ferenda, or moral postulates etc. The process of this kind of argumention is not circular, as it might seem. It is indeed a combination of rational steps of inference based on value suppositions, and the consequences are reached through a logical process using also cognitive elements (i.e. causal recognitions). The results of these deliberations are submitted to a new valuation. Suppose, e.g. that M is the necessary and sufficient means for achieving G. If the means M is valued so negatively that we prefer not - G to (M & G), then the goal will be cancelled. Often practical deliberations take the form of an analysis of different possible alternatives of the action programme or of practical conceptions. In the course of this, we develop the consequences of the various alternatives and make a choice of preference between these possibilities. The analysis of these alternatives should be a rational or even logical process and often consists of a purely cognitive argumentation, but the whole argumentation always embraces some elements of choice.

7.5 Partial and Relative Objectivity of Legal Reasoning

We are involved in an ongoing dynamic process of rationalization and relative objectivization of legal argumentations. I conceive this process as an interplay between logical-analytical steps and repeated acts of evaluation and choice. We start from our actual value convictions which are then checked and may be revised in the sequel by a process of analysis and repeated valuations of different possibilities and factual findings.

7.6 Extralogical Rationality in Lawyers' Reasoning

In addition to logical rationality we have to work in jurisprudence with certain criteria of reasonableness. Various ordering principles are to be applied according to the aims of the system under consideration, so that certain coherence postulates are measures for the rationality of a system of theses, of a narration, of a normative regulation or a system of actions. Only a useful kind of order is judged as rational.

Extralogical rationality also has a procedural aspect. It is reasonable to analyse practical standpoints in the light of experience and to take into consideration all possibilities of the field concerned. Combinatorial methods may be useful in this respect. Beside this so to say empirical approach in practical argumentation, it seems reasonable, at least from time to time, to call in question our own moral and other practical convictions. A certain amount of doubt and analytical scrutiny are constituent parts of human rationality. A rational man should be open to analyses of practical problems and he should in some way cultivate considerations about his practical convictions in private or/and with other persons. In this sense, rationality means also openness to procedures of practical analyses. It is reasonable to discuss our opinions with other persons, but consensus or dissent are not valid rational arguments in themselves.

NOTES

(1) Rawls, J.: 1971, _A Theory of Justice_, Harvard University Press, London, Oxford, New York; cf. also: Weinberger, O.: 1977, 'Begründung oder Illusion', _Zeitschrift für philosophische Forschung_, _31_, pp.234 - 251, reprinted in: Weinberger, O.: 1979, _Logische Analysen in der Jurisprudenz_, Duncker & Humblot, Berlin, pp.195-216.

(2) Cf. the various theories of life and of human behaviour which conceive spontaneous activity as the basic feature of all life-phenomena, e.g. Lorenz, K.: 1978, _Vergleichende Verhaltensforschung_, Springer-Verlag, Wien, New York; Roeder, K.D.: 1955, 'Spontaneous Activity and Behaviour', _The Scientific Monthly_, _80_, pp. 362-370; Weinberger, Ch.: 1983,

Evolution und Ethologie. Wissenschaftstheoretische Analysen, Springer-Verlag, Wien, New York.

(3) Cf. Weinberger, O.: 1983, Studien zur formal-finalistischen Handlungstheorie, Verlag Peter Lang, Frankfurt/M., Bern, New York, p.3f.

(4) In some communist countries it is explicitly ruled that all teaching has to be marxist.

(5) Cf. Lessing, G.E.: 1779, Nathan der Weise.

(6) "In der Logik gibt es keine Moral", Carnap, R.: 1968 (1934), Logische Syntax der Sprache, 2. ed., p.45.

(7) Schreier, F.: 1926, 'Über die Lehre vom "Möglichen Recht"', Logos, XV, pp.364-370.

(8) In any case, I believe that phenomenological analysis or Schreier's theory of the "Possible Law" does not allow us to overcome the distinction between considerations de lege lata and de lege ferenda. Even if we were able to describe a system of all possible legal systems, the essential difference between the two kinds of view would not vanish. In a consideration de lege lata we are bound to take the respective normative system for definitely given and providing fixed starting points for our judgements - incertainties being restricted to problems of interpretation -, whereas considering the legal system or parts of it de lege ferenda we have to valuate legal rules from different points of view, e.g. from moral, political standards etc. Having a list of all possible legal systems or of all possible rulings of a certain issue at our disposal, does not replace the need for the valuation of certain alternatives with the intention to plead for or against it.

(9) Perelman, Ch., Olbrechts-Tyteca, L.: 1970, Traité de l'Argumentation, 2nd ed., Edition de l'Institut de Sociologie, Bruxelles, p. 43.

(10) Aarnio, A.: 1977, On Legal Reasoning, Annales Universitatis Turkuensis, 144, Loisnaa.

(11) Kotarbinski, T.: 1965, Praxiology. An Introduction to the Sciences of Effective Action, Pergamon Press, Oxford.

(12) Cf. Alexy, R.: 1978, Theorie der juristischen Argumentation. Suhrkamp Verlag, Frankfurt/M.

(13) MacCormick, D.N.: 1984, 'Coherence in Legal Justification', in this volume.

(14) Ch.Weinberger, O.Weinberger: 1979, Logik, Semantik,

Hermeneutik, München, p.109ff., p.183.

ABSTRACT

In jurisprudence objectivity is aimed at in the following fields: decision making, justifying legal and social institutions, and legal science. In modern democratic societies people feel a need for objective justification of political and legal conceptions, since they feel responsible not only for the correct application of the law, but also for the content of the law itself and for the shape of social institutions. Philosophers have proposed three main ways towards an objective justification of 'Ought': religious faith, anthropological knowledge, or reason (logic itself or the phenomenological approach). Logic provides only relative objectivity. But rationality in legal reasoning is not restricted to logical relations. The author deals with the proposed sources of extralogical rationality (the way of life; procedural theories; the idea of coherence). In the last chapter he sums up his views on logic and extralogical rationality in lawyers' argumentation).

COHERENCE IN LEGAL JUSTIFICATION

Neil MacCormick

I. INTRODUCTION

Coherence in reasoning is one important test of its soundness as reasoning. It is a test which is not fully satisfied by mere consistency, in the sense of an absence of self-contradiction. What I say may be free of any internal inconsistency and yet fail to 'hang together' or to 'make sense' as a whole. The task of this paper is to elucidate this elusive notion of 'coherence', of 'hanging together', of 'making sense'. In the specific context of legal justification we find that it has two applications: to reasoning on matters of law, where what is in issue is 'normative coherence', and to drawing inferences of fact from evidentiary facts, where what is in issue is 'narrative coherence'. I shall consider these two applications in that order.

II. NORMATIVE COHERENCE

(a) The Meaning of Coherence

Why is it that a set of legal norms might sometimes appear incoherent, even when as a set they are not inconsistent? An example of such a set which I previously suggested(1) was if a statute laid down different speed limits for different cars according to the colour they were painted. Little did I realise when I figured that fanciful case that there was a real case rather like it. Now I am indebted to Mr.Juctice Ruggero Aldisert (Justice of the U.S.Court of Appeals, 3rd Circuit) for producing a real case more or less to the same effect. For some time ago the legislature in Italy determined that there should be differential speed limits for different types and makes of car(2).

Do such laws fail to make sense? And if they so fail,

A. Peczenik et al. (eds.), Theory of Legal Science, 235–251.

why do they so fail? My answer to that is that they fail to make sense if there is no common value which the enactment of such laws subserves. In our examples, is there no common value at issue? At least at first sight, it appears that there is not. Consider: there are three ends which statutes limiting driving speeds may promote, all of which we may suppose to be of serious social value: the safety of road users; economy in the use of fuel; and prevention of excessive wear and tear of road surfaces. If the colour of cars is purely a matter of taste, and many colours are available (as is true in Western Europe and North America at least), it seems doubtful whether any speed limit differential between differently coloured cars could possibly subserve effectively any such end as those envisaged above. Moreover, if people have bought cars prior to the colour-laws, it seems unfair that they should _ex post facto_ be treated differently according to the colour choice they made. So without subserving any value special to road safety laws, the colour-laws would in fact conflict with or subvert another value of importance in a very general way to legal systems.

Of course, we can imagine circumstances in which the colour-laws _would be_ coherent, and perhaps the Italian law did some coherent foundation. Perhaps there was a legislative intent rationally to relate differential speed limits to such objectives as economy and safety. Aldisert J. reports, however, that the car-drivers of Italy did not see it that way(3). They treated the differential speed limits as incoherent nonsense, and ignored them entirely. Desuetude overruled the act. May be the car-drivers judged wrongly. May be the lawmakers failed in persuasion rather than in coherent thought. But we need not go into that. Sufficient has been said to ground the suggestion, arising from these examples, that at least one aspect of normative coherence is a matter of the common subservience by a set of laws to a relevant value or values; and an absence of avoidable conflict with other relevant values (e.g. with justice, as in the above case).

Are there then other aspects of coherence? One candidate which comes to mind has to do with principles. We might say that a set of rules is coherent if they all satisfy or are instances of a single more general principle. If it is a principle that human life ought not to be unduly endangered by motor traffic on the roads, this will (help to) make sense of speed limit laws and many other parts of road traffic law taken together: but not of all possible speed-limit laws - judged by reference to that principle, car-colour differential speed limits will be arbitrary unless in some such expanded context as was imagined above.

The very fact that we can re-express in terms of common principles the coherence of a set of laws (road traffic laws) which we previously expressed in terms of common values poses the question whether appeal to 'values' is different in substance from appealing to principles, or only different in grammatical form. 'Safety on the roads' is a noun phrase; 'safety on the roads' conceived as a 'value' is the state of affairs signified by the noun phrase conceived as being a state of affairs which is a good or worthy purpose of human endeavour. 'That human life ought not to be unduly endangered by motor traffic on the roads' is a normative sentence which in virtue of its very general scope can be considered as a possible principle. It is actually somebody's principle or a principle of some normative system if some person assents to it as a practical norm for his/her own and others' conduct or, respectively, if it is an accepted or acceptable justifying norm for more particular and specific norms ('rules') within that system.

In short, if 'values' are, as I suppose, not merely the *de facto* purposes, aims, goals or ends actually pursued from time to time by individual persons or institutional agencies, but rather actually pursued or possible pursued states of being or of affairs which are conceived to be legitimate, desirable, worthy or even (the scale ascends by degrees) mandatory for pursuit as standing purposes, aims, goals or ends, then there appears to be an extensional equivalence as between 'values' and 'principles'. For any value V there is a principle according to which V either may be, or ought in the absence of countervailing considerations to be, or ought normally to be, or must in the absence of overriding considerations be, pursued or realised. Observance of principles is not an instrumental, but an intrinsic, means of realising values.

This conforms to a suggestion which I have recently put forward(4) that systems of practical reason necessarily require the framing of hierarchically related rules and principles of conduct exhibiting consistency over time and universalisability over cases, this being essential to the achievement of a rational order superimposed on what would otherwise be a chaos of particular purposes. Under such principles some actual recurring purposes can be generalised as standing aims legitimated (or made mandatory) by one's principles. So values are the product of a system of practical principles. This, if sound, accounts for the extensional equivalence of values and principles.

Nevertheless, the formulation of values (or virtues)like 'safety', 'health', 'considerateness', 'justice' or whatever may have a particular utility for the purposes of a critique of a system of rules or principles as hitherto (at any given time) formulated. By reflecting on what it is that we suppose to be, or are committed to treating as, good to bring about, we may be led to better or more general expressions of the principles of our practical systems or (yet more likely) to see new areas for their application. This perhaps accounts for the standing popularity of utilitarian and other consequentialist doctrines in ethics and legal and political philosophy.

I conclude that the coherence of norms is a matter of their 'making sense' by being rationally related as a set, instrumentally or intrinsically, _either_ to the realisation of some common value or values; _or_ to the fulfilment of some common principle or principles. At the level of the highest-order principles or values there is a further requirement of coherence: that, after allowance for the fulfilment of priority rankings of principles and/or values we consider that in their totality they express a satisfactory form of life(5), and one which it would be possible for human beings, as human beings are, to live. (This last statement is the best that occurs to me at the moment, though it seems unsatisfactorily vague). In short, the coherence of a set of norms is a function of its justifiability under higher order principles or values, principles and values being extensionally equivalent; provided that the higher or highest-order principles and values seem acceptable as delineating a satisfactory form of life, when taken together.

(b) Coherence in justification: an example

A well known decision in English criminal law, that of _Sweet_ v. _Parsley_(6) concerned the problem of interpreting a statute which provided:

> 'If a person - (a) being the occupier of any premises, permits those premises to be used for the

> purposes of smoking cannabis or cannabis resin or of dealing in cannabis or cannabis resin...; or (b) is concerned in the management of any premises used for any such purpose as aforesaid; he shall be guilty of an offence against this Act.'
> (Dangerous Drugs Act,1965, section 5)

The particular problem in the case focussed on paragraph (b), is to be precise, whether the offence of 'being concerned in the management of any premises used for any such purpose' requires or does not require guilty knowledge or intention or participation in the 'purpose' in question. Miss Sweet, a schoolteacher in Oxford, was tenant of a farmhouse outside Oxford. She sub-let rooms in the house to other persons. After a certain time, her car broke down, and she had to take rooms in Oxford. Thereafter she kept only one room in the farmhouse as her own, and sub-let all the other rooms, with a kitchen etc. retained for common use.She was able to visit the house only occasionally to stay over a night, collect rent, and check that the house was in reasonable condition.

In due course, the police discovered that cannabis was being smoked by residents in the house. They charged Miss Sweet, who was indubitably managing the premises, with 'managing premises used for [the purpose of smoking cannabis]", contrary to section 5(b) of the Act. The trial court found as a fact that 'she had no knowledge whatever that the house was being used for the purpose of smoking cannabis or cannabis resin'. Nevertheless, the court convicted her, being of the opinion that the Act was in terms which implied absolute liability, that is, liability without regard to a person's intention or knowledge as to the purpose in question, provided she or he were actually concerned in the management of the relevant premises. At the first level of appeal, this view was upheld: 'paragraph (b) in dealing with somebody concerned with the management of premises where cannabis is smoked contains an absolute liability; it does not depend on knowledge at all'(7), this being in contrast with paragraph (a), under which 'permitting' necessarily involves knowledge of what is going on.

On the final appeal to the House of Lords, this interpretation of the Act was rejected. There can be discerned at least three elements in the reasoning in favour of allowing the appeal and quashing the conviction: first, there is ambiguity in the Act: 'Is the "purpose" the purpose of the smoker or the purpose of the management?'(8) Either possible answer to that question being consistent with the express terms of the Act, either answer is permissible in law, given the requirement that rulings in law and particular decisions must not contradict established rules of law. Secondly, however, the consequences ('consequences-as-implications') of holding that the smoker's purpose suffices are unacceptable:

> 'The implications are astonishing. Parliament would not only be indirectly imposing a duty*) on persons concerned in the management of any premises requiring them to exercise complete supervision over all persons who enter the premises to ensure that no one of them should smoke cannabis, but Parliament would be enacting that the persons concerned in the management would be guilty of an offence if, unknown to them, someone by surreptitiously smoking cannabis eluded the most elaborately devised measures of supervision.'(9)

Such reasons - and many such points were taken by the five Law Lords - indicate the extreme undesirability, in their Lordships' view, of imputing to Parliament the intention that the word 'purpose' be read as referring to anything other than the purpose of the person concerned in management. What is said, of course is that Parliament 'cannot' have intended so unjust a result. But the grounds for the imputation of intention are the evaluation of the implications of the rejected interpretation; no independent recourse is available to the otherwise mysterious concept of 'legislator's intention'(10).

Thirdly, and also apparently essential to the justification of the ruling in law is what I call the argument from 'coherence':

*) A duty which, as Lord Pearce pointed out, (1969) 1 All E.R.356, would necessarily extend to 'the innocent hotel keeper, the lady who takes in paying guests, the manager of a cinema, the warden of a hostel, the matron of a hospital, the housemaster and matron of a boarding school' though 'the most that vigilance can attain is knowledge of their own guilt. If a smell of cannabis comes from a sitting room, they know that they have committed the offence. Should they then go at once to the police and confess their guilt in the hope that they will not be prosecuted?'

> 'A consideration of previous and analogous legislation removes any doubt that these words are intended to refer to such a special and limited class as I have described, one which quite clearly excludes such persons as the appellant.This legislation deals with other "anti-social" activities, such as the keeping of brothels, opium "dens" and gaming houses.'(11)

Thus for example:

> 'The Dangerous Drugs Act, 1920 dealt with opium. The relevant sections are reproduced in the Act of 1965 (s.8), and it is obvious that the provisions regarding cannabis are based on them. In dealing with the management of premises it seems clear enough that what is in mind is not the lessor of premises on which opium may come to be smoked, but a manager of what, if a noun is required, might be called "opium dens".'(12)

So, if a coherent view is to be taken of the legislation controlling such 'anti-social activities', it must be the view that none creates offences of strict or absolute liability. Reference was also made to the earlier decision of Warner v. Metropolitan Police Commissioner(13), from which, especially from the leading opinion of Lord Reid, there can be extracted a general principle as to the differentiation of offences of absolute liability from those requiring *mens rea*. As he re-stated that principle in *Sweet's* case, it is to the effect that there is a class of 'quasi-criminal acts', acts which 'in the public interest are prohibited under a penalty'. Being penalised in such matters involves no real moral stigma. By contrast, in the case of 'acts of a truly criminal character', 'a stigma ... attaches to any person convicted ... and the more serious or more disgraceful the offence, the greater the stigma'. It has then to be asked whether 'in a case of this gravity, the public interest really requires that an innocent person should be prevented from proving his innocence in order that fewer guilty men may escape'. That is: where an offence is created to regulate in the public interest a potentially dangerous activity of a specialised kind, and where conviction does not carry moral stigma, liability may be strict or absolute; where offences properly carry some stigma, *mens rea* ought to be required. This Lord Reid took to be the best - though not a perfect - rationalisation of the previous decisions on the question of absolute liability versus _mens rea_. In the former class of cases, in his view, the value of public safety is promoted without grave injustice. In the latter class of cases the injustice of a stigmatising conviction of a morally innocent person is normally and properly taken to outweigh any competing public interest in general safety and good conduct(14).

These considerations, together with the principle of interpretation that 'if a penal provision is reasonably capable of two interpretations, that interpretation which is most favourable to the accused must be adopted(15), completed the justification of the decision.

I hope it is clear how this reading of the reasoning in _Sweet_ v. _Parsley_ conforms to and illustrates what I said earlier by way of explaining my conception of 'coherence'. So far as concerns the coherence argument, the task the judges undertake is a two-fold one. First, the inquiry is as to the principles or values which as far as possible make sense of a relevant set of legal norms - statutes and precedents dealing with similar subject matter in the same field of law. These are partly to be found in the existing materials, partly to be constructed so as to establish a coherent view of the branch of the law, by showing it to be compendiously justifiable by reference to some 'underlying' principle or value or coherent set of principles and values which can be conceived as justifying the rules and rulings (the norms) under consideration. These are then applied to the purpose of justifying the actual ruling in the present case, as an analogous application of the same principles or values, and thus as coherent with the pre-established body of law. Where, as in _Sweet_ v. _Parsley_ (and, I suppose, as is almost invariably the case in codified systems of law) the problem concerns the interpretation of a statutory text, the rhetoric of such justification is to pose it as the intention of the legislator to legislate coherently. But there is an air of pious fiction about such rhetoric. It is the theory that legislators ought to legislate for a coherent body of law which justifies the imputation of such intention to the legislator; it is not an actual, independently established fact of proven legislative preference for coherent law which constitutes (as under voluntarist theories of law) a duty of the judge so to interpret the law in subservience to this real will of the real legislature.

(c) Why Coherence justifies

If we can only fictitiously treat a real legislative intention as that which constitutes 'coherence' as a relevant justifying quality in legal reasoning, we are left with the question why, and in what sense, coherence does justify. Is the theory that legislators ought to legislate for a coherent body of law a sound theory? More generally, is coherence a quality which legal norms ought to exhibit, and, if so, why?

To sketch a few considerations in virtue of which 'coherence' may be esteemed relevant to justification: first, it is agreeable to a certain conception of rationality in practical life, that which requires both universality and also the greatest possible degree of generality in practical principles. There are also reasons (fair notice to subjects, relative specifity in the law) why the law should be expounded at the level of relatively detailed rules. But these relatively detailed rules will be arbitrary if they are not also instances of more general principles, fewer in number than the number of the detailed rules, and more general in their terms.

More urgently, however, it is to be noticed that this implies a formalistic (formally rational) and relativistic sort of justification. Whatever the actual content of a legal system may be, the above considerations and their like imply that it is preferable that the system be interpreted and applied so far as possible under the supposition that its more detailed provisions can be conceptualised as derived from or instantiating (in the sense of being together justifiable under) some general principles; and also that the principles themselves as a set, making allowance for priorities and for different justificatory levels, are capable of being thought coherent in the ultimate sense suggested above at the end of I(a). Put in the alternative, the detailed provisions of the system ought to be interpretable as subserving a possible set of mutually compatible values. 'Coherence' can then be satisfied by a system which does subserve what those responsible for determining its content do suppose to be values (e.g. racial purity under various elements of National Socialist law), even although from another point of view - let us say bluntly a better one - these supposed values are truly evils.

On this ground, considerations of 'coherence' may be

considered to be only weekly justifying considerations. To borrow a term from David Lyon's recent paper 'Justifications and Easy Cases'(16) coherence concerns the derivability of a novel decision or ruling in law from the pre-existing body of law, not the ultimate defensibility of the decision or ruling from a moral point of view. Moreover, the constraint of coherence seems to be treated by lawyers as a relatively weak constraint, perhaps because it determines only what we might call the 'weak derivability' of a ruling or decision from the pre-existing law.

The reason why coherence determines only weak derivability of ruling or decision from established law is dependent on the fact that coherence is a desirable ideal feature of a system of law. As such, however, it may compete with other ideal features of law, like substantive justice (judged by appropriate criteria) and so forth. Yet it imposes a real and important constraint on judges if we interpret it in a negative sense: unless, by the coherence test, some ruling or decision is at least 'weakly derivable' from existing law, it is not permissible for judges in their judicial capacity to make such a ruling or decision, however desirable on other grounds it may be.

This finally enables me to contrast 'consequentialist' reasoning in the special sense alluded to above with coherence reasoning in law. For in the evaluation of the implications of rival rulings in a contested case judges and lawyers raise questions of justification in Lyons' sense of defensibility. Certainly, this is defensibility within narrow constraints posed on the one hand by the requirement of consistency and on the other hand by the negative requirement of coherence. The consequentialist judgment here is, however, a judgment of substance, not a 'formal' one in the way that judgments of coherence are. The issue is: what is the best principle on which to decide the case in hand? It is not: what is the principle which best explains the law as heretofore established by those responsible for establishing it?

III. NARRATIVE COHERENCE

'Narrative coherence' is my name for a test of truth or probability in questions of fact and evidence upon which direct proof by immediate observation is unavailable. Since almost all legal disputes, trials and litigations concern past facts and events, and since no past facts or events are susceptible of direct proof by immediate observation, narrative coherence is a test of great, indeed central, importance in the justification of legal decisions. For most legal decisions require findings of fact as well as applications of, or rulings on and applications of, the law.

Detective fiction, none better than the Sherlock Holmes stories, gives ample illustration of the force of 'narrative coherence'. The case of the dog that did not bark in the night is a good illustration. The horse has been taken from the stable by night. The trainer has been found dead on the Downs nearby. A suspicious-looking stranger has been picked up by the police and held on a murder charge. But Sherlock Holmes elicits from reliable witnesses the information that they did not hear the dog barking by night. 'The dog did not bark' and 'a stranger took the horse' are not propositions which contradict each other. Yet if, as a generalisation, dogs bark at strangers, then, under this common-sense principle, the dog's not barking becomes incompatible with a stranger's taking the horse, unless there is some further explanation, or some relevant exception to the general common-sense principle.

Compare real decisions at law, especially where the main evidence is circumstantial. Take _Rex_ v. _Smith_(17). The short summary of the evidence in the case given in the headnote of the report is sufficient for our purpose:

> The appellant was indicted for the murder of M. who had been discovered dead in her bath after having gone through a ceremony of marriage with him. At the trial evidence was given that subsequently to the death of M. two other women had died in their baths in similar circumstances after having gone through marriage ceremonies with the appellant. Evidence was also given of a consultant between the appellant and a solicitor concerning, _inter alia_, the effect in law of a voluntary settlement made by M., and whether the

trustees could buy an annuity without M's permission.

Not surprisingly, Smith's defence lawyers had objected at the trial to admission of evidence about the deaths of his two 'wives' subsequent to M. by the same misadventure as M. Not surprisingly, the defence had sought to exclude the solicitor's evidence. For reasons of no concern to us today, the Court of Criminal Appeal held that the evidence had been properly admissible, and therefore upheld the conviction.

But why was this evidence so damning? Why so vital to the defence to have it excluded? In a common-sense way, the answer is obvious. A man is to be pitied if he loses one wife by drowning in a bath, to be suspected if he loses two, and to be judged a murderer if he loses three. A man whose wife dies is to be pitied. A man whose wife dies shortly after he has checked to ensure that her death will benefit him financially is to be suspected as a possible murder. All this because unlikely misadventures which befall a person once can well be sheer accidents, but unlikely misadventures which recur three times in materially similar circumstances are not usually misadventures at all but the product of design. _A fortiori_ where there is not merely a possible motive (money) but where the motive has been checked out by the person whose possible motive it is. He had checked up to be sure about the money, so we can suppose this possible motive was an actual topic of interest to him, a matter actually before his mind.

(1) 'The first Mrs.Smith died in her bath, and Smith was at home at the time.' (2) 'The second Mrs.Smith died in her bath, and Mr.Smith was a home at the time.' (3) 'The third Mrs.Smith died in her bath and Mr.Smith was at home at the time'. (4) Before the first Mrs.Smith died, Mr.Smith checked up on the probability of his inheriting her money.'

These sentences (1) - (4) are not themselves contradictory of either (5) 'All the Mrs.Smiths died by sheer accident' or (6) 'Mr.Smith wilfully killed all the Mrs.Smiths in their baths.' Yet, in the absence of some further sentences disclosing weighty matters of exculpation, (6) coheres with (1) - (4) in a way that (5) does not. This justifies our concluding that we have less ground to doubt (6) than to doubt (5). Whether this puts (6) 'beyond reasonable doubt' so as to justify deeming it true for further evidence, such as was provided at Smith's trial) need not detain us.

Why is the story (1) - (4) plus (6) coherent in a way that (1) - (4) plus (5) is not? Why could that justify deciding to take (6) as being the fact of the matter?

The answer, I suggest, is that we treat the natural world as explicable in terms of explanatory principles ('laws') of a causal and probabilistic kind, and the world of human affairs as being explicable in terms of explanatory principles of a rational, intentional and motivational as well as a causal and probabilistic kind. Accidents which occur without human intervention have to be explained in a non-intentional and non-motivational causal or probabilistic way. The probability of the conjoint occurrence of the necessary causal conditions for any person's drowning in a bath is low. Even lower is the probability of these conditions recurring three times in the case of three persons successively enjoying the same relationship with a given fourth party. But the probability that a human agent can intentionally bring about the realisation of these necessary conditions is so high as to amount to certainty. And the probability that someone who has a strong motive to do this intentionally will do so is high. Given those explanatory principles, we can weakly derive (6) from the combination of (1) - (4) and the relevant explanatory principles. This is not a deductive derivation of (6) from the other set. Rather it is the case that (6) plus (1) to (4) belong within a single rational scheme of explanation of events; whereas (5) plus (1) to (4) does not; not without the supposition of further facts and auxiliary hypotheses.

So far, then, I have tried to do compendiously for narrative coherence what I did for normative coherence in sections II(a) and II(b), namely to show what I understand by narrative coherence and to illustrate very sketchily its application as a test justifying a decision about the *probandum* of a given murder case: that Mr.Smith wilfully killed his 'wife' M. in her bath.

I am left with my third question, why narrative coherence has justificatory force in making decisions about matters of fact. As to this, the idea of rationality again has a role to play. Our intellectual and our practical life is not a mere chaotic succession of Humean impressions and ideas. As both Hume and Kant recognized (though putting the point differently - perhaps less differently than is commonly supposed), we do not actually, nor should we if we esteem rationality a virtue, rest content with a mere succession of ideas and impressions. We construct explanatory principles of at least the various sorts mentioned above. We make our world an intelligible world for us. One of the conditions of intelligibility is the supposition that what we perceive is real. Another is the supposition that whatever is real is rationally related under some explanatory principle to whatever else is real. Therefore, whatever propositions about unperceived events fit into our explanatory schema in rational relationships with true propositions about perceived events are, under the second supposition, true propositions about the reality of the unperceived events. The difficulty, however, is that the suppositions have to be tentative, for three reasons: one is the revisability of our explanatory schemes; the other is the incompleteness of the information derivable from perception - there could be other relevant things that we failed to perceive, that we 'did not notice'. (Nobody, till Holmes came on the scene, had noticed that the dog had not barked); the third is the known delusoriness of some of our perceptions.

This lets us see both why the truth is sometimes stranger than fiction, and how it is that superseded explanatory principles come to be reckoned as 'fictions' or 'superstitions' by those whose world view is shaped by new explanatory principles. Think of our view of the Homeric Gods or the Ptolemaic spheres. We do not suppose there is no possible world or universe explicable in those terms (otherwise we would not understand how godly arbitrariness or Ptolemaic astronomy could count as an explanation at all). But we do suppose that no world or universe so explicable is our real one. It is our awareness that our successors will do for our explanations what we have done for our predecessors', and for the same good reasons as we have done so, that should encourage us to a proper tentativeness about our own explanations. But this tentativeness is not to be equated with scepticism.

Be that as it may. To sum up on narrative coherence: this provides a test as to the truth or probable truth of propositions about unperceived things and events. The test is of the explicability of the tested proposition within the same scheme of explanation as explains propositions considered true on the basis of perception. The relative probability of mutually inconsistent propositions relating to the same unperceived event (e.g. the drowning of a Mrs.Smith) depends on the number of other events which have to be supposed to have occurred to allow of coherence and on the extent to which further auxiliary explanatory hypotheses have to be resorted to achieve coherence. The most coherent story among mutually inconsistent stories is that which involves the lowest improbability by such a test. Few such stories allow of certainty as to the truth of the probandum in question. Such a test justifies beliefs, and thus justifies decisions about matters of past fact because (a) it is a necessary condition of the intelligibility of the phenomenal world; and because (b) rationality requires us to make the phenomenal world intelligible. This probably involves a transcendental presupposition that there is a noumenal world which is so ordered as to be perfectly intelligible, and that the phenomenal world perfectly replicates the intelligible structure of that noumenal world. So the propositions which satisfy truth conditions set within our schemes of explanation could be true about the reality of things. But we could never be sure that they are.

IV. CONCLUSION

My final question is whether normative and narrative coherence have anything in common. If my suggestions in each case are in some degree sound ones, then there is a genuine parallelism between that 'coherence' which justifies normative conclusions (or contributes to their justification) and that 'coherence' which justifies factual conclusions. A rational normative order is one which comprises a mutually consistent set of principles and values such that these principles and values propose an ideal pattern of a satisfactory way of life, and such that all the more detailed practical rules and principles within the system are justifiable under (and thus explicable with reference to), though not deducible from, the highest-order principles and values. A rational world-view is one which comprises a mutually consistent set of explanatory principles such that these principles delineate an intelligible pattern of events in a possible world, and such that they make intelligible (because explicable with reference to the principle) the events which our perceptions disclose to us. We could not have a rational normative order without having also a rational world-view. But we should not need a rational world-view, and should certainly not need ways of *deciding* about the reality of things, if we were not also active subjects desirous of practical rationality in our actions.

FOOTNOTES

(1) On normative coherence, chiefly *Legal Reasoning and Legal Theory* (Oxford, 1978) chapters 7 and 8; on narrative coherence, *id.* chapter 4, pp.89-92, and also 'The Coherence of a Case and the Reasonableness of Doubt', *Liverpool Law Rev.*, 2 (1980) pp.45-50.

(2) Ruggero J.Aldisert, book review of *Legal Reasoning and Legal Theory, Duquesne Law Rev.* 20 (1982), pp.383-398 and, in general, cf. Aldisert J.'s opinion in *Pfeifer*, v. *Jones and Laughlin Steel Corp*. 678 F 2d (1982) 453, esp. at 461.

(3) *Op.cit.*, p.395 'A few years ago Italian officials could not agree upon a speed limit for Italy's superhighway, the *autostrade*. They compromised on regulations that set speed limits according to automobile engine size. Thus a small Fiat was limited to 80 k.p.h., a larger car to 100 k.p.h., and so on. Each car owner was required to post on the rear of his vehicle a decal showing the car's assigned speed-limit. If the desired goal was road safety, the regulations seem absurd. Though internally consistent, they had no coherence. In practice no problems have resulted, however, because neither car owners nor police have paid any attention to the regulations'.

(4) In my paper 'The Limits of Rationality in Legal Reasoning', presented to the 11th World Congress of I.V.R., Helsinki, 1983; to be published in the proceedings of that Congress under the editorship of Aulis Aarnio; also, in a German version in MacCormick and Weinberger, *Die Theorie der institutionalistischen Rechtspositivismus* (Berlin, 1984, forthcoming).

(5) Cf. Aulis Aarnio, *On Legal Reasoning* (Turku, 1977) pp.126-9.

(6) [1968] 2 All E.R. 337; [1969] 1 All E.R. 347 (H.L.).

(7) [1968] 2 All E.R. 337 at 339, per Lord Parker C.J.;

the magistrates' finding of fact quoted above has the same source.

(8) [1969] 1 All E.R. 347 at 352, per Lord Reid.

(9) Id. at 355, per Lord Morris of Berth-y-Gest. On 'consequences as implications', see MacCormick 'On Legal Decisions and their Consequences', N.Y.U. Law Rev. 58(1983) 239-258.

(10) But see id. at 351, per Lord Reid: 'Speaking from a rather long experience of membership of both Houses, I assert with confidence that no Parliament within my recollection would have agreed to make an offence of this kind an absolute offence if the matter had been fully aired before it.' Does the counterfactual at the end support or confute my suggestion in the test?

(11) Id. 359 Lord Wilberforce.

(12) Ibid.

(13) [1968] 2 All E.R. 356 (H.L.).

(14) See [1969] 1 All E.R. p.350.

PARADIGMS OF JUSTIFYING LEGAL DECISIONS

Jerzy Wróblewski

I. INTRODUCTORY OBSERVATIONS

1. In contemporary language the term "legal decision" has many meanings. The core meaning seems to be that a decision is legal if it has some legal consequences and, therefore, is made by an authorized or competent agency.

In contemporary states there are many authorized agencies which are making legal decicions. In spite of the shift from the traditional non-interventionist State towards the contemporary interventionist one the role of judicial decision is looked at as the incarnation of the socially best legal, impartial, objective, just under the law, etc. decision(1). It is so in spite of the fact that the growing bulk of administrative regulation and administrative decisions seems to push the judicial activity from its earlier position. Judicial control of the administration is symptomatic for the place of judicial decision and the values it stands for. In several countries there are alternative ways of solving controversies replacing the judicial process, e.g. the arbitration or pre-trial proceedings. This seems to be a symptom of searching new ways of solving disputes which for some reasons are thought of as better than the court. The relevance of this phenomenon cannot be discussed here. It seems, therefore, that judicial decision stands as the best institution guaranteeing the highest standards of decision in terms of values listed above. This is the reason why we will take judicial decision as the topic of the present essay.

2. The judicial decision is a decision made within a framework of the procedural institutions of a determined type of legal systems operating in a socio-political and cultural context of a given society.

The rules of procedure single out some judicial activities as decision-making. We are not interested, however, in the procedures of concrete systems. From the theore-

A. Peczenik et al. (eds.), Theory of Legal Science, 253–273.

tical point of view a judicial decision is thought of as a choice of alternatives in the process of the application of law. Judicial decision, therefore, is either the "final decision" disposing of a case or a "fractional decision" determining some issues relevant for making this final decision.

As far as the legal framework of judicial decision is concerned, we take the systems of statutory law, as opposed to the systems of common law. This macrocomparativist typology is widely discussed but it is relevant for our purposes. The ways of justifying judicial decision are bound with the expectations of the role of the judge and the institutional bases of his activity, and here is the relevant difference between statutory law and "judge-made-law" ideas.

3. In contemporary culture judicial decision is expected to be a rational decision. In spite of all discussions concerning the concept of rationality it can be assumed, that the rationality of a judicial decision can be expressed in its justifiability(2).

Justification is thought of as giving arguments supporting a judicial decision. The justification of a judicial decision can be qualified as good or bad depending on its features and the criteria of the criticism. The decision is internally rational when it is based on the premisses it accepts. The premisses themselves, however, can be also analyzed and either approved of or rejected, and depending on that the decision is treated as externally rational or not.

In current practice of statutory law systems not all judicial decisions are in fact justified. This is so because of quite obvious pragmatic reasons. The law determines the situations in which the decisions ought to be justified and determines some of their elements ("minimum legal justification of judicial decision"). Judicial practice shapes some accepted ways of justifying the decisions, which can go beyond the requirements of law ("concrete legal justification of judicial decision").

There is the problem of a theoretical and practical paradigm of justification of judicial decision. These paradigms single out the basic elements of justification of judicial decision based on a theoretical analysis or practice of the ways in which the judge can justify his choices. The present essay will analyze this problem.

II. CONCEPTION OF JUSTIFICATION OF JUDICIAL DECISIONS

4. Justification of a statement gives arguments for its support. The arguments in question can be presented at least in three principal manners.

Firstly, one describes the arguments which are given in justification, and the result of this is a presentation of justifications given in some area of reasoning. Generalizing this kind of description one constructs a descriptive model of justification.

Secondly, one formulates the arguments which could be given for supporting some statements. We have to do with a theoretical approach which can eventually result in presenting a theoretical model of justification.

Thirdly, one singles out arguments which should be used to justify a statement according to an assumed pattern of a proper justification. These patterns, if presented in a more or less systematized matter, appear as the normative model of justification in a given area of reasoning.

These general observations cover all areas of reasoning where justification is given and/or expected. The same holds, of course, for judicial decision. One describes the justification eventually searching for a generalization of the "concrete legal justification" in some descriptive model. One analyzes the justification theoretically and one can construct a theoretical model of justification. Last but not least, one analyzes the patterns of due justification either formulated by law ("minimum content of legal justification") or in an ideology of judicial decision-making eventually including some theoretical ideas of rationality.

5. The most basic typology of justification consists in singling out an internal and an external justification. The former is defined as a justification of a conclusion, i.e. of decision, by its premisses according to the accepted rules of reasoning. These premisses in judicial decision cover the knowledge and the evaluations (preferences) of the decisionmaker. The latter deals with the correctness of these premisses(3).

From a methodological point of view there are several

types of justification(4). Justification _sensu largo_ gives reasons for any statement appearing in a discourse. Verification is a justification _sensu largo_ which deals with sentences, i.e. with statements which are either true or false in a determined language. Justification _sensu stricto_ is a justification _sensu largo_ of a statement which is not verifiable in the determined language. The concept of verification is linked with philosophical and logical assumptions. The verification depends on ontology, e.g. materialist, idealist, or materialistically or idealistically oriented culturalist ontology. The dependence on logic accepted in the language in question manifests itself in a verification based on alethic logic or on a non-alethic logic, provided that each logic is a formal calculus interpreted in the determined language.

Justification _sensu stricto_ deals with other argumentative techniques than verification. The most common name for them is a "non-formal logic", although there are several opinions against any use of the term "logic" in this context. In any case this area of reasoning uses arguments linking various statements in practical discourse. Their qualification is not in terms of truth but e.g. in those of good reasons, persuasiveness, etc.

6. Justification of judicial decision is, as a rule, treated not in terms of verification, but of justification _sensu stricto_.

Judicial decision should be justifiable: the decision-maker should be able to present arguments supporting it. These arguments usually refer to chosen directives and/or to evaluations of the decision-maker and, therefore, they hardly can be presented as verification, but in some rather exceptional and artificial situations, fulfil determined extremal conditions. I shall write about justification of judicial decision as a justification without analyzing whether a concrete type of decision is a verification or a justification _sensu stricto_.

III. THEORETICAL PARADIGM OF JUSTIFICATION OF A JUDICIAL DECISION

7. From the theoretical point of view any analysis of the justification of judicial decisions depends on the theory

of judicial decision-making itself.

This theory is based on an analysis of decision-making as an empirical cultural phenomenon. The theory in question depends also on the general theoretical framework in which the decision and its justification are placed. One can, e.g. use a decisional, an informative or a functional model(5).

The theory of judicial decision depends on the view of its justification as a kind of reasoning. And this reasoning is described within various theoretical frameworks. E.g. one treats it as deductive reasoning, as practical reasoning, as rhetorical and/or topical argumentation, as a linguistic game, as a transformation etc.(6)

In the present essay I will deal with the justification of a judicial decision from the meta-theoretical point of view. This means that I shall use a conceptual apparatus which is as little as possible theory-laden, without committing the analysis to any of the controversial explanations of justificatory reasoning in terms of a determined theory. I will try also not to depend too much on some generalized practice of justification, which can stimulate theory and which is basic in dealing with justification from the practical point of view (Cf. part IV).

These meta-theoretical reservations are, of course, only a matter of degree. One cannot deal with judicial decisions and their justification without taking some theoretical position. I hope, however, that the position implied by or presupposed in the present essay is so elementary, that it can be combined with most of the current theories of judicial decision and its justification.

8. The search for a theoretical paradigm of justification of judicial decision does not concern any heuristics of this decision.

We are dealing with a theoretical decisional model of the judicial application of law as a set of fractional legal decisions leading to the final judicial decision. The theoretical assumptions concerning the decision-making are simple:

a) The decision-maker has to make a choice of the alternatives of behaviour he is aware of, viz. he bases his decision on information he uses.

b) The choice of the alternatives of behaviour is made according to the preferences of the decision-maker as criteria of choice. These criteria form some more or less systematized aggregates; if there are some collisions between the preferences in question then some second order criteria are needed to solve them.

c) The use of the criteria of choice demand, as a rule, a knowledge of the decisional situation, evaluations, and more or less complex reasoning: owing to these factors the choices of different decision-makers concerning the same set of alternatives could be different.

d) The decision can be expressed in various ways, and especially in verbal or non-verbal behaviour. We are interested here only in the former case, because this is the case of judicial decision.

e) Depending on the circumstances some decisions are justified by giving determined supporting reasons. The justification, as a rule, is presented when the decision is controversial(7) or when the justification is a duty imposed by a rule in force. The former situation demands a justification before a certain audience or before the decision-maker himself in the case of post-decisional conflict. The second situation is typical for the "minimum legal justification" and will be dealt with below/point 19).

This elementary set of theoretical theses can be accepted as a rule, in contemporary theory of judicial decision-making. The only argument against them could be that some judicial decisions are simply "inferred" from the premisses without leaving any place for any choice. Against this argument one argues that according to theoretical analysis there are no such situations, or they appear as practically quite exceptional extreme cases(8). Moreover, treating decision as a conclusion in terms of logic implies several rather problematic conceptions of the involved logic and appears not as a description but as a fragment of a determined ideology of judicial application of law(9).

9. According to the decisional model of judicial application of law there are the following types of fractional judicial decisions:

a) the decision of validity, which states that the applied legal norm is valid and applicable (or- exceptionally - only applicable(10)) in case;

b) the decision of interpretation, stating that the norm to be applied has a determined meaning;

c) the decision of evidence stating that some facts of

the case are admitted as existing in the determined spatio-temporal dimensions;
d) the decision of the choice of consequences stating that according to the applied norm the proved facts of the case have determined legal consequences.

In the justification of fractional decisions one refers to the "sources of law" which are selected by the decisionmaker. Because of this I single out a source--decision, which is a sort of meta-decision in relation to the fractional decisions enumerated above.

The final judicial decision disposes of the case. This decision is prepared by and/or justified by the fractional decisions.

Each of the decisions in question is expected to be justifiable although the legal regulation of justification is interested mostly in the final judicial decision thought of as "the judicial decision".

10. The decision of validity states that a norm N is valid in the legal system LS in a determined spatio-temporal dimension ST.

The concepts related with the decision of validity are constructed by legal doctrine and used in legal practice.

Theoretically there are three basic concepts of validity, i.e. systemic, factual and axiological validity(11). The theoretical analysis identifies also another typology of the concepts of validity made by linking them with the singled out conceptions of legal system(12). In any case, to state that a norm N is valid one should use determined criteria of validity. These criteria are implied by an accepted concept of validity and can appear as the so--called rules of recognition(13).

Let us use the most common example of the systemic concept of validity related with the simplified legal system consisting only of enacted legal norms. A norm N is valid, i.e. belongs to the legal system LS in ST, if and only if: "(a)N is enacted according to a norm valid in LS; (b)N is not derogated by norms valid in LS; (c) N is consistent with norms valid in LS; (d) if N is inconsistent with norms valid in LS, then either it does not lose its validity according to the accepted rules of the conflict of norms, or N is interpreted in such a way that it ceases to be inconsistent with the norms in question"(14). The singled out conditions of systemic validity put together appear as the conceptual criteria which can be transformed into a correlated rule of recognition.

The theoretical analysis demonstrates that the criteria of the systemic concept of validity refer to evaluations at least when dealing with a use of some of the conflict rules and always when dealing with systemic interpretation for removing the inconsistencies or a choice of a fundamental norm(15). The justification should therefore be relativized to accepted values.

Analogous observations could be made concerning the axiological concept of validity mentioned above, but this is beyond the scope of the present essay. Using this one example I would propose the normal formula for justification of a decision of validity:

> The norm N is valid in legal system LS in spatio-temporal dimensions ST according to the rules of recognition RR_1, RR_2,.....RR_n and evaluations V^r_1, V^r_2,....V^r_n.

If some rules of recognition are not linked with evaluations, then the last part of the formula would be void.

11. Operative interpretation occurs when the law-applying agency has some doubts concerning the *prima facie* meaning of the norm he tends to apply in solving the case(16). The decision-maker has to determine the meaning of the norm in question in a manner sufficiently precise to use it in the decision. There are many sources of doubt, the most typical is the fuzziness of the legal language in which the norm is formulated, but there are also the doubts stimulated by the inconsistency of the interpreted norm with other norms of the system and/or the evaluation of the results of applying the norm in its *prima facie* meaning in terms of their instrumentality, justice, equity etc.(17). The interpretive doubt is stated purely pragmatically in the proces of decision-making and is expressed as the lack of *claritas*: whether a norm is clear or not depends, as a rule, on the evaluations of the decision-maker.

Interpretative decision is justified by reference to interpretative directives and evaluations. The directives in question are sometimes rather loosely formulated rules often referring to some evaluations, and they appear as argumentative or topical devices indicating how the interpretator should fix the meaning of the applied norm. There are many conflicting interpretative directives gathered together in sets called normative theories and/or ideologies of legal interpretation. The choice of interpretative directives is based on evaluations, if it is not indicated by the law itself. The same holds for normative theories of interpretation.

There are two levels of interpretative directives. Directives of the first level determine how one ought to ascribe a meaning according to the relevant context of the interpreted texts, i.e. the linguistic, systemic and functional context. The second level directives determine the way in which the first level directives are used (procedural directives) and how to solve the conflicts between the results of applying the first level directives (preferential directives).

The choice of the interpretative directives and, in many cases, their use depends on evaluations. This is typical of their argumentative and topical character, and, therefore, there are rather wide interpretative lee-ways. One can say that operative interpretation is, as a rule, an evaluative process.

Notwithstanding the varieties of interpretative heuristics(18) there is a normal formula of the justification of an interpretative decision:

> The norm N has the meaning M in the legal language LL and/or in the situation S according to the first level interpretative directives DI^1_1, $DI^1_2, \ldots DI^1_n$, and the second level directives DI^2_1, $DI^2_2, \ldots DI^2_n$, and the evaluations V^1_1, $V^1_2, \ldots V^1_n$; V^2_1, V^2_2, $\ldots V^2_n$.

12. A decision of evidence consists in stating that the facts of the case F exists in the spatio-temporal dimensions ST, on the strength of the material of presented evidence E which is analyzed, approved of or rejected(19).

There is, as a rule, a legal regulation of the issues relevant for the decision of evidence. The law in force determines the following issues: (a) the facts to be proved or which do not require any proof; (b) the admissibility of proofs; (c) the manner of accepting and/or rejecting the proofs; (d) the ways of controlling the decisions of evidence if any(20).

The legal regulation of proofs is important for the theoretical analysis because it shows the necessity of a reference to legal norms in the justification of decision.

The legal regulation of proofs is expecially important for determination of the rules of evidence. The comparative research supports the view that there are two types of normative regulation of the reasonings operative in the decision of evidence. Firstly, there is the so-called free evaluation of proofs in which the rules of evidence are the rules of the extra-legal sciences and of knowledge based on common experience. The rules in question are the empirical rules of evidence ERE. Secondly, there are legal rules of evidence LRE which determine what facts have to be treated as proved if other facts occur. The LRE are typical for the so-called formal theory of proofs. Contemporary statutory law regulation of proofs combines both theories, but LRE are exceptions in the prevailing domain of the ERE rules. The decision of evidence is, thus, justified by reference to ERE en LRE rules.

Further, the legal regulation is relevant because of the way of determining the facts of the case. The theoretical analysis demonstrates(21) that there are two opposite types of the determination in question: on the one hand the facts are determined descriptively, positively and simply (e.g. "a birth") and there are also, on the other hand, facts determined evaluatively, negatively and relationally (e.g. "faulty omission of a statutory duty"). For some determinations of a fact a reference to evaluation is needed, and, hence, this should be expressed in a justification of the decision of evidence.

The decision of evidence is based on the material of evidence. The decision-maker, according to the rules of evidence and his own conviction, accepts some of them and rejects the others. Even within the theory of "free evalution of proofs" which is aimed at determining the material truth(22), the decision of evidence should be internally consistent, i.e. it cannot contradict the accepted evidence and rules evidence. The normal formula of the decision of evidence is:

> The facts of the case F exists in spatio-temporal dimensions SF according to the evidence E_1, $E_2 \ldots E_n$ and the empirical rules of evidence ERE_1, $ERE_2 \ldots ERE_n$ and/or legal rules of evidence LRE_1, $LRE_2 \ldots LRE_n$, and/or evaluations $V^c{}_1$, $V^c{}_2, \ldots V^c{}_n$.

13. The decision of the choice of consequences consists in determining the legal consequences of the facts of the case according to the applied legal norm(23).

The range of the choice in question depends on the formulation of the applied norm. The extreme cases are that of no choice because the norm determines the only one consequence, and that of a complete freedom of the decision-maker. The former is exceptional, the latter is doubtful because the alleged freedom is moderated by the general principles and axiology of the legal system the applied norm belongs to. The usual situation is that of a decisional lee-way within some limits and, moreover, there are some directives of the choice of consequences formulated in law and/or by legal practice and/or legal doctrine.

The directives of the choice of consequences either determine the choice or only give some guidance enumerating the factors which have to be taken into account and/or aims which the decision-maker should follow. In any case the decision of the choice of the consequences has to be justified by the directives of choice DC. The directives in question are, as a rule, linked with the relevant characteristics of the facts of the case and depend on their evaluation. Evaluations are needed also for the choice of the directives of the choice of consequences and for their use.

The normal formula of the decision of the choice of consequences is:

> The facts of the case F with the characteristics C_1, C_2,...C_n have the legal consequences LC_1, LC_2,....LC_n according to the applied norm N and the directives of the choice of consequences DC_1, DC_2,....DC_n and/or evaluations $V^e{}_1$, $V^e{}_2$,....$V^e{}_n$.

14. In all justifications of the fractional decisions described above there are some references to the norms or directives which have to be accepted by the decision--maker. The rules referred to are treated as the arguments of justification. In a rather loose legal terminology they are referred to often as "sources" of law and/or of a decision. There is a variety of sources and the decision--maker should decide what sources he uses. We can treat the decision concerning the sources (the source-decision) as a general meta-decision in respect to the remaining fractional decisions.

Among the sources in the statutory law systems one singles out three categories of sources: the must-sources, the should-sources and the may-sources of law(24). The tripartition of these sources is based on some general features of the law and of the ideology of decision--making.

Each category of sources is complex and even between the must-sources there could be a choice if they are conflicting or indeterminate. A fortiori this holds for the two remaining categories of sources. If there is a choice then some evaluations are operative in making it. The formula of the justified source-decision is:

> The rules are used in arguments in justification according to the must-sources MS_1, $MS_2, \ldots MS_n$, and/or should-sources SS_1-$SS_2, \ldots SS_n$ and/or may-sources CS_1, $CS_2, \ldots CS_n$ and/or evaluations $V^s{}_1$, $V^s{}_2, \ldots V^s{}_n$.

Typical for justification of statutory law decisions is, that they must refer to statutory rules, i.e. statutes are the must-sources of decision. It depends, however, on legal regulation, whether only statutes are must-sources in this system, and what is the place of other enacted rules, if any. The source-decisions are needed as supplements to all of the fractional decisions.

15. The final judicial decision disposes of the case. It appears, therefore, as the combination of all the fractional decisions presented above.

There are many possibilities of presenting the justified final decision. I will use the simplest combination of the justified decisions taking the decision of the choice of consequences as the starting point.

The standard formula of the justified final decision is:

The facts of the case F having the characteristics C_1, $C_2, \ldots C_n$
which exist in spatio-temporal dimensions ST
according to the evidence E_1, $E_2 \ldots E_n$ and the empirical rules of evidence ERE_1, $ERE_2, \ldots ERE_n$ and/or the legal
rules of evidence $LRE_1, \ldots LRE_2, \ldots LRE_n$ and/or
evaluations $V^e{}_1$, $V^e{}_2, \ldots V^e{}_n$.
have the legal consequences LC_1-$LC_2 \ldots LC_n$ according to the applied legal norm N
valid in the legal system LS in spaciotemporal dimensions ST
according to the rules of recognition RR_1, $RR_2, \ldots RR_n$ and/or evaluations $V^r{}_1$, $V^r{}_2, \ldots V^r{}_n$
and having the meaning M in legal language LL and/or the situations
according to the interpretative directives $DI^1{}_1$,

$DI^1{}_2, \ldots DI^1{}_n$; $DI^2{}_1$, $DI^2{}_2, \ldots DI^2{}_n$ and evaluations $V^i{}_1$, $V^i{}_2, \ldots V^i{}_n$
and the directives of the choice of consequences DC_1, $DC_2, \ldots DC_n$ and/or evaluations $V^c{}_1$, $V^c{}_2, \ldots V^c{}_n$
taking into account the must-sources MS_1, $MS_2, \ldots MS_3$, and/or the sould-sources SS_1, $SS_2 \ldots SS_n$ and/or the may-sources CS_1, $CS_2, \ldots CS_3$ and/or evaluations $V^s{}_1$, $V^s{}_2, \ldots V^s{}_n$.

16. The normal formula of the justification of the final decision is based on the theoretical analysis of the justified fractional decisions. All the elements of the justification are relevant for supporting the decision, although one can present them in a more detailed manner than used in the present essay.

The crucial question of the present essay is, whether this theoretical analysis enables us to construct the theoretical paradigm of justification.

The reply depends on the conditions for using the term "paradigm" in this context. By paradigm we will understand here the essential features of justification, i.e. the features without which the decision could not be treated as a justified decision in the legal culture we are interested in(25). Then the normal formula of the justified final judicial decision can be treated as the basis of such a theoretical paradigm.

The paradigm in question is, however, more general or abstract than the normal formula presented above. It is, thus, less technically formalized, and is centered on the types of arguments which should be given and not on their theoretical analysis, which can be changed without changing the paradigm itself. This is the price of constructing a paradigm, which is characteristic of the legal culture and not of particular legal theories.

The theoretical paradigm consists of the following essential conditions of justification: (1) reference to the applied rules which are taken from the accepted sources, and are valid in the determined meaning; (2) reference to the acknowledged facts of the case; (3) use of the approved of rules of decisional reasoning linking sentences, evaluative statements, and decisions as their conclusions.

From the theoretical point of view each element of this paradigm is a necessary condition of internal justification of the final judicial decision in the historically evolved present legal culture in statutory law systems. This means, that any change of at least one of these elements amounts to a change in the basic theoretical insights of judicial decision-making and/or the accepted notion of justification.

The paradigm of internal justification is a formal paradigm in the sense that it refers to the necessary elements of justification without identifying their content. The identification in question depends on the theoretical constructs which are problematic and could be related to the legal regulation and/or to accepted practices of decision-making.

I leave open the problem of the paradigm of external justification of judicial decision. There are two ways of constructing it. The first way is a simple operation of transforming the internal justification paradigm by stating, that the decision in question is externally justified, if the singled out elements of the paradigm and/or the decision itself are justified according to the opinion of the critic or of a controlling agency. The second way describes the same situation by stating, that the elements in question and/or the decision itself are approved of by the relevant audience. The first way is a "justificatory way" leading with the chain of justifications to some ultimate premisses it is based on. The second way is an "explanatory way" explaining the approval of the premisses of internal justification by the fact or the hypothesis of a consensus(26).

IV. PRACTICAL PARADIGM OF JUSTIFICATION OF JUDICIAL DECISION

17. The practical paradigm of justification of judicial decision, if any, is based on the practice of justification accepted in the statutory law systems within our legal culture.

There are two types of justification of judicial decisions relevant for our argument. Firstly, the law regulates the content of justification when its formulation is imposed by law; this is the "minimal legal justification of judicial decision". Secondly, there is a practice of justifying decisions which determines what arguments are used in fact; this is "concrete legal justification of judicial decision"(27).

I will formulate my observations as hypotheses without referring to empirical comparative material of legal norms and decisions.

18. The minimum legal justification of judicial decision is determined by legal norms dealing with the procedural requirements of justification(28). Even the most cursory analysis of the regulation in question shows that the necessary elements of justification are the so-called "normative basis of decision" and the "factual basis of decision".

The normative basis of decision in minimum legal justification consists in reference to valid statutory norms which are applied in the decision.

The factual basis of decision consists of the statements concerning the facts of the case which are admitted as proved. Sometimes the presentation of reasons for accepting and rejecting evidence and the selection of facts influencing the choice of legal consequences determined in the decision, are required.

Taking this into account one can formulate the practical paradigm of the minimum legal justification of judicial decision in the following way: A judicial decision is justified if it is supported by the normative and the factual basis of this decision.

The paradigm is a very general one. It seems, that it cannot be more concrete because of the assumed features of the paradigm accepted here (point 16). It is patent that a change of this paradigm is not possible without changing our culturally bound ideas of justification of judicial decision which are expressed in legal regulation. This regulation seems to be correlated with the theoretical paradigm of justification, but is far less elaborated.

The paradigm reflects some relevant features of the concept of a proper application of law: the application of law is thought of as justified by valid legal norms applied to the proved facts. In other words, the decision ought to be in some manner bound by the law and be based on judicially determined truth.

19. The "concrete legal justification of judicial decision" is made in the practice of decision-making(29).

The way of justifying the decision depends, of course, on the "minimum legal justification of decision". But concrete justification usually goes beyond these minimum requirements. It includes elements which are legally not necessary but are treated as arguments relevant for improving justification. The variaties of these elements are expressed in the style of justification of decision.

The style of justification reflects, thus, not only the requirements of legal norms, but also the theoretical preçonceptions of the decision-maker anchored in legal cultural tradition. The styles of justification of judicial decisions are analyzed in comparative legal research and in the history of judicial decision-making of a country(30). The styles are differentiated e.g. in respect to the variety of used sources, the manner of using interpretative materials, the occurrence of references to the precedent decisions or to the scientific authorities, the extent of referring to the factual data concerning the functional context of decision-making, the expliciteness of evaluative arguments(31), etc. The styles of decisions epress the surface structure of concrete justification of judicial decisions. It seems that there are no reasons for searching a paradigm of this structure because it is the variety of styles that is important here for concrete justification.

20. The deep structure of concrete legal justification of judicial decisions opens, however, interesting vistas. It is my hypothesis, that in the deep structure of concrete justification there is the problem of formulating a rule of decision that will exactly fit the facts of the case. In other words, the deep structure consists in transforming the normative basis of decision into the rule of decision applying it to the factual basis treated as a settled datum.

The rule of decision(32) can be constructed of: (a) legal norms understood in their *prima facie* meaning; (b) legal norms interpreted in operative interpretation for the concrete case; (c) legal norms interpreted independently of the decided case; (d) principles of law; (e) extra-legal rules or evaluations. I have discussed the construction of the rule of decision elsewhere and I cannot deal with it here in detail. The central idea of this rule is that the normative basis should be transformed so that it would fit the facts of the case in such a way that the final judicial decision follows the rule of decision.

It seems that this search for a rule of decision can be linked with the theoretical paradigm of justification: the rule of decision would then appear as the combination of the decisions of validity, of interpretation and, in part, of the choice of consequences in the part containing the rules of choice of consequences.

If these observations of the deep structure of concrete justification are accepted, then the paradigm of this structure would be the following:

The facts of the case have the consequences C_1, $C_2 \ldots C_n$ according to the rule of decision RD.

The final judicial decision is justified by the decision of evidence and the constructed rule of decision governing the case.

V. CONCLUDING REMARKS

21. The paradigm of justification of legal decision has been discussed in reference to a justified judicial decision, because this type of legal decision is expecially important in our legal culture. Its importance is based on the opinion, that judicial decision implements in the highest socially possible degree such values as legality, objectivity, certainty, justice under the law etc.

The paradigm of justification of judicial decision is conceived as the set of conditions essential for its justification. The paradigm is bound with the determined legal culture which defines the notion of rationality expressed in proper justification.

In the present essay the theoretical paradigm of the justification of judicial decision has been constructed as the result of an analysis of justified fractional decisions.

The minimum legal justification of judicial decision is determined by law. In the contemporary statutory law systems the regulation expresses the basic expectations of our legal culture, and this opens the way to the practical paradigm of justification formulated above.

The surface structure of the concrete legal justification of judicial decision appears as styles of decision, and there are no reasons for constructing their paradigm. It is, however possible to state a paradigm of its deep structure. It consists in formulating a rule of decision which fits the facts of the case.

The change of a paradigm means by definition a change in legal culture, expressed in a new theory of judicial decision, theory of justification and the concept of rationality inherent in the expectation of justifiability of the decision. The change of the paradigm of justifying judicial decision is, thus, linked with profound changes in the legal culture as a whole. I have attempted to outline the paradigms inherent in the legal culture of our times, which are inherited from the deep rooted traditions of legal justificatory reasoning.

FOOTNOTES

(1) Cf. J.Wróblewski, Wartości a decyzja sądowa/Values and Judicial Decision/, Ossolineum 1973, chapt.III-IX.

(2) Cf. in general Rationality To-Day. La rationalité aujourdhui ed. Th.F.Geraets, Ottawa 1979. The relation of rationality and justifiability is, however, controversial: cf. A.Peczenik, The Basis of Legal Justification, Lund 1983, chapt.4; A.Aarnio, R.Alexy, A.Peczenik, "The Foundation of Legal Reasoning", Rechtstheorie 12, 1981, part.II.

(3) J.Wróblewski, Meaning and Truth in Judicial Decision, Helsinki, 1983[2], p.49-70. The internal justification is, in A.Peczenik's terminology, a "contextually sufficient legal justification" whereas the external justification is either a contextually sufficient or a "deep justification" depending on the way of justifying the premisses. Cf. A.Peczenik, The Basis, p.1.

(4) Cf.J.Wróblewski, "Verification and Justification in the Legal Sciences", Rechtstheorie, Beiheft 1, 1979.

(5) Cf. J.Wróblewski, Sądowe stosowanie prawa(Judicial Application of Law), Warszawa 1972, chapt. III.

(6) E.g. N.MacCormick, Legal Reasoning and LegalTheory, Oxford 1978, chapt. II, III; R.Alexy, Theorie der juristischen Argumentation, Frankfurt am Main 1978, part. C; Ch.Perelman, Logique juridique, Paris 1979[2], part II; G.Struck, Topische Jurisprudenz, Frankfurt am Main 1971; A.Aarnio, On Legal Reasoning, Turku 1977, chapt. II/2/; A.Peczenik, The Basis, chapt.2-4; A.Aarnio, R.Alexy, A.Peczenik, op.cit.;Metatheorie juristischer Argumentation, ed. W.Krawietz and

R.Alexy, Berlin 1983.

(7) E.G. Ch.Perelman, _Le champ de l'argumentation,_ Bruxelles 1970, p.203; Idem, "Justice and Justification", in: _Natural Law Forum_ 10, 1965, p.16.

(8) For the criticism of the l'école d'exegese cf. Ch.Perelman, _Logique juridique_, Paris 1979[2], part I, chapt.I; for the criticism of theoretical presuppositions of the positivistic ideology of the bound judicial decision cf. J.Wróblewski, Sądowe....chapt.XII/2/ and lit.cit. Especially the positivist thinking rejects any type of consequentialist argumentation, which appears as a kind of "second order justification" cf. N.MacCormick, op.cit.chapt.V; Idem, _On Legal Decisions and Their Consequences: From Dewey to Dworkin_, New York University Review 2, 1983.

(9) Cf.J.Wróblewski, "Legal Syllogism and Rationality of Judicial Decision-Making", _Rechtstheorie_ 1, 1974; Idem, Sądowe... chapt. X/8/; M.Taruffo, _La motivazione della sentenza civile_, Padova 1975, chapt. IV/1/.

(10) Cf. E.Bulygin, "Time and Validity", _Deontic Logic, Computational Linguistics and Legal Information Systems,_ vol.II, ed. A.A.Martino, North-Holland 1982.

(11) Cf.J.Wróblewski, "Three Concepts of Validity of Law", _Tidskrift, utgiven av Juridiska Föreningen i Finland_ 5-6, 1982 and lit.cit.; Idem, Sądowe...chapt.X/2/.

(12) J.Wróblewski, "Fuzziness of Legal System", _Essays in Legal Theory in Honor of Kaarle Makkonen_, XVI Oikeustiede Jurisprudentia 1983, Vammala 1983, p.319-322.

(13) Cf. H.L.A.Hart, _The Concept of Law,_ Oxford 1961, chapt.V/3/VI/1.

(14) J.Wróblewski, Three concepts...p.410.

(15) J.Wróblewski, "Operative Models and Legal Systems", _Artifical Intelligence and Legal Information Systems,_ vol.I. ed. C.Ciampi, Amsterdam 1982, p.220-229.

(16) For the features of the operative legal interpretation accepted in the present essay cf. J.Wróblewski, Meaning... p.1-48, 71-103; Idem, "L'intérpretation en droit; Théorie et idéologie", _Archives de philosophie du droit_ XVII, 1972; Idem, Zagadnienia teorii wykladni prawa ludowego/Problems of the Theory of Interpretation of the Socialist Law/Warszawa 1959, passim; Idem, Sądowe...chapt.VII.

(17) Cf. J.Wróblewski, Fuzziness...p.315-319, 322-326. Idem, "Legal Language and Legal Interpretation"/paper presented during the symposium "Normative Reasoning and Legal Interpretation" Turku August 1983.

(18) A very good example of heuristic of legal interpretation is in J.C.Cueto-Rua, Judifical Methods of Interpretation of the Law, Louisiana 1981.

(19) Cf.J.Wróblewski, Sądowe...chapt.VIII; chapt.X/4-5/; In the present account I am not discussing the problems of the relevance of evidence, cf. M.Taruffo, Studi sulla rilevanza della prova, Padova 1970.

(20) Cf.J.Wróblewski, "La preuve juridique; axiologie, logique et argumentation", La preuve en droit, ed. Ch.Perelman and B.Foriers, Bruxelles 1981, p.338-354. M.Taruffo, La motivazione...chapt.6; G.A.Micheli and M.Taruffo, "L'administration de la preuve en droit judiciaire", Towards Justice with a Human Face, Ghent 1977.

(21) J.Wróblewski, Facts in Law, ARSP 2. 1973.

(22) J.Wróblewski, Meaning...p.180-198.

(23) Cf.J.Wróblewski, Sądowe...chapt. IX/1,2/, X/6,7/.

(24) A.Peczenik, The basis...chapt.2.2.

(25) About the paradigm in legal science and especially of legal dogmatics cf. E.Zuleta Puceiro, "Paradigm Dogmatico y ciencia del derecho", Madrid 1981; idem, "Scientific Paradigms and Legal Change", /Essays in Legal Theory...; A.Aarnio, Philosophical Perspectives in Jurisprudence, Helsinki 1983, chapt. 10.

(26) E.f.A.Aarnio, R.Alexy, A.Peczenik, op.cit., p.437-444; J.Wróblewski, "Towards Foundations of Juridical Reasoning" Metatheorie..., p.247-252.

(27) J.Wróblewski, "Motivation de la décision judiciaire" La motivation des décisions de justice, ed. Ch.Perelman and P.Foriers, Bruxelles 1978, p.122-130.

(28) Cf.note 20.

(29) Cf.M.Taruffo, "La motivazione delle decisioni dei giudici"(in):Lo statuto dei lavoratori: prassi sindacali e motivazioni dei giudici ed. T.Treu, Bologna, 1977; J.Wróblewski, Meaning...p.157-179; Idem, La motivation...p.129 sq.; cf.note 30.

(30) Cf.J.Gillis-Wetter, The Styles of Appellate Judicial Opinions, Leyden 1960; G.Gorla, Lo stile delle sentenze. Ricerca storico-comparativa e Testi commentati, Roma 1968, 2 vol., Minin, Le style des jugements, Paris 1962[4]; K.N.Llewellyn, The Common Law

Tradition. Deciding Appeals, Boston-Toronto 1960.

(31) To an ideology of justification belongs the idea what arguments should be used for a "proper justification", e.g. the role of making evaluations explicit. Cf. e.g. the proposal of "Two-Level Justification Procedure", or in R.Wasserstrom, The Judicial Decision, (Stanford-London 1961, chapt.I) the proposal of "Two-Level Justification Procedure", or "second-order justification" (N.Mac.Cormick, Legal Reasoning..., chapt.V). The evaluation of consequences of possible decisions can appear, of course, as an evaluation singled out in each of the formulas of justifications of decisions given in the text of the present essay.

(32) J.Wróblewski, "La règle de décision dans l'application judiciaire du droit', La règle de droit, ed. Ch.Perelman, Bruxelles 1971.

MONISM, PLURALISM, RELATIVISM, AND RIGHT ANSWERS IN THE LAW

Sheldon L.Leader

When lawyers disagree over the right answer to a question of interpretation of the body of cases and statutes in a particular domain, what exactly is their disagreement about? Are the rival propositions they advance to be evaluated as true or false? If so, in what sense? If not then how is the evaluation to proceed?

This is one of those questions that busy practitioners are content to leave to idle, if edifying, speculation. But it can come back to haunt them when they find that their submission on a point has been dismissed by a judge with the words: "the proposition which counsel is advancing is most interesting, but unfortunately it is not the law." Such statements provoke the uneasy feeling that an issue has been fudged. Can the judge honestly say that he is fending off the 'interesting proposition' because of something he can claim to believe to be true about the body of the law that the defeated submission conveys less well? Given the both sides in the case knew all of the relevant statutes and cases, and given that neither of the rival submissions actually ignored any of them, then what other bit of truth was there left to disagree about?

1. THE TERMS OF DEBATE

There are several imaginable positions on this issue. First, there are those who can be called objectivists, holding that rival propositions interpreting a body of law are properly assessed as true or false according to criteria which those putting forward the propositions would in principle accept as valid and ultimately decisive of their disagreement. Within this group there are monists, who believe that this truth exists in the form of a single right answer to questions of law, even though that truth might not be discoverable and pluralists, who believe that

A. Peczenik et al. (eds.), Theory of Legal Science, 275–294.

the truth has the form of alternative answers, each reflecting equally well the subject matter. Monists and pluralists can be identified as objectivists because of their willingness to accept the same kinds of constraints on propositions if they would be true. For example, both would accept the possibility of appeals to a neutral terrain of institutional and personal history in searching for reflective equilibrium between moral theory and the totality of settled judgements. Similarly, both would accept, as do Professors Dworkin and MacCormick in their different ways, that normative coherence is a quality that correct answers in the law display(1).

Within monism, there is one imaginable species which can be called categorical, holding that there is always a single right answer to questions of interpretation, and another which could be called contingent, holding that there is very often a single right answer, depending on such background variables as the complexity of the legal system within which the question of law arises. Two species of pluralism can in turn be matched to these two species of monism: one is concerned to refute the categorical monist, holding that there will not always be a single right answer, or more extremely that there never will be, while the other contingent species focuses on the probabilistic claim in contingent monism, and argues that there is often no single right answer to questions of law, however complex the legal system might be.

Clearly, one cannot refute contingent monism with the argument that <u>sometimes</u> there are no single right answers to questions of law, since this is indeed what the contingent monist argues himself. That argument need trouble the categorical monist only. Contingent monism must instead be met with the claim that there is no single right anwer more often than this species of monism allows, and it is this claim that I associate with contingent pluralism.

Both monists and pluralists, as objectivists, could in turn be opposed to the relativist, who would hold that it is wrongheaded of monists of either variety to believe in a single right answer, and of pluralists to narrow the field to a group of right answers, since within this domain of questions someone who advances one proposition can never hope rationally to convince someone who advances a rival. The foundation for this relativism is sometimes an epistemological position drawn, inter alia, from the evidence provided by natural science of the relation between theories and facts. It can lead to scepticism about the possibility of one or several best fits between propositions of morality and law to the relevant history of a community, claiming that there is no neutrally describable terrain of such history to which appeal can be made.

I wish to argue for contingent pluralism. I will do so by way of a critique of the strongest case for contingent monism that we have been given so far: the theory of Professor Dworkin. The argument will not try to amend Dworkin's model of the ideal judicial decision, but will instead probe into its details. I shall claim that once these are properly fixed they display features which give greater support for contingent pluralism than for Dworkin's own brand of monism.

I shall also try to strengthen the attractions of pluralism by attacking the other end of the spectrum of positions: those inspired by epistemological relativism. Such a position often trades on some basic points about theories and observations which are deployed in a weak argument in the philosophy of science. There are lessons in this for the philosophy of law. Finally, I shall briefly indicate that support for pluralism is often wrongly taken to imply endorsement of positivistm

2. PROFESSOR DWORKIN'S MODEL: THE FIRST AND SECOND VERSIONS

Prof. Dworkin argues that in complex legal systems, such as the Anglo-American, there will usually be a single right answer to a question of law, even if there are no means of actually discovering it, and the occasions on which there are not will be very rare(2). He grounds this argument on the features of his model of adequate justification in judicial decisions. He writes, "One justification may be better than another in two different dimensions: it may prove a better fit, in the sense that it requires less of (the body of settled law) to be taken as mistakes, or it may prove a morally more compelling justification because it comes closer to capturing sound political morality"(3). The second dimension contains "the substantive ideals of political morality on which (the judge) relies in deciding whether any putative interpretation is to be preferred because it shows legal practice to be better as a matter of substantive justice"(4). In the theory, these two dimensions function in a broadly complementary way. The first dimension, the dimension of fit, has a prior claim to satisfaction, but above a certain threshold of fit the judge is entitled to look to the second dimension, the dimension of sound political morality. Above that threshold, that is, he is entitled to trade off a more comprehensive fit with institutional history against a proposition that is based on a more adequate moral and political theory(5).

How is this model linked to contingent monism? As it is elaborated and analogies are drawn with interpretation in literature, it looks as if it is grounded on what can be called the determinate features of institutional history. That is, it is claimed that when such history is sufficiently rich "the intersections and interdependencies of different legal doctrines will be so intense..."(6) that they will usually determine a single right answer as the goal of an adequate decision. "In a modern, developed, and complex system, the antecedent likelihood (that two rival propositions will display equally good fits) will be so rare as to be exotic"(7). On those rare occasions on which such history is indeterminate, then the second dimension can be invoked. "The second dimension - the dimension of political morality - supposes that, if two justifications provide an equally good fit with the legal materials, one nevertheless provides a better justification than the other if it is superior as a matter of political or moral theory; if, that is, it comes closer to capturing the rights that people in fact have. The availability of this second dimension makes it even less likely that any particular case will have no right answer"(8). In short, institutional history will usually be determinate if it is sufficiently dense, and when it is not then appeal can be made to the substantive merits of moral and political theory to break any ties within the first dimension.

It is important to be clear about the 'dimension of fit' in this argument. It does not stand apart from the requirement that a proposition of political morality be sound despite the wording of the foregoing quotations, but is instead _a_ way of assessing its soundness. It is one of Dworkin's central contentions that all propositions of law have embedded in them some theory of political morality, either trivial or controversial(9). The institutional morality of adjudication demands that a proposition of law be chosen for the fit it displays, as well as on its other merits.

The distinctive way in which Dworkin navigates between naturalism and conventionalism in legal and political theory can be appreciated only if we keep in mind the importance of this first constraint. In seeking to satisfy the dimension of fit, the judge reveals how the unique history of the community within which he works may ground a moral and hence a legal claim of right within it, which may not be available in another. It is a history making its own demands of consistency, forcing a judge to resist his inclination to follow his own preferred moral position on a contested issue. This insistence on adequate fit with a particular community history throws into relief the failure of some relativists to account for the weight of a judge's proper institutional scruple, and also shows us how to avoid the excesses of some naturalists, who would wish to base all appropriate answers to hard cases on the fit with moral facts transcending the contingent features of any particular society.

It is this delicate balance that Dworkin is trying to convey in the analogies he draws with interpretation in literature. In the earlier versions we were asked to consider exercises which pose questions about characters or events in a story for which answers were not explicitly provided in the text but which could be arrived at inferentially, by seeing which of several possible answers was most consistent with the body of information actually provided. The narrative consistency which such propositions display is the measure of their truth, Dworkin argues. There is nothing outside of the story to which they must correspond in order to be accounted true. On the other hand, the strength or weakness of the narrative consistency that a proposition exhibits is an objective quality that it possesses, which consultation of the information in the text will reveal(4a). In his most recent account the parallel with interpretation in literature (and other areas of art) is filled out, so as both to advance a distinctive view of interpretation in aesthetics and to draw out the parallels with law more completely. He asks us to imagine a work done by a series of individuals, one passing on the partially finished product to the next. Each individual in the series, like each judge in a chain of decisions forming a body of law, interprets the work in the very process of developing it.

Dworkin argues that there are two features of this interpretive activity, when carried out at its best. First, interpretation will obey the constraint of holding on to the identity of the work. It will respect the line between "explaining a work of art and turning it into a different one"(10). Secondly, adequate interpretation will reveal the way in which the work strives to be the best work it can be. The concrete form it takes manifests much about the virtues towards which it implicitly strives. It reveals its integrity as well as its identity. There are many distinct species of art work, each with its distinct qualities, into which any given piece <u>might</u> be fitted. But the need to preserve the work as a coherent whole as far as possible - to minimize the number of 'mistakes' in it - will rule most out. Thus, to interpret an Agatha Christie novel as a treatise on the meaning of death and not a mystery story renders a large part of the story irrelevant, and so is unacceptable. This is so even though we might say on independent grounds that such a philosophical treatises were more valuable art works than were mystery stories(11). Again, as with the two dimensional model of interpretation in law, we must interpret such works in the light of the soundest aesthetic theory we can find on substantive grounds, but it is the dimension of fit which carries the burden of guiding the selection of the most appropriate theory when more than one is available, and the denser the story is the more likely will it be that there is one such best theory.

So much for what we can call the first version of the theory. The closer we look at some of its details, however, it seems that a second is actually being offered. Its argument for predictive monism is implicit rather than explicit, and is differently focused. According to it, institutional history is an important constraint in the search for right answers, but it will <u>rarely</u> provide single ones. Instead, any prospect for the elimination of all remaining alternatives will usually be sought within moral and political theory itself, and now chosen squarely within the second dimension. The first dimension continues to be important in the effort to find a compromise between naturalism and conventionalism, but is no longer central to the monism.

Why is this? The answer lies in the phenomenon of

'thresholds' of fit. Clearly, the assessment of rival propositions of law is not quite the mechanical matter that some conceptions of the dimensions of fit might invite us to see it as. As Dworkin has repeatedly pointed out, one does not favor one proposition over its rival simply because it is consistent with a few more elements of institutional history than is another. Indeed, it is difficult to know how one could individuate parts of history sufficiently closely so as to know when something constituted one or two segments in the first place. Judicial practice and normal moral argument reflect this. They require no more than an implicit threshold beyond which it simply does not matter that one might show that one proposition justifies more of institutional history than does another(12).

In any given disputed interpretation of the law more than one proposition meant to resolve it will often be beyond such a threshold. Legal cases show this. One frequently finds that when rival propositions are advanced, particularly at appellate levels, more than one fits satisfactorily with the body of precedent. They have all passed the notional threshold. Notice that this would be true even if we were to cast our net as widely as does Dworkin's model judge, Hercules. That is, it would be true even if the judge were to look at all the areas of collateral law on an issue, as well as to consider and resolve any background institutional disagreements about e.g. the judicial function which may be affecting the present case, such as a disagreement over the weight to give to older as opposed to recent precedent, or over the proper way in which to construe the intention of the legislature. It is true that while several rival propositions may be above the threshold of fit with 'direct' precedents, a judge might find that one is far more widely supported in collateral areas then is another(13). It is also true that the proper resolution of background institutional disagreements which are embedded in rival propositions in a particular case may be evident because one position is so much more frequently supported in the history of the system.

This might incline Prof. Dworkin to say that a full satisfaction of the dimension of fit will usually yield single right answers in complex systems. That claim would be plausible, however, only if we jettisoned thresholds, and imagined that the ideal judge managed to produce a best fit with institutional history such that his answer yielded, precisely, the fewest mistakes in that history. We might never be able to reach this level of perfection, but we also know that the model functions as an ideal-type. However, if we construed the ideal-type in this way, we draw away much of its force in another. For, to reiterate a point that Dworkin himself makes, beyond the threshold of fit, a judge, including the ideal one, gives up the best fit with institutional history for the sake of adopting a principle which has greater substantive merit. This is one of the things that makes 'hard cases' hard.

This seems to be faithful to actual practice, and is not a feature we should expect to recede as we approach ever more perfect specimens of that practice. We have before us an idealized picture of adjudication, not a picture of a perfectly just society as well. If the society were perfectly just, then it would contain the perfect elaboration and ranking of principles, and the need for adjudication about those principles would become redundant. Its institutonal history would, ex hypothesi, manifest the best combination of principles and would therefore never have to be given priority over them.

What would happen in the perfect judicial world would be that Hercules would be able to make the correct trade-offs. He would manage to make just the right concession to the particular institutional features of the community within which he was adjudicating. But the activity of 'trading off' supposes that there is more than one proposition of law available above the threshold of satisfactory fit. Otherwise, there is nothing to trade. A judge who is adequately sensitive to institutional history does not trade between propositions, one of which is morally the best and clearly below any plausible treshold of fit, and another, its moral inferior, which is above the threshold. The point of the notion of a threshold is that it indicates the degree of fit a proposition of law must display before it is available to the judge. So, we should expect that the dimension of fit with institutional history will fail to yield a single right answer to a question of law as frequently as those questions are raised in 'hard cases'.(14)

We are well into a second version of the argument for contingent monism. Now it looks as if the second dimension of the model, the dimension of choice and/or elaboration of a theory of political morality on its own substantive merits, does not have a residual role to play, closing gaps that a complex institutional history occasionally leaves open, but instead occupies the center of the stage since those gaps will be more than occasional.

Here everything turns on one's confidence in the ability of a theory of political morality to deliver right answers to a problem of interpretation. Dworkin addresses the problem explicitly in the following terms. He tells us that apart from the position of the moral sceptic, which he advances independent (and convincing) reasons for dismissing(14a), there is none which need take seriously the possibility that rival propositions of law will each satisfy a moral theory so completely that they can be said to tie in their competition. The classical utilitarian, for example, would treat this as possible, but highly improbable. "He will recognise the theoretical possibility that two distinct sets of moral rules will have exactly the same pleasure-pain consequences over the long run; but he will think that the possibility is so small that it may be ignored in practice"15). For those rights-based theories recognising the right to equal concern and respect even the possibility of a tie is dismissed. "It is not ... easy to see how someone could accept the general idea of the equal respect theory and still maintain, not that he is uncertain which (rival proposition) is better, but that neither is. There seems to be no room here for the ordinary idea of a tie. If there is no right answer in a hard case, this must be in virtue of some more problematic type of indeterminacy of incommensurability on moral theory"(16).

Is this adequate? There is one fundamental difficulty. The foregoing argument is designed to show us that we have little reason to think that moral theory will be indeterminate because of the very nature of such theory. No qualifications are added about the workings of such theory in simple or complex legal systems. For the utilitarian,for example, exact ties in the pleasure-pain balance would be as rare in primitive legal systems as in complex ones. Similarly, the oddity, if it is an oddity, of thinking that there can be ties in the satisfaction of the demand for equal concern and respect is as much a feature of the articulation of that demand in primitive systems as it is in complex ones. Therefore, it is hard to see why we should attach any significance to the complexity of a legal system when grounding our expectation that there will be single right answers to questions of law within it. Yet this complexity is a variable to which Dworkin's general argument attaches central importance, opting as it has for contingent monism. In other words, reliance on the second dimension of sound political morality, as Dworkin characterises it, would shift his position about right answers to that of the categorical monist, who holds that there will always be right answers to any question of law whatever the complexity or simplicity of the legal system in which the question arises.

However, even if the categorical position were entertained it does not seem to offer much promise. For we should question the claim that the idea of a tie within the region of rights-based theories, embodying the value of equal concern and respect, is inappropriate. To fail to meet the requirement of showing equal concern and respect is a quite particular kind of moral shortcoming. It manifests the contempt that someone shows for someone else by systematically having his wishes count for less than those of another. But one can manage to satisfy that requirement while still falling short of having provided the single right answer to a moral problem. Thus, if I propose, in a case before the court, to hold trade unions liable for the acts of their shop stewards, and you propose to hold them immune, and if both propositions are consistent with the body of settled law to a satisfactory degree, will it be enough to invoke, in the second dimension, the requirement of equal concern and respect to settle the issue? It is quite imaginable that either solution, grounded as it is on competing conceptions of the role of trade unions in industrial society, could be adopted by individuals who could not be accused of systematically discounting the interests of one of the parties in the way captured by the injunction to show them equal concern and respect. The prospects for a single right answer are not improved if we flesh out the moral theory on which we rely as an instantiation of the demand for equal concern and respect. Take Rawls' theory as an example. It seems that the decision procedures he specifies are not likely to yield single right answers in given situations. On his own account, his theory "..is often indeterminate; it is not always clear which of several constitutions, or economic or social arrangements would be chosen (from the original position). But when this is so, justice is to that extent likewise indeterminate. Institutions within the permitted range are equally just... they are compatible with all the constraints of the theory"(17).

Is Rawls saying something about his theory in general which would not apply to the judge in particular? Is something added by the particular impingement of institutional history? That is doubtful, because an adequate theory on his account already pays due regard to the unique features of a community's history of judgements on concrete issues, as it does for an individual. This is just what his requirement of reflective equilibrium is about. The possible options for the parties in the original position are narrowed by the need to find a best fit with the bulk of their settled and undistorted convictions, adjusting the rest to fit with the theory which best accounts for those. As Dworkin has shown, this model of reasoning for individuals can be expanded to understand the proper fit of alternative moral theories with convictions in a community, which it is the task of the judge to discover. But what he discovers are the solutions within the 'permitted range' to which Rawls refers.

3. THE CASE FOR CONTINGENT PLURALISM

In effect, the argument for contingent pluralism has just been given. As a legal system develops an ever longer history of settlement of a type of dispute, and widens in the range of disputes it settles, the probability rises that fewer and fewer alternative propositions of law will fit the body of relevant precedent equally well. The range is further narrowed with the search for the soundest moral-/political theory to deploy. So, 'strong' judicial discretion shrinks, as we would expect, as the full body of precedent lengthens and the variety of legal doctrine intersect and overlap. What does not seem justified is the conversion of this judicial search for right answers in law into something like a realist version of the search for truth in science: a view that there is at bottom a single truth of the matter. Instead, even for the perfect judge, there will often be the need to choose between several equally valid truths.

A few further points should be added, to clarify the position advanced and to bring out its relationship to broader issues in legal theory. First, I am not arguing that Dworkin's model endorses pluralism, as that term normally figures in moral and political theory. Normally, we think of pluralism as a doctrine about the nature of ultimate values. It claims that between competing social ends, such as the advancement of liberty and equality, or the preservation of democracy or efficiency, there is no global principle which will guide the correct balance(17a). But the version of pluralism advanced here is not making that claim. It does not, for example, shy away from the possibility of a global theory which among rivals emerges as the single best theory with which to work. Thus, in the competition between utilitarians and rights-based theories it does not claim that there is no ultimate way to choose between them. It would not feel <u>obliged</u> to resist Dworkin's claim that the most plausible moral and political theory with which to both characterise and carry on within Anglo-American legal systems is a rights-based theory. It also accepts that this theory is 'global' in the sense that it provides the framework of political morality within which all particular questions can receive a partial answer, insofar as they raise moral issues. It would, as he does, limit the range of the theory to the solution of problems of social relations and would not pretend to specify some dominant personal end for individuals. But what it denies is that such a global theory can provide a unique set of right answers in any system. This is not just because facts are complex, or situations are inevitably vaguely described. It is also because it is wrong to think that principles, and the moral theories they embody, usually have a single correct set of elaborated features, or can usually be arranged in the single correct pattern of relative weights.

Here, it is best to draw a distinction. We can distinguish between local theories and global ones. A local theory gives an account of some particular realm of activity such as punishment or promise-keeping, or some particular feature of an institution such as trade union's prerogatives over individuals or the proper basis of liability in contract, or finally, some particular constitutional feature of the legal and political system such as the proper relationship between legislature and the courts or the relative weight of old as proposed to recent precedent.

Contingent pluralism, then, tells us that we will often have more than one correct set of weightings and elaborations of local theories and principles in satisfaction of the demands of a global rights-based theory. It claims that this is quite compatible with the belief that the theory is for the moment the single best theory, or indeed with the belief that even if we were eventually to reject that theory in favour of another, that there is in principle and at bottom a single best global theory even if we could never articulate it. This species of pluralism is not obliged to pronounce for or against such a view. It can remain agnostic about pluralism at the level of global theories, being content to affirm it at the level of local ones. It is for this reason that it can draw on Rawls as an example of a pluralist political theorist even though he is not a pluralist at the most basic, global level.

4. OBSERVATION, THEORY, AND EPISTEMOLOGICAL RELATIVISM

Someone who abandons monism at the level of local theories might, unless he is careful, succumb to the attractions of relativism. This may be because he is impressed by a particular kind of argument about interpretation in law, science, and literature. Professor Stanley Fish has given us a recent version of it in a commentary on Prof. Dworkin's picture of the 'dimension of fit'.

He writes: "It is of course tempting to think that the more information one has ... the more directed will be one's interpretation: but information only comes in an interpreted form, it does not announce itself. No matter how much or how little you have, it cannot be a check against interpretation, because even when you first 'see' it, interpretation has already done its work"(18). For Fish it seems to follow that disagreements between lawyers, like disagreements among interpreters of a text, cannot be settled by reference to the existing law (or story, etc.).

Now there is one simple way in which this familiar argument needs to be filled out before it will yield a conclusion like that of the last sentence in the quotation. Someone who believes in the possibility of neutral appeals to the body of e.g. institutional history can accept that there are no theory-free observations against which theories are tested. He need not, that is, rely on what Lakatos called dogmatic or naturalistic falsificationism to make his case(19). One can accept that the statement 'this needle has moved five degrees to the left' is one which it is possible to make only in the light of a theory relating to the workings of the instrument, as is also true of the statement 'the temperature of this gas has risen 5° because its molecules have vibrated more quickly'. However, to move from the general observation that any theory will always confront observations-as-interpreted, as does Fish, to the conclusion that _therefore_ there is no common body of interpreted observation on the basis of which one can assess the claims of rival theories, is too quick. If true, it would be because we have good reason to expect mutual dependence, at all levels of a theory, of the meanings of the terms used in that theory. One can certainly accept that there is such mutual dependence between the implicitly defined terms of a theory, such that e.g. 'force' is defined in terms for a certain relationship between 'mass' and 'acceleration'. But that there be in principle and of necessity any equivalent mutual dependence between the meaning of terms in the statement about the movement of the temperature needle and the statement about the behavior of gas molecules is not plausible. It seems more reasonable to think that even if the two statements are theory-laden, they are laden with different independent theories. The theories are 'independent' in that at any given point of enquiry we can draw a serviceable distinction between background theories, which for the moment we are not contesting, and those in the foreground, which we are. Rival background theories may well explain the movement of the temperature needle differently, but nothing obliges us to enter and settle _that_ contest in the process of settling the contest _about_ the cause of the rise of temperature in the particular gas before us.

The same point can be carried over into the law. It is certainly true that the body of precedent in a given system is describable in some of its parts in the light of one proposition of law, in a way that another rival proposition does not (and sometimes cannot) entertain. This is clearest when a fact in a particular case emerges as material in the light of a subsequent distinction drawn in a later case which before that case was not seen by anyone as material. Here, the very description of the body of precedent will vary retrospectively with the proposition of law being advanced, as when a particular proposition about breach of contract is one which counsel tries to limit to commercial contracts, saying that it should not apply to contracts of employment and relying on the fact that the leading cases establishing the proposition were, indeed, commercial cases(20). It may not have been an important point to anyone at the time, but come to be seen later as crucial.

But this accounts for only a small area of judicial reasoning. There remain e.g. those cases in which the material facts were similar to those in the case at hand but in which the weight assigned to a particular principle was different, and which was arguably a 'mistake'. Here the description of the constituents of that case forming the precedent it sets do not vary with the particular proposition of law being advanced: only the evaluation of those constituents does. This portion of the terrain of precedent is theory-neutral. This would seem to be so even if one were also to agree that the very conception of precedent, fixing its elements and their respective priorities, is itself a product of a moral/political theory. It is not a product of the same moral/political theory giving us a particular conception of e.g. a principle about breach of contract.

It is certainly true that several moral/political theories touching on several areas of law are often simultaneously in context in a particular case. It is also true that any particular conclusion for or against the plaintiff or defendant will be a product of the combination of positions taken. The combination may even be such that each of its elements was necessary to the outcome, such that were any of them different then the outcome would have been different. But, as with complex causes of events, so here, there is no reason to think that just because elements of a complex set of antecedents are all in this way mutually dependent that there is no way of testing rival hypotheses as to whether or not a particular element is indeed part of that set. This is true for enquiries into matters of fact (was C one of the causes of an event?) or normative ones (is P properly part of an adequate set of moral propositions which provide the best justification of institutional history?) In both cases one can hold some variables constant, assuming provisionally the truth of the theories explaining them, while imagining others to vary, permitting the assessment of rival theories of that variable.

Here, it is important to distinguish the features of an ideal state of affairs from actual ones. If we looked at the ideal set of local theories accepted by Hercules in a given legal system we might believe that each portion of his account of the law was internally related to his account of the rest. Thus the full description/best justification of liability in contract would have to include the description/best justification of the totality of institutional history. If this were so, it would be hard to imagine evaluatable rival hypotheses providing rival justifications of a portion of that history. But of course we have reached the point in this ideal state of perfect knowledge when the notion of rival hypotheses is itself redundant. The mistake, however, is to let this view of the imagined end-point of enquiry affect our understanding of the nature of the appropriate steps to take along the way. Our actual theories are more approximate than they are in this ideal state of affairs, allowing us to draw boundaries between one portion of a system, which is the province of one local theory, and other portions relating to other local theories. This allows us to imagine variations within one portion only, since others can be held constant, and so to evaluate rival local theories of that portion. This is all an objectivist in legal theory requires, and it provides the objective foundation for contingent pluralism.

5. CONCLUSION

Professor Rupert Cross once voiced pluralist objections to Professor Dworkin, and approval of Professor Hart, in these terms: "Hart's account of judicial behaviour in hard cases is to be preferred to Dworkin's if only because the latter is too perfectionist, not merely in the sense that few judges are equipped to begin to fulfil the Herculean task proposed for them, but also because it is unlikely that its performance would produce one uniquely correct answer to many judicial problems. For example, it is difficult to see how any scheme of abstract and concrete principles could provide a coherent justification for all the relevant precedents, cited and not cited, in the *Spartan Steel* case .... The besetting weakness of the rights thesis is its tendency to exaggerate the certainty of the law in hard cases"(21).

Cross seems to be right in his scepticism about unique results, but for the wrong reasons. Dworkin never pretended that any judge, even the ideal one, could find a unique interpretation justifying all relevant precedents. His theory of 'mistakes' is meant precisely to show that no line of precedent could be expected to be entirely consistent. Instead, the task of the ideal judge is to minimise the number of mistakes to which his own interpretation will inevitably consigne some of the decisions. But, on the argument that has been advanced here, even if this task is carried out as well as Hercules might, it would still be unlikely to yield the single right result - even in combination with the substantively best moral-/political theory.

However, Professor Cross, along with others, seems wrong to let a conclusion about the ineradicable uncertainty in adjudication lead him to side with Hart. On the face of it, his seems to be a reasonable reaction. For it might be thought that the foregoing argument for pluralism endorses positivism for the following simple reason. On pluralism's own account, the judge is often faced with a plurality of equally sound solutions to a problem in interpretation between which he must choose without the confidence that the settled law or the best moral-/political theory determine the single result which he must do his best to discover. Clearly, the choice he ends up making will establish what the law is among several possible and equally valid outcomes, and this portion of the law is 'the law' just because it has been chosen. It is the law because of the authoritative institution source to which we can point, while the other possible solutions are not.

The role of appeals to the pedigree of rules and principles, and the width of judicial discretion in the interpretation of those rules, are thereby heightened. But is is a mistake to think that they are heightened in a way that should console the positivist. It is no part of Dworkin's argument that appeals to the pedigree of rules and principles play no role in determining what the law is at a given moment. He is rather concerned to show that this is not sufficient to fix the line between what the law is and what it is not. He is able to concede that a full description of the law on a point can and will be different before and after a given decision fixing an interpretation of it. This is an event in institutional history like the others, of which account must be taken. Dworkin's concern is rather with the monistic feature of Hart's argument from the pedigree of rules, whereby all sources of law are traceable back to a single rule of recognition as a guide. Instead, the sources of law are as various as are the factors which give principles their weight and character. "We could not bolt all of these together into a single 'rule', even a complex one ....", Dworkin objects(22).

The pluralist can agree, while regretting that one species of monism is trying to replace another. He can also agree that insufficiencies in Hart's picture of judicial discretion flow from his conception of the rule of recognition. Many of the principles which the pluralist agrees form part of the law escape the purview of any master rule that a judge can meaningfully consult in a hard case, but they are also properly understood as the fruits of judicial discovery.

However, he insists that often the process of judicial discovery will be directed towards a set of equally appropriate local theories and principles. Within the set the judge must choose. In this way, the pluralist claims to be better situated than is the monist to convey that element of legitimate legislation which lawyers know in their bones to be occasionally a proper feature of adjudication.

NOTES

(1) cf. D.N.MacCormick, 'Coherence in Legal justification' (This Volume, p. and R.M.Dworkin 'No Right Answer?' in Law, Morality and Society eds. P.M.S. Hacker and J.Raz (Oxford: 1977) p.78ff.
(2) 'No Right Answer?' version in 53 N.Y.U.L.R. 1 at 30-31. (1978).
(3) Taking Rights Seriously (London:1977) p.340 (Hereafter TRS).
(4) '"Natural Law" Revisited' XXIV U.Florida L.R. 165 at 171 (1982).
(5) Ibid.
(6) Supra n.1 p.84.
(7) Supra n.2 p.30.
(8) Supra n.2 p.30-31.

(9) 'Law as Interpretation' 60 U. Texas Law Review 527 (1982).
(9a) Dworkin, supra n.1 p.78ff.
(10) Supra n.9 p.531
(11) Supra n.10 p.532.
(12) TRS p.68, 341.
(13) See e.g. R. v. Lemon [1979] 1 All England Law Reports 898, especially the dissent by Lord Diplock, Read v. Lyons [1947] A.C. 156 at 170-1, per Lord Macmillan.
(14) Professor Dworkin seems himself to reflect this view in a short passage in a recent article where he writes "It should be apparent ... that any particular judge's theory of fit will often fail to produce a unique interpretation... two principles may each find enough support in the various decisions of the past to satisfy any plausible theory of fit. (my emphasis). Supra n.9 p.545.
(14a) e.g.Supra n.4 at 173 ff. I do not think that his distinction between internal and external points of view can be extended to deal with the points presently being raised, but that is an issue that cannot be treated here.
(15) Supra n.2 at 31.
(16) Ibid.
(17) A Theory of Justice (Oxford:1971) p.201. This is because there is often no uniquely correct way of construing what counts as a gain to the least advantaged since there is no uniquely correct way of construing primary goods as self-respect (cf.p.362).
(17a) cf. eg. B.Barry, Political Argument (London: 1965) Ch.III.
(18) 'Working on the Chain Gang: Interpretation in Law and Literature' 60 U.Texas L.R. 551 and 554 (1982).
(19) 'Falsification and the Methodology of Scientific Research Programmes' in Criticism and the Growth of Knowledge, eds I.Lakatos and A.Musgrave (Cambridge University Press: 1970) pp.93-103.
(19a) E.Nagel, The Structure of Science (London: 1961)p.91-100.
(20) As in eg. Thomas Marshall v. Gunile [1979] Ch. 227 at 238 where counsel for the defendant sought to distinguish Decro Wall International SA v. Practitioners in Marketing Ltd. [1971] 1 WLR 361.
(20a) As counsel in Thomas Marshal v. Guinle (supra n.20 at 241) was obliged to say about Wm Robinson and Co.

v. Heuer [1898] 2 Ch. 451 and Warner Bros. Pictures Inc. v. Netsan [1937] 1 KB 209.

(21) Precedent in English Law, (Oxford: 1977) p.220-221.

(22) TRS p.40.

DISCOVERY AND JUSTIFICATION IN SCIENCE AND LAW

Martin P.Golding

In this paper I propose to discuss a certain account of the nature of judicial decision-making by bringing to bear on it a distinction that was long held to be canonical in the philosophy of science, namely, the distinction between the context of discovery and the context of justification. In recent years this distinction has come under attack from two rather different directions. Although I shall not be able, here, to evaluate the validity of these attacks as far as the philosophy of science is concerned, I think that they provide useful perspectives from which to study the judicial process. The particular account of judicial decisionmaking is associated in Anglo-American countries with the disparate group of thinkers known as Legal Realists. I am not sure as to how widely this view is still held, but for a reason that will later become apparent, I suspect that it has more proponents than is openly admitted.

Let me begin with a classic source for this account. It is the often-cited 1929 article by Judge Joseph C. Hutcheson with the title "The Judgment Intuitive: The Function of the Hunch in Judicial Decision" (14 Cornell Law Q., 274). A judge, says Hutcheson, "really decides by feeling and not by judgment, by hunching and not by ratiocination, such ratiocination appearing only in the opinion. The vital motivating impulse for the decision is an intuitive sense of what is right or wrong in the particular case; and the astute judge, having so decided, enlists his every faculty and belabors his laggard mind, not only to justify that intuition to himself, but to make it pass muster with his critics". The hunch is characterized by Hutcheson as "that intuitive flash of understanding that makes the jump-spark connection between question and decision..."

It is not entirely clear to what aspect of judicial decision that Hutcheson's remarks are addressed. Judges may be called upon to decide at least three kinds of questions: 1) questions of law, 2) questions of fact, and

A. Peczenik et al. (eds.), Theory of Legal Science, 295–305.

3) questions involving the application of legal rules or standards to complicated or unusual fact-situations. It is in relation to the last of these question-types that Hutcheson's account is most plausible, that the "vital motivating impulse for the decision is an intuitive sense of what is right or wrong in the particular case".

Hutcheson's account, nevertheless, seems to have been understood as applying also to questions of law. Jerome Frank, who cites Hutcheson with approval in his book _Law and the Modern Mind_ (1930), speaks quite unrestrictedly of the "dominance of the conclusion" in judicial thinking, and he sets this account within a broader philosophical context. Frank maintains that everyone's ideas and beliefs fall into two categories: those based upon the direct observation of objective data and those based upon subjective factors such as personal desires and aims or, in a word, values. The latter condition the ideas and beliefs of judges no less than of ordinary people. Frank does not explain what he means by "the direct observation of objective data", but he is firm in claiming that judicial decisions are infected by bias and that opinion-writing is an exercise in "rationalization". (It should be added, however, that in many places Frank so qualifies his position as to negate it). The implication of this approach is that in order to explain a given judicial decision ineliminable subjective factors must be taken into account; the biography of the judge will be an essential part of the explanation.

This approach was intended to throw into question the so-called _objectivity_ of judicial decision-making; it questions this objectivity not merely in actual practice but also as a feasible ideal. The fact that subjective factors play a role in judicial decision and are essential elements in the explanation of such decisions purportedly undermines the possibility of judicial objectivity. The fully "mature" judge is merely one who is aware of his biases, but that awareness makes him no more objective than the unaware judge. For in either case the "dominance of the conclusion" prevails, and the judge's ratiocination will be directed towards justifying a conclusion that is chosen, at least in part, because of privately held values.

Now if it is a fact that judges' opinions give a false picture of the process whereby they arrive at conclusions on questions of law, how far does that fact undermine the objectivity of judicial decision-making on questions of law? A response to the Realist position may be found by drawing a distinction between the process of discovery and the process of justification along the lines of the distinction in the philosophy of science. (This response was suggested by Richard Wasserstrom in his 1961 book, The Judicial Decision, and I have myself employed it in a 1984 book, Legal Reasoning). This distinction was central to the logical empiricist approach to the epistemology of science. To the context of discovery, it was held, belong the factors that stimulate the generation and formulation of theories and hypotheses. These factors may include creative genius, religious beliefs, dispositions inculcated by experience in research, dreams and so on. To the context of justification, on the other hand, belong the factors that determine the truth or acceptability of theories and hypotheses: experimental testing, consistency, fertility, simplicity, scope, and so on. The fact that the idea of the benzene ring came to Kekule in a dream is neither of probative nor disprobative weight. The truth or acceptability of that idea is a matter of the context of justification, and it is there that the objectivity of science resides, for the criteria of truth or acceptability are independent of the scientist's personal predilections and values.

The analogy of the judicial situation with that of science is apparent. Hunches, biases, and values may well affect the process by which judges come to formulate their conclusions on questions of law, but all this belongs to the process of discovery and in no way undermines the objectivity of judicial decision. For contrary to the Realists, the objectivity of the judicial process is to be located in the process of justification, that is, in the "rationalizations" judges give in support of their conclusions. The crucial issue is whether the reasons given are adequate to establish the conclusions, and not whether the conclusions were products of hunch, bias, or value-predisposition.

Now it seems to me that this response considerably weakens the Realist's claim to have undermined the objectivity of judicial decision-making. But I'm not sure that it disposes of that claim entirely. An examination of some attacks on the discovery/justification distinction in the philosophy of science will be useful here. These attacks came from two different directions, for two implications were drawn from the distinction. It was held, firstly, that there can be no "logic of discovery". Thus, Mill's Methods, for instance, are merely procedures that suggest hypotheses for subsequent submission to controlled experimental testing. Secondly, it was held that there are two different kinds of studies <u>about</u> science: (a) a study of the psychological, social and cultural factors that contribute to the acceptance or rejection of knowledge claims, and (b) a study of the standards that ought to be employed in accepting or rejecting theories and hypotheses. Plainly, (a) is related to the context of discovery and (b) to the context of justification. The scope of the philosophy of science was restricted to the second type of study. The attacks on the distinction proceed by claiming either that there at least is something like a logic of discovery or by claiming that in important kinds of circumstances the decision to accept or reject a theory is never completely a function of so-called objective standards but is also in part determined by psychological, social, or cultural factors.

The claim that there is something like a logic of discovery clearly does not threaten either the objectivity of science or of law either, if the analogy drawn above holds. Yet a brief examination of this claim, which has been put forward by Norwood Russell Hanson ("Is there a logic of scientific discovery?" in Feighl and Maxwell, eds., <u>Current Issues in the Philosophy of Science</u>, 1961, pp.20-35) and Herbert Simon, may shed light on judicial justification. Hanson argues that philosophers of science have been too much enamoured by the "Logic of the Finished Research Report" and have overlooked the fact that scientists do supply reasons for proposing an hypothesis, for thinking that a given hypothesis is at least plausible. Inquiry into the grounds of plausibility, Hanson maintains, is a conceptual, not a psychological inquiry. His basic point is that it often will be reasonable to anticipate that an hypothesis that is proposed as an explanation for a certain phenomenon will be of a certain <u>type</u>, and this is not dependent on hunches.

According to Hanson, by the time a law gets fixed into the standard hypothetico-deductive system, the original scientific thinking is over, but the H-D analysis is silent on the reasoning that determines which type of hypothesis is likely to be the most fruitful to investigate. The hypothetico-deductive "plot" has three steps:

1) the suggestion of an hypothesis, 2) the deduction of observation statements, which turn out to be true, thus 3) establishing the hypothesis. All this fits in well with the idea of a "finished research report", but science is concerned with problematic phenomena that call for explanation. Scientists, Hanson maintains, also reason towards hypotheses, in the process of which they eliminate hypotheses of the wrong type. In this process they draw heavily upon analogies between their own problematic situation and laws or theories that have already been established. These analogies of course do not establish their own hypothesis as true - the logic of discovery is not a manual - but they do render it plausible and fruitful to propose. The traditional analysis is mistaken in holding that the origination of an hypothesis is to be treated as a matter of indifference to the scientific enterprise.

Now a judge's opinion, which gives the reasons for a conclusion on a question of law, is similar to a "finished research report", but if there is anything in law like Hanson's notion of reasoning towards plausible hypotheses, which would directly confute the Realist account and thus reinstate judicial objectivity, it would have to be at the level of the thinking that leads the judge to propose his conclusion as one that is plausible. And it is here that the similarity with the so-called logic of discovery, as Hanson himself understands it, breaks down. For it seems to be Hanson's view that the reasons that render a given hypothesis as plausible are different in kind from the reasons that render it acceptable or true. In the judicial situation, however, it would appear that the reasons a judge would have to have for thinking that a given answer to a question of law is plausible are substantially of the same sort as the reasons that purport to establish the answer as correct, for "plausibility" here means plausibility within the particular legal system in which the judge is operating. There are other writers, however, who do not accept Hanson's own understanding of the logic of discovery. Carl R.Kordig, ("Discovery and Justification", Philosophy of Science, vol. 45, 1978, pp. 110-17), for instance, maintains that three distinctions are needed: 1) initial thinking, 2) plausibility, and 3) acceptability. Logic is not essential to the first, but is to the second and third. There is no fundamental difference, according to Kordig, between the reasons for 2) and the reasons for 3), though acceptability requires more than plausibility.

This approach seems to have a better fit with the judicial situation.

But what could it mean, in the judicial situation, to say that acceptability requires more than plausibility? In drawing an analogy between law and science it must be kept in mind that there is an important point of disanalogy between them: there is in law no experimental test for conclusions of law. The "test" of a conclusion of law, to put it rather briefly, is whether it can be inferred from premises that are true or acceptable within the particular legal system, and a conclusion that is valid in one system might not be valid in another. In view of these considerations it might be said that a conclusion can be regarded as plausible when the judge has reasons that go toward establishing it as correct but he does not yet know whether these reasons are themselves correct within the system; the judge therefore needs to supply a subsidiary argument to establish the premises and thereby render the conclusion acceptable.

If we look at the matter in this way, it could be said that it is not merely the conclusion that is being proposed as plausible but that so also are the supporting reasons. This approach, in fact, allows us to construe the judge's opinion, the finished research report, as having a logical structure that is somewhat symmetrical to the hypothetico-deductive structure of explanation: the conclusion of law is parallel to the explanandum and the supporting reasons (the justification) are parallel to the explanans. And in order for the supporting reasons to be plausible they must satisfy criteria not unlike some of those that the explanans must satisfy: they must be relevant, not *ad hoc* (i.e., yield more than the particular conclusion), and perhaps they also should not be inconsistent with widely held legal or ethical points of view. But there is no exact analogue in judicial reasoning to anything like Popper's thesis that scientific explanation is explaining the known in terms of the unknown and of course there is no experimental test for the judge's supporting reasons. These reasons have to be established as correct by a subsidiary argument. Therefore, even if we can blunt the Realist position by distinguishing the process of discovery from that of justification, and even if there is in law something like a logic of discovery - as I think there undoubtedly is - the problem of judicial objectivity reemerges at the level of subsidiary arguments.

A few words should now be said about Herbert Simon's approach to a logic of discovery because it is helpful in suggesting a step toward the solution of this problem. Simon ("Does scientific discovery have a logic? "Models of Discovery, 1977, pp. 326-37) denies that it is necessary to postulate an "irrational element" or "creative intuition" in order to explain discovery. The possibility of a normative theory of discovery has been dismissed because it is supposed to depend on a solution of the problem of induction, but this supposition, says Simon, is erroneous. There can be norms of procedure, and thus criteria for evaluating processes for the discovery of scientific laws, because such norms may be derived from the goals of scientific activity. A process is said to be logical if it is efficacious or efficient for accomplishing the goals for which it was established. We may be able to show that a process should have certain properties A, B, C, if it is to be efficacious for attaining goal Y. Simon thinks that he can give rules for the detection of patterns in phenomena, subsequent to which detection these law-like patterns can be subjected to experimental testing.

The issue of whether all this amounts to logic of discovery may be put aside for our purposes. Instead, let us ask, for judicial decision-making, questions similar to those dealt with by Simon. Is it necessary to postulate an "irrational element" in order to explain either a judge's conclusion of law or the justification of the conclusion? Are there goals of the judicial process from which norms of legal discovery or justification may be derived? Since these questions begin to broach the second type of attack on the discovery/justification distinction, I want to turn to that now. Let me note in passing, however, that there is an ambiguity in the notion of explanation that is relevant to our problem. It will be brought out in the discussion that follows.

Probably no attack of the second type is as well known as Thomas Kuhn's. Kuhn (The Structure of Scientific Revolutions, 2nd ed., 1970; The Essential Tension, 1978) denies that there can be a study of the standards for accepting or rejecting scientific theories that is completely independent from the psychological, social, or cultural factors that contribute to such acceptance or rejection. He does not deny that there are certain characteristics that a good theory must have, but claims that they often will be insufficient to account for theory choice. Thus, the abandonment of one theory in favor of another "cannot be resolved by proof".Kuhn therefore holds that the discovery/justification distinction fails to constitute even a plausible or useful idealization. Subjective factors, he maintains, are ineliminable elements of decisions to adopt a scientific theory. One can explain why particular men made particular choices at particular times, but for that, says Kuhn, one must go beyond the list of shared criteria of a good theory to "characteristics of the individuals who made the choices." It seems to me that Kuhn's doubts over the discovery/justification distinction extend beyond so-called revolutionary science to normal science as well. He argues that textbooks oversimplify the choices scientists make; the distinction makes these choices seem unproblematic, while in fact these are often good reasons for each possible choice.

Now I am not concerned, here, to evaluate this attack, but merely to stress its similarity to the Realist critique of judicial objectivity. What is not entirely clear, however, is how far the Realist could admit that there may be good reasons on each side in the answers to questions of law without considerably undermining his own position.For the Realist would then be maintaining, as Kuhn does in respect to science (appearances to the contrary), that there merely are limits to judicial objectivity, not that there is no objectivity. To me, at any rate, it seems difficult to hold these more limited theses, in law or science, without acknowledging a distinction between the contexts of discovery and justification, though it would be a distinction of degree.

We can now turn back to the questions raised at the end of the exposition of Simon. Is it necessary to postulate an "irrational element" in order to explain judicial conclusions or justifications? Are there goals of the judicial process from which norms of legal discovery or justification may be derived?

Let us consider the first question but substitute the less pejorative term "subjective" for the word "irrational". On this question the Realist takes a position similar to that of the Kuhnian; both hold that biography is an essential part of the explanation of decisions. Now it appears to me that there is a sense in which this claim is true. Since the decision, or choice of a conclusion, is in one sense an event in a particular judge's biography, the explanation of that event will contain details of that biography, including the values held by the judge. To put the matter a bit differently, we might want to know the "reason why" a judge decided a question of law in a particular way, just as we may want to know the reason why a certain event - say, the collapse of a bridge - occurred. Reasons in this sense are explanatory reasons. A full statement of the explanatory reasons, furthermore, will have to contain much more than details of biography, for social and cultural factors will be required in order to explain why the question of law came up when it did. But we may also want to know what a judge's justifying reasons for a conclusion were. If we call these an explanation of the judge's decision, they clearly constitute an explanation in a different sense of the term; and an explanation in this sense typically will not include a judge's privately held values as privately held values but rather the legal reasons from which the conclusion of law are inferred.

Two issues nevertheless remain: (a) Is there a way of showing that a judge's particular justifying reasons are good legal reasons without bringing in the judge's values or other extra-legal considerations? and (b) How are we to understand the situation in which there are good legal reasons on each side of a question of law? Only a very brief and incomplete discussion of these complicated issues can be given here.

It is useful to consider whether there are goals of the judicial process from which norms of justification, if not norms of discovery, can be derived. One of these goals, I think, is simply the resolution of disputed questions of law by means of reasoned decisions on such questions. Conclusions of law, therefore, should be supported by justifying reasons. Moreover, judicial justification is not justification in the abstract. Rather, it is a process in which the judge is attempting to justify his decision to the losing party, to other persons who may be affected by the result, and to the legal community at large. The reasons for the conclusion must therefore, in some sense, be acceptable to this group; the explanans of the decision, so to speak, must satisfy a criterion over and above those mentioned earlier as criteria of plausibility. The fact that there is a social component to judicial justification does not have the extreme relativistic consequences that it may have in other justification situations. (In law, of course, a justification is also relative to the system for which it is supposed to hold). If values enter into a justification, they do not enter into it as personal predilections. The values must have some purchase on the community to which the justification is addressed. Although these considerations do not supply us with a way of demonstrating that some justifying reason is a good legal reason, they at least suggest that the problem may not be insoluble.

The problem of how to understand the situation in which there are good reasons on each side of a question of law may be more difficult. For here a judge will have good justifying reasons for his conclusion but, unless the reasons on one side are in some sense "better" than those on the other - and what could that mean? - he will not be able to show his conclusion to be uniquely correct. Contrary to Professor Dworkin, (Taking Rights Seriously, 2nd ed., 1978), I think that this situation not infrequently obtains in Anglo-American law, or so it appears to many observers. For this reason, I suspect, the Legal Realist account of judicial decision has more proponents than is openly admitted. Since a conclusion of law in this situation cannot objectively be shown to be uniquely correct, the most we can do is to explain, give explanatory reasons, as to why it was chosen. Still, I do not think that there is good cause to be depressed by this outcome. For while the judge's conclusion cannot be shown to be uniquely correct, it can at least be shown to be reasonable. And this may be all that we should ask for.

ABSTRACT

According to an account of judicial decision-making that was widely held by American Legal Realists, judicial decisions are a function, in part, of the biases and value-predispositions of the judge, and the opinion gives a false picture of the process of decision; the opinion is a "rationalization" of a choice that is reached on other grounds. This account is intended to undermine the objectivity of judicial decision.

In this paper the distinction between the context of discovery and context of justification - a distinction that has been canonical in the philosophy of science until recently - is considered as a possible response to the above account. Two recent attacks on the distinction are examined: 1) that there is a kind of logic of discovery (Hanson and Simon), and 2) that the values of scientists are ineliminable from explanations of their choices on the acceptance of theories (Kuhn). The process of judicial justification is discussed in relation to these two attacks. It is argued, at the end, that the Realist account has some plausibility with respect to judges' conclusions on questions of law in cases in which good legal reasons can be found to support answers on each side.

REASONS AND CAUSES IN CONNECTION WITH JUDICIAL DECISIONS

Gunnar Bergholtz

Some legal realism, part of sociology of law, students of the so-called psychology of judges and some common sense views seem in one way or another to assert that the views of the judge in questions of law or of fact may on the whole depend on his person, i.e. his personality traits, attitudes or something of the sort(1). There is reason to ask if these modes of reasoning hit the target and entail the whole truth.

In collegial courts there are now and then dissenting opinions. The following lines are based on an investigation of some dissenting opinions on questions of law in the Supreme Court of Sweden. Court-decisions with dissenting opinions can be useful in this respect, because they make explicit differences between judges ways of reasoning.

As an illustration I have more or less randomly chosen some cases from Swedish-international law. The reason for this choice is that I have some knowledge from that legal field and also because most questions around international law are in Sweden resolved by case law. The dissenting opinions in these cases therefore ought to lend themselves better to examination than cases of traditional interpretation of statutes. The result of the examination of these cases is as follows(2). It is possible to give clear arguments for or against all divergences of opinion. Those arguments are to be found in precedents and legal writing. The reasons given for stands taken can be traced back to legal-normative clusters of ideas or "theories". The different lines of reasoning are thus clearly intersubjectively structurable.

As another illustration of a rational justification I will, in a few lines, describe a case from another part of Swedish-international law. It is a case from the Svea Court of Appeal (SvJT 1958 rf. s. 13) without dissenting opinions and with a very simple structure of reasoning. The facts of the case are as follows. A Swedish national rented a flat in Germany. She did not pay the rent and

A. Peczenik et al. (eds.), Theory of Legal Science, 307–314.

after some time she returned to Sweden. The question presented to the court was whether a debt of rent from Germany could be claimed before a Swedish court. Swedish law on the subject mentioned only debts of rent for property situated on Swedish soil and consequently an answer was neither to be found in Swedish law nor in international agreements. Nevertheless the court found it possible to try the claim. The reasons given were that the former tenent was a Swedish national with habitual residence in Sweden and with all her assets located there. The German landlord thus naturally wanted a decision enforcable in Sweden. In its decision the court also pointed out that it was very doubtful if German courts could have tried the case. If the Swedish court had refused jurisdiction, most probably the landlord could not have had his claim taken up anywhere.

I mean that this case, in all its simplicity, is a paradigm of practical legal rationality and I mention it to show how difficult it is to believe that the judges could have been governed by something else than the reasons given in the decision.

Usually a distinction is drawn between justification and explanation of legal decisions. Justification deals with the legitimation of legal decisions, explanations on the other hand are normally either sociological or psychological cause-explanations, e.g. as follows:

"Le rapport entre la rationalité et l'explication psychologique ou sociologique de la décision judiciaire est tout á fait autre. Chaque décision a une explication psychologique et une explication sociologique, qui depend seulement de l'etat de la connaissance des faits empirique. On explique la décision quand elle est rationelle ou quand elle n'est pas(3)."

It is not unusual to give the explanation in form of a correlation between judges' attitudes at large and their standpoint on legal questions.

Now I want to pose a question, which, perhaps, can also be seen as an empirical hypothesis testable in principle: Is it possible to perceive the avowed <u>reasons</u> for legal decisions as <u>causes</u> of these decisions?

My intention is not to verify that this is so, but just to suggest that the view is plausible and to try to clarify some presuppositions for posing the question the way I do.

Justification in connected with the so-called internal

perspective, the lawyers viewpoint as a lawyer, and the external viewpont is connected with that of an observer, e.g. a sociologist. The internal perspective deals with the connection between the avowed legal reasons as premisses and the judgment as conclusion. There can be disagreement about the question if p & q are conclusive reasons for the judgment R, or, if reasons r & s & t are conclusive reasons for judgment R'.

The external perspective can deal with causal explanations of the behaviour of the judge. Among the causes one can find the judge's attitudes at large including his likes and dislikes of certain kinds of persons, of some political developments and his general beliefs about persons and societies. It is often taken for granted that the general attitudes of the judge influence his acting as a judge - perhaps in a decisive way(4). This opinion must of course assert a causal connexion between attitudes and behaviour (see note 5).

According to my view a somewhat different position is possible. Through his legal knowledge and his experience the judge must have acquired a "legal attitude". This attitude contains components of beliefs and a disposition to act. Researchers of attitudes are, generally in agreement about the concept of attitude containing cognitive elements(5). The disposition to act may be a "simple one" and the same for different judges, even if they disagree about what to do (the legal action). That is the case, when if a judge beliefs that an alternative R is backed by stronger legal reasons, he will choose R, or, if he thinks that he, as a judge, ought to choose R, then he chooses R. Differences between judges in their decisions will then depend on differences between them concerning the belief component. Judge D1 believes that p & q are conclusive reasons for R. D2 may believe r & s & t to be conclusive reasons for R'.

The following differences between (a) my view and (b), for example, Glendon Schubert's, (see note 4) can then be said to exist.

1. The disposition to act

a) Judges have a common disposition to act according to what they think is right or in accordance with valid law. The judge chooses solution R in spite of his disliking it because it is the legal solution.

b) The actions of judges are influenced by a lot of dispositions to act, different from one judge to another depending on what they like an dislike etc.

2. The component of belief

a) The beliefs that de facto influence the actions of judges are such that their contents in the internal perspective are relevant as reasons. When a judge passes judgment his action is caused essentially by his belief that p & q & r are conclusive reasons for R. Here p, q, r according to the internal perspective can be defended as relevant reasons. This view seems to be that of the Danish philosopher of law Alf Ross(6).

b) The judge is influenced by a lot of beliefs which are "legally irrelevant", i.e. the content of the beliefs cannot, according to the internal perspective, be defended as reasons. The Norwegian Vilhelm Aubert has stated such a view in disputing Alf Ross's thesis(7).

I think it is possible to say that the judgment is caused by the judge's beliefs, which in the internal perspective are seen as good legal reasons. It is also possible to maintain an epistomological equality between "legal" and "non-legal" (2a and 2b) beliefs in a theory of causal connexion.

By this I mean that it is doubtful if more scientific or somewhat different grounds can be shown to hold for a causal connexion between non-legal (2b) beliefs and judicial behaviour than can be the case between legal beliefs (2a) and the same behaviour.

Neil MacCormick, in his "Legal Reasoning and Legal Theory", argues that justification has some connexion with the outcome of a case(8). In this context MacCormick just seems to mean, that the judge, to the best of his abilities, gives his "real" reasons for the decision(9). On the other hand he does not say that the judge's belief in these reasons act as causes. However, such a view seems to follow from his reasoning. My analysis of cases (note 2) in this way verifies MacCormick's analysis. Such an analysis as MacCormick's shows at least that the avowed legal reasons for legal decisions - precise and legally valid and testable - cannot as a rule be given to hide other more mean reasons.

Can arguments be put forward to defend the view, that judges have a common disposition to act as judges and are acting on beliefs that in the internal perspective are seen as relevant reasons?

Often judges are in total agreement on both reasonings

and outcomes of cases while laymen disagree wildly. A complete and rationally defensable justification does not come from heaven of from a magician's hat. Very often a lawyer can predict that agreement or disagreement will show itself in some cases and also what the disagreement will be about and what reasons will be used to defend the different positions. If you ask for causes, which question is the most natural to ask: Is the judgment and the reasoning behind it caused by the judge's personality or by his legal knowledge? If you try to predict the outcome of future cases, the reasons used in former cases of the same kind are very useful. If you take if for granted that the judges are influenced by several dispositions to act and by a lot of "legally irrelevant" beliefs, you will have difficulties to explain the prevailing consensus among judges.

One possibility could be that those of the judge's beliefs that in the internal perspective are good legal reasons are one type of causes and that those causes used in sociological and psychological explanations are of another type. This view is impossible only if you suppose that the judgment is caused by certain causes and holds that the judgment is totally independent of its avowed legal reasons.

When providing a scientific causal explanation there is a requirement not to introduce ontologically or epistemologically spurious entities. If two presuppositions are reasonable, namely that there are not any relevant ontological differences between 1a & 2a and 1b & 2b above and that there is also between them an epistomological equality in a theory of causal relations, the view that the judge's belief in the avowed legal reasons for a legal decision can be seen as causes of that decision will be more plausible.

In this respect, it is appropriate to ask if there really are any decisive differences between the frequent "sociological" variables and variables with a more "legal" content including beliefs used as reasons in the justification of legal decisions.

In an article in Law and Society Review ('The legal/extra-legal controversy: Judicial decisions in pretrail release' by Ilene H. Nagel, Law and Society Review, Vol. 17, No. 3 1983 p. 481) the author seems to deal with related questions and says among other things:

"It is also the case that if influential legal variables are excluded from our models, there will be a distorted picture of both the relative importance of extra-legal variables and the legality of judicial decision-making in general. One implication is that sociologists of law must be able to work with legal source material as well as empirical data."

The author ends the article with the following statement:

"The legal/extra-legal controversy has long dictated a model of research that has failed to give due weight to either side of the debate. The complexities of law have often been ignored, and the extra-legal category has been narrowly and selectively defined. My research suggests that both components should be refined and reexamined as part of the continuing process to understand the bases for the decisions that lead to court outcomes."

NOTES

(1) For an overview of the research in USA; see Hans Joachim Koch, Zur Analyse richterlicher Entscheidungen, especially pp. 42-120.

(2) The differences in the investigated cases can be characterized as follows:

(1)NJA 1978 s 126 This case does only state a difference in reasonings for the same outcome. The majority avowed a legal principle and the minority made use of what can be called a "concrete judgment of the situation", if you like an authority reason versus a substantive reason.

(2)NJA 1953 s.4. Two legal principles compete and one gives another outcome than the other.

(3)NJA 1969 s. 310. The difference concerns different arguments for the same outcome. It deals with the relative weight of different legal reasons.

(4)NJA 1979 s. 242. Two legal principles compete. One is defended with authority reasons and the other with substantive reasons.

(5)NJA 1974 s. 629. The dissenting opinion is grounded on a consequentialist-argument of fundamental nature.

6)NJA 1973 s. 486. A general teleological reading of the law stands here in oppostition to a black-letter reading. A distinction made by Robert Summers (Robert

S. Summers, "Working Conceptions of 'The Law' in Law and Philosophy, Vol. 1., No.2 P. 263) between different "working conceptions of law" comes to mind. The majority applies a 'reason-conception" and the minority a 'rule-conception".

(3) Jerzy Wróblewski, Communication au C.N.R.L., le 11 janvier 1975.

(4) See for example, Judicial Behaviour: A Reader in Theory and Research, ed. by Glendon Schubert. On p. 446 You can read among other things the following:
"For present purposes, we need only summarize briefly the kind of answers that the materials of chapter III suggests. Judges differ in their attitudes because they have come to accept some beliefs, and reject other, as a result of their life experience. What a judge believes depends upon his political, religious, and ethnic affiliations; his wife; his economic security and his social status; the kind of education he has received, both formally and informally, and the kind of legal career he has followed prior to his becoming a judge. His affiliations, marital and socio-economic status, education, and career will in turn be largerly influenced by where he was born, and to whom, and when. From the point of view of predicting judicial decision-making, it is not assumed that judicial decisions are a direct function of judicial attributes. Decisions are a direct function of attitudes; and attitudes are a direct function of attributes; thus, decisions are affected by attributes only through the mediation of intervening attitudal variables."

(5) On attitudes; see Robert Audi, "On the Conception and Measurement of Attitudes in Contemporary Anglo- American Psychology", Journal for the Theory of Social Behaviour, 1972 p. 179. On the theory of connexions between attitudes and behaviour see especially pp 193-195.

(6) Alf Ross, On Law and Justice, pp. 47-48:
"Valid danish law can be defined as the normative ideology that de facto is at work, or can be imagined to be at work in the judge's mind, because it is percieved by him as socially binding and therefore is effectively followed. The sign of this efficiency is, that the hypothesis presupposing this ideology can be used to explain the judges verbal behaviour, his legal

decisions as connected in the context of meaning and motivation with other social acts which according to the content of the normative ideology conditions the reactions of the judge (acts of legislation, other official of private legal acts, actions in general)".

(7) Vilhelm Aubert, "Conscientious objectors before Norwegian military courts", in Judicial Decision-making, ed. Glendon Schubert, p. 202:
"One thing is certain: no valid theories of the decision process can be framed if one is limited exclusively to the reasons that the judge himself states when he gives an opinion. His articulated premisses for judgments are of course important, but it would seem a questionable approach to interpret judicial opinions in psychological terms in order to yield a cue to the judge's mind and motivation (as for example, Alf Ross does several places in his treatment of these questions)".

(8/9)Neil MacCormick, Legal Reasoning and Legal Theory, pp. 16-17.

CHAPTER 4

TECHNICAL RATIONALITY IN THE LAW

SYNOPSIS

Aleksander Peczenik

This chapter takes up technical rationality, that is, adequacy of means to ends, and its relevance in legal reasoning.

Zirk-Sadowski discusses four dichotomies, that is, external vs. internal rationality, formal vs. material rationality, theoretical vs. practical rationality and technical vs. hermeneutical rationality. Combining the dichotomies, he obtains 16 types of rationality. He eliminates some of them as contradicting his basic philosophical assumptions and considers the following seven types: (1) internal-formal-technical-theoretical; (2) internal-material-technical-theoretical; (3) internal--material-technical-practical; (4) internal-material-hermeneutical-theoretical; (5) internal-material-hermeneutical-practical; (6) external-material-hermeneutical-theoretical; and (7) external-material-hermeneutical-practical. He then assumes that legal interpretation is governed by the ideal of a rational lawmaker, who chooses the best means to preferred ends. Legal interpretation may thus be governed by three first types of rationality, all of them technical. However, since legal interpretation is a practical activity, only the third type remains. The author thus concludes that rationality of legal interpretation is internal, material, technical and practical.

Aart de Wild assumes that legal dogmatics (that is, the evaluatively-analytical study of the substance of law, developed at European universities) constitutes a mixture between a description of valid law and its evaluative interpretation. To make this discipline intersubjective and comparable to science, three courses of action are

A. Peczenik et al. (eds.), Theory of Legal Science, 315–318.

possible: (1) splitting dogmatics into a descriptive (scientific) part and a normative part; (2) changing dogmatics so that it will study empirical relations between legal norms and other social factors; and (3) modifying philosophy of science so that it will cover a normative discipline. The author rejects the first possibility as incompatible with the social function of legal dogmatics. He recognizes the third possibility as legitimate but, in this paper, argues in favor of the second, which -in his opinion- would be directed to the practice of the Bar and could make proposals for the enactment of new legal norms, without having a normative character.

A question occurs immediately, how a non-normative science can make such proposals. This seems to be possible only if the proposals are conceived as technical, that is, as suggesting optimal means to goals embodied in the legal system.

Technical, goal-means, rationality is also a subject of Robert Summers's paper. At first, he repeats his theory of judicial reasons in common law cases. There are either authority reasons or substantive reasons. The latter derive their justificatory force from normal, political, economic, institutional, or other social consideration. Two fundamental Goal reasons have the following features. The judge predicts decisional effects, establishes the fact that the predicted state of affairs affects or instantiates public interests, and concludes that it serves or disserves a social goal needed in the community. Neil MacCormick has criticized Summer's theory and pointed out that the courts do not predict the likely behavioral (that is, causal) consequences of their decisions. They rather formulate a general principle that covers the case under consideration and check whether its logical (not causal) consequences are acceptable or not. Summers now argues that the general principle may be formulated in many ways, each having different logical consequences, some acceptable, some not. What decides the case is rather an evaluation of likely (behavorial) consequences in future legal cases and even some out-of-court behavorial consequences. Summer's paper, and his analysis of an interesting example, make thus an account of goal-means reasoning in the law much more detailed and precise.

Technical rationality plays also a central role in Per Olof Ekelöf's teleological construction of statutes. His central idea is as follows. One distinguishes between the interpretation and the construction of a provision. One performs the former in the same way as, for example, an interpretation of a sentence in a newspaper article. However, such an interpretation is applicable only to establish the meaning of the provision in ordinary cases, belonging to the core of this meaning. Ekelöf regards the practical effect of this interpretation as the purpose (function) of the provision. One then construes the provision in extraordinary cases in such a way that the construction serves the same purpose. From this starting--point, Ekelöf criticises Bydlinski's doctrine of statutory construction and MacCormick's theory of normative coherence, cf. supra in this volume. Bydlinski recommends teleological construction of statutes as a correction of results obtained by other methods, literal, systematic, logical and historical. Ekelöf asks rhetorically why not employ the teleological method from the beginning. MacCormick bases his normative coherence on such principles as, e.g., safety, health, considerateness and justice. Ekelöf thinks, however, that such principles are of no use in applying a statute in concrete cases. One shall perhaps mention that Ekelöf's idea that one shall derive the goal of the statute from the statute itself is very influential in Swedish courts. It seems to fit the Swedish version of the statute-positivism.

Enrico Pattaro presents the philosophical system of Guido Fasso. Technical rationality plays the central role in this system. Human reason imposes rationality on the irrational multiplicity of reality. This task is performed by scientific laws and by practical laws. All practical laws are technical rules. For example, given the scientific law governing a phenomenon, if you want to produce the phenomenon, you shall assume its scientific law as a rule for your behaviour. If you, for instance, wish to melt platinum, you shall bring it to a temperature of 1780° C., because the scientific law in question says that platinum melts at this temperature. All will, in opposition to sheer instinct, is a rational volition of means in view of an end. All human values are relative to ends. They constitute law, in a broad sense. Law is thus science in imperative form. All human institutions, human history, human activity etc. are "legal" in this sense, that is, they are governed by technical goals-means rationality.

The rationale of many attempts to reduce legal rationality to technical rationality lies in the ambition to make it value-free. One cannot rationally choose the ends, only the means to the ends. Mats Mattsson, however, gives an example of the great difficulties the technical rationalists meet in the law. He starts from critizing Hayek's and Raz's idea that the rule of law is the same as predictability of decisions. Although a similar idea has been used by many Swedish Legal Realists, the legal practice, nevertheless, restricts the use of such words as "legal security" (Swedish _rättssäkerhet_), roughly corresponding to "the rule of law". The restriction is not value-free. The core of the use of the words "legal security" is thus as follows: Legal security is considered to be in danger first of all in the fields with clear principles, where the legislator authorizes public bodies to make any or doubtful exceptions from the main principle. Possible exceptions from a main principle must be predictable. Since these principles cannot be established in a value--free manner, the conclusion is that Legal Realists, in spite of their programme of value-freedom, had to rely on valuations, not merely on technical rationality. On the other hand, not all kinds of legal situation involving moral values have importance for legal security. One needs both predictability and valuations.

LEGAL RATIONALITY AMONG DIFFERENT TYPES OF RATIONALITY

Marek Zirk-Sadowski

1. EXTERNAL AND INTERNAL RATIONALITY

The question about rationality of axiological decisions is very often felt to be a paradoxical problem. Rationality and the choice of values have nothing in common for those advocates of axiological relativism who claim that there is no possibility to justify a commitment to any moral, religious, scientific etc., practice. Such forms of human life create their own standards by which we may judge them. Only as the advocates of a given system of values may we, committing ourselves to its notions, make this system understandable for us. The commitment determines the correctness of a system but there is no justification for the commitment itself. Only deterministic explanations, e.g. our psychic features, the surroundings we were brought up in may eventually give reasons for the commitment but they do not justify it.

The attitude of this type, suggestive as it is, gives rise to a number of doubts. First, it divides human ways of life into separate systems. From the point of view of this conception the obvious fact of our ability to synthetize human reality is paradoxical. Second, it does not explain why passing from one system to another is possible. How generally does a change of an outlook on life take place? And third, it assumes the existence of some ultimate voluntaristic commitment everything follows from. After all, commitment, by nature, cannot be differentiated; so we must assume that there is one commitment which somehow justifies the acceptance of an infinite number of systems of human ways of life. The three counterarguments or objections mentioned seem to be the most essential. They result in the fact that the conception of commitment is of little use for a lawyer in justifying his axiological decisions since it offers a sharp caesura between rationality and irrationality in the choice of systems of values. The rivalling theory is the one which assumes the

A. Peczenik et al. (eds.), Theory of Legal Science, 319–333.

existence of a continuum between the rational and the irrational and practically means the existence of many kinds of rationality which could be arranged along the axis connecting rationality with irrationality.

The division into external and internal rationality is the basical differentiation in the interval: the rational-the irrational(1). The thesis that it refers to such ways of thinking whose character is systematical is the assumption hiding behind it. If a set of notions about the world can be arranged in a system, even in the weakest sense of the word, we may speak about a cause for this differentiation. External rationality comprises the set of criteria guiding a given subject at the choice of the system of values or notions. So it decides which system suits a given person. For instance, from the whole set of moral norms functioning in a given society a judge chooses a certain group of norms which he regards as the rules of community life.

However, this criterion may have a double interpretation. The question arises whether we say that the judge bases his choice on external rationality only when he does not refer to a system of notions of a higher order than the systems he chooses from, or whether we use this notion also in the situation where he applies criteria contained in the system but of a higher degree. In other words, whether external rationality is connected with the choice under conditions of the absence of a superior system, or not. In the latter case external and internal rationality would be relative notions and they could be replaced by the notion of, for instance, a net of systems. So the first interpretation, which treats external rationality as a set of criteria of the choice from systems for which no metasystem exists, is more interesting. However, this does not mean a necessity to choose from systems under conditions of some "epistemological vacuum". In the environment of a system various factors act which justify the choice of it, but their nature may not be systematic. Thus to refer to them in the criticism of the justification from the point of view of external rationality does not mean a reference to a metasystem. So the environment of a system within the framework of which the criteria of rationality (internal) are obligatory has a character close to the form of life Wittgenstein wrote about. The form of life has no systematic character but may manifest itself in the form of various systems, e.g. of values.

Of course, the choice of one out of the two versions of this problem is also conditioned by the philosophical views of the person who makes it since the decision must be made whether the situation is possible at all, if the choice is made in the absence of a system. We deal with rationality which we connect with the context of justifying decisions. Philosophical assumptions which decide about one of the interpretations mentioned concern the question about the existence of a theoretical system following Platon or Hegel. So if we choose the interpretation of external rationality which assumes the lack of such metasystem we proclaim ourselves against such philosophies. However, this does not mean that we resign any philosophical assumptions such as, for instance, the assumption about the existence of the form of life since it turns out that the accepted version of differentiating between external and internal rationality must assume determined philosophical attitudes. Since it does not want to refer to the notion of commitment it rejects apriorism in favour of rationalism. However, it must be such rationalism which proves to us in some way that it is rational. After all, every rational motivation must have some reasons. Of course, the danger of the vicious circle appears, or the necessity to admit that irrational decisions are the basis of the rational attitude. However, in the latter case a rationalist is not able to criticize those who, on an irrational basis, choose something different from rationality.

A lawyer obviously cannot accept such an attitude as the final basis for the justification of his decisions. He needs a philosophy of law which will allow him to avoid the dangers described. Every decision made on the basis of law must undergo intersubjective evaluation.

This problem becomes particularly important in the case of axiological decisions where the judge has decisional lee-way and is limited only by the demands of rationality. So it depends on a solution of the problem of the vicious circle in the construction of rationality whether it is external or internal. According to the accepted philosophical conception we shall evaluate the judge's reaction to this decisional lee-way differently.

2. PRACTICAL AND THEORETICAL RATIONALITY; FORMAL AND MATERIAL RATIONALITY

Up till now we have mainly discussed external rationality. The division into material and formal rationality also concerns internal rationality. Since internal rationality is connected with system approach there is a trend to define it formally, as a rule. In jurisprudence it is expressed in various constructions of deontic logic.

As long as it is difficult to give any ultimate justification for the decision made within the framework of external rationality, internal rationality requires referring to some procedures, standards accepted by a definite audience which give a possibility to defend the achieved results of reasoning. In this connection in literature emphasis is put on the necessity of dividing argumentations which are regarded as internally rational into firstly argumentations about which we may say that conclusions following from them are "derived" in the sense that they cannot be impaired from the point of view of the procedure of truth and falseness, and, secondly, argumentations whose conclusions are only "indicated"(2). In this case rationality is a category broader than truth and falseness because the second group of argumentations allows of connecting descriptive sentences, norms and evaluations.

Differences between the two kinds of argumentations may be considered from different points of view. The most frequently mentioned pairs of distinctive features are as follows: 1) certain-uncertain, 2) close-open (close are those which have a precisely determined moment terminating the argumentation, open are these where such a moment is not determined), 3) true-plausible. So it turns out that ranking this very great class of argumentations in the category of internally rational reveals the fact that by rationality we understand a set of many different values(3).

It is greatly emphasized that a rational discussion differs from others by the fact that its aim is not to mislead its participants. It is claimed that this rationality demands of every participant to use the arguments he himself believes in. If a rational discussion requires a quotation of arguments then such values as freedom of the access to the discussion, freedom of discussion, exclusion of an element of pressure must be fulfilled(4). Since the thus formulated rationality is a set of notions, we include a lot of argumentational schemes into the group of rational argumentations. There is the agreement as to the necessity of separating formal logical argumentations whose correctness is determined according to the criterion of truth and falseness. However, in the interval we are interested in, the truth is not the only demand of rationality and sometimes it does not even need to be identified in order to determine the argumentation as rational. Only in logical argumentations does this criterion appear but logic is understood widely as intensional or extensional. The rest do not derive a conclusion in the sense logical argumentations do, but they point to it, e.g. argumentations based on norms. In spite of this they are not deprived, as we have said, of the character of rational argumentations.

Thus it turns out that the rationality we call practical is a complex set of values, part of which we have mentioned above. Legal argumentations per analogiam, a contrario, a fortiori may serve as examples of this type of argumentations. This set of values consists not only of widely understood praxiological values but also of ethical ones (freedom, honesty,etc.).

Two main tendencies may be observed while studying argumentations of the practical type. The first one may be called axiological and it refers to legal-natural conceptions. The dependences occurring in rationality are to be a manifestation of a construction of logical nature(5). The second tendency which may be called "pragmatic" refers to the conventional activity. It assumes that the set of rules of the practical discourse which is to contain pragmatic rules of conduct of the disputants may be reconstructed. They ensure rationality but do not give sufficient certainty of the obtained conclusion of the discussion. Therefore enumeration of rules defining the practical discourse and its rationality is never exhaustive and does not aspire to the title of the code of practical reason. The theory of practical discourse in this approach creates the basis for constructing the theory of legal discourse. As has already been mentioned R.Alexy's work is the most developed theory of legal discourse against the background of practical discourse. The fact that rationality is defined in a way close to conventional activity is a defect of the construction of this type, since it brings about difficulties already described in analysing the dilemmas connected with justifying external rationality. The fact that for the conventional activity which we here call practical discourse there are no, as has been remarked upon in literature, accurate criteria which enable us to state whether it was accomplished in a felicitous manner or not, is an additional difficulty. In the background there appears the problem of formal and material rationality. Rationality is formal when its criteria are exclusively syntactic and, as in mathematical logic, independent of the contentual aspect of notions. Contrary to them the criteria of material rationality refer to the semantic aspect of notions, i.e. their contents.

3. TECHNICAL AND HERMENEUTICAL RATIONALITY

Technical rationality manages that actions run according to the scheme of means and ends. The ends must be determined exactly enough to choose optimum means for the realization of them. He who can determine the aim of the activity, and chooses the best means he knows or has at his disposal for its realization, acts rationally. From the point of view of such rationality there is no difference between the driving of nails into the wall and the creation of laws by the law-maker since this type of rationality refers to the explanatory and not to the understanding attitude. It assumes that acting in natural order we can - on the basis of anancastic sentences - formulate instrumental directives.

The scheme of technical rationality is adequate for some domains of the life of man and particularly for man's contact with nature. However, it turns out to be a counter-factual scheme in relation to many human acts in the sphere of culture(6) where the dependence of means and ends is too weak to serve for formulating instrumental directives.

Schemes of actions which do not take place according to the above presented scheme, but are nevertheless regarded as rational, we shall call schemes of hermeneutical rationality.

This type of rationality was first elaborated in the theory of organization. Charles E.Lindblom(7) deserves the credit for it. Before we start discussing this conception we must remark that the differentiation of the two types of rationality is valid when we treat them as means of description. This remark is substantial also for the reason that the creators of models based on technical rationality frequently treat it as a normative tool designed mainly for the criticism of the existing patterns of actions. According to Lindblom the technical (in his terms synoptical) model of rationality is applicable only if a subject who wants to act rationally had absolute clarity about his ends, if he possessed all necessary information, if all participants in the decisional process professed the same values, if, at last, all necessary sources were equally easy to utilize. Since these elements are very difficult to gather together in one concrete decisional process, this model is of low heuristic value because it does not occur in practice, and it is of low normative value as well, because things which cannot be realized cannot be postulated either. In this connection Lindblom suggests the analysis of the criteria of rationality which he determined as the model of mutual adjustment of partners.

Lindblom observes that an a priori accumulation of all necessary information is useless and even superfluous. Information is introduced into the decisional process as the result of pressure, counter-pressure and negotiations such as the mutual adjustment of the partners requires. Complexity of the world which surrounds us, and the great degree of uncertainty connected with it, cause that the interests of partners are very often contradictory. Owing to the vagueness of their goals partners can arrive at a mutual policy. If they tried to be explicit about their goals at the very beginning they would immediately deprive themselves of the possibility to cooperate and communicate. In the technical model vagueness of goals makes its application impossible, contrary to the hermeneutic model which assumes the mutual adjustment of the partners. In this model it is also assumed that the partners recognize different values, and therefore associating on the basis of non-questioning these values is easier than establishing common aims.

So how is progress in action possible? According to Lindblom it is possible owing to a common experiment which allows of understanding the solved problems as well as discovering the sources necessary for solving them. Therefore the model which we called technical according to the cited authors has an a priori character and pretends to the role of the code of practical reason. On the other hand, the model assuming the mutual adjustment of the partners is of an posteriori character.

The critics of the hermeneutical model argue that in this model we may find A.Smith's fundamental idea, the idea of a hidden hand which manipulates the market processes so that their particular participants competing with one another act on behalf of the general interest thinking that what they do serves their own particular interests. In this model rational logic of means and ends does not exist. The set of norms, the observance of which is necessary for this mutual adjustment of all the partners taking part in decisionmaking to be possible, is the only constant catalogue of rules. Hence in some reconstructions of the criteria of rationality we may even meet ethical norms, e.g. the ones concerning the freedom of discussion.

Both the technical and the hermeneutical model do not occur in practice in their pure form, of course. There are the spheres of justification of human activity where it is difficult to find the criteria of technical rationality based on means and ends. In the sphere of culture this scheme has a counter-factual character as a rule since it is stressed that in this sphere human activity lies not only in the processing of the objects of reality but also in the common constituting of meaning, and the activity which assumes understanding cannot be structuralized in justification. Understanding somebody's activity is its cultural identification, grasping its culturally determined sense(8). For this reason dialogue, discussion, argumentation cannot be grasped in the scheme of technical rationality.

4. LEGAL RATIONALITY AGAINST THE BACKGROUND OF THE PRESENTED TYPES OF RATIONALITY

The considerations presented above have revealed the fact that we have to do with a multiplicity of types of rationality or, in other words, that the criteria of rationality may be classified in different ways(9). In this approach the general differentiation of rationality and irrationality creates the interval in which we may place different criteria of rationality, one of them forming the stronger, the other the weaker type of rationality. So rationality is a gradable feature. In our considerations at least four divisions of rationality have come into view. The division into external and internal rationality is the first one, performed according to systematicality of the criteria of rationality. We say that rationality is internal when the criteria it consists of refer to some system of norms, values or theorems. External rationality appears when its criteria are connected with no such system.

The second division differentiates formal and material rationality. Rationality is formal when its criteria are exclusively syntactic and, as in mathematical logic, are independent of the contentual aspect of notions. Contrary to them the criteria of material rationality refer to the semantic aspect of notions, i.e. their contents.

According to the role of truth in the criteria of rationality we distinguish theoretical and practical rationality. Theoretical rationality as a test for rationality treats the possibility to refer to true statements in the sense of the classical theory of truth. Truth may be established extensionally or intensionally. On the other hand, practical rationality regards the criterion of truth as one of many possible criteria of rationality and in the extreme version it allows of the lack of this criterion.

The last division comprises technical and hermeneutical rationality. The role of the argumentation of means and ends is the criterion here. In technical rationality he acts rationally who is able to determine the goal of an action and chooses the best means for its realization. Hermeneutical rationalization is characteristic of the activity in the sphere of meanings and the process of commonly identifying the sense of cultural objects is submitted to its evaluation. In the process of establishing meanings, e.g. in the course of argumentation, argumentation of means and ends is only, at most, one of the moments of argumentation or it does not appear at all.

As follows from the presented divisions every conception of rationality must consider at least four problems: 1. systemacity, 2. formality, 3. the role of argumentation of means and ends, 4. the role of truth. We do not claim that a solution of these problems will be enough for the construction of the notion of rationality but these are the problems which must necessarily be solved, anyway. So their solution gives at least a partial definition of rationality.

Thus we have achieved four divisions of rationality: 1. internal and external (symbolically I/E), 2. formal and material (F/M), 3. technical and hermeneutical (symbolically S/H), 4. theoretical and practical (symbolically T/P). We should be aware that each of these divisions is of a dychotomic character. It is undoubtedly an idealizing operation because in reality the features which constitute every one of them are gradable. Just as we, in the division into the rational-the irrational, have established the existence of a continuum of criteria which approximate either one or the other extreme of the division, so could each of the four classifications be thus characterized.To simplify our considerations we assume that passing from one kind of rationality to the other is of a step character.

While the idealizing character of these divisions may be observed to the least degree in divisions 1 and 2, the bases of divisions 3 and 4 are felt more as gradable features. For instance, the decision about the dychotomic character of the division into technical and hermeneutical rationality includes in the group of hermeneutic criteria rationality of such argumentations as are based on the relation: means and ends but in which identification of means and ends is accomplished on the basis of axiological criteria but not on causal or functional ones(10).

Dychotomy of these divisions has some definite advantages in spite of necessary idealizing assumptions. In each of the suggested divisions we may easily isolate weak and strong rationality. The facility of intersubjective verifiability of the criteria introduced by each of the formerly distinguished eight kinds of rationality is the criterion of this division which we put upon them. From this point of view internal rationality may be regarded as stronger than external , formal stronger than material, technical stronger than hermeneutical, theoretical stronger than practical rationality(11). Respectively, in relation to each division we illustrate this dependence with the following symbols:

$$\overset{+}{I}/\overset{-}{E},\ \overset{+}{F}/\overset{-}{M},\ \overset{+}{S}/\overset{-}{H},\ \overset{+}{T}/\overset{-}{P}.$$

As we have pointed out, the characterization of rationality must embrace at least four mentioned features which are at the same time components of our divisions. This means that rationality may be described by combinations of the above mentioned kinds of rationality and in every arrangement one element of the four divisons, e.g. I, M, H, P must appear. Since the character of each division is dychotomic the number of possible combinations amounts to 2^n where n is the number of divisions. This gives 16 possible arrangements:

	+- I/E	+- M/F	+- H/S	+- P/T	number of pluses
1	I+	F+	S+	T+	4
2	I+	F+	S+	P-	3
3	I+	F+	H-	T+	3
4	I+	M-	S+	T+	3
5	E-	F+	S+	T+	3
6	I+	F+	H-	P-	2
7	I+	M-	S+	P-	2
8	I+	M-	H-	T+	2
9	E-	F+	S+	P-	2
10	E-	F+	H-	T+	2
11	E-	M-	S+	T+	2
12	I+	M-	H-	P-	1
13	E-	F+	H-	P-	1
14	E-	M-	S+	P-	1
15	E-	M-	H-	T+	1
16	E-	M-	H-	P-	0

It is easy to observe that particular fours have been arranged from 1 to 16 to the number of strong types of rationality which occur in a given four (denoted: +). In this way we have obtained 16 arranged types of rationality, internal, formal, technical and theoretical ones being the strongest of them. For instance, the deterministic conception of physical dependences based on the interpretation of mathematical models corresponds with it.External, material, hermeneutic and practical type to which we may assign the attitude of common sense in philosophy is the weakest.

The constructed scale of types is very weak because the dependence stronger (+) weaker (-) holds only inside each dychotomic division but gives no basis for differentiating the fours with the same number of pluses.

The above list is of course only theoretical. In reality we do not meet the majority of the distinguished sixteen types of rationality for the reason that they demand the acceptance of definite philosophical assumptions. Let us consider this problem now.

Arrangements which cannot be accepted because of the accepted philosophical assumptions are the types of rationality containing the following pairs: EF, ES, FH, PF. Acceptance of EF requires the assumption that it is possible to construct some formal ontology, and, moreover, refer to something following Platon's world of ideas. Acceptance of ES requires in turn the existence of some final aim which cannot be a means for other aims. Acceptance of FH requires the introduction of an assumption about a possibility to construct a formal hermeneutics. PF the assumption about a possibility to construct a formal code of practical reason. Since we do not recognize the assumptions mentioned we reject in our arrangement these types of rationality which contain the above mentioned arrangements, i.e.: 2, 3, 5, 6, 9, 10, 11, 13, 14. The following types come into play:

1	I+	F+	S+	T−
4	I+	M−	S+	T+
7	I+	M−	S+	P−
8	I+	M−	H−	T+
12	I+	M−	H−	P−
15	E−	M−	H−	T+
16	E−	M−	H−	P−

Seven types of rationality have been left, and only two of them comprise external rationality. It is characteristic that the strongest and the weakest types of rationality remain.

Against the background of these considerations, what does the problem of rationality constituting the deep structure of legal interpretation look like?(12) According to this conception an interpreter always assumes that the law-maker's knowledge and preferences meet some idealizing demands which follow from the rule of rationality. Namely we assume, that if a given person assents to consistent and systematical knowledge and cherishes transitive and asymmetrical preferences, this subject takes up from possible activities the one which leads to the highest preferred (in the light of this knowledge) state of affairs. So a rational law-maker does not create inconsistent or unnecessary norms. Even if this construction is completed with the construction of the perfect law-maker, and it is, thus, admitted that the law-maker's knowledge comprises convictions regarded by the interpreter as right, different interpreters may ascribe different preferences to the law-maker and apply different interpretational directives as the result. This follows from the fact that the construction of the perfect law-maker does not decide about what the law-maker's preferences are.Therefore it is necessary to use the normative theory of interpretation.

A rational law-maker is thus a part of the construction of the perfect law-maker and the latter is, on the other hand, a part of legal culture(13). Differences in understanding of the perfect law-maker are, according to L.Nowak, not only a result of ascribing different knowledge and evaluations to him. There are also differences connected with dissimilarity of inferential rules of different epochs or different legal cultures. The statement follows that rationality we discuss here is internal because it refers to knowledge, evaluations and formal schemes of a given legal culture.

It is also rationality based upon the rule of means and ends. The rational law-maker, because of his knowledge and preferences pursues his best ends, choosing the best of means he knows.

So we may at once determine the considered rationality from the point of view of our typology as internal and technical rationality. In our considerations these features occur in three types of rationality: I,F,S,T; I,M,S,T; I,M,S,P. Although rationality of the law-maker does not determine the contents of his knowledge or preferences it is, however, not formal in the sense we have accepted since we have accepted formality in a very strong sense such as in axiomatic logical calculus. However, the described construction assumes that legal language is simply a section of natural language which decides about the way of understanding the linguistic competence of the law-maker(14). So the question is about formality of the construction in a weaker sense than the one we accept. For this reason we must number this rationality among the type of material rationality. This is the price we must pay for the dychotomy of the accepted classification.

So two more types of rationality come into play: I,M,S,T and I,M,S,P. The role of the criterion of truth in the set of criteria of rationality decides about their choice. It is not clearly determined in the construction discussed. We may determine it on the basis of the consequences which are brought about by the construction of the rational law-maker. As we have said, it follows from this that the law-maker does not create inconsistent or unnecessary norms, i.e. the ones that result from the previously created norms, or normalize situations which cannot, in the light of empirical knowledge, take place. There is no evidence that in the deep structure of legal interpretation the statement is contained, that the determination of inconsistency of one norm in relation to another or of its superfluity, is based upon the criterion of truthfulness. On the contrary, the evaluative moments which come into play lead one to suppose that inferential rules belong to the criteria of practical rationality. If so, then rationality making the deep structure of legal interpretation belongs to the I,M,S,P type. So it is rationality internal in relation to legal culture, based on the criterion of means and ends. The choice of ends and means which lead to it is made on the basis of the criteria concerning contents and it may be preceded by axiological decisions. However, we should remark that in the theory of interpretation which refers to this conception the stress is put on the first two features of the type of rationality mentioned.

NOTES

(1) J.Wróblewski: Meaning and Truth in Judicial Decisions.Helsinki 1979, p.63.
(2) Cf.I.Tammelo: Modern Logic in the Service of the Law. Wien-New York 1978, p.150 ff.
(3) The examples of reconstruction of rationality which are in my opinion sets of values of different kinds are I.Tammelo's works "Modern Logic...." op.cit., p.149 ff.and R.Alexy's Theorie der juristischen Argumentation, Frankfurt am Main 1978.
(4) Cf.R.Alexy, ibid., p.239 and the following.
(5) I do not discuss in detail the problems connected with this type of rationality because they are widely discussed in literature against the background of various conceptions of the law of nature.
(6) Speaking about culture I rather mean symbolic culture here.
(7) E.Lindblom: The Intelligence of Democracy, New York 1965.

(8) M.Ziókowski uses such definition of understanding in his work _Znaczenie, inteligencja, rozumienie_ ("Meaning, intelligence, understanding"). Warszawa 1981, p.19.

(9) K.O.Apel suggests different division in "Types of Rationality Today", "The Continuum of Reason between Science and Ethics" in _Rationality Today_. Ottawa 1979, pp.318-346.

(10) Cf.J.Wróblewski: "Evaluative Statements in Law". _Rivista Internazionale di Filozofia del diritto_, LV/III, 1981, pp.618, 619.

(11) I do not justify every division in detail because we decide here that cultural conditioning is easily perceptible.

(12) L.Nowak: _Interpretacja prawnicza_ (Legal interpretation). Warszawa 1973, Ch.III.

(13) Ibid., p.151.

(14) Ibid., p.153.

PARADIGMS OF LEGAL RESEARCH
Empirical Science and Legal Dogmatics

Aart Hendrik de Wild

1. INTRODUCTION

It is quite common to characterize legal dogmatics as the discipline that consists of the interpretation and systematization of (valid) legal norms. It is also quite often recognized that legal dogmatics is connected with other kinds of legal research such as sociology of law, history of law, philosophy of law and so on(1). Legal research thus comprises activities of many kinds. In this paper I would like to pay some more detailed attention to that branch of legal research that is called legal dogmatics, or in German 'Rechtsdogmatik'. In particular I am interested here in the question whether or not legal dogmatics is comparable to empirical social-scientific research.

2. SOCIOLOGY OF LAW AND LEGAL DOGMATICS

Legal dogmatics is, as I said, connected with many other kinds of legal research. But is it also comparable to other branches of legal research such as c.q. sociology of law or history of law? Let me take sociology of law as an example.

Sociology of law is purely descriptive and explanatory; it is an empirical science. It would however be misleading to characterize sociology of law as a discipline that aims at giving 1) a description of the (legal) norms valid in a certain legal system, and 2) an identification of the valid sources in the legal system one is studying.

In characterizing sociology of law in that way, one would give, in my opinion, a very narrow description of the subject of study of the sociologist of law. There is however a tendency among sociologists of law to restrict oneself to the behaviour of the legal officials of the

A. Peczenik et al. (eds.), Theory of Legal Science, 335–344.

system one is studying, or more general to what I would like to call 'norm behaviour'. By 'behaviour' in this connection I mean verbal behaviour (officials, such as judges, lawgivers etc. apply norms, express norms, enact norms and so on, thus showing verbal behaviour) as well as non-verbal behaviour. In this connection H.L.A.Hart is perhaps a good example. In his Concept of Law(2) Hart declares that his study could also be labeled as an essay in descriptive sociology of law; he defines 'law', as is well known, in terms of primary and secondary (legal) rules, and in Hart's opinion one can come to know these rules by studying the (verbal and non-verbal) behaviour of the legal officials.

It is however by no means necessary for the sociologist of law to restrict himself to the study of norm behaviour. Donald Black e.g. in his The Behaviour of Law(3) gives a description of what he calls 'law' without taking norm behaviour into account. It is even questionable whether norms or norm behaviour can, as a matter of fact, give causal explanations of the facts of social life in which a sociologist of law is interested. Empirical research shows that the role of legal rules as significant independent variables in the explanation of social phenomena is rather limited(4).

The tendency in the sociology of law to concentrate on norm behaviour and to recognize only the behaviour of legal officials as a relevant subject of study, is in my opinion due to the fact that many sociologists of law are under the influence of the perspective of the practical jurist or the legal dogmatist: both the practical jurist and the legal dogmatist of course consider their behaviour (both verbal and non-verbal) as important in the legal system. An explanation of the fact that quite a lot of sociologists of law are, in their definition of their subject of study, influenced by the perspective of the practical jurist, could perhaps be found in the circumstance that many students of the sociology of law are also legal dogmatists or practical jurists.

2.1 Some differences

It is however important to realize that it is (a priori) by no means plausible that the concepts used by participants in the legal system (like those used by legal dogmatists or judges) are also adequate concepts to describe processes taking place in the legal system. On the contrary; I am inclined to think that this is not the case.

The participants in a legal system (lawyers, citizens, judges and so on) use legal rules, principles etc. to classify their own behaviour, behaviour of other participants, to approve of behaviour, to justify behaviour and so on; these rules, principles etc. are thus reasons or motives for behaviour. But this reflects the situation viewed 'from within', from the internal perspective(5). If I want to justify my behaviour to another participant in the legal system, or if I want to 'explain' my behaviour to another participant in the legal system, I could do that by invoking legal norms or other rules: I acknowledge or recognize (to use a phrase of H.L.A.Hart) those rules and use them. But the circumstance that I, as a participant in the legal system, invoke norms, legal rules, vis à vis other participants in the legal system, does not mean that an (empirical) investigator of the legal system (i.e. a person who is studying, among other things, the behaviour of particpants of the legal system from the viewpoint of the spectator, i.e. not from the internal, but from the external point of view) should be bound to my internal point of view when such an investigator tries to describe and explain my (legal) behaviour. A scientific reconstruction of e.g. judicial behaviour is not eo ipso the behaviour seen 'through the spectacles of the participant'.

A possible difference between the point of view of the legal dogmatist, of the judge or, in general, of the participant in the legal system on the one hand, and the point of view of the investigator of behaviour in the legal system on the other hand, might be illustrated from the discussion whether or not 'law' (a legal system) is a system of rules (Dworkin)(6).

If someone, like e.g. Dworkin, wants to see principles (in contradistinction to rules) as justificatory grounds for judicial decisions, I am tempted to ask whether or not one is blurring, or at least neglecting, the distinction between the perspective of the decision maker, the judge himself, and the perspective of the investigator of the process of judicial decision making. When a judge, in the process of making up his mind, invokes principles or standards, or if he gets, with the help of principles, new ideas for the formulation of a new rule (e.g. because of the fact that he does not like the 'existing' rules and that he does not want to apply them in his case), then this does e.g. not mean that those principles or standards also give a justification of the newly formulated rule, not to mention an explanation of that new rule. The reconstruction of the decision making behaviour of the judge as applying rules is not equivalent to using those rules in the process of discovering a solution for a particular case. It is e.g. also possible that in the perspective of the judge himself, the judge thinks that he is not applying rules (he might even not be aware of any applicable rules), but principles; nevertheless, the possibility remains that an investigator of the decision making behaviour of the judge, as e.g. the sociologist of law, could reconstruct the behaviour of the judge as purely rule governed. Reconstructing behaviour with the help of scientific explanatory hypotheses and theories is something different from bringing about behaviour!

3. THE CONCEPT OF TRUTH

The foregoing means that e.g. judges or legal dogmatists can quite truthfully say from their 'internal' point of view that e.g. law is not merely a system of rules, but that at the same time e.g. a sociologist of law from his 'external' point of view can equally truthfully say e.g. that law is a system of rules.

The concept of truth for the participant in the legal system (i.e. from the internal point of view) is a concept of truth different from that of the empirical investigator of the legal system.

The proposition 'A is a valid norm' can be uttered both by the legal dogmatist on the internal level and by the sociologist of law on the external level; both the legal dogmatist and the sociologist of law can meaningfully say that this proposition is true or false. Nevertheless, the differences are great. The proposition of the legal dogmatist is uttered in a normative context; though this statement is in a descriptive form and though the participants in a juristic debate have discussions about the 'truth' or 'falsity' of the proposition, this proposition is 'normatively loaded' on the internal level. If a legal dogmatist says that 'A is a valid norm' he does not mean that he has drawn a conclusion from empirical data in a manner comparable to the ways of reasoning of the sociologist or natural scientist. What he does mean when he says that 'A is a valid norm', is (e.g.) that A has been acknowledged by the Supreme Court, that A gives good or at least acceptable solutions for the cases A is applicable to, that A allows for a better systematization of the other 'valid' norms, that the judge ought to apply A, and that therefore A 'is' valid. The legal dogmatist can only express himself in a descriptive way because of the fact that legal dogmatists have certain values in common (both methodological and axiological): because of the fact that these values are shared, there is no explicit reference to them. Except the case that there is, for a certain disputed legal issue, no set of shared values among legal dogmatists, there will be no reference to these values and it will look as if the proposition 'A is a valid norm' were a descriptive proposition. In fact however such a proposition is a mixture of descriptive and normative claims. The mixture-character will normally only come to the surface in cases where there is no unanimity about the values involved.

The situation with the sociologist of law is totally different. The claims of the sociologists of law, when putting forward phrases like 'A is a valid norm', are purely descriptive and empirical. Of course, the sociologist of law also has some values in common with his fellow sociologists of law, but these values have only to do with the methodological rules to be followed in the investigations and not with the content of law, or, in other words, these shared values are not axiological. The behaviour of the sociologist of law as a sociologist of law is not governed by legal norms or other non-methodological value viewpoints; the behaviour of the legal dogmatist however, is, in a sense, governed by legal norms.

4. SPLITTING UP LEGAL DOGMATICS IS UNDESIRABLE

A conclusion of the foregoing might be, that even if one were to characterize legal dogmatics and sociology of law in such a way that these disciplines would have in common that both sciences aim, among other things, at giving descriptions of the valid legal norms (which, by the way, is, as I argued before, a false supposition, since,among other things, it gives a too narrow characterization of sociology of law) even in that case, legal dogmatics and sociology of law (or any other discipline studying law from an empirical point of view) are non-comparable. Legal dogmatics is only quasi-descriptive.

Should we then not try to split up legal dogmatics into two parts, one descriptive (and 'scientific') and the other evaluative (and 'unscientific')? After all, a mixture of description and evaluation is neither epistemologically nor logically preferable. The descriptive part would gain in intersubjectivity, and perhaps also the evaluative part, since the values would then be explicitly stated.

Nevertheless, a splitting up of legal dogmatics into these two parts, is in my opinion not to be preferred.

Legal dogmatics fulfills an important ideological and political role in our societies. Legal dogmatics makes critical comments on judicial decision making, statutes etc. and is the most important part of the curriculum of the law student in the law school. If one tried to educate our law students in a 'scientific' way by restricting the education to the descriptive part and thus leaving the evaluation of e.g. judicial decision to the 'unscientific' practice after the law school-period (and to the 'unscientific' practice outside the 'scientific' law school), we would be taking the wrong road: it is very undesirable to educate people for, among other things, evaluating behaviour of other people, without discussing the process of evaluating during the academic training. This would be tantamount to making the future judges, lawyers, lawgivers and so on, blind to the normative, evaluative and political aspects of their respective jobs.

5. CHANGING THE PARADIGM

There is however a way of making legal dogmatics comparable to other sciences, to empirical sciences, without losing the political and evaluative functions of legal dogmatics.

This would however involve a change of paradigm in legal research. I am referring here to a proposal made by Hans Albert in his 'Erkenntnis und Recht'(7).

According to Albert the legal dogmatist restricts himself to two kinds of questions: first the _identification_ of the sources of the law, and secondly the _interpretation_ of these sources. Legal dogmatics is, in the context of these two questions, structured around the interpretation of texts; it thus treats legal norms _as if_ they were constituting an _autonomous cosmos_ of norms. The legal norms are treated as if they were _revealed_ by the legal sources. Legal dogmatics is thus structured around the '_revelation model_' of truth: the legal sources reveal what is valid or not. Legal dogmatics is, thus, as a matter of fact, a _normative_ science, which tries to give _knowledge_ about the normative order, the '_Sollen_' .

Legal dogmatics thus neglects in the opinion of Albert - and I think he is right on this point - the fact that legal norms are social facts, cultural acquisitions; law is one of many social factors influencing each other. In others words, there are relations between the sphere of the _Sollen_ and the sphere of the _Sein_, and these relations can be studied _empirically_. If we want to study law 'scientifically', and if we want to pay attention to the fact that law does not only play a role in the sphere of the _Sollen_ but also in the sphere of the _Sein_, then we have to treat the concept of 'legal norm' _also_ as a phenomenon from the realm of everyday social life, and study it with the help of suited methods and concepts.

This does not mean that legal dogmatics should turn into sociology of law. A legal dogmatist or a legal science which would specialize in studying empirical relationships between the _Sein_ and the _Sollen_ could (among other aims have as its aim the formulation of proposals for interpretation of the normative objectives recognized in valid law; such a legal science could of course also make proposals for the enactment of new legal norms. Such a legal dogmatics, or legal science, would be directed to the practice at the Bar _without_ having a normative character. This legal science would not be 'dogmatical' in the classical sense, but would work with hypotheses into which the relevant knowledge from the domain of the social sciences would be incorporated. This conception of legal science could perhaps be characterized as a discipline which conceives law as a means for 'social engineering'.

To my mind the study of the empirical relationships between the _Sein_ and the _Sollen_ can only be of _practical_ significance if legal norms are important means for a social behavioural technology. It is however an open question whether or not legal norms can be important in the realm of social engineering. As I pointed out before, the relevant empirical research until now has shown that the role of legal rules as significant independent variables in the explanation of social phenomena, is rather limited. Nevertheless, I think legal science should try to follow this path; it makes legal dogmatics comparable to other empirical social sciences, less evaluative and more intersubjective, without losing its relevance for the practice at the Bar. The question whether or not this path will be profitable, can however only be answered by empirical research.

6. CHANGING THE MODELS

There is still another way of making legal dogmatics comparable to empirical social sciences. In the foregoing I discussed the possibility of adapting legal dogmatics, or of changing the paradigm of legal dogmatics, in order to make it more intersubjective and less evaluative, so as to make the concept of truth (in the sense of this concept as used by empirical scientists) applicable within the domain of legal dogmatics.

It is however also possible to make legal dogmatics comparable to the paradigms of empirical research, leaving the paradigm of legal dogmatics intact, but by changing or adapting the models as used in the philosophy of science to characterize empirical science.

Aleksander Peczenik has, in his recent book *The Basis of Legal Justification*, given a survey of several models from the philosophy of science and discussed the question whether or not these models are applicable, be it in an adapted form or not, to the analysis of the structure of legal dogmatics. His conclusion is that both the models of Lakatos(8) (research programmes) and of Sneed(9) are applicable to legal dogmatics. I myself have argued *in extenso*(10) that there can be constructed a normative counterpart of Lakatos' model, which I have called the model of '*law development programmes*', and which is applicable both to the analysis of the process of judicial decision making and to the analysis of the structure of legal dogmatics. Perhaps also the views developed by Laudan(11) are applicable to legal dogmatics as is argued by Villa(12). I would like to conclude by saying that we should in our research follow the two paths I have indicated above. Two important questions for further research are thus:

(1) To what extent can the paradigm of legal dogmatics be changed that it becomes an empirical discipline _without_ losing its relevance for the practice at the Bar, i.e. without abstaining from practical advice in matters evaluative?

(2) In which ways can the models used in the philosophy of science be adapted so that they become applicable to legal dogmatics in its present form?

To answer these questions, both conceptual (logical and epistemological) and empirical research is needed. Trying to anwer these questions can only deepen our insight into the structure of legal dogmatics.

NOTES

(1) See e.g.Aleksander Peczenik: _The basis of legal justification_, Lund, 1983, p.118.

(2) First impression London 1961. See the preface to the first impression.

(3) Academic Press, 1976.

(4) See e.g.J.Griffiths: _Is law important?_, Deventer, 1978 and the literature quoted by him.

(5) See for an interesting characterization of the difference between the internal and external perspective: A.Wedberg: 'Some problems in the logical analysis of legal science', in: _Theoria_, 1951. See also my paper 'Progress in legal science' delivered at the 11th World Congress on Philosophy of Law and Social Science (Helsinki, august 1983).

(6) R.M.Dworkin:'Is law a system of rules?', in: _Essays in legal philosophy_ (ed. R.S.Summers) Oxford, 1968, p.2-60. See also: R.M.Dworkin: _Taking Rights Seriously_, London, 1977.

(7) H.Albert: 'Erkenntnis und Recht. Die Jurisprudenz im Lichte des Kritizismus', in: _Rechtstheorie als Grundlagenwissenschaft der Rechtswissenschaft. Jahrbuch für Rechtssoziologie und Rechtstheorie_ (H.Albert u.a. Hg.), Düsseldorf, 1972.

(8) I.Lakatos:'Falsification and the methodology of scientific research programmes', in: _Criticism and_

the growth of knowledge (I.Lakatos and A.Musgrave eds.), Cambridge, 1970.

(9) J.D.Sneed: The logical structure of mathematical physics, Dordrecht, 1971.

(10) See A.H.de Wild: The rationality of the judicial decision (in Dutch: De Rationaliteit van het rechterlijk oordeel), Deventer 1979. See also my paper 'Progress in legal science', delivered at the 11th World Congress on Philosophy of Law and Social Philosophy (Helsinki, August 1983), and my 'Legal reasoning from the perspective of the philosophy of science' in: Methodologie und Erkenntnistheorie der juristischen Argumentation (A.Aarnio, I.Niiniluoto, J.Uusitalo, Hgs.), Berlin, 1981, p.229-234.

(11) L.Laudan: Progress and its problems, London, 1977.

(12) Vittorio Villa: 'Theories of Natural Science and Theories of legal Sciences. Models and Analogies', paper delivered at the 11th World Congress on Philosophy of Law and Social Philosophy (Helsinki, august 1983).

ABSTRACT

In this article I discuss the question to what extent legal dogmatics (in German 'Rechtsdogmatik') is comparable to empirical social sciences such as e.g. sociology of law. Though it might be argued that there are some problems common to legal dogmatics and sociology of law, the differences are many and fundamental. Legal dogmatics and sociology of law differ both in methodology and semantics. The concept of truth as used by the legal dogmatist is only quasi-descriptive. Nevertheless the paradigm of legal dogmatics can be changed in such a way, that it becomes comparable to an empirical science, without losing its relevance for the practice at the Bar. In this way legal dogmatics also increases its intersubjectivity. Empirical research, however, shows that there are some limitations on changing the paradigm of legal dogmatics in the manner indicated.

Finally the possibility of changing the models, as used in the philosophy of science to characterize empirical research so as to make them suitable for enlightening the structure of legal dogmatics, is touched upon.

GOAL REASONS IN COMMON LAW CASES -- ARE THEY PREDICTIVE?

Robert S.Summers

The two most important kinds of reasons to be found in judicial opinions in common law cases are authority reasons and substantive reasons. Common law authority reasons consist primarily of appeals to the binding force of previously decided cases -- to precedent. Such authority reasons derive their justificatory force from two main sources: (1) the substantive reasons behind the precedent itself, and (2) the applicability of further substantive reasons that support the doctrine of binding precedent.

What are substantive reasons? Substantive reasons derive their justificatory force from moral, political, economic, institutional, or other social considerations. If I am correct, there are two fundamental types of substantive reasons: goal reasons and rightness reasons. A goal reason is a reason of a type that derives its justificatory force from the fact that, at the time it is given, the decision it supports can be predicted to have consequences that serve a good social goal. A good rightness reason does not derive its justificatory force from predicted goal-serving consequences of the decision it supports. Rather, a rightness reason draws its justificatory force from the way in which the decision accords with a socio-moral norm of rightness as applied to the parties' past interactions or to a state of affairs resulting from those interactions. Rightness reasons ordinarily invoke concepts of fault, equity, due respect, diligence, fairness, etc. -- concepts that originate in justificatory practices outside the law.

Substantive reasons, in my view have primacy in the common law. By "primacy", I mean that substantive reasons are more "foundational" than authority reasons, not that they are numerically more common, nor that they are always of over-riding significance. Substantive reasons are more "foundational" than authority reasons in four major respects. First, most of the justificatory force of precedents -- of authority reasons in the common law -- derives from the substantive reasons that figure in or lie behind

A. Peczenik et al. (eds.), Theory of Legal Science, 345–355.

the precedents, including the doctrine of precedent itself. Second, a judge cannot soundly interpret and apply the usual common law precedent without taking into account the bearing of the substantive reasons that figure in and lie behind the precedent.Third, in the American common law system, cases frequently arise in which there is no directly applicable precedent, or the applicable precedents are in conflict. Here substantive reasons must be decisive. Fourth, judges have power, within limits, to over-rule precedents. In such cases, the judges usually invoke substantive reasons to criticize the precedent being overruled and to formulate the content of the new precedent to take its place. In all these ways, then, substantive reasons have more foundational primacy than authority reasons

If I am correct, goal reasons are of major importance in common law cases. We must therefore try to understand the construction and evaluation of goal reasons. In 1978, I published the fruits of an extended study of a wide variety of examples of goal reasons I had identified in common law cases(1). I concluded that these cases reveal a number of features of the ideal goal reason -- one with maximum justificatory force. These features are as follows(2):

(1) All facts relevant to the construction and evaluation of the reason are indisputably true.

(2) The judge has an unquestionably reliable basis for predicting decisional effects, and the prediction itself is highly certain.

(3) The judge accurately characterizes the predicted state of affairs in all relevant respects.

(4) The predicted state of affairs affects or instantiates public interests - not merely the personal interests of the parties or of the judge; thus, the future state of affairs clearly serves (or disserves) a possible social goal.

(6) The predicted increase in goal subservience is plainly needed in the community (or the predicted decrease in goal subservience is plainly prejudicial to the community).

(7) The reason is well formulated.

Although it is abstract and idealized, this model is not a personal or utopian contrivance that I have conjured up without reference to the realities of common-law decisionmaking. Most aspects of the model are derived from the goal-reasoning practices of common-law judges. In this respect, the model represents a rational reconstruction of actual judicial practice.

In my 1978 account of goal reasoning in the common law, I claimed that judges commonly predict the out-of-court behavioral consequences of projected decisions either to serve or disserve social goals. For example, on my account, judges _predict_ whether holding newspapers liable for merely negligent as distinguished from malicious or reckless defamation of candidates for political office will lead newspaper editors to reduce the flow of information about candidates and thus diminish democracy -- a flourishing democracy being a desirable social goal.

Professor D.Neil MacCormick of the University of Edinburgh in the first Dewey Lecture at New York University delivered in October of 1982 has criticised my account on two grounds(3). First, he suggests that when common law judges typically engage in goal reasoning, they do not _predict_ the likely consequences of decisions. Second, he suggests that "in the main" the judges are not interested in the out-of-court _behavioral_ consequences of their decisions.

Professor MacCormick also offers what he consideres to be a positive alternative account of consequentialist reasoning in the common law. I will briefly respond to his two criticisms of my positive account, before turning to his claimed alternative. As for his criticism that judges do not, contrary to my view, actually predict or try to estimate the likely consequences of a projected decision (or rule based on the decision), I can only say that I have found what I take to be a good deal of judicial language and other evidence that at least American judges, when purporting to give goal reasons, do commonly seek to estimate the actual likely effects of alternative decisions. Whether this occurs so often that it can be said to be _typical_ when judges engage in goal reasoning I cannot say, but I would claim that it is very common.

As for Professor MacCormick's other criticism, namely that in giving goal reasons, judges are not, contrary to my account, commonly concerned with out-of-court _behavioral_ consequences of their decisions, here again, I have found what I think to be a good deal of evidence that at least American judges are commonly concerned with such consequences. Certainly many opinions include language that so indicates.

Where does this leave the dispute between Professor MacCormick and me, so far? The dispute seems to me to be largely over the facts and over the proper interpretation of those facts. He believes the facts are different from what I believe them to be. Of course, it is also possible that there are some differences between British and American common law judging here, and that we could both be right about what commonly occurs in our own countries. To try to settle our dispute, we would need at least to sit down together and examine a good many cases jointly. But the dispute is of real importance, for surely we need to understand how far goal reasoning is predictive and how far the objects of those predictions consist of out-of-court behavioral consequences. To understand such things is to understand important aspects of a general type of justifying reason -- namely, certain of the sources of its justificatory force.

I now turn away from Professor MacCormick's criticisms of my account to his claimed positive alternative account of goal reasoning. He says, as we have seen, that usually consequentialist reasoning in the common law is neither predictive nor behavioral. Rather, in his view the judges are concerned only or in the main with "possible" consequences of their decisions and with the "normative status" of those possible consequences as such. Professor MacCormick calls this "hypothetical consequentialism". Now, his claimed alternative account is very different from my positive account. He believes that his account is "in the main" true both of British and American common law and if he is right, my positive account is wrong, for I also claim that my account is in the main true of much if not most goal reasoning at least in American common law.

First, I will set out a detailed example Professor MacCormick uses to illustrate his claimed alternative account. From this, we can see rather clearly just what his account involves. I quote from Professor MacCormick's Dewey Lecture(4):

> "Do you recall the English case of Regina v. Dudley & Stephens? The case concerned the prosecution for murder of two shipwrecked mariners who saved their own lives by killing the cabin boy and eating him after they had been adrift in the lifeboat eight days beyond the end of their food supply and six since they had finished their last drinking water. The defense tried to set up a

principle of necessity as a justification for killing in order to save one's own life. But said Lord Chief Justice Coleridge:

> 'It is not needful to point out the awful danger of admitting the principle which has been contended for. Who is to be the judge of this sort of necessity? By what measure is the comparative value of lives to be measured? ...It is plain that the principle leaves to him who is to profit by it to determine the necessity which will justify him in deliberately taking another's life to save his own... [I]t is quite plain that such a principle once admitted might be made the legal cloak for unbridled passion and atrocious crime'.

Notice that the 'awful danger' apprehended here is an awful danger concerning what will follow logically from the principle if adopted. The only statable way of framing a suitable exception for necessity in the law of murder will be a ruling in law, or in Lord Coleridge's term a principle, which necessarily authorizes people whose lives are in pressing danger to judge whether they should kill another innocent victim of the same danger in order to save themselves. The alarming or unacceptable quality of the principle is exhibited by attending to its logical implications as a possible governing norm for future cases...

Of course, what is not certain is whether the hypothetical test cases will arise at all, whether passing judgment with reference to them will alter or affect behavior patterns in the community. Such questions about the outcomes consequent on adoption of a given ruling, questions about what Rudden has called "behavioral consequences", are indeed open only to conjectural answers. There are plenty of studies, although many more in respect of statute law than case law, of past legal changes that indicate how difficult and disputable it is to establish what have been consequences and outcomes of, or responses to, any given change in the general rules of law. All the more conjectural must be fore-

> casts about the likely impact of novel rules or rulings on the way people will conduct themselves. No wonder, then, if judges rest but lightly on such predictions".

This concludes Professor MacCormick's example, and I now resume my own discussion at this point. I would like to pause to pay tribute to Professor MacCormick before I respond. May I say that I believe the way he resorts to examples in legal theory is most salutary and sets a very high standard for us all. Wittgenstein taught us that "contempt for the particular case" is a great source of error in doing theoretical work. One finds no such contempt in Professor MacCormick's writings.

Now, to turn to his treatment of the famous case of Regina v. Dudley and Stephens(5). Let me put one matter aside at the outset. I do not deny that what I would call in my own general theory a powerful "rightness reason" lurks behind the general law against killing -- killing is generally wrong. I concede that in the opinion of the Dudley court this general rightness rationale exerts powerful pressure against the recognition of an exception for necessitous killing. But Professor MacCormick does not choose to analyse the case frontally along these lines. Rather, if I read him correctly, he treats the case as one resting on goal reasoning -- on consequentialist analysis.

Professor MacCormick points out that in the common law (civil and criminal -- and much of the law of excuse in criminal law is common law) judges often reach decisions by considering the consequences for future cases of the alternative possible decisions in the case before the court. One technique is to hypothesize cases under some alternative rule and if they would be unacceptably decided under that projected rule, then at least the court has a reason not to adopt that alternative, even though this may involve doing an injustice in the case at hand. Thus, for example, it might be that on the facts of the Dudley case, the defendants could not help doing what they did and also only killed the cabin boy as a very last resort, both factors that would make it unjust to punish the defendants in that case, yet but because of the consequences of recognizing a general defense of necessity for future cases, the court might refuse to recognize such a defense.

Now, I agree with Professor MacCormick that the Dudley case may be just such a case, but for quite different reasons from those he gives. That is, I do not interpret the Dudley case as he does. He claims that the consequences of recognizing a general defense of necessity that the court takes into account do no consist in the main of behavioral consequences but consist rather of what he calls "logically necessary implications" for future cases of recognizing the defense, and since he finds the logical consequences contrary to what is right, he interprets the decision rejecting the defense to rest merely on such "logical" (or "juridical") consequences. As he says(6):

> "The only statable way of framing a suitable exception for necessity in the law of murder will be a ruling in law, or in Lord Coleridge's term a principle, which necessarily authorizes people whose lives are in pressing danger to judge whether they should kill another innocent victim of the same danger in order to save themselves. The alarming or unacceptable quality of the principle is exhibited by attending to its logical implications as a possible governing norm for future cases."

Now, the court here did not in the actual opinion say that "the only statable way of framing a suitable exception for necessity in the law of murder will be a ruling in law ... which necessarily authorizes people whose lives are in pressing danger to judge whether they could kill another innocent victim of the same danger in order to save themselves". Indeed, this way of phrasing the claimed implication of recognizing a defense of necessity almost makes it sound as if certain persons are being invited to take life ("authorized") and to do so as a matter of cold calculation.

In truth, it is possible to frame a defense of necessity in the law of murder so that it is not necessarily so alarming or unacceptable as Professor MacCormick assumes. Such a principle might read as follows:

> "If it is found beyond a reasonable doubt that the defendant took the victim's life as a last resort, and could not have helped doing what he did, and that the choice of who was to be killed to save the defendant and any others was fairly done by lot, then the defendant may be acquited."

It is true that the Dudley court did not explicitly articulate any such principle as the one I have just hypothesized. But then neither did the court adopt Professor MacCormick's general wording. Indeed his wording does not take the form that defenses in the criminal law usually take. My wording is nearer to that form. At the same time, my wording is realistically conceived, at least in the sense that it is naturally suggested by the very facts of the Dudley case. There is evidence that the defendants in the case did kill only as a last resort and that they probably could not have helped doing what they did. There is no evidence that the cabin boy was selected to die by lot and thus also no evidence that he participated in any such process of drawing lots. Thus this narrow defense of necessity would not have acquitted these particular defendants, unless perhaps the requirement of choice by lot be eliminated.

Now, let us apply Professor MacCormick's analysis to this projected defense (or principle). His analytical procedure here is to "attend to the logical implications" of the projected defense "as a possible governing norm for future cases". By this procedure, would he concluded that, as a mere matter of logic, this norm would have what he calls an "alarming and unacceptable quality"? I think it would be something of a distortion merely to say unqualifiedly, as Professor MacCormick might, that this norm "necessarily authorizes people whose lives are in pressing danger to judge whether they should kill another innocent victim of the same danger in order to save themselves". At the very least, there is much more to the story. As a result, I fail to see how as a mere matter of logic, it follows necessarily that such a norm would have an alarming and unacceptable quality. Even if we drop the requirement of choosing by lot, and thus make the norm more clearly applicable to the circumstances of the defendants in the case, I still fail to see why, on balance, such a norm necessarily would have an alarming and unacceptable quality, overall. Here is what else there is to the story. It is not obviously just to punish someone who could not help doing what he did; persons who take life in such circumstances may be said not to act responsibly or at least to act with diminished responsibility (rightness reason). Moreover, allowing such persons a defense can also be justified as a concession to human frailty (rightness reason). Our criminal law does not make such concessions. Further, it cannot be said that defendants who act in circumstances to which the above defense would apply are deterrable by the prospect of incurring penal sanctions, and deterrence is a major function of the criminal law (goal reason). Another function of the criminal law is to express moral denunciation. In my view, defendants who can, on the facts, claim the above defense are not morally blameworthy and thus not fit objects of denunciation.

So far, then, I believe I have shown that it is not possible to rule out the foregoing, realistically conceived, defense of necessity using the type of "logical" or "juridical" consequentialism that Professor MacCormick says is not only at work in the _Dudley_ case, but it is typically(?) at work in common law goal reasoning. But I also want to try to show two further things. I want to try to show that it _is_ possible to justify _not adopting this very defense_ on the basis of the kind of predictive behavioral consequentialism that I claim is common in American common law cases. I also want to try to show how it can be argued that the court in _Dudley_ was in fact to a substantial extend making just such predictive behavioral arguments when it actually refused to recognize the defense of necessity. If I can show that the goal reasoning in _Dudley_ was predictive, this will be especially significant, for the fact pattern in that case is not of the recurring type in which one would expect to find much predictive reasoning.

Even if, in the abstract, we can justify recognizing a defense of necessity along the lines I suggested above, a court might still justifiably refuse to adopt the defense in the end. First, it can be predicted that the defense would be difficult to administer consistently and faithfully to the facts of the actual cases that do arise. That is, the defendants would usually be the only witnesses to the taking of life involved, and thus would be in a position to manufacture false evidence. As a result, courts and jurors would have special difficulty in differentiating between genuine cases of necessity and cases in which the necessity really rests on false evidence. It is also true that the difficulty of consistently administering the defense would be all the greater because different courts and jurors would not agree over time about whether a killing really was, conceptually, a "last resort" and about whether it was by defendants who could "not have helped" doing what they did. I believe the court in _Dudley_ had such reasoning as this in mind when it said: "Who is to be the judge of this sort of necessity", and when it said:

"It is plain that the principle leaves to him who is to profit by it to determine the necessity which will justify him in deliberately taking another's life to save his own".

Reasons of this kind are goal reasons and involve predictions of likely consequences in future legal cases. These consequences are behavioral, although they occur in court. Because they relate to the institutional apparatus of courts -- here their fact-finding capability and their capacity to function in a generally consistent fashion -- I classify these reasons as institutional goal reasons in my own general scheme for analyzing types of reasons(7).

But I also believe that a goal reason involving out-of-court behavioral consequences can also be given to help justify refusal to recognize the defense of necessity I have hypothesized. Partly (but not only) because of the force of the foregoing institutional reasons, it can be argued that recognition of the defense of necessity might influence some prospective defendants to take life as a first resort and in circumstances in which they could have helped doing so, all the while believing that if they were caught, and if worse came to worst, they could manufacture false evidence that would lead to their acquittal under the legally recognized defense of necessity. After all, the law of murder is one of those few branches of the law the leading principles of which tend to be generally known to many lay persons. Now, I believe that the *Dudley* court did think in just these terms, when it said that the defense of necessity, "once admitted might be made the legal *cloak* for unbridled passion and atrocious crime". Here the reference is plainly to predicted out-of-court behavioral consequences. Professor MacCormick notes this language in the *Dudley* opinion but characterizes it as a mere "hint" of predictive behavioral consequentialism(8). I think it is more than a hint.

If I am right, two kinds of predictive goal reasons, an incourt institutional reason and an out-of-court behavioral reason, then, go far to justify judicial refusal to recognize even a carefully formulated general defense of necessity to the crime of murder. If I am right, the *Dudley* case refusing to recognize a defense of necessity to the crime of murder rests mainly on just such reasons and not on mere logical implications of a projected defense of necessity, however formulated. Further, I believe this to be characteristic of goal reasoning at least in American common law cases. Such reasoning is predictive, and very largely in-court institutional or out-of-court behavioral or both.

Even if I am wrong about the extent to which predictive goal reasoning of these kinds actually influenced the court, and even if Professor MacCormick is right that in the actual Dudley case the court's reasons are of the logical sort he stresses, it would remain open to me to make the further claims: (1) that the reasons I identify and stress would be alone sufficient to justify refusal to recognize the defense, and (2) that however much Professor MacCormick's analysis fits the reasoning of Victorian judges, many judges today at least in America would find it more congenial in such a case to rely on the kinds of reasons I identify and stress.

In closing, I should say that this topic raises two further issues I hope to treat later, namely, the extent to which Professor MacCormick's non-predictive "logical consequentialism" applies to what I call rightness reasons, and the matter of the various significantly different ways judges reason when they test proposed decisions against likely future cases.

NOTES

1) Summers, Two Types of Substantive Reasons: The Core of a Theory of Common-Law Justification, 63 Cornell L.Rev.707 (1978).
2) Id.at 748.
3) MacCormick, On Legal Decisions and Their Consequences: From Dewey to Dworkin, 58 N.Y.U. L.Rev.239 (1983).
4) Id. at 250-51.
5) Regina v. Dudley and Stephens, L.R. 14 Q.B.D. 273 (1884).
6) MacCormick, supra note 3, at 251.
7) Summers, supra note 1, at 722-24.
8) MacCormick, supra note 3, at 252.

TELEOLOGICAL CONSTRUCTION OF STATUTES

Per Olof Ekelöf

1. BYDLINSKI ON THE CONSTRUCTION OF STATUTES

A recent book about the construction of statutes is *Juristische Methodenlehre und Rechtsbegriff* (1982) by F.Bydlinski, professor of civil law at the University of Vienna. He describes *Wertungsjurisprudenz* (*wertkritische Methode*) as the art of application of statutes which at present is the most fashionable one in Germany and Austria(1). This method is in essence "objective-teleological" and considered as a modification of the *Interessenjurisprudenz* of Philipp Heck(2). In contrast to Heck's theory, the purpose of a provision should not merely be deduced from the state of affairs prevailing at the time when the provision came into force.

However, Heck meant that his theory about the application of statutes should be used generally. Bydlinski, on the contrary, recommends that - at least in principle - all the various means of interpretation should be used in a fixed order of priority(3). The judge should first interpret the statute according to its wording. If this does not bring about an acceptable result, he passes on to a systematic-logical and then - if necessary - to an historical interpretation. Only as a last possibility should one resort to the objective-teleological method. But how is the *télos* or in other words the purpose of a provision to be ascertained? Bydlinski means that this should be done on the basis of the "idea of justice" and such general principles as *Rechtsfriede* and *Gerechtigkeit*(4).

But what makes a particular literal interpretation or a systematic or historical one acceptable? As regards this we must consider that all provisions are more or less vague and that consequently their field of denotation can be divided into a core and a penumbra (fringe) of the meaning. To the core belong such cases as the provision quite obviously covers, while a case in the penumbra represents a semantically indeterminate meaning. Current

A. Peczenik et al. (eds.), Theory of Legal Science, 357–362.

usage is not ever so fixed that one can ascertain whether a particular case is covered or not by the meaning of the provision. Bydlinski thinks that a method is practicable only if it eliminates the vagueness of the provision(5). And the ranking order cannot be used as a basis of an application per analogiam(6).

However, Bydlinski emphasizes that an acceptable interpretation must fulfil another condition, too. The result of - for instance - an historical interpretation is unacceptable if it disagrees with the objective-teleological method. This test must be used as Begleitkontrolle for all other methods of interpretation(7). But would it not under these circumstances be simpler to employ the objective-teleologic method from the beginning?

It could be of interest to compare Bydlinski's theory with another teleological method, to which I have drawn up an outline in my essay "Teleologische Rechtsanwendung" in Österreichisches Zeitschrift für öffentliches Recht 1958 pp. 174-189(8). There I distinguish between the interpretation and the construction of a provision. By the former one establishes the contextual meaning of the provision. This will be done in the same ordinary way as in the interpretation of - say - a sentence in a newspaper article. However, the interpretation will not be conducted further than to establish cases of ordinary types, belonging to the core of the meaning of the provision(9). The practical effect of its application on these cases I regard as the purpose (function) of the provision. On the basis of the function, thus established, I then determine the borderline between a restrictive (narrow) and an extensive (broad) application of the provision in its penumbra-field. And in the same way I solve problems of analogy and reduction. One must "reduce" a provision when an application would be in opposition to its purpose, even when the case at hand belongs to the core of the provision's meaning(10). With this method one constructs a field of application, which may more or less diverge from the meaning of the provision and can be considered as an "explication" of it.

According to this theory the relevant purpose is established in close connection with the provision and not on a sweeping estimate of the demands of justice. At the end of my essay I emphasize that my theory cannot be used to supplement provisions in force with a completely new norm, but only to correct semantical shortcomings in the legislation. The former should be a task for the legislature and not for the courts.

However, it seems to me that this opinion is becoming antiquated. Recently the changes in Western countries have been so rapid that the legislatures cannot adapt the written law to the new state of affairs speedily enough. As a doubtful remedy broad provisions, referred to as "general clauses", are being used(11). In Sweden there are statutes, where the whole material regulation merely consists of one such clause! Bydlinski has a whole chapter about gesetzübersteigende Rechtsfortbildung which certainly is intra jus but extra legem. In this connection, too, he recommends the idea of justice as guidance(12). To my mind, the formulation of a norm with a more comprehensible content is a most urgent task for jurisprudence.

2. COMMENTS TO PROF. MACCORMICK'S PAPER

Now, I want to compare the opinion of professor MacCormick about normative coherence with my own theory about teleological construction of statutes, to which I referred in the previous section. According to MacCormick a set of norms is coherent, if they "make sense together", in the signification that they "subserve a common value". He also believes that the application of the norms on concrete cases should be founded on this value. He demonstrates this opinion by an illustrative legal case, in which the courts had to apply the following paragraph in the Dangerous Drugs Act of 1965:

> "If a person - (a) being the occupier of any premises permits those premises to be used for the purpose of smoking cannabis---; or (b) is concerned in the management of any premises used for any such purpose as aforesaid; he shall be guilty of an offence against this act".

Relevant in the legal case at hand was the provision b and the question was, whether this provision was applicable even if the manager - living in another part of the country - did not know anything about the cannabis smoking in the premises. In the provision a the word "permits" plainly shows that such a knowledge is a prerequisite to the liability. But as you see the wording in b is different and its sense may be that the manager has strict liability

The courts found also that this was the case and convicted the manager. However, the House of Lords was of a different opinion and acquitted the accused. According to MacCormick, the judgment in the last instance was based on "a coherent view of this branch of the law", and "on an analogous application of the same principles or values and thus as coherent with the pre-established body of law". The applicable principle is "the general principle as to the differentiation of offences of absolute liability from those requiring mens rea". On offences similar to those described in the Drugs Act, there is in English law no strict liability but mens rea is required. Certainly the learned lords expressedly founded their judgment on "the intention of the legislator". But in MacCormick's opinion "there is an air of pious fiction about such rhetoric". And obviously the House of Lords had not made any historical investigation into what the originators of the paragraph had intended in the past.

The method that I recommend in the article mentioned above seems to resemble the method of MacCormick based on "normative coherence". In both methods valuations play an important role and he and I both reject an interpretation founded on the "intention of the legislator". But there are also differences which can bring about varying practical effects when applying a provision. According to MacCormick the relevant value is of a general character like "safety, health, considerateness and justice". And you can also use general principles based on such values or "on analogy from provisions in other statutes, in their character similar to the statute at hand".

On the contrary, according to my teleological method you have to establish narrowly which quite restricted function the provision b fulfils in the framework of other provisions in the same statute. In the provision a the interest of the occupant is - quite clearly - secured by the condition that he had knowledge about the cannabis smoking. It stands to reason that a manager, who is not living in the premises, shall have the same privilege. Consequently, the coherence of the two provisions presupposes that also such a person is liable only if he knows of the cannabis smoking. This is a reasoning simpler than the one I describe in my paper and which is adopted to more complicated problems of application. But you see that my construction of the provision b is based not on the purpose of any broad principle but on the paragraph's practical function in the framework of the Drugs Act. The provision b shall be applied so that the practical effect of the statute will be realised as well as possible.

I think broad principles of law are of no use in applying a statute on concrete cases but only in applying precedents. In the arguments of the courts you do not always find a norm but only some substantial facts and their legal consequences. Then you get a good guidance for the future by inducing a principal from the precedents.

As a reason for this opinion MacCormick maintains that ordinary people only grasp an application based on general principles of law. But I think that people can very well grasp a teleological argumentation, if this is clearly recorded in the judgment. Unfortunately, this is often neglected by the courts. Even if the real reason for his judgment is teleological, a judge often does not want to announce this, but prefers to found the tenor on factual circumstances, as on "the plain meaning of the statute" or on "the intention of the legislator" - although this is a pure "show justification". I think that Bergholtz in his paper "Reasons and causes in connection with judicial decisions" (published elsewhere in this volume) has not observed this practice of the courts.Certainly, the theory of the American realists that the courts are influenced by what is ironically called "stomach jurisprudence" is an illusion - as Bergholtz shows in his paper. But despite this you cannot be sure that the arguments in the judgment also are the real causes for the judge's decision about the outcome of the case.

Finally, I want to say a few words about what Professor MacCormick calls "narrative coherence" which "provides a test to the truth or probable truth of propositions about unperceived things and events". As application of provisions and precedents should be based on "normative coherence", evaluation of evidence should be founded on "narrative coherence". Certainly, such an evaluation is not only based on the perception of facts but also on laws of nature and commonplace generalisations. And between these there must be "coherence". But the criterion for this coherence seems to me to be quite different from that of normative coherence. As we have seen a set of norms is coherent, if they have a common purpose, aim, goal. On the other hand, laws of nature and commonplace generalisations shall represent a consistent view of the world and ought consequently not to be contrary to each other. When "evaluating" evidence we do not consider something as convenient or inconvenient but estimate approximately the probability of a course of events. MacCormick says "that we could not have a rational normative order without having also a rational worldview". That is right but this does not imply that the application of norms has something in common with the evaluation of evidence.

NOTES

(1) Pp.123-139.
(2) Pp.131 and 542.
(3) Pp.557 and 560.
(4) P.133.
(5) P.561. cf.pp.
(6) Pp.557 f.;467 f.
(7) P.565.
(8) See also my essay 'Teleological Construction of Statutes in _Scandinavian Studies in Law_ 1958 pp.75-117.
(9) Cf.Bydlinski pp.543 and p.549.
(10) Bydlinski neither distinguishes between restrictive and reductive applications nor between extensive application and analogy (pp.480 and 475).
(11) Pp.481-500.
(12) Pp.486 f.

REASON, LAW AND HISTORY

Enrico Pattaro

1. THE ISSUE

The purpose of the present paper is to illustrate the following passage by Guido Fassò.

"Sociality, (a) which is equivalent to being legal, (b) is the necessary form of human history (c) so that the latter can be human and history. (d) Sociality means being legal just because it is multiplicity, (e) and because multiplicity implies reason (f) which puts an order in it and organizes it in a system: (g) theoretical and practical reason at once, (h) which, in its being practical cannot be but legal, (i) that is, involving the statement of abstract rules. (j) Law is, as it was once called, ratio: (k) reason, that is, scheme, (l) abstraction: (m) in practical life law is the same as science is in theoretical life. (n) In fact, law is nothing but a translation of science into an imperative form"(1).

2. WHICH KIND OF REASON

Let us start from "multiplicity" and "reason". Idealism used to distinguish between reason and understanding. Understanding was assumed to deal with multiplicity, or better to produce it. Multiplicity is not true reality, it is appearance. Understanding produces pseudoconcepts, its outcome is science, which is not true knowledge, but a kind of practical, economical activity. This was, for instance, Benedetto Croce's view(2).

Reason, on the contrary, was assumed by idealism as true knowledge. Reason knows the concrete unity of the real, the absolute, the pure concept, the concrete universal. The outcome of reason is philosophy or speculative logic as it was conceived of by idealism.

Reason, unity, true knowledge, true reality, the abso-

A. Peczenik et al. (eds.), Theory of Legal Science, 363–372.

lute, the unconditioned, philosophy: this was the metaphysical top of the idealistic conception.

Fassò cut out the metaphysical top of idealism, and topless idealism revealed the following(3).

(i) Concrete unity of the real, the absolute, the unconditioned, the concrete universal, true reality beyond the appearance and so on are not matters of human knowledge. They are a matter of concern for religion.

(ii) Idealistic reason conceived of as a process of identification between subject and object, ego and nature, thought and reality, philosophy and history and so on, is an illusion. Processes of that kind are possible only in mysticism, which is however the very negation of reason.

(iii) Reason is reduced to what idealism called understanding. It deals with multiplicity and produces science. Science does have many of the limits indicated by idealism, but it happens to be knowledge: it is the _only_ possible kind of human knowledge.

Correspondingly, concrete multiplicity of the real is the only possible object of knowledge. Reality is an endless flow of concrete particulars, a heterogeneous _continuum_ of original unrepeatable events (Heraclitus). In space and time there is not a pair of perfectly identical phenomena. Such a reality is irrational, but it is not beyond reason (_viz._ understanding) as the idealistic concrete unity of the real is. It precedes human schemes of thought, it does not transcend them. It is grasped through a common empirical experience, not through an exceptional mystical experience as the concrete unity of the real is.

The concrete, irrational, changing multiplicity of the real reminds one of the Kantian matter of knowledge consisting of various, disordered sense impressions(4). Human understanding, that is reason, puts an order in such multiplicity by abstracting, generalizing and issuing scientific laws ("naturalistic laws", as Fassò calls them).

Scientific laws are sentences announcing relations among empirical concepts of phenomena. Such relations are necessarily established by the human mind, which cannot avoid putting an order among empirical phenomena (cf. Kant, instrumentalism, Popper). In fact, in this way the human mind tries to attain a unity out of the scattered data of experience: unsuccesfully, however, because the unity of the real - as we have seen - is beyond the schemes of the human mind.

Scientific laws are built up by Man in his eternal

effort to reduce multiplicity to unity, becoming to being, contingency to necessity, the changing to the lasting.

Such an effort is the effort to make rational the irrational. Since the only means of knowledge Man has is reason, then he tries to force any object of his knowledge into the schemes of reason. On the other side, knowledge is the indispensable basis of any action. Mere experience of immediate particular data is not knowledge, nor is it a sufficient basis for behaviour. Every action implies a scheme of action: we act according to schemes of action even when our action is not a conscious one(5).

What I have just said affects points (b), (e), (f), (k), (e) in Fassò's mentioned passage.

As for (e) and (f) Fassò assumes a correspondence between multiplicity and reason. Here he is echoing Kantian themes. However, he drops the Kantian distinction between schemes and understanding, on the one side, and between understanding and reason, on the other. According to him reason, schemes and abstraction are the same (point k and l in Fassò's mentioned passage).

Moreover, according to Fassò, knowledge is necessary only in the sense that Man cannot avoid using schemes and cannot behave without schemes. This does not mean that knowledge is objective and universal (universally valid) as it is according to Kant. On these issues Fassò is closer to Hume and Popper than to Kant. Man expects that between two phenomena there is always the same relation he has observed between two other similar phenomena. This expectation however cannot be justified. It is just a subjective belief.

As I have said before, Fassò, while refuting the idealistic tenet of a speculative, absolute, universal knowledge, preserves the idealistic doctrine about the limits of scientific knowledge. Reason, which produces scientific knowledge, is the only (and unavoidable) guide Man has in his dealing with reality: it is, however, a guide of relative validity, defective and fallible.

As for (b), the assumption that Man uses reason necessarily in the process of knowing and acting does lead to the conclusion that law is the necessary form of human history on the condition that other assumptions are made, namely: that law is reason and human history is the whole of Man's actions.

In order to introduce such assumptions let us now consider how the relation between theoretical and practical reason is conceived of by Fassò.

3. SCIENTIFIC AND PRACTICAL LAWS

Fassò maintains that scientific laws and practical (imperative) laws, albeit mutually irreducible one to another, can be turned one into another. They are mutually irreducible because, while scientific laws belong to knowledge, practical laws belong to will. They can, however, be turned one into another as follows.

(i) Given the scientific law of a phenomenon, if you want to produce the phenomenon, you shall assume its scientific law as a rule for your behaviour. E.g.: "Platinum melts at 1780°c." (scientific law); "if you want to melt a piece of platinum, you shall bring it to a temperature of 1780°c." (practical law).

(ii) Given any practical (imperative) law, its enactment implies an issue having an aim: an issuer who has assumed that if such a law is obeyed, then his aim will be attained. This is to say that any practical law is a prescription to act according to a scientific law of a given phenomenon the realization of which is aimed at by the issuer of the prescription. E.g.: "Bring this piece of platinum to 1780°c" (practical law) implies that the issuer aims at getting the piece of platinum melted and has assumed that "platinum melts at 1780°c." (scientific law)(6).

4. REASON AND MORALITY

Among the consequences attached to such views are the following ones. All practical laws are technical rules because they put forward scientific laws in an imperative form. There is no categorical imperative. Any will, in its being distinguished from mere instinct, is rational. In other words, any will is the volition of a means in view of an end. Rationality, here, consists in the establishing of a relation between a phenomenon taken as an end and another one taken as a means to this end. The choice of the so-called final or ultimate ends cannot be but irrational (emotional) as they are not chosen as a means for a further end. The value of any practical law is just the value of that scientific law which has been turned into an imperative form through the practical law: it is the relative value of reason which is abstract, fallible and defective.

Fassò assumes that true morality and values are absolute. The absolute, being beyond reason, can neither be an object of knowledge nor of volition. Moral volition is impossible because any volition requires a means-end law. The absolute has no law, transcends any law, be it theoretical or practical. As a consequence human values and morality are not, properly speaking, values and morality(7). They are, Fassò maintains, legal values and law: any practical activity is rational, which prevents it from being moral; law is science in an imperative form, it is rationality in practical life; any practical activity cannot be but legal (points g, h, i, j, m, n in Fassò's mentioned passage)(8).

Despite these assumptions it would be misleading to say that Fassò reduces morality to law. In fact he does the following. He distinguishes sharply between true morality and any human activities. True morality, in his view, is divine, it belongs to a religious mystical élan experienced by very few exceptional men, for instance by Jesus Christ. Such an experience is not only rare and exceptional, but also unlasting as it is not compatible with the needs of any historical institutions. Churches themselves, albeit originally inspired by the example of such exceptional religious men as Jesus Christ, do not have and *cannot* have anything to do with that true morality experienced by those to whom they attribute their origin, faith and devotion. On this subject Fassò wrote a very important, learned and beautiful book, *Christianesimo e società* (Christianity and Society), which was largerly debated in Italy(9). Apart from such true morality, which is impossible to apply in human affairs, Fassò maintains that any human activities are in the final analysis legal.

5. HISTORY AS INSTITUTIONAL REALITY

Fassò names "institutions" all relations among human (historical) phenomena. For example, not only the family or the state, but also a queue waiting for a bus or a couple making love, are institutions(10). A human (historical) phenomenon is any phenomenon where a man is involved. On the other hand, both knowledge and will imply establishing a relation between two or more phenomena. As a consequence, any relations Man puts between two or more phenomena by knowing or willing them is an institution provided that they are human phenomena. In fact, Fassò goes further. As knowing and willing are human phenomena themselves, also a relation Man puts between two or more natural phenomena by knowing or willing them is an institution. At this very point, if you demonstrate that all institutions are legal, you will also demonstrate that all human activity is legal, that "history is legal experience", to use Fassò's expression.

Before considering this final conclusion let us remind that human history is made equivalent by Fassò to human activity, and notice that human activity seems reduced to knowing and willing. As for knowing, Fassò seems to leave out of it only immediate experience of sense-data, immediate feelings, appetites, impulses, and mystical élan (true morality included). In his opinion, however, moving away from this immediate basic experiences, if only a few steps, is ipso facto involved in schemes of thought, and, therefore, in reason and knowledge, irrespective of one's awareness of it.

As for willing, Fassò assumes, as we have seen, that any scientific law can be the object of volition and thereby turned into a practical law. A scheme of thought, on the other hand, is considered by him as involving a scientific law, that is a relation established between (at least) two objects or elements of an object. As a consequence, any unit of knowledge is the potential object of an act of will. Correspondingly, any volition involves the use of a scheme of thought, consciously or unconsciously assumed by the willing person as a scheme of action.

At the roots of history, there are then (irrational) sense-experiences, feelings, appetites, impulses, possibly mystical élans too. But history in itself is (rational) human activity, that is, knowledge and will, which -as rational- cannot be but knowledge and will of relations among phenomena, knowledge and will of institutions: history is an institutional reality(11). What I have just said affects points (b) and (c) in Fassò's mentioned passages.

6. LAW AND SOCIALITY

Fassò considers the features commonly assumed as peculiar to law and especially to statutory law, namely: generality and abstractness, imperativeness, coercion, organization, intersubjectivity.

As for generality and abstractness, they are characteristics typical of any schemes of thought or action. In this respect, then, no difference is recognized between the law of a state and any scientific or practical law. As for imperativeness, as we have seen, Fassò assumes that every scientific law can be turned into an (imperative) practical law and viceversa. In this respect too, then, no difference is recognized between the law of a state and any other practical law. As for coercion, every imperative implies some degree of it. In this respect, there are differences between statutory laws and, say, a practical law issued by an individual. They are, however, differences in degree, which are obviously important in a scientific consideration of law, but irrelevant to a philosophical consideration of it. As for organization, the same remark applies: there are differences in degree between the sofisticated organization of a modern state and the very simple relations occurring among primitive men. Such differences, however, are not, and must not be, relevant to a philosophical consideration of law. As for intersubjectivity, it is the relation established through a law between two or more persons. As we have seen, according to Fassò, every scientific or practical law establishes a relation between two objects. Fassò goes further assuming that every practical law establishes a relation between two persons. In my opinion, a difficulty arises here, because it is logically possible to think of our world as inhabited by only one man: in such a case no intersubjectivity is possible. Fassò, however, explicitly maintains that even the first and the last men on earth, albeit being alone, think and act in terms of relations to other men, that is in terms of sociality. On such assumption Fassò states in the aforementioned passage (a) that sociality "is equivalent to being legal" and (d) that "sociality means being legal just because it is multiplicity".

In conclusion, all practical laws are legal and all relations they establish are legal institutions, because all practical laws are (i) general and abstract, (ii) imperatives, (iii) coercive to some degree, (iv) moreover they establish a relation, which is (v) a kind of organisation and (vi) at least potentially, intersubjective. Moreover because of (vii) sociality is equivalent to being legal.

8. REASON, LAW AND HISTORY

Let me quote another passage by Fassò. "Though imperative, law is always a product of the abstraction by means of which our reason isolates and schematizes the infinite, flowing aspects of human history, just as it (the reason), when dealing with equally infinite and flowing aspects of nature, isolates and schematizes them in scientific constructions. With regard to nature these constructions of our understanding preserve an indicative form (even if they, too, turn into the imperative form as soon as their object becomes an aim for human action), while we are accustomed to consider them also and above all in their normative version with regard to history. In both cases, however, the logical phenomenon is the same, just as its outcome is the same laws and abstract rules, which catch generalia, _id quod plerumque fit_, or, which is the same, _id quod plerumque debetur,_ in the Heraclitean stream of the concrete particulars. In fact, it makes no difference and it is somehow incidental whether those abstract _generalia_ are arranged in the _is_-form or in the _ought_-form"(13).

Idealism identified history with philosophy, meaning that it is absolute reason and true morality in process. Fassò shows and maintains that history is legal experience: it is law, that is scientific, fallible reason in the imperative form. Theoretical reason is science, practical reason is law. The absolute, that is morality, is not of our world.

There is a strong moral liberal stand at the basis of

Guido Fasso's philosophy of law. "I leave history -he wrote- and society, which is Man's dimension, to Man's instruments: to experience, and to a reason not pretending to possess the absolute truth and then to know the absolute value, but working out experience-data, organizing them in building up human, imperfect, historical truths, and pursuing its own values on their basis. I exclude from history the absolute values all kinds of rethoricians want to introduce in it, the absolute values of those who erect dogmas by taking their own beliefs as the universal model of knowing and behaving because of their velleity of lowering into history the superhuman life of the Absolute, of which, in their little humility, they presume to be the interpreters"(14).

NOTES

(1) G. Fassò, La storia come esperienza giuridica (History as Legal Experience), Giuffrè, Milano, 1953, pp. 103-104 (my translation and letters in parentheses).

(2) Cf. B.Croce, Logica come scienza del concetto puro (Logic as Science of Pure Concept), Laterza, Bari, 1947, pp. 7 ff., 20 ff., 30, 43 ff., 114 ff., 159, 212 ff., 219 ff., 337 ff., 346, 356, 359, 360, 390; Filosofia della pratica (Philosophy of Practice), Laterza, Bari, 1957, pp. 7, 33, 40, 69 ff., 92, 125, 250 ff., 254, 260 ff., 327 ff., 342, 347 ff., 403, 404. Cfr. E.Pattaro, Sull'assoluto. Contributo allo studio del pensiero di Guido Fassò (On Absolute. A Contribution to the Study of Guido Fassò's Thought), in G.Fassò, Scritti di filosofia del diritto (Legal Philosophical Writings), edited by E.Pattaro, C.Faralli, G.Zucchini, I, Giuffrè, Milano, 1982, pp.LXXVI-LXXVIII, note 69, where Croce's and Gentile's views are compared with I. Kant's.

(3) Fassò's critique of idealism is illustrated in my Sull'assoluto, pp.XXXVIII-LXI, and was the main topic of my oral presentation of the present paper at the Lund conference.

(4) Cf. I. Kant, Kritik der reinen Vernunft, 2nd ed., in I. Kant, Werke, edited by E. Cassirer, III, verlegt bei Bruno Cassirer, Berlin, 1923, pp. 34, 113 ff., 156.

(5) Fassò, La storia come esperienza giuridica, pp. 98,

104, 105, 40-41, 50.

(6) Example (*i*) is given by Fassò who took it from Benedetto Croce. Example (*ii*) is mine: in this case, Fassò uses another, more complicated example taken from Croce too, needless to be considered here.

(7) Fassò uses capital letters when speaking of values and morality in the "proper sens".

(8) Cf. Fassò, *La storia come esperienza giuridica*, pp. 39 ff., 133 ff.; *Legge naturalistica e legge pratica* (Naturalistic and Practical Laws), in Fassò, *Scritti*, I, p.165 ff.; Pattaro, *In che senso la storia è esperienza giuridica: l'istituzionalismo trascendentale di Guido Fassò* (What is is Meant by History as Legal Experience: Guido Fassò's Transcendental Institutionalism), in *Rivista trimestrale di diritto e procedura civile*, 1983, p.404 ff.; *Sull'assoluto*, pp. XXVIII ff., LXII ff.

(9) Fassò, *Cristianesimo e società* (Christianity and Society), Giuffrè, Milano, 1956, 2nd ed. 1969.

(10) Cf.Fassò, *La storia come esperienza giuridica*, pp.59 ff., 96 ff.,

(11) Cf. Pattaro, *In che senso la storia è esperienza giuridica*, pp.413-423.

(12) Cf. Fassò, *La storia come esperienza giuridica*, p.101.

(13) Fassò, *ibidem*, pp.103-104 (my translation).

(14) Fassò, *Cristianesimo e società*, 1st ed., p. 157 (my translation).

THE RULE OF LAW IN LEGAL REASONING

Mats Mattsson

This paper will deal with some problems in connection with "the Rule of Law", a concept frequently used by F.A. von Hayek. To the legal dogmatician it is the judicial relevance of the concepts that is important and in my opinion von Hayek's concept of the Rule of Law, being adopted for use within the field of political science, is less useful in judicial contexts. After some brief notes on Hayek and the Rule of law I will discuss a similar Scandinavian concept. Finally, I will argue for a judicially relevant and useful Rule of Law concept.

1. HAYEK AND THE RULE OF LAW

Hayek has in several contexts made attempts to define the meaning of the Rule of Law(1).

"government in all its actions is bound by rules fixed and announced beforehand - rules which make it possible to foresee with fair certainty how the authority will use its coercive powers in given circumstances, and to plan one's individual affairs on the basis of this knowledge(2)"

On the whole I do agree with Hayek that predictability is closely associated with the Rule of Law. I consider, however, Hayek's definition to be too wide to be of practical use in legal contexts.

Hayek's idea has been further developed by Joseph Raz in his paper "The Rule of Law and Its Virtue(3)". Raz's concept seems to be better adopted to legal reasoning. In the following I will therefore direct my criticism at Raz although it might just as well have been directed at Hayek.

Raz considers the Rule of Law to be a formal concept, in no need of ethical undercurrents. Thus legal rules may have any content as long as court decisions are predictable. Such decisions are divided into two categories: predictable and unpredictable ones. Now Raz seems to mean (if

A Peczenik et al. (eds.), Theory of Legal Science, 373–382.

I understand him correctly) that it is the rules that conflict with the Rule of Law if decisions actually made lack predictability.

As I see it the positive ring that the term "the Rule of Law" actually possesses does not accord with the concept being used only in formal and neutral contexts. In addition it is unfortunate for Raz to suggest that predictability should concern decisions actually made, since many decisions do not necessarily follow from preexisting rules. The rule-creating activity of the courts, for instance when they establish precedents, is hard to predict. The rules that regulate this activity should then be in conflict with the Rule of Law. Another consequence of Raz's reasoning seems to be that the rules that a judge applies run into conflict with the Rule of Law if a judge makes incorrect decisions. Consequences of that kind should preferably be avoided since it would otherwise be difficult to see the advantage of using such a concept in legal dogmatics. In my opinion predictability, in legal dogmatics, should concern the normative range of applicability of a rule and not decisions actually made. By "normative range of applicability" I mean the range of normatively possible applications of a rule (often illustrated in hypothetical cases). But these are normative possibilities, not empirical probabilities.

2. THE RULE OF LAW AND "RÄTTSSÄKERHET"

There is a historical connection between the concept of the Rule of Law and the Scandinavian term Rättssäkerhet. There is no good equivalent in the English language of the Swedish term Rättssäkerhet. Some authors have used "legal security" and others "the Rule of Law(4)".
The Enlightenment philosophers, perhaps especially John Locke, greatly influenced the development of the German concept of "Rechtsstaat". These ideas were then partly codified in the Bürgerliches Gesetzbuch for the German Reich, in 1896. This code of laws strongly influenced Swedish law during the first decades of the 20th century. A prominent representative of Swedish private law, Tore Almèn, argued for legal security (within contract law), in 1916, by referring to German sources,(5) and so did the Swedish philosopher Axel Hägerström, in 1917(6). In both

of these cases the German influence expressed different ideals as to how the "Rechtsstaat" should function(7).

In spite of several similarities between the Rule of Law and legal security, they also differ to a certain extent. The problematic thing about many American and English concepts of the Rule of Law is that they are mainly used in political or ethical rather than legal debates. In American law especially, the borderline between judicial and ethical reasoning is quite vague. In Scandinavia legal positivism and, since the 1920s, legal realism have centered their interest upon a legal reasoning which should be purely scientific, that is, free from value judgments(8).

3.LEGAL SECURITY (9) AND LEGAL DOGMATICS IN SCANDINAVIA

Hägerström made a strong distinction between doctrines "*on* morals" and "*in* morals". The latter he considered to be nothing but emotional expressions, not accessible to empirical criticism. Such "unscientific" sentences should be banished from legal science(10). This probably explains why Scandinavian lawyers are, more than American ones, anxious to make a clear distinction between ethics and law. A sharp line of demarcation between ethics and law gives rise to many difficulties. For the present, however, I will confine myself to pointing out the deep traces that this distinction has set in the development of legal conceptions. Having this in mind it is no wonder that attempts have been made to use the concept of legal security in a judicial-technical sense without any emotive undercurrents whatever. In fact this has at least partly failed, a point to which I will return later.

The dilemma was posed for the legal realists by their way of classifying recommendations, for the resolution of hard cases, as emotive and unscientific. If legal dogmatics is to be useful and capable of acting as a guide for law in action it must, however, discuss hard cases. Any legal dogmatician who is also a legal realist will then be forced to regard one part of his activity as scientific and another as unscientific. The traditional legal realist, though, was too much of a lawyer to avoid discussing hard cases. The legal realists also ran into trouble when the rules within a certain legal field needed to be systematized or justified. In such a situation a traditional legal dogmatician often introduces new principles in order to make it easier to survey the field in question. The principles may also help to explain the relative weight that individual rules carry within the system. It is obvious that value judgments must necessarily play an important part in such an activity. And since the legal realists (for instance Knut Rodhe) partly accepted the traditional method, one may say that they introduced value judgments by, so to speak, the "back door".

In the 1950s legal realists began to take an interest in questions of emotive character. I think this can be traced back to P.O. Ekelöf's book on teleological construction of statutes(12). Academic dissertations should, according to Ekelöf, not only describe the law _de lege lata_ but also put forward recommendations _de lege ferenda_. If not, legal dogmatics is only confronted with technical or practical problems which in a sense are quite easy to solve. Ekelöf's aim was to make legal dogmatics more fruitful to lawyers. The author of an academic dissertation is supposed to use his special competence in giving recommendations. Furthermore, he tried to bring some order in the method of how to give recommendations. Recommendations, as how to solve hard cases, should be based on the purpose of clear and typical cases, derived from the plain meaning of the words in the statutory probition. (Ekelöf classified the method as a teleological construction of statutes)(13). This, I think, inspired legal science to be more inclined to give recommendations, thus being unscientific in Hägerström's (and the legal realist's) sense.

In later years legal security has mainly been used as an argument against certain types of arbitrary and unpredictable authority decisions. a conceptual analysis of legal security was made in 1957 by Eckhoff(14). Olivecrona(15), Ross and Eckhoff all seem to think that legal security is valid in varying degrees of importance in different fields of law. Questions regarding legal security have, for example, often been raised in public law. This is probably due to the special type of problems that is characteristic of this legal field, something which was pointed out by Eckhoff in the above-mentioned article, written in 1957.

4. THE TERM LEGAL SECURITY AND ITS USE IN JUDICIAL CONTEXTS

On the basis of an investigation of the language used in connection with the term legal security in Swedish legal doctrine, I have drawn the following conclusion:" In fields with clear principles, where the legislator authorizes public bodies to make many or doubtful exceptions from a main principle, legal security is considered to be in danger." This is often expressed as:"possible exceptions from a main principle must be predictable". Almén's above-mentioned use of the term legal security may serve as an example of this. The main principle of the law of contract is pacta sunt servanda. From this main rule the statutes admit several exceptions. Almén commented on those exceptions and was of the opinion that legal security was not in danger provided that the exceptions were clearly specified and that the range of interpretation was not too wide(16). Another field in which judicial questions of legal security are of great importance is social welfare law. This is very much so in Sweden today. In this field one has, in Scandinavia, long discussed questions of legal security in connection with decisions to take minors in charge for social care. It is quite clear that what is best from the child's point of view should be put in the foreground. This aphorism, however, is too devoid of content to serve as a judicial principle. The difficult problem is to decide whether the social welfare authorities or the parents should be given the right to decide what is best for minors. The main principle within family law is that the responsibility for the child and its education lies with the parents. Within social welfare law, however, important exceptions are made from this main principle. The authorities have been given extensive powers to intervene and the statutes concerning those interventions are very vague indeed(17). This could perhaps explain why social welfare authorities have coercively taken charge of so many minors(18). Lawyers have long discussed those problems but lately several examples of remarkable cases(19) with dubious outcomes have been given great attention in the Swedish media.

The fact that lawyers have, in the situations described, used the concept of legal security, makes it possible to analyse or at least to illustrate the difficulty in those situations. The previous pages have made it evident that Scandinavian lawyers have been trying to keep law and ethics apart, due to the strong infuence from the legal realists. In spite of this lawyers have recommended new principles ans suggested solutions to hard cases.

It has long been a virtue, even a necessity for a theoretically inclined lawyer to avoid conceptual obscurity. Quite recently, however, the concept of legal security has encountered a new sort of crisis(20). Some persons, less versed in the use of legal method, have experimented with conceptions of legal security so vague, that it is virtually impossible to understand what they are talking about(21). Let me say that even if you do not agree with the legal realists on the question of how to draw a clear line between ethics and law, you should join them in their search for clarity of thought and precision. A good way of achieving this is to adapt one's concepts, in this case the concept of legal security, to the legal problems one is trying to solve, and not to social or political problems, even if one cannot ignore them completely. Legal security as understood by Almén is so clear and instructive that it would be a bad mistake to refrain from using it solely on the ground that it has some value connotations. However, it would be even worse for legal science if its spokesmen started to use the concept of legal security in any kind of legal situation involving moral values. The present well-established order, in itself perhaps insufficient, might then be tragically lost.

FOOTNOTES

(1) For a full survey see Gottfried Dietze, "Hayek on the rule of law," (in Essays on Hayek) New York, 1976.
(2) The road to serfdom, London, 1944, p.54.
(3) Joseph Raz, The authority of law, Oxford, 1979, pp 210 ff.
(4) Other possibilities are "legal predictability" and "legal certainty". I will just give a few examples of how "Rättssäkerhet" has been translated into English.
- Torstein Eckhoff, Rettferdighet og rettssikkerhet-justice and the rule of law, Oslo, 1966, p 397 "the Scandinavian term 'rettssikkerhet' is "in some cases, but not always, synonymous with the English expression 'the rule of law'".

- Bent Christensen, "Efficiency and the rule of law in public administration", _Scandinavian Studies in Law_, 1973, p.53: "The original Danish title of the paper was 'Effektivitets- og retssikkerhedshensyn i forvaltningen'. The translation into English of all the words in the title except one has been easy. But no English word conveys exactly the meaning that lawyers and administrators connect with the word 'retssikkerhed'. The literal translation of the word 'sikkerhed' is security, safety or safeguard, The word 'ret' is etymologically the same as the English 'right' in e.g. 'the right method', but as a legal term 'ret' means law both in the sense of, e.g. the law of the land and the legal position of some legal person. In Anglo-Saxon debates about public administration the term here chosen - rule of law - is normally used in the same sense as 'retssikkerhed' in Scandinavian debate. But other words, e.g. 'justice', may be used as a contrast to efficiency when the organization and methods of public administration are being discussed."
- Åke Frändberg,"Some reflections on legal security," in _Philosophical essays dedicated to Lennart Åqvist on his fiftieth birthday_, ed. Tom Pauli, Uppsala, 1982,
p 41 " The term 'legal security' (Sw. 'Rättssäkerhet') has, in fact, many connotations. Lawyers often use it as synonymous with 'legal predictability', and so shall I in this paper."
- Folke Schmidt & Stig Strömholm, _Legal values in modern Sweden_, Stockholm, 1965. The authors seem to use "the Rule of Law" as equivalent to "Rättssäkerhet" see pp 26 ff. It is also interesting to note C. D. Broad's translation of Axel Hägerström's _Till frågan om den objektiva rättens begrepp l. viljeteorin_ ((Uppsala, 1917) into English. This book is included in Hägerströms book _Inquiries into the nature of law and morals_, Uppsala, 1953. The term "Rättssäkerhet" was used twice in the Swedish edition, p 19 and p 24. Compare these pages with p 77 "to secure certainty in the law" and p 84 "legal security" in the English edition.

(5) Tore Almén, _Lagen om avtal och andra rättshandlingar på förmögenhetsrättens område_, Stockholm, 1916, pp 172, 216.

(6) Hägerström, op. cit., pp 77, 84

(7) The word "rättssäkerhet" was, to my knowledge, used for the first time in legal reasoning, in 1853-54, by the lawcommittee of the Swedish parliament (riksdagens lagutskott). The committee discussed whether the courts could try the substantive constitutionality of statutes. The committee arrived at the conclusion that the courts did not have that right because that could "endanger the 'rättssäkerhet' (legal security)". I think one can trace a direct influence, in the reasoning of the committee, from the Enlightenment philosophers.

(8) See the editor Karl Olivecrona's preface to Hägerström's Inquiries..., p XXI "Hägerström's writings on legal philosophy are strictly scientific. He does not discuss any problems of valuation; he is only concerned with questions of fact. It would therefore be vain to seek in this volume for any attempt to establish guiding principles of legislation, judicial practice, or social organization. In his more popular writings Hägerström sometimes gave expression to his own valuations; but such occasional utterances are, of course, to be distinguished from his scientific propositions."

(9) I have for practical reasons, in this paper, chosen the term "legal security" to stand for "Rättssäkerhet". Personally I think that "the Rule of Law" is the best translation but in this paper I have used the term "legal security" to avoid its being mixed up with Hayek's and Raz's concepts.

(10) Inquiries..., p XI "From these brief indications it is evident that Hägerström must deny the existence of objective values. So-called value-judgments, according to Hägerström, are judgment only with regard to the verbal form. If we say, e.g., that a certain action is desirable, the property of being desirable is ascribed to the action. But no such property can be discovered in the action as belonging to the context of reality; nor can the property in question be identified with the fact that the action is desired. The utterance springs from an association of a feeling of pleasure with the idea of the actuality of the action. No real judgment lies behind the sentence; nevertheless, the sentence takes the indicative form. This form of language is the ground

for the objectification of values.
The objectification of duties is explained in a similar way. What lies behind the sentence: "this is my duty" is the association of the feeling of conative impulse with the idea of an action; but the verbal expression of this association is a sentence in the indicative form. In his inaugural lecture (1911) on the truth of moral ideas. Hägerström drew the conclusion that there can be no science of duties; only the actual ideas of duties can be the object of scientific investigation."

(11) Inquiries...., pp XI-XII "Briefly, no science of the Ought is possible. The subject matter of legal philosophy will be the analysis of legal concepts actually used, such as the concepts of rights and duties, of the state, etc.; th investigation of ideas concerning justice and the purposes of the law; the study of the actual function of legal institutions such as, e.g. punishment. Obviously, legal philosophy as a science will not be sharply differentiated from what is now generally called sociology of law. But while field investigations are the proper domain of sociology, legal philosophy is chiefly concerned with conceptual and psychological analysis."

(12) Per Olof Ekelöf, Ar den juridiska doktrinen en teknik eller en vetenskap? Lund, 1951.

(13) For a closer account of this theory see Ekelöf, "Teleological construction of statutes", in Scandinavian Studies in Law, 1958, pp 75 ff. See also the paper by Ekelöf, discussed at the Conference on Legal Theory and Philosophy of Science, Lund, 1983, "Bydlinski on the construction of statutes", elsewhere in this volume.

(14) Eckhoff, op. cit. pp 384 ff.

(15) Karl Olivecrona, Rättsordningen, Lund, 1976, pp 279 ff.

(16) Almén, op. cit., p 216

(17) Gunnar Bramstång, Omhändertagande, förebyggande åtgärder och flyttningsförbud (Social care of minors, preventive measurements and removal prohibition), Lund, 1983, p 14. From the statutes it is not clear who is to prove what and which evidence is needed. Page 8 "The author advocates both the maintenance of strong requirements within the process of evidence evaluation as well as the adoption of a restrictive

statute interpretation style. Such recommendations gain further credit while considering not only the general aspects of individual legal security, but before all the fundamental ideas of the new legislation, i.e., voluntariness, mutual understanding and concern for individual integrity." Note that Bramstång is using the term "legal security."
In all areas where the law regulates the freedom of private persons the principle of legality is the predominant legal principle. Exceptions from this principle are treated in the same way as was the case with pacta sunt servanda. cf. Almén above. For a closer account concerning the importance of the principle of legality, see Hans Thornstedt, "Legality and teleological construction in criminal law," Scandinavian Studies in Law, 1959, pp 219 ff.

(18) It is quite clear that the problems have strong political undertones. As to the main question, who is to decide what is best for the child the "collectivists" tend to think that social authorities are more suited to do so than parents, while the "individualists", of course, are of the opposite opinion. The judicial technicalities, such as the incidence of the burden of proof and the question of how strong the evidence need to be are, of course, closely related to the main issue. This example also illustrates the difficulties you run into when trying to separate ethics from law. It certainly is difficult, in social welfare law to formulate a basic principle without taking a standpoint on controversial political or ethical questions.

(19) Brita Sundberg-Weitman, Saklighet och godtycke, Lund, 1981, passim.

(20) Hans Ragnemalm, "Yttrandefriheten och integriteten", in Rättsfonden Nr 5, "Om Våra Rättigheter, Uppsala, 1980, pp 51 ff.

(21) Mats Mattsson, Staffan westerlund och rättssäkerheten", in Svensk Juristtidning (Swedish Law Journal), 1981, pp 459 ff.

CHAPTER 5

SOME SPECIAL TOPICS
CONCERNING RATIONALITY AND LEGITIMACY IN THE LAW

SYNOPSIS

Aleksander Peczenik

Alejo de Cervera points out that legal reasonings are often incomplete. When one tries to make them complete, they appear to be circular. He gives 17 examples; let us quote two of them. One assumes that the law must be just. On the one hand, the concept "law" must have been elaborated to some degree before "justice" could possibly be referred to it. On the other hand, the same data shape our concepts of "law" and "justice". The other example is, as follows. One says that some legal concepts are vague and therefore there is uncertainty in the operation of subsumption. On the other hand, the concepts are vague precisely because there is some degree of hesitation - by the judges - in the proces of their elaboration. One has thus two sequences, from vagueness to uncertainty and from uncertainty to vagueness. To break the circle, one must tell which one is stronger. The author concludes, for example, that the sequence which begins in the uncertainty is more fundamental than the one which begins in the vagueness. His conclusion is that one shall always look for the fundamental sequence, and let all legal reasoning rest on such a sequence.

The question of the proper starting point for legal reasonings is related to the problem of legitimacy of legal systems. Both problems result from the search for certain foundations for the law and for legal reasoning. Mats Flodin thus starts his paper from stating that, according to Hans Kelsen, all legal thinking presupposes the basic norm, that is, the following rule: Norms regulating official use of force should be enacted under the conditions and in circumstances prescribed by the first

A. Peczenik et al. (eds.), Theory of Legal Science, 383–384.

constitution in history and by the norms enacted in accordance with it. The basic norm is open to the objection that it lacks any justification. Albeit contrary to Kelsen's intentions, one possible justification could rely on contractarian theories but these are based either on fictions or on false theses. Flodin suggests, instead, that a rational reason to obey the law consists in the fact that other people tolerate it. This gives the law-maker an implicit power of tolerance. This justification is weak and stronger may be needed. However, it seems to remove the basic objection of irrationality of the basic norm.

Roland Schött deals with a historical topic, concerning conservative ideas developed by Sir Edward Coke. At the end, he approaches, however, a related problem. One needs an external justification of legal assumptions in philosophical or political terms. This can be done either in a rationalist manner, e.g. through appeal to "obvious" principles of natural law, or more empirically. The first course, chosen e.g. by Hobbes, is controversial. Reason alone can provide systematization of data, but not the data themselves. In agreement with Coke's standpoint, one can obviously find the data in the historical development of society. It is a striking coincidence, concludes the author, that some modern philosophers of law, e.g. Aulis Aarnio, express similar views. According to the latter writer, the external justification in the law consists in rational acceptability which is a historical and social matter of development of our forms of life.

Yemima and Hanina Ben-Menhem discuss quite a special topic. They compare Popper's criterion of refutability, deciding about the scientific character of hypotheses, with an analogous criterion, found in Talmudic law. Biblical law provides that one who bears false testimony is to suffer the injury which he intended to inflict upon the person against whom his false evidence was directed. This principle is, however, restricted to plotting witnesses only, not to all false witnesses. A plotting witness is such that one can strongly falsify his testimony, that is, rule out that he was in good faith. The Talmudic law states that a testimony which is not strongly falsifiable is inadmissible. The authors compare in details this criterion with Popper's and conclude that the former is in many respects superior but unfortunately not transferable to the philosophy of science.

AN UBIQUITOUS PARALOGISM IN LEGAL THINKING

Alejo de Cervera

It is the purpose of this paper to arrive at the thesis that the operation with legal norms all too often indulges in a kind of thinking which includes the risk of what we may call the paralogism of the neglected sequence. Many controversial, and even untenable, attitudes denote a failure of reasoning which is to be accounted for. Both the danger of the paralogism, and the paralogism itself if at all real would seem to be particularly visible in civil law by reason of its being centered on the concept "norm", a circumstance which is accompanied by a paramount attention to concepts.The common law would seem to be affected also, but only in the measure it goes beyond the attention to precedents.

We may begin by considering the feeling of unsatisfactoriness apt to be felt on many occasions when thinking about the law. It raises the question of whether everything which is said that is needed to understand the law will consequently improve our use of the law. Let us look at several examples.

1. EXAMPLES

1.1. Concerning legal matters

1. To say for instance that a norm is, by definition, something general is undisputably correct. But on reflection is has an odd ring.Is everything said that is needed? The saying strictly implies that, in the process for the elaboration of norms, there is first a concept (like a Kantian a priori) with the traits of generality and imperativeness, called "norm", only to be actualized in some specific would-be norm if generality is included. We would be facing a norm only then.

2. It is very frequently heard that the law must be just; that legal norms must be just. Together with the sense of

A. Peczenik et al. (eds.), Theory of Legal Science, 385–399.

undisputability, the same odd ring appears here. As in the example above, it is legitimate to wonder whether this is all we need to bear in mind. Such an assertion implies or leads to conclude that first there are the concepts of "norm" and "legal norm" (each like a Kantian a priori), to which being just belongs, only to be actualized if justice is included.

3. When investigating the relationship between law and society, it is most usual to proclaim that the law is a means of social control. This is again undisputable. But at times, the law is even seen as the Deus ex machina of order. Law is referred to as first in time, society coming thereafter, and depending thereon. This sounds strange. It is interesting to note that we find such an emphasis predominant in common law legal thinking.

4. In the teaching of law it is usually said, correctly, that certain rights and obligations are the adjuncts of a certain institution ("deposit is a contract...." "mandate is a contract..." etc.). But it is somehow implied that the connection begins with the institution: somehow, there is an institution (like a Kantian a priori concept), and therefore, as a consequence, certain rights and obligations are established by the law. Certainly, we need to understand the operation of law better.

5. Concerning the practice of law, it is usually held that as soon as a legal nexus (a contract of lease, for instance) is established between the parties, some rights and obligations are ascribed to such parties. The correction of this contention is supported, a) by the fact that a legal nexus is usually established following the model of some institution to be found in the law, and b) by the fact that legal institutions may be used and may be neglected.

Such a relationship, by itself insignificant to us, deserves attention because, by reason of affinity with the one referred to in the preceding paragraph, it might suggest that the latter does not need more comment; that institutions are first; that everything begins with the institution; it might even suggest that there could be institutions that direct the possibilities of the legal thinking activity. But the evident affinity between the two relationships should not obscure the fact that the connection between a legal institution and the attendant rights and obligations remains one thing, but that the connection between a legal nexus and the rights and obligations ascribed to the parties is something quite different. The former may exist without the latter. Moreover, there always is a time-sequence between the two relationships when one of them is eventually established. Then, the affinity does not warrant the suggestions above.

6. It is commonly asserted that to each and every people belongs the right of self-determination, a relationship of correspondence to which there are no objections. But it is implied that the connection between a people and the quality of self-determination is to be understood as beginning in the people, never in said quality. Consequently, the first thing to do is to eventually establish that some group is a people, to invest it afterwards with the right to self-determination. This is strange.Even though there certainly is a movement of the legal thinking activity from the group to the right, we know of many groups which determine themselves without any previous declaration of their being a people.

7. Vagueness of legal concepts and uncertainty in the operation of subsumption are also in a relationship of correspondence. But is is usually seen in a sequence running from the vagueness to the uncertainty, as though there was nothing prior to the concepts, there usually being no express warning against this assumption. That is, the concepts are vague and then there is uncertainty, only in this time sequence. This is puzzling. Of course, there is such a sequence. But vague concepts are made to appear as prior, as it were out of the blue. Everything suggests here that things that matter are left unsaid.

8. Uncertainty as to legal rules concerning their commands or their scope and perplexities of the courts are usually seen in a sequence which starts in the legal rules. This view concerning legal norms demands and is demanded by the view above concerning vague concepts. The vagueness of concepts explains in part the uncertainty concerning legal norms. There is a close parallelism between this example and the preceding one, even in terms of the priority granted to concepts and of the suggestion of things left

unsaid.

9. If it is question of a legal institution the elaboration of which is still in the first stages, the decisions to be taken to promote said elaboration may lead to actual bewilderment. Ever since the "creation" of the Special Drawing Rights, there is a lively controversy concerning their "nature". Are they money? Are they credit? Evidently, the legal thinking activity starts here in the concepts "money" and "credit", moving in the direction of the Special Drawing Rights, only in this time-sequence, as though such concepts were previous to the whole operation (like a Kantian _a priori_). And this suggests that the near hopelessness of the controversy might be alleviated if something that is left unsaid is taken into consideration.

10. As is well known, it belongs to the orthodoxy of the civil law to affirm that to decide the outcome of a controversy the judge looks first at the facts and then at the law, directly, looking among its norms, and through their concepts, for the norm best suited to subsume the facts. The implication is that everything gyrates around the operation of subsumption that is, around concepts. Here again, a feeling of perplexity is legitimate. Both contentions have been qualified, even negated, so many times that they are difficult to accept. There must be things that are left unsaid.

11. According to one of the basic tenets of the "Pure Theory of Law", the concern for the law should circumscribe itself to the field of normativity, of validity of the law, a completely " formal" endeavour, leaving aside any concern for "contents", which is left to others, and consequently for the empirical and historical _data_ which influence contents. This attitude is a consequence of another tenet which expressly recognizes something merely implied in some of the above examples; namely, that there is a concept which is previous to any operation of the law, called "law", to which any concern for legal evolution or experience, is alien. By the same token, any result concerning contents, as attained by a non-jurist, is of no relevance for the jurist; apparently, no jurist who entertains such preoccupations deserves the name, and he should renounce to it _motu propio_. In the final result, the concept "law" determines what is legal. Likewise,

there is a concept called "State", which determines the actions to be considered as pertaining to the State. There are many others with the same characteristics.

In this example, apart from our general perplexity facing the "concepts first" thesis, and recognizing that to leave to others the concern for contents is perfectly legitimate, the ensuing results raise a particular perplexity because it implies a movement in circle. On both accounts, we have to have recourse to something which has been left unconsidered.

1.2. Concerning non-legal matters

We may also find in fields outside the law, both in connection and unconnected with the law, many parallelisms to the above examples, suggesting that we are on solid ground.

12. Schumpeter, in his classic work _Capitalism, socialism and democracy_ (1942), traces the existing and widespread antagonism to the market system to the inability of the bourgeoisie to rule. This may be true. But the implication rings particularly odd, namely, that there are individuals who are born bourgeois, and therefore are incapable to rule. Is there anything lacking in Schumpeter thesis?

13. In the socio-economic sciences we find many who say that the market is bound to favor the rich as against the poor. Again, this may be true. But the implications ring as odd as in the previous example. In effect, such a contention is predicated on the assumption that first there are people who are, as per essence, poor, and who happen therefore to be hurt by the market; and people who, also as per essence, are rich and thereby favored by the market. Something must have been left unsaid here.

14. In the political, philosophical, religious, etc., discussions, it is quite common to explain opinions, pronouncements, attitudes, etc., by labelling the individual: he says what he says because he is a socialist, a positivist, a calvinist, etc. In effect, many hold certain opinions as a consequence of a previous position-taking. But the implication is quite indigestible, namely, that to be something pointed out by the label is first, and to hold a

certain opinion or attitude is only a consequence. The odd ring is still more evident if, more dramatically, we understand the implication as saying that first everybody is born a socialist, etc., to have ipso facto and therefore some opinion or other. There must be something very wrong here, in terms of circumstances left uncommented.

15. The same odd ring is to be found on the occasion of many preoccupations which are not bound to lead to the legal regulation of controversies or tensions. Let us think of a very common type of reasoning in which some consequences are attributed to a concept which refers to a profession or dedication, a variation of the preceding example. Let us think, for instance, of the proposition that theologians care very much about those who are deprived. Leaving aside the improper use of the word "deprive", such a proposition is quite defensible. But it leads to think that the first thing is to be a theologian, and then and therefore, the individual is prompted to care about those who are not economically well endowed. Again, something have been left unconsidered.

16. It is quite common to hold that "political science is value-free", and that political scientists should refrain from indulging in any concern for values. The situation has the structure we saw when commenting on the "Pure Theory of Law". Apparently, any "political scientist" is disqualified as such if he indulges in the concern for values, or if he is failing to make any contribution to political science. This sounds odd, notwithstanding the legitimacy of leaving such a concern to others.

17. In a very popular television program, certain animals' behavior is attributed to the fact that they are lemurids, because they are lemurids. The phrase is understandable but, evidently, implies something absurd. Apparently, some animals are first lemurids and then behave in such and such a way. The program would seem to assume that if any of those animals would happen to behave in a different way, it would conceivably be possible to mend the animal's ways by reminding the animal that, as a lemurid, he is expected to behave accordingly; or, simply, that he is a lemurid.

2. THE NEGLECTED SEQUENCE

We have seen that the same odd ring, the same feeling that something important is left unsaid, appears in connection with all of the above examples.Many more examples could be given, both from inside and from outside the law. As a consequence, the conclusion is legitimate that it is a most general occurrence.

Now, the question is to find out whether in effect, in each case, there actually is something which has been left unsaid, and whether its consideration is apt, or necessary, to dissipate the odd ring, thus improving our understanding of the law. The answer is in the affirmative on both accounts. As a matter of fact, it is rather easy to show that in all of the preceding examples there still is another sequence which happens to be neglected. To this effect, some of the considerations that follow might appear more cogent than others. But their joint impact leaves no doubt as to the presence of the neglected sequence.

1a. With regard to the generality of norms, it is to be kept in mind, for any further concern about norms, that the concept "norm" is also to be elaborated to some degree or other; and that such an effort is made in view of several legal propositions in some elaboration or other, in order to control them. From some elaboration of several such propositions the thinking activity proceeds to some correlative elaboration of the concept "norm", in a process which, in retrospect, is seen as running from the propositions to the concept.

2a. It is clear that, on the one hand, the concept "law" must have been elaborated to some degree before "justice" could possibly be referred to it. On the other hand, "justice" must have been abstracted, to some degree or other, on the basis of the very same legal propositions borne in mind to elaborate the concept "law". There are two processes here - the one to elaborate the concept, the other to abstract "justice" - running against the grain of the movement which ends in the contention that the law must be just.

3a. Concerning the relationship between law and society,

and whatever could be said of many statutes and law-making decisions, the thesis that the law is a means of social control implies an invitation to neglect that there is also a movement - of the legal thinking activity - from society to the law, a movement starting from society. Thus, such a thesis happens to provide too narrow a view of the relationship in question here. It is not necessary to be an admirer of custom law to accept that the law-making activity also proceeds in view of social patterns and depends on them.

4a. Regarding the relationship between an institution and the rights and obligations attendant to it, while it is true that we are legitimated to say that to a certain institution correspond certain rights and obligations, it is no less true that the institution is elaborated,to some degree or other, under the influence of the progressive elaboration of those rights and obligations, including a process of selection.

5a. As to the relationship between a legal nexus involving more than one person and the rights and obligations ascribed to them, it is now obvious that if parties actually had the possibility of following legal models, it is by reason of the efforts of other parties, repeated many times over, which have coalesced into the sanction of the legal model - of the institution - by the law.

6a. It is impossible that any legal thinking activity should proceed only from a people - a real group - to the right of self-determination. There must also be a march of the legal thinking activity from the latter to the former, a process running in such a direction. It is impossible that the right of self-determination be proclaimed in favor of all peoples without the actual and more or less occasional happening that some groups, in their relations to other groups and peoples, arrive at self-determination. There must be a sequence running from a situation of self-determination to the characterization of a group as a people.

7a. The sequence running from the vagueness of concepts to the uncertainty as to subsumption is obviously not the only one in the relationship involved. There is also a movement running from the latter to the former.It could not possibly be otherwise. No concept is possible - vague or not - without some elaboration. If it is vague to some degree, there must have been the same degree of hesitation - by the judges - in the process of its elaboration.

8a. The close parallelism between the preceding relationship and the one between uncertainty as to legal rules and the perplexity of the courts is evident anew. The sequence which starts in the legal rules is obviously not the only one. There also is the process, followed by the legal thinking activity, which runs from the hesitations of the courts to the uncertainty of the rules.

9a. Concerning the problem of the Special Drawing Rights, the situation would seem to be hopeless if it does not dawn upon us that there also is a process running from what is usually subsumed under the concepts "money" and "credit" to the concepts themselves, a process to be reactivated when the Special Drawing Rights are dealt with, exactly as it is reactivated as to other concepts whenever any other newly established or refurbished institution is dealt with. To be sure, the problem boils down to one of subsumption: whether to subsume them under the concept "money" or under the concept "credit". But then, the Special Drawing Rights should be taken into account to go on with the elaboration of said concepts, until a subsumption is called upon excluding the other. It is even possible that the circumstances should demand the elaboration of some other concept or concepts.

10a. Undoubtedly, on many occasions, the judge makes his decision by looking for a certain result. Aware of the law, he looks again at the facts, trying to propose to himself a result which the facts make desirable. In this case, the use of a certain norm is only a rationalization of an outcome previously reached with the desired result in mind. This contention is held quite frequently in the common law, even to the point of proclaiming that such is the case on all occasions.

11a. As against the "Pure Theory of Law", it is often held, correctly, that legal concepts are the result of centuries of observation of, and activity or meddling with, historical <u>data</u>, in the mass of which the legal thinking activity has carved out said concepts. So, a particular set of finalities has inspired such a carving out thus prompting the elaboration of the legal concepts we find in a legal system. Some other set of finalities would have resulted, or has already resulted, somewhere, in some other conceptual outcome.

12a. As to Schumpeter, the following complement is needed: there are individuals who are incapable of leading others. This trait, together with other correlative ones, makes them congregate, join other individuals who seem to share the same traits; or to be included, by the would-be outsiders, in a group called by many bourgeoisie.

13a. The conclusion according to which there is the sequence referred to in point 13 needs to be complemented by the assertion that there is another sequence operating precisely in the opposite direction. There must be a process, followed by the thinking activity, running from the market to the economic situation of the individuals. At the historical beginning, a beginning which has repercussions as to every individual, each one of us is allowed, or invited, to exploit every personal talent and capacity. Those who do not do well in the market - in the rat race - are the poor ones.

14a. In political, etc., labelling, there is also another sequence which runs on the same track and precisely in the contrary direction. In effect, there also is the process which begins with the adoption of some opinion, belief, or notion, which leads to, if there is repetition and it is accompanied by other opinions, and they integrate a more or less harmonic constellation of opinions, to the adoption of some label, or to the acceptation of some labelling made by others.

15a. There is also here a process which runs contrary to the one referred to in point 15. There undoubtedly is a mindset to which the concern leading to the cultivation of theology belongs.

16a. Concerning the concept "value-free political science", there must be a movement leading to the concept, along which "value-freeness" has been introduced. Just as it happens concerning the concepts considered by the "Pure Theory of Law". It could have been left out. We could have elaborated another concept with the concern for values as an ingredient.

17a. Concerning any animals, our thinking activity also moves from - among other characteristics - the animals' behavior as point of departure, and elaborates the more or less adequate concepts. The discovery of other animal forms of life, manifested in a different behavior or otherwise, forces biology to a constant re-elaboration of concepts. Exactly as it happens with the concepts "money" and "credit".

3. THE FUNDAMENTAL SEQUENCE

We have seen that to any of the sequences we have exemplified corresponds another one which runs on the same track but precisely in the contrary direction. If we try now to get out of the vicious circle implied in any of the examples, that is, if we do not succumb and remain content with a simple explanation of any of the sequences by its correlate, we should pay attention to something that has already made itself visible in some of the examples, namely, that one of the sequences is first in time, and stronger or more important than the correlate. Let us ratify such an impression.

If we take for instance the relationship in which law and society are, and realize that prior to any society and to any law there are the physico-natural things, and that they are nearer to society than to law, we will conclude, without any interest for giving credit to the historical school, that the neglected sequence is more important than the non-neglected one. No doubt, as a function of the degree of its elaboration, the law controls and influences the social patterns. But the stronger sequence starts from the social patterns.

In the relationship between a certain institution and the corresponding rights and obligations, the movement running from the selection of the rights and obligations is certainly stronger than the one with the contrary direction. The process through which legal concepts (institutions) are elaborated has been described many times, and alluded to in several of our examples, implying the rejection of any notion of the priority of concepts.

It will be sufficient to say here that on the basis of such a selection the elaboration of the concept begins. Gradually, and in a second stage, it will be possible to use the institution for particular finalities.

As to the relationship between uncertainty and vagueness of concepts, it is not necessary either to render hommage to the Historical School to hold that the sequence which begins in the uncertainty is the stronger one. Certainly, only by reason of the hesitations they provoke concepts are characterized as vague. We would not pronounce any judgment as to their uncertainty or lack of it if we had not tried the use of the concept in view of a case, real or hypothetical. Regarding the relationship between the uncertainty of the legal norms and the perplexities of the courts, the same reasoning leads to think that the neglected sequence is more important than the correlative one.

Without entering into the investigation of more examples, the conclusion is legitimate that, in all of them, one of the sequences is stronger than its correlate, and that the stronger sequence is the neglected one; the fact notwithstanding that it may be open to discussion which is the stronger sequence. Let us call the stronger sequence fundamental.

4. THE PARALOGISM, AND THE REMEDY

We are now in a good position to understand the odd ring commented upon. The fundamental sequence is neglected, meaning that it is not recognized as fundamental, meaning in turn that the non-fundamental sequence is held to be the fundamental one. There are other examples where, regrettably, the fundamental sequence is ignored with more or less deliberation, implying or not that the correlative non-fundamental sequence is the fundamental one. And still others where there is unawareness of the fundamental sequence, where, consequently, the correlative non-fundamental sequence is held to be the only existing one.

In all of those examples, the possibility is open for the strange conclusions we are now familiar with, precisely on account of the neglect of the fundamental sequence. Had we not neglected it, the conclusions would have been impossible because, as was made clear, they are incompatible with such a sequence. As therefore such conclusions have been reached on the basis of a defective insight, we are facing what we may call a paralogism, the paralogism of the neglected sequence - in the best of cases. It follows, that we may now say that all of the conclusions reached in our examples without taking due account of the fundamental sequence are paralogical conclusions. Some of the conclusions dealt with here are evidently defective (with awareness or not of the paralogism); some other are defective, but not evidently; on these occasions, either we discover the paralogism, or the situation cannot be improved.

Our paralogism is more to be feared in the civil law than in the common law. It so happens that the civil law preferably follows, and rests on, the non-fundamental sequence, by reason of the importance it warrants to concepts. Whereas the common law is more aware of the fundamental sequence on account of its going from one case to another.

Of course, the possibility of our paralogism is not a paralogism. We may neglect the fundamental sequence and still not be guilty of the paralogism; the risk of the paralogism has remained just a risk. Consequently, the paralogism is not to be found on all occasions; far from that. But it is to be found in all fields; this is its ubiquity. The risk, though, is to be found whenever and wherever the fundamental sequence is neglected.

As to the remedy, if paralogical conclusions are incompatible with the attention given the fundamental sequence, the remedy against the risk of such conclusions is well at hand: To always look for the fundamental sequence, and to let all legal reasoning rest on such a sequence.

5. EXPLAINING THE FREQUENCY OF THE RISK OF OUR PARALOGISM

We already know of several reasons. For instance, the lack of any visible damage when the neglect of the fundamental sequence is routinely for didactic purposes. The visibility of the movement running from legal concepts to their actualizations leads also to such a neglect; let us bear particularly in mind the very naturality of many operations with the law, referred to in point 5; it explains the easiness and almost carelessness with which civil law legal thinking indulges in such a neglect. Etc.

But there is also the support the neglect receives from a persistent and venerable Platonian "realism of concepts", with which the neglect has the distinct affinity of which we are already aware. As we may now note, most of our paralogical conclusions postulate the priority of concepts whenever a concept is involved. Such a "realism", very popular in the 19th century, especially in neokantianism, which makes the attitude of neglect very persistent and venerable, deserves some comments to put things in a convenient perspective.

Even though the fundamental sequence is first in time, it is not as easily seen as is the non-fundamental one. Knowledge of the latter is, on account of this, previous to that of the former. We have seen how, for this very reason, and following the non-fundamental sequence, concepts end by being thought of as previous in any operation with the law. As this view is supported by the Platonian "realism of concepts", presumably in turn a result of the non-fundamental sequence, the vindication we have made of the fundamental sequence amounts to negating such a realism.

But Platonism, despite all, has a way of promoting the neglect of the fundamental sequence. It is quite clear today that there is a distinction to be made between concepts, essences, and categories, all of them undistinguished in the Greek philosophy, the Platonian entity corresponding to categories and essences, but not to concepts. As is well known, the distinction is strongly promoted by the phenomenological school. However, the distinction is not widely accepted as yet, partly on account of the difficulties it raises. And as long as the distinction fights for its consolidation, the neglect of the fundamental sequence will be supported by the realism to be found anyway - we are told - in essences and categories, through the attribution to concepts also of such a Platonian entity.

We might now be on the brink of understanding the role of the "Pure Theory of Law" in legal thinking. It might be that the whole effort of this school implies the adumbration, in the historical and empirical <u>data</u>, of the properly so-called "categories". In effect, this school accepts and proclaims that there are originary things, previous to all legal endeavours. Only, it conceives of them in the manner of the Kantian *a priori*; hence, its use of the word "concept".

POWER OF TOLERANCE - ON THE LEGITIMACY OF A LEGAL SYSTEM

Mats Flodin

1. INTRODUCTION AND SUMMARY

In this paper I will discuss the character of the relations between the legal system and the society as a whole. The concept of a legal system can be the object of investigation without significant attention(1) to the problem of its legitimacy(2). A well-known example of this is H.Kelsen's pure theory of law, where the mere identity (social cognizance) and efficiency(3) of the system constitute the content of the source of validity, the basic norm.

A number of theories use either a real or a hypothetical social contract as a necessary basis for the legitimacy of a legal system, as for instance, in the theories of Grotius, Hobbes, Rousseau and Rawls.

I will use another way of approaching the problem. In Swedish contract law there is a concept called "power of tolerance" (toleransfullmakt), meaning that a person (or a group of persons) can voluntarily interfere in other persons' affairs with legally binding consequences as long as it is being tolerated by that person or group. The power of tolerance may be part of a combination of explicit or implicit power, but the significant content is the tolerance of the person or group who grant the power to the person or group who initiate a power or exceed an existing power.

In analogy with this I will say that the essential characteristic of the legitimacy of a legal system is its being tolerated in the community where the system operates. It follows that the fact that the tolerance expires, which causes the discontinuity of a legal system, cannot be a failure to fulfil an obligation or a breaking of a contract. It also follows that power of tolerance has to be maintained continuously as a type of social function. The aim of this paper is to describe the phenomenon of interrelation between the legal system and society with

A. Peczenik et al. (eds.), Theory of Legal Science, 401–411.

the help of a legal concept and also a weak sense of legitimacy(4).

2. CONCEPTUAL BASIS AND DEFINITIONS

To begin I give some necessary definitions of some concepts and terms used in this paper.
1) The notion of a legal system divides itself into three elements, a. a body of rules, b. social institutions treating these rules as valid and c. total or almost total recognition in society of element 1 as "law" and element 2 as "rulers".
2) Society means all individuals and organisations.
3) Contract means a legally or morally binding agreement.
4) Legitimacy means conformity in relation to a presupposed standard, such as, for example, natural law or a contract (binding agreements)(5).
5) Efficiency means the influence of the legal system as a whole on social behaviour; not only the following of a single rule(6).
6) Tolerance is the other side of efficiency(7). It is a social attitude among the members of a society, which may or may not exist at a certain time.
Imagine the following scale: Acceptance - tolerance - rejection. Acceptance means an active, positive (or at least not negative) opinion of the legal system. Tolerance implies a passive, not necessarily value-influenced attitude towards the legal system. Rejection implies an active negative opinion against the system.

3. CONTRACT THEORIES

As a comparison with the idea of power of tolerance I will give a short sketch of four well-known contract theories, the first three based on natural law.

H.Grotius' (1583-1645) concept of a social contract, in De iure belli ac pacis (1625), relied on the idea that the members of society in early history reached a state of reciprocal enmity and therefore yielded a degree of their freedom to a ruler. The form of government could be absolutism, republic or restricted monarchy depending on the historical circumstances.

The members of society have, through the social contract, given up their right to take back the power they have left to the ruler, and therefore have the duty to obey him(8,9). Grotius realized that no explicit agreements in the dimension of social contracts could have occurred. He thought that the members of society could enter into a social contract by remaining passive to the wielding of power, a tacit approval, making even future generations bound to the contract.

T.Hobbes (1588-1679), in De cive (1642) and Leviathan (1651), saw the social contract as a necessity because of the egoism and anti-social behaviour of human beings. In order to gain security, the people have given total power(10) to a sovereign (monarch or republican assembly)(11) who has not taken part in the agreement and cannot be responsible if he breaks the contract(12). However, if and only if the sovereign fails to provide the sufficient security, the people no longer have a duty to obey him(13). And so, they can overthrow him and choose another sovereign. If he, in turn, succeeds in giving his subjects the necessary security, the new contract will be valid.

J.J.Rousseau (1712-1778), in Du contrat social (1762), thought of the social contract as a means to defend the personal integrity and property of the participants in the contract(14), the people themselves. The social contract was an agreement(15) among the people to yield their personal power to the whole community(16). No one was to have personal power over another participant. The common will would be the sovereign(17). Everyone should take part in legislation and ruling, but as this would be impossible, proxies were to be elected for these tasks. All decisions should be approved by the people. More important ones should rely on referendums. The people would have a total right to overthrow the rulers, without breaking the contract, because the contract was not meant for the rulers, but for the whole people.

The contemporary philosopher J.Rawls has a notion of a hypothetical social contract in his book A Theory of Justice (1972) p.11:

"...not to think of the original contract as one to enter a particular society or to set up a particular form of government. Rather, the guiding idea is that the principles of justice for the basic structure of society are the object of the original agreement".

p.12:
"The choice which rational men would make in this hypothetical situation of equal liberty, assuming for the present that this choice problem has a solution, determines the principles of justice".
In justice(18) as fairness the original position of equality corresponds to the state of nature in the traditional theory of the social contract.
p.13:
"Then, having chosen a conception of justice, we can suppose that they are to choose a constitution and a legislature to enact laws..."

Grotius, Hobbes and Rousseau think of real or ideal social contracts, based on the necessity of following a natural law. Rawls's social contract is a pre-legal hypothetical(19) agreement on moral issues. Consequently, the concept of a social contract has several divergent meanings.

4. SOME REMARKS ON SWEDISH CONTRACT LAW

In Swedish contract law, a legally binding contract can arise from the tacit factual cooperation of two parties. This requires the activity of at least one party and the passivity of not more than one party.

The concept of a legal power presupposes three parties: 1) the one who grants the power, 2) the authorized agent, and 3) the third party.

Normally, legal power comes from an initiative of the party who grants power, such as an order to perform tasks of a certain kind, an employment contract, etc. The purpose of the agreement between parties 1) and 2) is the performance in relation to the third party.

The concept of legal power in Swedish law contains, no matter if the power is founded explicitly or by passivity, both the legally relevant facts constituting the power and the authority to act based on those facts.

The term "power of tolerance" was introduced in Swedish contract law by the law professor K.Grönfors(20). There is no special paragraph constituting such a power, but in the preparatory materials of the codification it states: "if, for example, someone repeatedly appears as an authorized agent for someone else, without the objection of this person, such passivity can under certain circumstances give birth to a power also for the future to undertake legal actions of this kind".

Through precedents, the legal power of tolerance has established itself as a relatively stable concept, with approximately the content foreseen in the preparatory materials(21). In practice, however, the power of tolerance rarely appears in isolation. In most cases it is combined with explicit powers, which are being exceeded.

So, these circumstances are important for the power of tolerance: 1) the attitude of the one who grants the power; knowledge of the actions of the authorized agent together with omission to protest; 2) the behaviour of the authorized agent; taking action on his own initiative in the legal sphere of the first party; 3) the impression of the third party that the agent has legal power to act the way he does.

Further, any significant change in the course of action of the authorized agent or the absence of tolerance from the first party, as well as the lack of the impression of the third party will end the power of tolerance automatically. In Swedish law, it is conceptually possible to recall a legal power.

5. THE POWER OF TOLERANCE OF THE LEGAL SYSTEM

Now, let us make an analogy inferring from the micro-level of the contract law situation to the macro-level of the legal system and the whole society.

The legal system will, if tolerated or accepted, represent the society as a whole in the intervention into the affairs of the members of society. The individual member has a double role in this interrelation; 1) as a more or less influential part of the whole society in which he lives, and which grants the power to the legal system, 2) as a "third party" in relation to society and legal system.

If a legal system has been operating for some time, dealing with public affairs including international relations, and there has been no reaction from the society that could be characterized as intolerance, then a power of tolerance exists for future actions of the same kind. A condition, of course, is that the legal system cannot resist the clear utterance of intolerance or rejection when it reaches a certain level. How high this threshold is can vary considerably from one society to another. A clear case of social intolerance that lead to change in the legal system was the Iranian revolution in the late 70's(22). Huge street demonstrations with about half a million participants and pressure from religious authorities resulted in the mainly non-violent overthrow of a military-supported dictatorship of a superpower satellite. Important parts of the rule-system and the old social institutions operating on them were replaced with legal rules of religious origin and social institutions under supervision of the clergy, according to the wishes uttered in the demonstrations.

A case of weaker utterances of rejection were the events in Poland in 1980-81. The political conditions resembled those in the pre-revolutionary Iran, but in Poland the intolerance lead to an opposite change; the body of rules were applied much stricter and some personal changes in the ruling institutions took place. After some time the new situation in Poland stabilized.

The social intolerance as a macro-phenomenon (of which we are speaking here) seems to exist in an all-or-nothing fashion. If tolerance exists, then the situation will stabilize. Tolerance is the ultimate cause of stability.

The extent and limits of the power can be very divergent. Different societies have varying possibilities for uttering rejection and intolerance, depending on the form of government. This is more a question for political science, and will not be discussed here; neither the moral question of when a legal system should be intolerated, or the problem of choice of rejection method.

As far as legitimacy is concerned, I say that there is a reason to obey a legal system which is tolerated by the society. This reaon is not at all a sufficient reason, but it is a necessary one. It is a necessary condition for the existence of the legitimacy of a legal system. No weaker sense of legitimacy is possible, without being equal to or weaker than the notion of efficiency.

The legal system has also no legitimacy of its own; only in relation to its performance of the tolerated task. When a legal system tries to change its course of action by its own initiative, it will be a gamble with the power of tolerance.

Of course there can be a degree of positive social

acceptance of the legal system. If this acceptance goes far enough, it will cause the disappearance of the legal system(23). If no one would break the rules, the power-exercising institutions would have nothing to do. A total acceptance is the opposite of intolerance, which also ends the legal system. Thus, the attitude and opinion of the legal system must lie somewhere in between these extremes, but not at them.

Less acceptance in different ways, such as, for example, voluntary participation in voting, running for offices etc. gives a foundation for the thought of a social contract in some societies, but then only to the legal system as it is at the time of these utterances and always combined with the power of tolerance.
This figure may help to explain the theory:

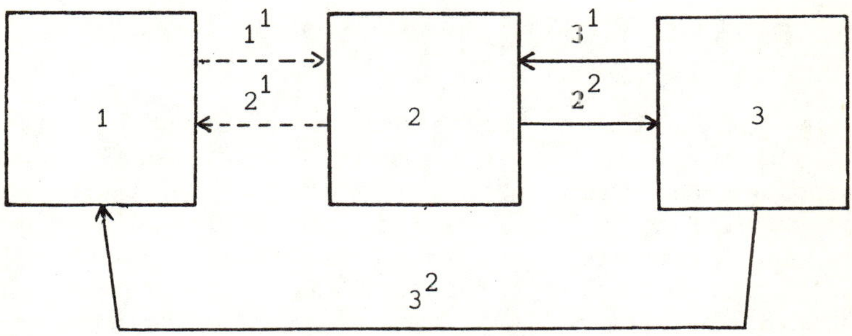

2 is the legal system
3 is the individuals and organisations of the society as separate elements
1 is the society as a collective agent(24)
the arrows with broken lines show potential of action.
the solid arrows show continuous phenomena.
1^1 is potential intolerance of 1
2^1 is a shift in the character of action by 2
2^2 is the efficiency of the system
3^1 is the cognizance of 2 as law and rulers
3^2 is the sumtotal of actions and attitudes in 3 with relevance to the presence of tolerance in 1(25).

The contract theories presuppose real or hypothetical

participation from the members of society, tie them to the obligations of the contract, with little or no right to break it.

The social contract, even though it is a fiction to some writers, is used as a logical precondition for the justification of the existence of a legal system, the external perspective.

The power of tolerance is open to automatic change in the legal system and society; the power of tolerance has a continuous existence, can expire without notice and implies little obligation to the members of society. As an external justification of a legal system and as a starting point for the internal justification, the notion of power of tolerance will be a more flexible alternative, and also a more appropriate description of the structure of the ruler-society phenomenon.

6. THE POWER OF TOLERANCE AND RATIONALITY

One of the topics of this conference is rationality of legal thinking. According to Hans Kelsen, legal thinking presupposes the basic norm, that is, the following rule: Norms regulating official use of force should be enacted under the conditions and in circumstances prescribed by the first constitution in history and by the norms enacted in accordance with it(26). This basic norm is open to the well-known objection that, from the external perspective, it lacks any justification. Albeit contrary to Kelsen's intentions, one possible justification could rely upon contractarian theories. Unfortunately, the idea of social contract, as already stated, is either a false historical description or a fiction. On the contrary, the power of tolerance can constitute such an external justification. It is not a fiction because we actually tolerate the legal system. This fact of tolerance is, as said before, a rational reason to obey the legal system. This justification is obviously weak, and stronger ones may be needed. However, it seems to remove the basic objection of irrationality of the basic norm.

NOTES

(1) H.Kelsen, *Reine Rechtslehre*, Wien 1960, p.213: "Von einem juristischen Standpunkt aus gesehen ist es gleichgültig, ob diese Änderung der Rechtslage in Wege einer gegen die legitime Regierung gerichteten Gewaltanwendung, oder durch Mitglieder dieser Regierung selbst, ob sie durch eine Massenbewegung des Volkes oder durch eine kleine Gruppe von Individuen herbeigeführt wird".

(2) H.Kelsen, *Reine Rechtslehre*, Wien 1934, p.128: "Wenn die Reine Rechtslehre eine Legitimierung des Staates durch das Recht ablehnt, so nicht etwa darum, weil sie jede Legitimierung des Staates für unmöglich erklärt. Sie leugnet nur, dass die Rechtswissenschaft eine Rechtfertigung des Staates durch das Recht oder - was dasselbe ist - des Rechts durch den Staat zu leisten vermag".

(3) H.Kelsen, *Reine Rechtslehre*, Wien 1960, p.214-215: "Wäre die Revolution nicht erfolgreich, das heisst: würde die revolutionäre - nicht nach der alten Verfassung zustande gekommene - Verfassung nicht wirksam werden, würden die von ihr vorgesehenen Organen keine Gesetze erlassen, die von den in diesen Gesetzen vorgesehenen Organen tatsächlich angewendet werden, sondern würde d[illegible] Verfassung in diesem Sinne wirksam blieben, läge kein Anlass vor, an Stelle der alten eine neue Grundnorm vorauszusetzen. Dann würde die Revolution nicht als ein neues Recht erzeugendes Verfahren, sondern - nach der alten Verfassung und dem auf ihr beruhenden noch als gültig angesehenen Strafgesetz - als Verbrechen des Hochverrates gedeutet werden. Das hier zur Anwendung kommende Prinzip bezeichnet man als das der Effektivität. Das Prinzip der Legitimität ist durch das Prinzip der Effektivität eingeschränkt."

(4) This paper is an excerpt from the manuscript of a coming dissertation with the title "On valid law and valid moral". The notion "power of tolerance" belongs to a part of the dissertation where universalizable structural and procedural aspects of law are discussed.

(5) H.Kelsen, op.cit., p. 280: "Das Prinzip, das eine Norm nur von den zuständigen, das ist durch eine höhere Norm hierzu ermächtigten Organ zu setzen ist,

ist das Legitimitätsprinzip. Es ist wie schon festgestellt durch das Effektivitätsprinzip eingeschränkt". (According to international law to which Kelsen also refers, the legitimacy of a legal system is due to its being factually effective in a certain area. But then the legitimacy of international law remains to be explained, which leaves the problem unsolved).

(6) H.Kelsen, *General Theory of Law and the State*, Cambridge, Mass. 1949, p. 41.

(7) The existence of tolerance within the social institutions of the legal system will not be discussed here.

(8) H.Grotius, *De iure belli ac pacis*, engl. transl., New York 1964, p.139: "That as a general rule rebellion is not permitted by nature".

(9) H.Grotius, op.cit., p.155 ff: Rebellion is under certain circumstances allowed against a very bad ruler or an ursuper.

(10) T.Hobbes, *Leviathan*, reprinted Oxford 1952, p.136 (90): "Whatsoever the Soveraigne does is unpunishable by the Subject".

(11) T.Hobbes, op.cit.,p.131 (87): "...which is a Common Power, to keep them in awe, and to direct their action to the Common benefit. The only way to erect such a Common Power,......is, to confere all their power and strength upon one Man, or upon the Assembly of men,...".

(12) T.Hobbes, op.cit., p.134 (89):"Secondly, Because the Right of bearing the Person of them all, is given to him they make Soveraigne, by Covenant only of one to another, and not of him to any of them; there can happen no breach of Covenant on the part of the Soveraigne; and consequently none of his Subjects, by any pretence of forfeiture, can be freed of his Subjection."

(13) T.Hobbes, op.cit., p.170 (114): "The Obligation of the Subjects to the Soveraigne, is understood to last as long, and no longer, than the power lasteth, by which he is able to protect them".

(14) J.J.Rousseau, *Du contrat social*, reprinted Paris 1933, p.24: "Trouver une forme d'association qui défende et protège de toute la force commune la personne et les biens de chaque associé, et par laquelle chacun, s'unissant à tous, n'obéisse pourtant qu'à

luimême, et reste aussi libre qu'auparavant".

(15) J.J.Rousseau, op.cit., p.23.

(16) J.J.Rousseau, op.cit., p.25.

(17) J.J.Rousseau, op.cit., p.28.This interpretation is made by T.Strömberg, Rattsfilosofins historia, Malmö 1979, p.39.

(18) J.Rawls, A Theory of Justice, Cambridge Mass. 1972, p.10: "The concept of justice I take to be defined, then, by the role of its principles in assigning rights and duties and in defining the appropriate division of social advantages".

(19) J.Rawls, op.cit., p.12n, where he points out that Kant's concept of original agreement is hypothetical.

(20) K.Grönfors, Ställningsfullmakt och bulvanskap, Stockholm 1961, p.338ff.

(21) K.Grönfors, op.cit., p.349.

(22) M.Flodin, Om begreppet gällande rätt (paper), Lund 1979, p.12.

(23) H.Kelsen, Reine Rechtslehre, Wien 1934, p.68: "Ja es muss sogar die Möglichkeit einer Diskrepanz zwischen der normativen Ordnung und dem ihr zugeordneten Bereich des tatsächlichen Geschehens bestehen. Denn ohne eine solche Möglichkeit hat eine normative Ordnung überhaupt keinen Sinn."

(24) L.Lindahl, Position and Change, Uppsala 1976, p.67: "An agent is either an individual person or a so-called collective agent".

(25) T.Strömberg, Inledning till den allmänna rättsläran, Lund 1978, p.52, uses the expression "social convention" in a somewhat similar meaning.

(26) H.Kelsen, Reine Rechtslehre, Wien 1960, p.203-204: "Zwangsakte sollen gesetzt werden unter den Bedingungen und auf die Weise, die die historisch erste Staatsverfassung und die ihr gemäss gesetzten Normen statuieren".

SIR EDWARD COKE'S LEGAL CONSERVATISM

Roland Schött

In the history of ideas, the social and professional background of those who have made contributions to the formation of the respective ideologies cannot be disregarded. They must, furthermore, be considered with regard to their respective nationalities and historical periods. However, one is the look-out for features in common between for instance Bolingbroke and Disraeli(1) in the case of British conservatism. It is possible to make a distinction between practical men, on the one hand, and philosophers, on the other, although a few, like Burke, fit both categories. Among the practical men politicians naturally constitute the main group, though judges may also be included, as they are of special interest in discussing conservatism with respect to legal matters, given the authority they exert, and the pragmatic and casuistic nature of their office. These qualities notwithstanding, the way the judges interpret their role may have far-reaching ideological implications in the matter of legal reform.

Sir Edward Coke, or the Lord Coke as he was usually called after being appointed in 1613 Chief Justice of King's Bench(2), has been paid many tributes. A Luther of the Law(3), the living embodiment of the Common Law(4), the head of our jurisprudence(5), he is perhaps most famous for his quarrel with James I, and for being one of the architects of the Petition of Right in 1628(6). His great influence (not least through his writings), even after his death in 1634, on generations of jurists, in America too, has been duly appraised and stressed(7). Coke rose to power and wealth in the reign of Elizabeth, born in Norfolk of poorish gentry stock and educated at Trinity College, Cambridge, and at the Inner Temple in the Law(8), thriving as a practising lawyer and with skill holding public office; he was elected speaker of the House of Commons and appointed Attorney General already in his early forties.

In those days, law and politics, together with religion, would seem to have been more than usually inter-

A. Peczenik et al. (eds.), Theory of Legal Science, 413–424.

twined; and Coke, as a prominent member of the English ruling elite(9), was a typical representative of the age: dynamic, ruthless, ambitious. Above all he might be described as a patriot, in love with England and her ancient ways, while at the same time looking ahead towards the future with inquisitive and fearless eyes, full of confidence, as was befitting for a late Renaissance age, in general mentality so very different from the outlook of the medieval era(10). There was no feeling of internationalism coupled with this discovery of a new world(11), our modern one and its possibilities opening up before enterprizing man. It was rather the opposite: fierce competition, in particular between nations. Indeed, the strong spirit of a national revival, a consciousness of and an interest in England and her customs and capabilities, appears to have been generally characteristic of the age in question(12). This nationalism seems to be the first and perhaps fundamental point to bear in mind about Coke.

In the field of law this nationalistic tendency is about as evident as anywhere else. The law of the land, the common law, with its ancient and medieval roots, enjoyed its golden age in the sense of its formative period during the thirteenth century(13). After that, subject to a strong and, from a Cokean standpoint, alien pressure, it gradually decayed; for instance, the continuous reporting of cases petered out; and the whole tangle of illogicalities, its very disorder(14), made its survival in a new Renaissance age uncertain. This alien pressure was, of course, the Roman civil law, not to mention its canon law, originating from the same source: papal Rome, with, in addition, powerful princes only too eager to lend their services in the name of Counter-Reformation. There were also, as an integral part of the alien pressure directed against England and her national common law heritage, the doctrines prevalent on the Continent, advocating strong, centralized, autocratic government(15). These doctrines suited those crowned heads admirably. Thus, the historical liberties, the rights of English subjects, protected by the existence and operation of the common law, though never formulated as an abstract concept, were obviously in danger.

The Tudors, for all their centralizing ambition, understood, it has been argued(16), the national spirit of the English immanent in the common law. Henry VIII's break with Rome, followed at last by Elizabeth's acceptance of the Spanish challenge in a bold bid for English seapower, made it imperative as an urgent matter of national survival to check further "continentalization". For Spain had come to represent the most deadly aspect of what I have here termed alien pressure; a violent show-down had sooner or later to be faced(17).

In this connection one may ask whether England could not have expanded as well with Roman law, but that argument can, I think, be persuasively countered by again stressing the unavoidability of a general major clash with Spain. The conflict of interests and objectives went too deep, and the state of perpetual skirmish, the policy of neither war nor peace that Elizabeth pursued for so long, would, it may be argued, not necessarily have removed the pressure, nor made it less deadly. Allied to this external-threat-factor-argument there is, or rather was, the utter vulnerability of the little up-start island country before it could be known that the high gamble that it nevertheless must have meant to take on Spain would come off so successfully, and its gains later be consolidated. This point emphasizes the element of uncertainty. Viewed in this way, the almost xenophobic nationalism of the English makes at least some sense.

What all this amounts to is, I should say, that the general situation in the latter part of the sixteenth century in England, according to a commonsensical interpretation of the historical panorama of the age, led or at least contributed towards a tremendous boost for all that Coke stood for: puritanism, patriotism, the economic and political ascendancy of the gentry, and, as to the law, the common law's finding eventually an equilibrium with equity(18).

The question may be put whether Coke, the Puritan gentleman with his narrow sort of nationalism, the champion of the common law, inbred in the working of it, looked beyond the actual quarrels over jurisdiction towards a more principled position as to the proper function of the law and of the judiciary in government and in the life of the state. One of the main theses of Professor White appears to be that he only gradually and belatedly arrived, in the late 1620's, at a position of any generality constitutionally, as a parliamentary champion of "rights"(19). Accepting that, maybe one could hazard the conjecture that he at least proved to be something of a spiritual father of the long parliament of the 1640s, and of the "glorious revolution" of 1688, his views having thereby been borne out by history, and from then on proven a viable and integral part of Anglo-American legal and political heritage(20). This is saying quite a lot, and, of course, there is the break in the constitutional continuity represented by the American revolution, introducing a more clear-cut division of powers, those of the US Supreme Court standing out as something that Britain lacks, where Parliament, as is often emphasized, rules unhindered in the sense that there are no limitations to its legislative capacity(21). Nevertheless, even if this holds good in theory, the English judiciary in reality appears to enjoy a very considerable prestige and a good measure of independence(22); and the constitutional arrangements that have brought this about can be traced all the way back to 1688(23). So, the avowed influence of Coke could not quite unreasonably be qualified as a contribution to the saving of the judiciary in common law legal cultures from having become or standing the risk, typically, of becoming so to speak the hostages of either the executive or the legislature, or both(24). This would appear to be the second point to recall about Cokean legal quasi-theorizing. He may here well have been driven to a greater or lesser extent by professional jealousy, or by various baser ulterior motives. Even if so, that does not have to make the impact less real, or his theory, as far as it goes, less selfsupporting, whatever the point of view from which criticism is advanced.

So much for the position of the judiciary(25). Over legal reform or policy-making, the redressing of grievances in law, Coke was likewise opinionated, explicitly so in detail, more implicitly so in theory. It is in this regard tempting to confront his thinking with that of his great rival, Francis Bacon(26). On the surface of it, Coke was simply appealing to history in his dogged defence of common law institutions and procedures, whereas Bacon was bringing that great wisdom and subtlety to bear on the interest of the Royal prerogative as interpreted by the new king from Scotland, James Stuart. Neither Coke nor Bacon had though, if one tries to do credit to them both by not pushing convenient simplifications too far, any ready-made answer to the complex matter of legal change, the multifold considerations that there must be taken into account. But Coke's method, if it can be called that, in dealing with these problems on a level above the minutiae of the concerns in question, be it a case before the bench or a matter for legislative consideration, seems to have been one of historical empiricism that presages the critics of the great French revolution(27). It is, I believe, what Professor Skinner calls Coke's classic formulation of the theory of continuity(28). On the other hand, Bacon's admittedly substantial and indeed almost awe-inspiring contributions to the problems of modern man, just then newly awakened, and of his social as well as political surroundings, brilliant as they are, smack of constructivist intervention, based upon abstract philosophical speculation(29). Herein lies, it may be argued, the nub of Coke's conservatism.

Moreover, Coke's conservatism was not rampant. Especially the way in which he conducted himself as a parliamentarian would seem to support that(30). Instead of focusing on the obstinate judge that so angered King James, or talking about Coke's being motivated by class-interest, viz. that of the gentry, both of which characterizations may have something to say in themselves, the more fruitful approach to Coke would seem to be that of recognizing his constructive and reasonable side, which he also showed in his attitude towards the changing nature of the society and of the law(31). Viewed in that light, labels such as libertarian or radical(32) even, with their associations to the political terms of property rights, whigs(33) and _bourgeois_ yet to come, may more easily be attached to the man; but on balance it seems fairest to refer to him as a conservative with a small "c", rather of the Lord Salisbury brand, given above all the pragmatism and moderation that were parts of Coke's mental disposition.

As to the actual structure of Coke's method or theory, there exists a near-contemporary criticizing scrutiny carried out by Hobbes(34). Since I find it revealing, I propose here to give a resumé of it; and in defence of Coke rope in Professor Gray(35). Hobbes's Philosopher in the famous dialogue starts off by saying that "... the common law itself is nothing else but reason ... Equity is a certain perfect reason, that interpreteth and amendeth the law written ... and consisting in nothing else but right reason"(36). To which the Lawyer, the student of the common law, answers, citing Coke's _Institutes_, to the effect that it is just an artificial sort of perfection of reason in law, the law itself being the product of "many successions of ages". At this the Philosophers comes back, stressing the authority of the legislator as upholding the reason, and the discussion gets increasingly technical. At a later stage the Lawyer is accused of trying to undermine the King's authority(37). And the Philosopher tries to undermine Coke's by showing how often the historical and legal sources simply do not support his famous sayings(38). It all gives the impression of a guarded mockery of Cokean positions, which makes the Lawyer cry out in exasperation: "But I know that neither history nor precedent will pass with you for law"(39). In another instance(40) he is really in difficulties; and Hobbes makes him say: "It is somewhat hard, but it has been so practised time out of mind". And so it goes on.

What makes Hobbes' criticism unfair is, I think, that the totality of Coke's various positions in no way seems to add up to a Jurisprudence proper, a complete theory standing as it were on its own(41). Coke had no preconceived general Theory of law and state. What he had, and used, was what may here be conveniently termed an internal perspective, accepting the formal language in which this sort of discussion is habitually conducted. By that phrase I mean that he, presumably, identified himself with a given historical legal heritage, seeing it from the inside, as it were. This, in order to be serviceable, would seem to necessitate a process of supplementary underpinning, which could be done either philosophically or politically. As has already been indicated by the aforesaid, and also by the title of this paper, I believe that the latter alternative applies in Coke's case(42).

Professor Gray, in his seemingly balanced and sympathetic assessment of Coke after all as a kind of legal philosopher, elaborates the first step about the historical heritage by inspecting Coke's claim to artificial reason. His analysis follows, interestingly, I think, two lines: the one is the old judge's emphasis of legal imagination; the other is that its use, the creative adjudication, should be subject, not to an abstract ideal, but to the values that define the national identity, the ancestral choices that made England England(43). And dr. Wingfield-Stratford makes the same point even more rhetorically when he says that "these men (the Puritan upholders of the common law) for all their impracticable formality, were standing for England, in the sense that the Law, with all that was good and bad in it, represented what was deepest rooted in the heart of the nation"(44).

This may not save anything of the splinters after the Hobbesean barrage taken as such, but it helps one better to understand, in my view, how misconceived it was. Further, it points to the spiritual kinship with both the more philosophically inclined conservatives and those who restrict themselves to conservative dogma(45). By studying the creed they have in common, we stand a chance of learning more about a social interplay, the outcome of which is our present and our future law(46).

A few additional words may be justified concerning Coke's internal perspective as a lawyer. Much later, especially in the United States and in Scandinavia, the so-called legal realist movement criticized this internal point of view as an irrational expression of emotive attitudes. To answer this objection, one needs as has been remarked above, an external justification of legal assumptions in philosophical or political terms. This can be done either in a rationalist manner, for instance through appeal to "obvious" principles of natural law, or more empirically. The first course, typical for natural law philosophers of Coke's time, including Hobbes, is quite controversial. Reason alone can provide a valuable systematization of data, but never the data themselves. In agreement with Coke's standpoint, one can obviously find the data in historical development of society. It is a striking coincidence that, in view of the most recent development of legal philosophy, the ultimate yardstick of external justification in the law consists in rational acceptability which is "a historical and social matters .... a matter of the development of our forms of life"(47).

NOTES

1) Cf. Gilmour, I., Inside right. A Study of Conservatism, Hutchinson of London 1977, p. 49.
2) Chief Justice of England, in common parlance.
3) Hill, C., Intellectual Origins of the English Revolution, Oxford 1965, pp. 259 ff.
4) Wingfield-Stratford, E., The History of British Civilization, London 1930, p. 483.
5) Bowen, C.D., quoting Sir Henry Spelman in The Lion and the Throne. The Life and Times of Sir Edward Coke 1552-1634, London, 1957, p. 462.

6) See, inter alia, White, S.D., Sir Edward Coke and the Grievances of the Commonwealth, Manchester 1979, with, too, references as to pertinent literature.
7) For instance by Thorne, S.E., Sir Edward Coke 1552-1952 in Seldon Ssociety Lecture, March 17th 1952, published by Bernard Quaritch, London 1957, and Holdsworth, Sir W., A History of English Law, Vol. 5, 3rd ed. 1945; reprint, London 1966, pp. 499 ff. Also appreciative is Allen, Sir C.K., Law in the making, 6th ed., London 1958.
8) This phrase has been borrowed from Professor G.R. Elton in his article on Coke in Biographical Dictionary of British Radicals in the Seventeenth Century I, ed. by R.L. Greaves and R. Zaller, Brighton 1982, pp. 156 ff.
9) White, op.cit., p.3.
10) Cf. Barbara Tuchman's portrait of De Coucy in A Distant Mirror. The Calamitous 14th Century, Alfred A. Knopf, Inc., 1978; and Penguin.
11) Rowse, A.L., The Expansion of Elizabethan England, Macmillan 1955; Papermac 1981.
12) Many a work could here be cited. Perhaps is Lytton Strachey's way of putting it, in the beginning of his famous book about Elizabeth and Essex, unsurpassed for its economy and literary beauty. Very useful is Rowse, A.L., whose piece of scholarship mentioned in note 11 was published after a first volume. The England of Elizabeth. The Structure of Society, 1950/1981. See, too, Zagorin, P., The Court and the Country. The Beginning of the English Revolution, London 1969, pp. 1 ff., and likewise Prestwich, M., Cranfield. Politics and Profits under the early Stuarts. The Career of Lionel Cranfield, Earl of Middlesex, Oxford 1966, pp. XV ff.
13) According to David, R., and Brierley, J.E.C., Major Legal Systems in the World Today. An Introduction to the Comparative Study of Law, London 1968, p. 278. The reader may wish to consult further with Plucknett, T.F.T., A Concise History of the Common Law, 3rd ed., revised and enlarged, London 1940, a more recent work being Milsom, S.F.C., Historical Foundations of the Common Law (1969).
14) The author acknowledges having borrowed these telling characterizations from Wingfield-Stratford, op.cit. p. 479. Cf. that author's sub-title on the same page:

The Revival of the Common Law; and Plucknett's op. cit., p. 50: The Supremacy of the Common Law, which, together with The Stuart Constitution 1603-88. Documents and Commentary, edited and introduced by J.P. Kenyon, p. 105, relates to a later stage.

15) In this respect, see Berkowitz, D.S., Reason of State in England and the Petition of Right, in Staatsräson. Studien zur Eines Politischen Begriffs, edited by Roman Schnur, Berlin 1975, pp. 165 ff., and Skinner, Q., The Foundations of Modern Political Thought I & II, Cambridge University Press 1978. For further treatment of the general "crise de conscience" following in England upon the climax of the Elizabethan age and culminating, eventually, in the civil war, see Wingfield-Stratford, op.cit., pp. 484 ff., and Hill, op.cit., pp. 4 and 287 f., with reference to Allen, J.W., English Political Thought, 1603-44 (1938), pp. 386 ff. The discussion among historians concerning this complex topic, started by E.J. Hobsbawm's paper in Past and Present, in 1954, should also be noted here. Dacre's (then H.R. Trevor-Roper) analysis in the same publication, in 1959, is famous, and indeed germane to our issue; Swedish readers may here conveniently consult Rystad, G., ed. Problem I Aldre Historia. Europeiskt 1600-tal, Studentlitteratur 1977.

16) By an older generation Swedish historian, Stavenow, L., Den Stora Engelska Revolutionen I det Sjuttonde ÅRhundradets Mitt, Gothenburg 1895, pp. 6 ff. and pass. This rather appears to be a legio kind of interpretation.

17) This armchair foreign policy assessment of the options open to the English leadership visavi Spain echoes, I understand, much of the argumentation, or rather the political spirit, of the war party of the days in question, whose headquarters and source of inspiration was Essex House and its master, the Earl being its active spokesman at court, when not commanding task forces, Anthony Bacon its chef de cabinet, manipulating the threads of secret diplomacy. Burghley and Elizabeth, in contrast, for long personified restraint and caution. This division between activists and passivists generally appears to be intelligible when discussing British foreign policy, cf. Blake, R., DISRAELI, University Paper-

back, London 1969, pp. 574, f., referring to Southgate, D., The Passing of the Whigs, 1832-1886 (1962).

18) David, R., and Brierley, J.E.C., op.cit., section 283; according to Wingfield-Stratford, op.cit., p. 482, "The equity jurisdiction of Chancery had been formed to remedy some of the notorious delays and injustices of the Common Law Courts".

19) White, op.cit., pp. 22 ff., 192, and 222 ff.

20) Scruton, R., The Meaning of Conservatism (1980), pp. 60 ff., where this point is made in politico-ideological terms, according to conservative dogma -in Scruton's interpretation. It is not inconsistent with Coke's views, to say the least.

21) For a discussion of this, see Fuller, L., The Morality of Law, New Haven 1969, pp. 113 ff. The details of the constitutional position as such do not interest us here, but, as Berkowitz points out, it is indeed something of a paradox that the Commons and the Lords, in standing up to the Crown and winning, contributed to the fact that eventually only a omnicompetent Parliament survived the wreckage, that, fallaciously, as it developed, their triumph was seen as the absolute supremacy of the law, op. cit., p. 212.

22) Hailsham, Lord, The Door Wherein I Went, Collins , London 1975, pp. 245 ff.

23) Near-breaks occured in the field of representation and suffrage reform in 1832, 1867, and 1906

24) Df. the title of Bowen's biography, referred to in note 5 above.

25) See, too, White, op. cit.,pp. 76 ff.

26) For instance as depicted in the foreword by R. Lockyer in Bacon's The History of the Reign of King Henry VII, The Folio Society, London 1971, pp. 10 f., and in Bowen, op. cit., pass., and Francis Bacon. The Temper of a Man (1963), III:1.

27) Burke, Coleridge, de Maistre.

28) In History and Ideology in the English Revolution in The Historical Journal, Vol. VIII, 1965, No. 2, p. 152, note 5.

29) Cf Hayek, F.A., Law, Legislation and Liberty. A New Statement of the Liberal Principles of Justice and Political Economy, Vol. I, The University of Chicago Press 1973, p. 74

30) Most markedly, this appears to have been the case in

the early 1620s; cf. above in the text with note 19, referring to Professor White.

31) Cf. Skelton, A.N., *Constructive Conservatism*, With an introduction by Sir R. Horne, William Blackwood, Edinburgh and London, 1924, and Young, K. *Sir Alec Douglas-Home*, J.M. Dent & Sons Ltd, London 1970, pp. 28 ff.

32) Cf. note 8 above.

33) Many authors, Disreali among them, have referred to the revolution of 1688 as the confirmation or foundation of the Whig oligarchy in England.

34) *A Dialogue between a Philosopher and a Student of the Common Laws of England*, ed. by Sir W. Molesworth in *The English Works of Thomas Hobbes*, Vol. 6, London 1840.

35) Gray, C., *Reason, Authority, ans Imagination: The Jurisprudence of Sir Edward Coke* in *Culture and Politics. From Puritanism to the Enlightenment*, ed. by P. Zagorin, University of California Press 1980, pp. 25 ff.

36) Molesworth, op. cit., p.4.

37) Molesworth, op. cit., pp.62 f.

38) See Thorne, op. cit. stressing the consequence that, yet, Coke's influence on posterity became so formidable; cf. also Bowen's works and Wingfield-Stratford, op. cit., p. 481.

39) Molesworth, op. cit., p. 109.

40) The spirited conversation has then reached the subject of "felony; the land".

41) Cf. Gray, op. cit., and Plucknett, op. cit., p. 51 with references.

42) Cf. Elton, op. cit., pp. 157 and 158.

43) Gray, op. cit., p. 39.

44) Wingfield-Stratford, op.cit., p. 483.

45) Cf. Gilmour's selection, op.cit., Part Two. Roger Scruton's account, op. cit., is of the latter sort. Sir Ian's treatment of the constitutional crisis, in his view chiefly brought about by Labour in recent years, is in many ways of interest to issues touched upon in this paper (Chapter 5 to Part Three).

46) More specifically, the Swedish reader may wish to take stock in Fredrik Lagerroth's paper *Rättskontinuitetens problem i Sverige 1809 sett i belysning av den rena rättsläran* in *Festskrift tillägnad Nils Herlitz*, Stockholm 1955, pp. 204 ff.

47) Cf. Aarnio, A., On Truth and the Acceptability of Interpretative Propositions in Legal Dogmatics, in Rechtstheorie, Beiheft 2, Berlin 1981, p. 49.

POPPER'S CRITERION OF REFUTABILITY IN THE LEGAL CONTEXT

Yemina Ben-Menahem, Hanina Ben-Menahem

In this paper we compare two criteria; Popper's criterion of the refutability of scientific hypotheses and an analogous criterion, found in Talmudic law, for the admissibility of legal testimony. Popper's criterion and the methodology based on it faces philosophical as well as pragmatic difficulties. We survey the major objections to Popper's criterion and argue that they do not apply in the legal context. The paper is both a result and an illustration of the interaction between philosophy of science and philosophy of law.

The two criteria we wish to discuss are the following:

*Any testimony that is not refutable is not admissible.

*Any conjuncture that is not refutable is not scientific.

We shall begin with a brief discussion of Popper's criterion and proceed to the legal analogue.

Popper shares with other methodologists of science, including the logical positivists, the idea that testability is the hallmark of science. But it is Popper's special insistence on refutability (rather than say, verifiability or testability) that has to be explained. There are two dimensions to the explanation, one logical, the other normative.

First, Popper was motivated by his wish to solve the problem of induction, which has always been considered the main obstacle to a rational scientific method. There is no valid inference from the truth of a finite number of observation statements to the truth of a generalization or a law. But it is the law that interests the scientist. If the general law is in principle unverifiable, then to make verifiability the criterion of science would mean that the most interesting part of science is non-scientific, which is absurd. Popper's move from verifiability to refutability purports to neutralize the problem of induction. According to Popper no inference to the truth of a general

A. Peczenik et al. (eds.). Theory of Legal Science, 425–435.

law is required in the scientific process. The inference that is required is <u>deductive</u> rather than inductive and hence perfectly legitimate; it is the rule known as <u>modus tollens</u>, which leads from the falsity of a particular instance to the falsity of the generalization.

The other reason for Popper's insistence on refutability goes far beyond logical considerations. Popper regards any trace of dogmatism as a threat to rationality. For him, rational thinking means critical thinking. In order to overcome the temptation to succumb to dogmatism he sets the negative ideal - falsify - rather than the positive ideal - verify (although the end, of course, is a theory which has withstood the severe test and has not been falsified). That these psychological or normative considerations go beyond what can be justified by logic is evident from Popper's treatment of what he calls 'the conventionalist stratagem'. A conventionalist stratagem is an attempt to save the refuted hypothesis by modifying it, modifying the auxiliary hypotheses, explaining away the counter evidence or using any other method to harmonize the hypothesis with the data. Such a move cannot be rejected on logical grounds, and Popper is fully aware of this: "The only way to avoid conventionalism is by taking a <u>decision:</u> the decision not to apply its methods. We decide that if our system is threatened we will never save it by a conventionalist stratagem.(1)"

We can now formulate Popper's requirement of refutability in somewhat different terms. When putting forward a scientific conjecture a double risk is involved: a) The logical risk; the conjecture has to be of the appropriate logical form, a form that enables its refutation. b) The normative risk; the scientist must be willing to take a risk in the sense that if his conjecture is refuted he will abandon it rather than attempt to save it.

Let us now turn to the Talmudic analogue. Biblical law provides that one who bears false witness is to suffer the injury which he intended to inflict upon the person against whom his false evidence was directed(2). The rationale of this principle, which occurs in other ancient legal systems (e.g. LH s.4) is, among other considerations, to deter potential false witnesses from carrying out their malicious intentions. However, it fails to distinguish between an ill-intending witness, who deliberately gives false evidence, and one who in good faith gives incorrect testimony based on some misleading impression such as wrong identification. Cases of mistaken identity are known to be "alarmingly frequent"(3). Jerome Frank observes that "perhaps erroneous identification of the accused constitutes the major cause of known wrongful convictions"(4). An English committee came to the conclusion that "evidence as to the identity based on personal impressions, however, _bona fide_, is perhaps of all classes of evidence the least to be relied upon"(5). These observations show how likely it is that innocent persons will be convicted on the basis of wrong identification. However, the very same observations also lead to the conclusion that false testimony is not necessarily culpable; indeed in many cases it is not. The principle of _talionis_ in the case of false witnesses has therefore to be mitigated by some device which would restrict its applicability to those instances where the _mens rea_ of the witnesses is beyond doubt. This goal is achieved in Talmudic law by introducing a distinction which is unique to Talmudic law, that between _plotting witnesses_ and _false witnesses_, and by restricting the applicability of the principle of _talionis_ to the former case only.

Consider the following case: A and B give evidence that C murdered D on such and such a day in such and such a place. Later, E and F come along and testify that on the day on which the alleged crime was committed, A and B, the eye-witnesses, were elsewhere with them and could not possibly have seen the crime, (what is called an alibi of the eye-witness)(6). Now in this case, if we believe E and F, it is evident that A and B are false witnesses with malicious intention, that is, plotting witnesses; wrong identification is ruled out. Implicit here is the psycho-legal assumption that a mistake in dating the event, as opposed to a mistake in describing the event, can safely be ruled out. The Talmudic presumption concerning timing mistakes is that 'A man is to be treated as not liable to make even the slightest mistake'(7).

In our example A and B are plotting witnesses and the Biblical principle of _talionis_ would be applied to them. In all other cases of falsified testimony, for example when it has been proven that the alleged murderer was elsewhere at the time of the crime, the false witnesses will not be punished - though their testimony is dismissed - in case they have in reality seen a crime but mistakenly identified the defendant as the murderer.

Note that parallel to the distinction between a plotting witness and a false witness a distinction has to be introduced between a strong and a weak type of falsification. A testimony is strongly falsified only if the witnesses have been proven to be plotting witnesses; in all other cases of falsification we shall say that the testimony has been weakly falsified. Note further that strong falsification results in punishment whereas weak falsification only results in dismissal of the evidence.

As pointed out, Talmudic law restricts the applicability of the Biblical principle of talionis to the case of a plotting witness, thereby protecting the bona fide erring witness. From the point of view of the defendant, however, it is essential that an ill-intending witness should be deterred by subjecting himself to refutation and thereby to punishment. But unless the witness specifies time and location of the alleged crime it is impossible to charge him with plotting.Bearing these two considerations in mind, i.e., protecting the bona fide witness on the one hand, and protecting the defendant on the other, a demarcation criterion is required to determine what kind of statement qualifies as admissible testimony.

The Talmudic criterion is the following: A testimony which is not strongly falsifiable is inadmissible(8). (Of course, this is a necessary but insufficient condition; other conditions such as legal competence have to be satisfied).

The two elements that have been distinguished in Popper above can be detected here as well. First, the formulation of the testimony itself must be such that it includes time and location of the event upon which evidence is given. This specification ensures that the testimony is strongly falsifiable. Second, the witnesses themselves must be liable to punishment should it be proven that they were not present at the scene of the alleged crime, otherwise it cannot be said that they have actually risked themselves. If, for some technical reason they will be exempt from punishment, their testimony is inadmissible even if it is of the appropriate logical form. Here is an example:

In Talmudic law one's legal competence to bear witness depends on one's legal status. If one is a slave one lacks legal competence to appear in court as a witness. Now suppose two babies, the master's and his slave's, have been switched around in such a way that we cannot tell which is which. The Talmud ponders whether both together, when grown up, could testify to an event they had both observed, since one of them surely has legal competence to testify. This possibility is rejected, however, on the basis of the following argument: Should it be proven that they were plotting witnesses, neither of them could be punished, for either of them could claim 'I am the disqualified witness, hence I am not liable'. It follows that these witnesses did not risk themselves and consequently, on the basis of the above cited principle, their testimony is inadmissible(9) even if it is of the appropriate logical form.

Let us now confront the two refutationalist criteria. In both cases we have noticed the existence of two elements: The logical element dictating the structure of the testimony or the conjecture and the punitive (or normative) element according to which the witness (or scientist) has to bear the consequences of his endeavour.

To appreciate the significance of the emphasis on refutability in the legal context, let us juxtapose the Talmudic principle and the requirement of corroborative evidence.As an example, Section 6 of the Evidence Ordinance 1924 issued by the Mandatory Authorities of what was then Palestine reads as follows: 'No judgment shall be given in any case on the evidence of a single witness unless such evidence is corroborated by some other material evidence which, in the opinion of the court is sufficient to establish the truth of it'. Corroborative evidence is evidently based on a different conception as to what grants credibility to a given piece of testimony. The requirement of refutability, as opposed to that of corroborative evidence, squares with the conception that a person is to be believed unless it can be proved he has lied.

We must not, however, lose sight of the differences between the two demarcation criteria. The motivation for the requirement of refutability is different in the two cases. As we saw above, Popper was led to his criterion by the realization that a typical scientific conjecture is in principle unverifiable. This consideration does not hold in the legal context. The typical legal testimony is composed of particular statements which can be both verified and falsified by other witnesses. The emphasis on refutation in this case derives from the need to subject the witness to strong falsification. Lack of confirmation of his testimony is not sufficient to establish his guilt and verifiability is therefore an inappropriate requirement.

Popper does not specify what counts as a refutation and we may conclude from his writings, that any type of refutation is sufficient to overthrow a scientific conjecture. The Talmudic criterion, on the other hand, spells out what counts as a refutation: The witness is falsified only if it is proven that he was elsewhere at the time of the crime. Thus, a testimony as to the colour of the defendant's shirt would meet Popper's but not the Talmudic criterion. This difference can be explained in terms of the difference in motivation we have just observed. The Talmudic principle strives to establish the guilt or the innocence of the witness and this can only be done via strong falsification. Falsifying the colour of the defendant's shirt would not show the guilt of the witness (though it will dismiss the evidence), since a bona fide mistake (wrong impression, bad memory) cannot be ruled out. The problem we have touched upon here - what counts as refutation, is at the root of some interesting difficulties we discuss later.

Popper's objection to the conventionalist stratagem is paralleled by the Talmudic prohibition against retracting one's testimony, that is, adjusting one's testimony to fit counter evidence(10). Note however, that Popper's objection is based on purely normative considerations (as he himself admits) and is indeed the most controversial part of his methodology. It has been argued by many of his critics that contrary to his demand that we abandon the refuted hypothesis, modification and elaboration of a theory in face of counter evidence is a vehicle to progress. On the other hand, the prohibition against retracting one's evidence is self evident. The scientist suggesting a new hypothesis is in a different position than the witness. The scientist's hypothesis has yet to be tested by experiment and its being a modification of a previous hypothesis does not, logically speaking, interfere with the process of testing its validity. Giving evidence is not to be viewed in terms of suggesting a hypothesis but rather in terms of establishing its truth. Allowing the witness to readjust his testimony would therefore almost be analogous to allowing the scientist to readjust the readings of his instrument and would thus frustrate the legal process.

Turning to the main arguments raised against Popper's refutationalist philosophy of science, we shall examine their applicability in the legal context.

a. The holistic argument. Pierre Duhem, as early as 1905, has pointed out that there is hardly ever a test involving a single scientific hypothesis. Typically, when a hypothesis of a mature science is tested, a whole complex of auxiliary hypotheses is tested as well. It follows that when a prediction fails we may only conclude that the complex as a whole cannot be true. No inference to the falsity of the individual hypothesis we initially set out to test can be justified by logic alone. Thus, according to the holistic argument, falsification is, practically speaking, just as problematic as verification. Faced with counter evidence, the scientist has to decide which components of the complex he will regard as falsified, and this decision is necessarily based partly on extra-logical and extra-empirical considerations which have come to be known as conventions. In short, falsification is, according to the holist, inconclusive and thus tentative.

Does this argument hold in the legal context? As you may recall, the only issue at stake as far as strong falsification is concerned, is the truth of the statement asserting the presence of the witness at the scene of the crime. This is an isolated, well-defined statement, the testability of which is beyond reasonable doubt. If there are any isolated testable statements which only an extreme sceptic would doubt, this is one. (Note that falsifying the witness does not entail the assertion that the facts about which he testified have never occurred. Indeed, the person against whom the testimony was directed can still be convicted on the same indictment on the basis of other, non-falsified testimony). We may therefore conclude that the strong type of falsification gets as close as possible to the falsification of an isolated statement and is not threatened by the holistic argument. In other words, strong falsification is, practically speaking, conclusive, and the punishment of the strongly falsified witness can be rationally justified.

Does the holistic argument apply to weak falsification? Here we have to be more cautious. It is true that in the case of direct evidence the holistic argument will again seem unconvincing. A typical piece of such direct evidence is composed of a series of observation reports, each of which can be viewed as independently falsifiable (this man was driving the car, the speed was 120 km/h, etc.). But there are more complicated cases as well. The evidence may be circumstantial evidence or expert's evidence. Here the interdependence between the several components of the evidence and the dependence of the evidence on background theory is much more significant. Though even here the holistic argument may be considered less powerful than in the case of an advanced scientific theory, it cannot be ignored altogether as in the case of strong falsification.

b. The idealist-relativist argument. Like the previous argument, this one purports to show that the picture according to which scientific hypotheses can be conclusively refuted by objective facts is inadequate. However, the relativist is more radical than the holist in his philosophic outlook. Whereas the holist recognises the existence of objective 'facts' or 'experiences' but claims that a typical scientific theory is underdetermined by those facts, the relativist denies the very existence of objective facts. According to him the commonsense distinction between fact and theory or observation statement and theoretical statement is untenable. He claims that all facts are theory-laden, so that what constitutes a 'hard fact' from one perspective may not constitute a fact or not the same fact from a different perspective. Clearly, if any fact is theory-laden then the view that theories are refuted by objective facts is far too naive. Experimentation, on this view, is an intratheoretic affair, not an objective transtheoretic process which may serve to compare different theories with regard to their objective truth values.

How powerful is this argument in the legal context? It has to be conceded: The extreme relativist will find the 'facts' reported by the eye-witness just as vulnerable to his argument as any other facts. However, on a more moderate version of relativism it can be argued that the theory-dependence of facts is a matter of degree. Thus, what the physicist sees in a cloud-chamber may be more affected by his theoretical frame of reference than what a typical witness sees when one man is killing another. Further, the extent to which conceptual frameworks of individuals can differ varies from one domain to another. The physicist's framework is indeed different from that of the layman when a sophisticated experiment is concerned, but there is little ground for assuming that the physicist will actually see something different than the layman when offered a cup of coffee. Even if it is granted that theoretical frameworks affect what one sees, the framework of the two persons in our last example as well as in most other daily occurrences appears to be the same.

From the point of view of moderate relativism, strong falsification is again different from weak falsification. The dependence of fact on theory when what is at issue is the presence of the witness at the scene of the crime can be taken by the moderate relativist to be minimal. Hence the strong falsification of a witness is, practically speaking, an objective and conclusive procedure. On the other hand, the falsification of circumstantial evidence or expert's evidence is vulnerable to the relativist's argument and could not be taken by him to be conclusive.

c. <u>The psychological argument.</u> According to this argument, there is in the realm of knowledge a kind of <u>horror vacui</u> - a theory is not to be overthrown even in situations which the refutationalist would regard as refutations unless a viable alternative exists. A famous example of a theory adhered to in spite of well-known 'anomalies' is Newtonian mechanics. The known anomalies were not considered refutations of Newton's theory until Einstein's theory of relativity succeeded in accounting for them in a more satisfactory manner. Popper's methodological maxim according to which a refuted theory should be abandoned is viewed as totally unrealistic (and hence inadequate) by the proponents of the psychological argument. Before considering the application of the argument in the legal context it should be noted that Popper would find this argument easy to meet. He would admit that the conservatism of some scientists has occasionally prevented abandoning a refuted theory but he would insist that it is precisely these dogmatic tendencies which his methodology was meant to combat.

If we now turn to the legal context, we immediately realise that there is no analogue to the psychological argument. It may be true that judges are sometimes victims of a similar type of 'horror vacui' and cling to their conception of a case in spite of accumulating counter evidence. But there is no doubt that court procedure must be such that conservatism and dogmatism on the part of the judges should be minimalized. Here, evidently, Popper's position is the only reasonable one. Falsified testimony cannot be relied upon just because no viable alternative exists. This goes both for strong and weak falsification.

d. <u>The evolutionary argument.</u> This argument draws our attention to the fact that a scientific theory often has to evolve from a preliminary set of hypotheses into a sophisticated comprehensive theory. To enable this evolutionary process, the theory should be modified and adjusted rather than abandoned in the face of counter evidence. No idea would ever develop into a full-blown scientific theory if it were abandoned when it first encoutered a difficulty. This argument, reasonable as it is in the scientific context, again has no analogue in the legal context. The process of developing a theory has no counterpart in the process of giving testimony. Hence, the witness cannot be granted permission to modify and adjust his testimony.

e. The varying weight argument. According to this argument it is unreasonable to assign the same weight to each component of a scientific theory. Should we regard a theory as refuted when one minor prediction fails, or should we be more tolerant and abandon the theory only when one of its major elements, its core, is refuted? If the latter is the case, how are we to distinguish minor from major elements? Newtonian mechanics can again provide an example. Were the discovered anomalies central enough to undermine the whole theory? The problem is a serious one. Popper's refutationalism is in danger of becoming an empty slogan unless one specifies exactly what counts as a refutation. The same problem exists in the legal context. Is the whole testimony refuted when one of its minor components is refuted? At what point do we dismiss the testimony? At what point do we charge the witness with perjury? As we have seen, the Talmud tackled this fundamental problem. The Talmud takes the most extreme position concerning the credibility of the evidence. Any refutation counts as a refutation in the sense that it is sufficient to dismiss the evidence. The Talmud goes to the other extreme, however, when the guilt of the witness is at issue. Here only strong falsification counts.The rationale in both cases is the same. Convicting a person, whether the defendant or the false witness, should be on a firm factual basis and any doubt should operate to his benefit.

To conclude, Popper's falsifiability criterion encounters philosophical as well as pragmatic objections. We have tried to show that none of these objections, with the exception of extreme relativism, can be raised against the Talmudic criterion of strong falsifiability. The philosophical debate concerning Popper's criterion enables us to provide a rationale for the Talmudic distinction between strong and weak falsification, a distinction which has attracted the attention of scholars but has nevertheless remained obscure. At the same time the discussion of Talmudic law has provided us with a relatively unproblematic application of the falsifiability criterion. Unfortunately, it does not seem possible to transfer the Talmudic distinction back into the scientific context. Reaching the goal of a rigid methodology of science seems as far as ever.

NOTES

(1) Popper, K.R.1959. The Logic of Scientific Discovery, London, p.82.
(2) Deut.19:19
(3) Wall, P.M.1965. Eye-Witness Identification in Criminal Cases, p.7.
(4) Cited ibid.
(5) Ibid.
(6) Gooderson, R.N. 1977. Alibi, London, p.165.
(7) Babylonian Talmud, Pesahim 11b.
(8) Babylonian Talmud, Sanhedrin 41a.
(9) Palestinian Talmud, Yevamoth 11, 5.
(10) Babylonian Talmud, Kethuboth 18b.

CHAPTER 6

CRITICISM AND DEVELOPMENTS IN PARTICULAR AREAS OF THE LAW: PROPERTY, CONTRACTS AND TORTS

SYNOPSIS

Lars Lindahl

This chapter is devoted to criticism and developments in some particular areas of the law.
Contract law and the choice between competing theories in this field is the subject of Marsha Hanen's paper. Against the classical theory, according to which promise and agreement is the central element in contract around which everything turns, it has been argued both that it is reminiscent of nineteenth century liberalism and that it is inadequate for dealing with a number of problems (for instance, mistake and impossibility) arising in this area of the law. As opposed to the classical doctrine, other theories, purportedly more "realistic", have been proposed such as economically orientated theories on allocation of risks and losses, and so-called "relational" theories of contract. In her paper, Marsha Hanen suggests that the various theories may be found to be at cross-purposes since some of them may be primarily aiming at moral justification while others are more concerned with explanation or prediction. (It is, however, also maintained that none of these three functions of a theory can be wholly disregarded by an acceptable theory of contracts). The author specially emphasizes the similarities between theory choice in contract law and theory choice in the field of science. In each case, there are a number of constraints with which any acceptable theory must comply, but there need not be any unique right or correct theory. In the law, as in science, the choice between several right or correct theories should be made by comparing the theories considered with respect to their coherence and comprehensiveness.

A. Peczenik et al. (eds.), Theory of Legal Science, 437–439.

As a legal scientist working with contracts, torts and insurance, Jan Hellner raises the question of the relevance of legal theory to the kinds of problems facing modern legal science. The author emphasizes that outside the realm of traditional legal dogmatics there are other problems of great importance to legal science in modern society. Firstly, there is the problem of analyzing various alternative means for achieving societal ends, where the author describes how, in contemporary Sweden, traditional contract law has been supplemented by a number of new means of control and how, in connection with torts and insurance, legal science is increasingly aware of the variety of ways of rendering compensation to victims. Secondly, there is the important problem of ends and values to be pursued in connection with legal rules and proposals for reform. Criteria for the evaluation of legal rules have been proposed by the "law and economics" movement but the legal scientist feels at a loss as to whether purely economical considerations are conclusive. The author ends by urging legal theorists to consider in what ways they can contribute to these kinds of problems, confronting the legal scientist.

The "law and economics" movement is represented by Göran Skogh's paper, where the main topic discussed is the relationship between contracts and trade-laws. Assuming a legal system where contracts are publicly enforceable, the author first presents a number of (purportedly testable) hypotheses on how contract formation or contract formulation is influenced by a number of factors such as the value involved in potential conflict, the possibility of having recourse to private enforcement, the difficulties involved in the evaluation of the product etc. There follows a discussion of how non-mandatory trade-laws will influence transaction costs in contract formation, the author maintaining that trade-laws will have the effect of reducing transaction costs if the laws reflect what the parties would have agreed on ex ante, had they made an explicit agreement on the matter. In a final section, the author presents some putative objections to such mandatory trade-laws not conforming to what would be agreed on if the mandatory law did not exist. It is argued that there will always be "ways around the law" though at higher transaction costs and it is claimed that if the legislator intends to benefit one of the parties (for instance, a consumer party) he will not achieve his end since the other party will have recourse to higher prices or will avoid bargains with such parties as are intended to be favored by the law. In the author's view, the most efficient way to protect consumers in trade is to encourage competition, e.g. by securing free entry to the market.

THEORY CHOICE AND CONTRACT LAW

Marsha P.Hanen

The underlying problem to which this paper is directed is the question of what we can learn about the nature of theories and theory choice by looking comparatively at science and law. The notion of law as science in western legal theory, which can be traced back at least to jurists of the eleventh and twelfth centuries(1), finds expression also in legal views more familiar to us because closer to us in time. I have in mind not only continental legal science of the nineteenth century but also the development (around 1870) of the case method of teaching law at Harvard by the first dean of the Law School, Christopher Columbus Langdell. Langdell conceived of law as a science whose laboratory was the library and whose data were the reported cases. He believed that there were relatively few fundamental legal doctrines and that an understanding of them could be reached by studying a fairly small number of reported cases. In the preface to his first casebook, on Contracts, which appeared in 1871, Langdell wrote

> "It seemed to me, therefore, to be possible to take such a branch of the law as Contracts, for example, and, without limits, to select, classify, and arrange all the cases which had contributed in any important degree to the growth, development, or establishment of any of its essential doctrines;..."(2)

Langdell's picture was one of the growth of legal principles in an evolutionary way from darkness to light using almost exclusively judge-made law; and the role of the law professor, as distinct from the judge, was to build up a coherent system from the data of reported cases.

It is interesting that this picture of legal science is one that suggests that individual branches of the law rather than the law as a whole can be formulated as theories; and clearly there are few areas of the law that lend themselves as well to representation as a theory as does contract law(3). I want in this paper to examine the

A. Peczenik et al. (eds.), Theory of Legal Science, 441–454.

nature of theory in the context of contract law with a view to trying to shed some light on the relation between theory choice and the development of those presuppositions and values that underlie particular legal outcomes.

How do we choose among different theories of contract? Are there reasonable grounds for so choosing or is the choice entirely non-rational? And how are such theories related? Are they to be viewed as incompatible alternatives, at most one of which can be correct though all might be incorrect or might each be correct in its own sphere and each complement the other? Is there anything we can learn from recent controversies in philosophy of science, especially those concerning scientific realism, that would help us to resolve current theoretical disputes about contract law?

It is usefull to begin with a recent version of the classical theory developed by Charles Fried. For Fried, promising is the moral basis of contract law.

> "There exists a convention that defines the practice of promising and its entailments. This convention provides a way that a person may create expectations in others. By virtue of the basic Kantian principles of trust and respect, it is wrong to invoke that convention in order to make a promise and then to break it"(4).
>
> "To enforce a promise as such is to make a defendant render a performance (or its money equivalent) just because he has promised that very thing"(5).

The natural measure of damages on this view would be expectation, for, by hypothesis, the obligation was undertaken freely, and so the promisor is bound to the performance of his promise (or at least to putting the promisee in as good a position as if he had performed). But this has often seemed harsh, particularly in cases where the promisee has suffered no harm by relying on the promise. In such cases the proper measure of contract damages has been taken to be the reliance or harm suffered by the promisee, or sometimes the actual benefit received by the promisor before he reneges. The question is whether the fact that we do use these non-expectation measures undermines the theory of contract as promise. Fried, naturally enough, claims not(6). Sometimes, he says, we shift to harm as a measure because it is hard to find the monetary value of the expectation; sometimes the fairest measure of expectation just is the amount needed to correct the harm caused by reliance; and sometimes harm suffered is the proper measure of damages, but on a tort theory rather than as a substitute for the enforcement of a promise. If Fried is right to claim that a shift from expectation as a measure of contract damages provides no counterexample to the theory of contract as promise, the question becomes whether there are any facts that do undermine the theory(7).

Besides the question of the proper measure of damages, there are a number of problem areas for traditional contract doctrine which raise serious questions about whether contract can really be conceived of as essentially a matter of promise or agreement; and these pose a fundamental problem about the status of apparent counterexamples to the theory. Mistake is one such difficult area which is a useful focus for our inquiry(8). Is the fact that classical contract doctrine doesn't deal well with mistake an argument against the conception of contract as freely--entered-into bargain? In the classic 1864 case of Raffles v. Wichelhaus(9), for example, Fried claims that there is no agreement on a crucial aspect of the contract, and so it is not possible for the court to enforce the will of the parties. The buyer had refused to accept delivery of a shipment of cotton arriving in Liverpool from Bombay on the ship Peerless, as he had contracted to do, because the Peerless he had had in mind had sailed in October, while the seller's ship Peerless has sailed in December(10). Fried claims that there had been a misunderstanding regarding an important aspect or fundamental term of the arrangement, namely the ship on which the goods were to arrive. If contract presupposes agreement and there is no agreement then it just is not clear what the contract requires, and so the court's decision must be based on other grounds; but what other grounds, given that the only basis for decision is supposed to be an agreement?

A possible answer would be simply to say in all such cases that, there having been no genuine agreement, there can be no contract and so there is nothing to enforce and no damages to be contemplated, but courts virtually never say this. It used to be held that, in cases of mistake, the parties are to be held to the literal contract, however harsh this might seem(11), perhaps on the ground that the only (or best) evidence of what the parties intended is to be found in what they have said in writing, and to this they must be held. But the harshness of this recourse has had little general appeal, and so classical contract law has turned to other devices.

One is the doctrine of presumed intent. What has worried many writers about this appeal is its fictional character. Clearly, where there is no agreement, courts are imposing one that may bear little relation to what the parties would have done had they been fully aware of the situation. The less evidence one has about what the parties <u>actually</u> intended, the more likely it is that the decision will reflect something other than what they intended -- perhaps what the ordinary reasonable person would have intended in the circumstances. But this resort to an "objective" standard of interpretation is also, quite plainly, to impose an external standard on the parties. The solution is reasonable and perhaps even the only one available in the circumstances; but there is no reason at all to think that it is correctly described as a matter of agreement or will.

In a case such as <u>Raffles</u> we are faced with having to reconcile the law's intervention, which amounts to imposing a result that cannot be what the parties intended, with the classical doctrine. But what else <u>can</u> we do, given that some decision must be made? It seems impossible in such cases genuinely to claim one is enforcing an agreement, since there <u>is</u> none; indeed "one may ask what it would even mean to give effect to 'the will of the parties' in a case where the parties had no convergent will on the matter at hand"(12).

According to Fried, mistake and impossibility are species of contractual accident, and, as with other accidents we can either let the loss lie where it falls, or shift it according to some appropriate principle. Clearly, then, the appeal in these cases will at least sometimes be to a principle we ordinarily regard as characteristic of <u>tort</u> law: compensation for harm. Restitution for benefits conferred, or sharing of loses where both parties are harmed, neither benefits, and no-one is really at fault are additional principles to which courts often appeal. It follows that not every issue that arises in trying to work out the requirements of a contract can be solved by an appeal to the underlying promise: sometimes we must have recourse to tort or other non-promissory principles. Fried is prepared to bite the bullet here and admit this fact: there are gaps in contract law that must be filled in by other principles. But Fried insists that it does not follow that contract is pervaded by tort.

Now this is a very odd argument. Gaps in contract are filled by other legal principles, but how can we tell whether a principle is non-contractual? Only by question--begging means, it would appear; for it sounds as though Fried wants to say that whatever is not promissory is also not contractual, and this is, of course, precisely what is at issue. Fried insists on a narrow construction of the notion of contract, according to which much of what ordinarily falls under that heading must be classified otherwise. What needs to be asked is whether this is merely a matter of semantics or whether something important hinges on the classification.

Fried touches on but does not develop an aspect of this problem in a brief discussion of parallels with general legal theory. He refers to the legal theory of Ronald Dworkin, according to which there is a single right answer in every civil case and thus the law can have no gaps, and a will theory of law must be incorrect. However, it dows not follow that a will theory of contracts must be incorrect or that contractual arrangements can have no gaps. Gaps in contract are filled by appeals to general principles of law, including tort law, so as to leave no gaps in the law conceived of as a complete system(13).

Fried thus accepts the Dworkinian account of the law as a seamless web while insisting on a will theory of contract derived from principles of liberal individualism. But there is a tension between these two strands in his thought, and one consequence of his insistence on a relatively narrow construction of the notion of contract is to place much of what would ordinarily go under that label in some other sphere. But, as Patrick Atiyah tells us, this very circumscribed conception of contract bears little relation to what has been going on in the law in recent years. He writes:

> "...the circumstances in which it is felt desir able that promises should not create obligations are today far wider than is generally recognized in the literature on promising and ... there are a great many circumstances in which no explicit promise is given, but in which obligations are imposed because of the recognition of entitlements in respect of expectations, reliance, or unjust enrichment ideas...

> We thus find a decline in the belief that the individual has the right to determine what obligations he is going to assume, and an increased strength in the belief that the social group has the right to impose its own solution on its members, dissent as they may"(14).

That classical contract theory is procrustean and unrealistic in just the way Atiyah suggests is by now a standard complaint of those who hold that there is far more to contract behavior than is dreamt of in the classical philosophy. One possible alternative can be seen if we look at the handling by the Supreme Court of Canada of the case of The Queen v. Ron Engineering & Construction (Eastern) Ltd. - a case that also raises the problem of mistake. This case arose out of a builder's tender of $ 2,748,000 along with a deposit of $ 150,000. The builder was the low bidder. However, within an hour after the tenders were opened the builder advised the owner that the tender was in error and should have included an additional amount of $ 750,058. The contract was awarded to the next highest bidder at a price of $ 3,380,464. The builder commenced an action for the return of the $ 150,000 deposit. The trial court dismissed the builder's action. The builder appealed to the Ontario Court of Appeal which allowed the appeal and granted judgment to the builder for $ 150,000 plus costs. The owner appealed to the Supreme Court of Canada which allowed the appeal, set aside the judgment of the Ontario Court of Appeal and restored the judgment of the trial court(15). I take it that our intuitive sense of justice accords with the middle decision.

The Supreme Court took the position that, once the contractor submitted the tender (which was to be irrevocable for sixty days) his deposit was forfeit because he refused to execute the contract when requested to do so. The Court insisted that there was no mistake because the contractor "intended" to submit the tender that it did. One interesting commentator on this case -- John Swan -- finds this unhelpful, for the "assumption of the parties' 'intention' has no counterpart in the real world; it is nothing more than a device to hang a legal argument on"(16). But, one wants to ask, what is the relevance of this appeal to the "real world"? Is it merely to say that the evidence available to us is compatible with the parties' having had any one of a number of intentions, and there is no way conclusively to decide which was the actual intention? If so, how is this situation different from almost any situation in law (or in science) where our conclusions necessarily go beyond the evidence? There is no special reason to think that intentions are any more intrinsically inaccessible than are other phenomena which are underdetermined by the data(17).

Swan claims than an analysis based on risk allocation is more substantive than one based on intention and that, to focus on risk allocation is to focus on a "problem that can be seen as part of the real world, the world of contractors and owners who are participating in a business that must provide some background to the transaction in question"(18). But what needs explaining is why this world is more real than the world of intentions or the world of moral obligations created by promises. What is plain is that whatever analysis we choose will presuppose certain values, and these will often involve more than a simple assessment of costs and benefits. For example, the Court in this case tells us that the "integrity of the bidding system must be protected where under the law of contracts it is possible so to do"(19), but they do not explain how this goal required the result that the builder's deposit be forfeit, nor do they give an account of how this goal should be weighed against competing ones.

Swan's quarrel with the Supreme Court in this case turns on their being, in his view, too wedded to abstract doctrine and insufficiently sensitive to the need to engage in risk allocation and related economic determinations. But clearly this emphasis on risk allocation and determining who should bear the loss rather than on the reasons for the loss or the expectations of the parties would not be congenial to Fried, and so mistake cases seem to illustrate some of the differences between an essentially economic approach to contract and one based rather on the morality of promising. It is not that an appeal to economic factors is necessarily incompatible with classical liberalism; rather, I wish to suggest that a focus on risk allocation and related matters is a quite different focus from one on the morality of keeping promises once they are made.

There is an additional factor that is touched on in Swan's account but that requires further elaboration, and that is the matter of the context of the contract. Swan's suggestion is that "courts have to investigate the context of the contract when the context provides important information on the parties' expectations and understandings"(20), but many would argue that the context always provides such information and indeed that the notion of context needs to be broadened to include the entire social context of contract making. This is no doubt a great part of the point of yet another alternative -- a relational theory of contract. Patrick Atiyah, for instance, in Promises, Morals and Law argues that morality (and hence promising) is fundamentally a social phenomenon, and that the bindingness of promises arises not from the promise itself but from certain other features of the context of the promise such as the rendering of benefits or acts of detrimental reliance. But the point is more pervasive, for Atiyah does not shrink from tying the whole theory of promising and of contract to changing ideology. He speaks, for example, of the wide gap which his book

> "may have opened up between a legal theory of promising and much modern philosophical work on the theory of promising. It will, [he thinks], be apparent that the philosophical tradition in this sphere is still heavily dominated by writers who reflect liberal values. Liberal values no doubt still have many attractions, but it is simply misleading to think that the moral beliefs of our generation are congruent with these liberal values. In particular, the value of free choice, which underlies the high regard with which the nineteenth century thinkers viewed the institution of promising, has declined greatly in contemporary society. By contrast, paternalism, which is a negation of the ideal of individual free choice, has become a much more important aspect of social policy. No contemporary lawyer can be unaware of the extent to which changing values have weakened, and in some areas quite destroyed, the belief in freedom of contract which was so prominent a feature of nineteenth--century thought. Any theory of promising which fails to take account of movements in opinion of this magnitude is in danger of ending up as an obsolete theory. Far from explaining the moral basis of promising in any absolute or eternal sense, such theories turn out to be firmly rooted

> in the ideology of nineteenth-century Western thought"(21).

If we take Atiyah seriously here, does this mean that there is no room for the idea of a true or correct theory of contract but only for a theory that is in constant flux and subject to changing ideology in such a way as to make rationality, stability and predictability quite impossible? Or is there some middle ground that would allow us to take account of both the need for rational justification and the facts of social change? One naturally thinks of recent disputes in philosophy of science and writers who have wanted to preserve the notion of science as a rational enterprise(22). In a related vein, some commentators have thought of Atiyah and perhaps even more obviously Ian Macneil(23) as offering a new paradigm of contract in some Kuhnian sense according to which contract is to be viewed as a social institution which has the function of projecting exchange into the future. On this view contract involves a number of relations among parties with the role of promise being only a fragment of any contractual relation if it is even necessary at all.

The question that arises is whether, if indeed the classical promising, the economic, and the relational views of contract represent competing paradigms, there is any way of choosing among them on rational grounds; and in addressing this question we must also ask to what purpose the theory is directed. In particular, is the theory meant to be explanatory, predictive, justificatory, or some combination of these? Currently popular analyses of contract law (and this applies equally to other areas of private law such as torts) can also be classified in another way -- as primarily moral, economic or sociological. Fairly obviously, the focus of Fried's "contract as promise" theory is on the moral and justificatory function, that of Swan's analysis of the _Ron Engineering_ case is on the economic and predictive function, whereas Atiyah and Macneil focus (in rather different ways) on sociological and explanatory functions. Clearly, to explain is not to justify, and the theorist who explains a legal decision by showing that it leads to economic efficiency is not _thereby_ showing that it is the _right_ decision, either morally or legally. But this is too quick, for an adequate justificatory theory must be based on an adequate explanatory one in at least the sense that it must take account of how the law actually functions in society. Fried, I think,

means his theory to meet this condition; the more economic and sociological accounts focus on prediction and explanation as <u>constituting</u> justification, to the extent that justification is thought to be required. Perhaps, then, the various theorists have been at cross purposes in pressing their theories against one another and insisting on their own as best: There may be no conflict in that each theory may have a different focus from the others and may be designed to answer different questions. Even in the case of conflicting scientific theories, it is not clear that one just must be right and the other wrong.
Nelson Goodman, for example, maintains

> "that many world versions -- some conficting with each other, some so disparate that conflict or compatibility among them is hardly determinable -- are equally right. Nevertheless, right versions are different from wrong versions: relativism is restrained by considerations of rightness. Rightness, however, is neither constituted nor tested by correspondence with a world independent of all versions"(24).

If this is correct, then an intelligible account of theory choice in science does not require realism; and it does not follow, any more that it does in morals or the law, that such choice is arbitrary. Choice, where it is required, will be based on some notion of maximal coherence, given an appropriate fit with whatever we take to be the data. This is important because there is a tension in so much of what has been written about theory in contract law between a conception of theory change being simply a result of changing social circumstances that the law must regulate, and a conception of theory as the true or most realistic account of the decisions of the courts, or even a correct working out of the moral principles that underlie the legal doctrines. But these ways of looking at theory present us, I suspect, with a false dichotomy. In science, there may be several theories at a given time that fit the facts and the choice among them may have to be made on pragmatic or other not purely logical grounds. Nevertheless, this goes no way at all toward establishing the anarchist conclusion that any belief is as good as any other or that the notions of objectivity and truth are entirely dispensable.

Several morals suggest themselves for a theory of contract. First, valuable though the moves toward incorporating social context and facts about historical and social change into an account of contract are, one should not suppose that it follows that anything goes. In particular, the oft-repeated suggestion that we must look to the "real world" of business practice to determine how contracts are actually made should not be taken to mean that whatever is done in the world of commerce is acceptable, or that legal decisions must alway promote economic efficiency. An exhortation to look at the "real world" is often meant to be read as a justification for practices that we consider, or ought to consider, unacceptable. Legal judgments can be mistaken not only in relation to actual business practice or actual precedents or legal doctrines, but also in relation to what those practices or precedents or doctrines ought to be, given the values that society is committed to fostering.

Second, if a theory of contracts is conceived of as an account that justifies settled law, then it is just not obvious that there is, at a given time, only one such theory that is acceptable. It may be that both a classical and a relational view will "save the phenomena" and each may serve best in different contexts. But if we are to make a choice between them we should be clear that this choice is likely to be on grounds of which theory provides the best and most coherent account for some purpose rather than which is not falsified by any of the legal decisions. The move to a relational view is thus probably more like the move to a new paradigm in science than it is a move occasioned by a Popperian falsification of the classical view.

Third, it is misleading to see the choice of one view as somehow based on the fact that it is more realistic than the other. For one thing, it is just unclear what realism comes to in this context. Is it a metaphysical claim about the way the world is, a claim about how courts operate, about the nature of commercial practice, or even a claim about human psychology? All of this needs to be clarified if the notion of a realistic analysis is to do any work in the selection of a theory of contract.

It is interesting that discussions of this issue are often directed by a denigration of the idea of legal fictions such as the appeal to the intentions of the parties in a case where mistake is a problem. Indeed, legal fictions have had a bad name at least since Bentham, and this negativism stems, I suspect, from a mistaken analysis both of the nature and function of fictions in the law(25) -- an analysis that in turn results from a failure to recognize that fictions, metaphors, models and analogies are pervasive in all areas of thought, including science and law. There is no problem with fictions and metaphors as long as what they tell us is correct or appropriate; and indeed it is hard to imagine how very much can be said without their use. Many authors(26) have pointed out that such devices are not mere stylistic embellishments, external to the literal point of what is being said but rather that they are essential to that point itself.

I am not, of course, advocating any particular analysis of contract in terms of any particular fiction, any more than I earlier wanted to hold any particular brief for an analysis of mistake in terms of intention. But the problem with that analysis is not that it is fictional, or not in accord with the real world, but that it does not yield a sufficiently comprehensive or coherent or integrated theory. And theories are not, nor are they meant to be, mirrors of reality, regardless of whether they are intended to serve an explanatory, predictive, justificatory, or even some other function.

NOTES

(1) See, for example, Harold J. Berman, "The Origins of Western Legal Science", Harvard Law Review 90 (1977), 894-943, and Harold J.Berman, Law and Revolution, Harvard University Press, Cambridge and London, 1983, especially chapter 3.
(2) This passage is quoted in Lawrence M.Friedman, A History of American Law, Simon and Schuster, 1973.
(3) Indeed it is this very fact that allows Grant Gilmore to argue so coherently (in his The Death of Contract, Ohio State University Press, 1974) that the theory of contract that originated with Langdell and culminated in the work of Holmes and Williston is now dead.
(4) Charles Fried, Contract as Promise, Harvard University Press, Cambridge and London, 1981, p.17.
(5) Ibid., p.4.
(6) Ibid., especially chapter 2.
(7) See, for example, Fried, chapter 3 and 4.
(8) Mistake traditionally includes frustration and impossibility. Fried discusses these issues on pp. 58-64

of Contract as Promise.

(9) Raffles v. Wichelhaus, 159 Eng. Rep. 375 (Ex. 1864).

(10) Commentators suggest that the reason for the buyer's insistence on the earlier Peerless was that the price of cotton had fallen substantially between the two arrival dates, possibly owing to American cotton's again being available on the Liverpool market once the port of New Orleans had been captured by the north in the American Civil War. See Fried, pp.59-60.

(11) Fried, pp.64-67.

(12) Ibid., p.63.

(13) Ibid., p.69.

(14) P.S.Atiyah, Promises, Morals and Law, Oxford (Clarendon Press), 1981, p.130.

(15) The Queen v. Ron Engineering & Construction (Eastern)Ltd., (1981), 119 D.L.R. (3d)267, 35 N.R. 40 (S.C.C.).

(16) John Swan, "Case Comment", U.B.C. Law Review, (15:2) 1981, p.460.

(17) The notion of theories as underdetermined by the data is discussed in many places in the literature of philosophy of science. See, for example, F.Suppe, The Structure of Scientific Theories, Second Edition, University of Illinois Press, 1977.

(18) Swan, op.cit., pp.461-2.

(19) Ibid, p.462.

(20) John Swan, "The Allocation of Risk in the Analysis of Mistake and Frustration", in Barry J.Reiter and John Swan (eds.) Studies in Contract Law, Butterworths, Toronto, p. 232.

(21) Atiyah, op. cit., p.215.

(22) See, for example, Barry Barnes, Scientific Knowledge and Sociological Theory, Routledge and Kegan Paul, London and Boston, 1974; and W.H.Newton-Smith, The Rationality of Science, Routledge and Kegan Paul, Boston, London and Henley, 1981.

(23) See Ian Macneil, "The Many Futures of Contracts", Southern California Law Review (47), 691-816 and The New Social Contract, Yale University Press, New Haven and London, 1980.

(24) Nelson Goodman, "On Starmaking", in Symposium: Nelson Goodman's Ways of Worldmaking, Hackett Publishing Company, 1980, p.39.

(25) For an interesting discussion of this matter, see Lon Fuller, Legal Fictions, Stanford University Press,

1967.

(26) See, for example, Max Black, Models and Metaphors, Cornell University Press, Ithaca, N.Y., 1962; Mary Hesse, Models and Metaphors in Science, The University of Notre Dame Press, Notre Dame, Indiana, 1966; and W.H.Leatherdale, The Role of Analogy, Model and Metaphor in Science, North Holland Publishing Company, Amsterdam and Oxford, 1974.

TRENDS IN LEGAL SCIENCE RELATING TO CONTRACTS AND TORTS

Jan Hellner

1. INTRODUCTORY REMARKS

I dislike the term "paradigm", which is both pretentious and unclear. Although the subject of this paper has considerable affinity with what is sometimes discussed under the heading of "paradigm", I prefer to use the more modest term "trend". It indicates that the paper deals with some tendencies that are found in actual legal science and which are not mutually exclusive. Few or none of them can be called revolutionary or even entirely new. To a large extent the trends are explained not by new insights into permanent problems but by the evolution of new institutions that are regulated by law and thus belong to the legal sphere. However, the total impact of the new types of research is considerable and calls, in my opinion, for an examination of whether legal theory has considered sufficiently the development of legal science.

The reason for my interest in the subject is mainly that, like many others, I want to understand what I am doing, what the aim of my research is, what requirements can be made on such research, what arguments are valid, etc. When reflecting on these matters, a legal scientist cannot limit himself to his own work. He must see what others are doing and how, explicitly or implicitly, they answer the questions that he raises with regard to his own work. On the other hand, probably no one - and certainly not myself - can claim to have a complete understanding of all legal science. I therefore confine myself to that part of the law with which I have some familiarity. That happens to be the law of contracts and of torts. For obvious reasons, I shall generally focus the attention on Swedish and other Scandinavian legal research. Even a limited part of the field of legal science can provide a basis for a more general discussion, if the inherent limitations are observed. This limited experience can at least be used for falsifying unwarranted generalizations.

A. Peczenik et al. (eds.), Theory of Legal Science, 455–468.

With this starting point, there are two types of work that may be relevant. The first one is that of recent research dealing with substance rather than with method. Sometimes proposed research, rather than research whose results have already been published, will be referred to(1). The second type is that of theory dealing with methods, techniques and aims of legal research. This theory generally has a wide range and is not limited to the small field just mentioned. I shall deal very little with legal theory. My knowledge of it is limited, and most of it does not contemplate the issues with which I am particularly concerned.

2. THE COURTS' POINT OF VIEW

There is a traditional type of legal science which is often called Rechtsdogmatik in German, rättsdogmatik in Swedish. A literal translation into English would be "legal dogmatics", but the expression carries, I believe, overtones either of arrogance or of expounding a faith. I prefer to speak of "traditional legal science", thus avoiding the difficulties of finding a suitable characterization. This kind of science can most easily be identified by the material which it uses. It deals with statutes and decrees, reports proposing legislation, bills to parliaments, international conventions, decisions by courts and public authorities, usages, contracts (especially standard form contracts), and similar material. So far it is fairly easy to describe it. When we ask how the material is treated, the difficulties begin. It is sometimes stated that the object is to describe "valid law" (gällande rätt). But then we pass per obscurum ad obscurius, since we must find out what "valid law" means and whether it is true that traditional legal science describes valid law. I prefer to look at the way in which legal science deals with the material mentioned.

There is a well-known view, prevailing in Continental Europe, according to which the main task of legal science of the traditional kind is to interpret and construe statutes. This view seems to conform to the use of the term Rechtsdogmatik. But even if this view may have been appropriate for legal science at one time (there was e.g. a dominating école de l'exégèse in France), it is not applicable to most modern work on contracts and torts. Admittedly, works dealing particularly with the construction of statutes can be found and are important, especially for use by practitioners. Commentaries on statutes, for example, will naturally concentrate on questions that relate to the construction of the statutes(2). There are also modern monographs that deal principally with the interpretation of statutes. But these works are a minority(3).

What then do modern works in the field of contracts and torts deal with? I shall mention a few examples.

Let us start with works that resemble those just mentioned, i.e. works that aim at analysing questions that may arise before the courts and are to be decided principally with the help of such materials as statutes and precedents. Many writers - including myself - will start by trying to formulate questions, then decide what material is relevant for answering these questions, and finally answer them, with more or less conviction, or decide that it is impossible for a legal writer to provide an answer that is not too arbitrary(4). The most important part of this procedure is in my opinion the finding and the formulation of the questions. This is done to some extent through the use of statutes but to a much greater extent by procuring knowledge of practical problems, including such as have been the subject of court decisions, by a study of previous writings, e.g., those emanating from other countries with a similar legal structure, and by a systematic survey of the subject matter. This part of the research involves a preliminary analysis which aims at identifying the important questions. The next step, i.e., the choice of relevant material, is also important. It is rare that the choice is obvious, and the discussion in legal theory about "sources of law" will in practice provide little guidance except in regard to the simpler cases. The third step is perhaps the one that provides the most scope for an individual choice of method by a writer, but I will not discuss it further.

The procedure indicated above may seem pretty obvious, although the choices involved in it are often disputable. However, it is surprising how differently problems present themselves when the procedure mentioned is chosen in preference to the one of starting from the wording of a statute and looking for situations in which the application of the statutory rules seems dubious(5). There are an innumerable amount of important problems to which the statutes do not provide answers. Swedish statutes often deal with a limited number of questions and lack the all--embracing character of Continental European codes(6). Moreover, even if a certain matter is covered by a statutory rule, it does not always give an answer even to simple questions. As an example, the Swedish Tort Damages Act of 1972 may be mentioned. Many of the rules of this Act are essentially empty formulas, which give little guidance for dealing with the problems that arise(7). They may serve as a check list of relevant circumstances,but this does not indicate how these circumstances should be dealt with(8). On the other hand, the Act contains a number of rules for questions regarding details which have only limited interest(9). Legal science dealing with tort law will therefore to a great extent have to choose both its problems and its solutions independently of the Act.

At the same time, the indirect influence of statutory rules is often considerable and is reflected in legal science.The United Nations' Convention on Contracts for the International Sale of Goods, of 1980, states (article 7): "Questions concerning matters governed by this Convention which are not expressly settled in it are to be settled in conformity with the general principles on which it is based...". This is, I believe, a principle which is also often applied in works dealing with national law. However, since this paper is not focused on the use of statutory material in legal science, I shall abandon the subject (somewhat reluctantly).

3. CONTROL OF STANDARD FORM CONTRACTS

The works now indicated can be said to conform to the traditional pattern of legal science by dealing with questions of law as they appear from the point of view of the courts, in particular the ordinary courts that adjudicate litigation between private parties. In modern contract law, standard form contracts play an increasingly important role, and dealing with them incites legal science to follow new paths.

Since standard form contracts are often oppressive to the inferior party, the focus has been on control of the contracts. There are a number of means for such control, which is mirrored in the treatment by legal science(10). In Sweden (as in other countries), the most important among them are the use of mandatory rules in legislation on contracts, advance control by public authorities, methods of interpretation of the contracts by the courts, and the setting aside of contract terms as unfair, by virtue of a so-called general clause in the Contracts Act (sec 36). A method whose practical importance is increasing in Sweden is the prescription of certain terms (or the omission of terms) as a condition, e.g., for obtaining permission to build a house or for receiving a loan of public money on favourable terms.

As a consequence of the existence and the interplay of these methods, works dealing with contract law must take account not only of what the courts do but of the other methods as well. The main approach is thus no longer that of a court having to reach a decision on litigation between private parties,but that of a private party surveying the various means of protection that are at his disposal or, more appropriately, the protection of consumer interests in contracts as a general issue. In this way the traditional borderlines between private law and public law are crossed. Even the distinction between arguing de lege late and de lege ferenda is blurred. There is of course an important practical difference between resorting to legislation - which is in many ways the most cumbersome method available - and using means of control that are easier to manage. The choice between legislation and other methods is therefore a complicated issue(11). But this fact does not justify making more of the distinction between legislation and other means than practical considerations justify.

It should be pointed out that legal science working from the perspective now mentioned - which may of course be applied to matters other than the control of standard form contracts - may still be called "traditional" in a wide sense. The material is that of statutes, decrees, decisions, etc., which was mentioned previously. We have not yet entered the fields of sociology of law, or of law and economics. Nor do we face, as far as I can see, any new problems regarding method, relevance of arguments, etc.

4. EVALUATIVE ANALYSIS OF LEGAL RULES

Control of standard form contracts between enterprises and consumers can hardly be discussed without entering into research that analyses legal rules for the purpose of evaluating them. This is not the only example of such analysis. In my own work, I find numerous instances of investigations of this kind. How should consumers be protected in insurance relations?(12) Are the general rules on sales suitable for consumer relations?(13) How do the Uniform Law on International Sales and its successor, the UN Convention on International Sales, hold up under an evaluative analysis, and in particular under a comparison with national bodies of sales law?(14) My own interest in these issues may be explained by the fact that I have been deeply involved in preparing legislation, but I am certainly not alone.

Attempts at evaluation will lead us forward to problems on a higher level. There is a great amount of work discussing, sometimes with appreciation, sometimes with pronounced scepticism, the value of control of consumer contracts(15). Here we encounter the arguments associated with "law and economics", which no doubt constitute one of the most important innovations in modern legal research(16). It is possible to arrive at the same problems from the other direction. Those whose primary object is to discuss law and economics in general see the law of contracts, especially that of consumer contracts, as part of their field of inquiry(17). But here we face the fundamental problem of deciding how far we are willing to let us be guided by economic considerations.

It is also possible to investigate consumer protection, and other issues of contract law, with the methods of sociology of law, but as far as I know, comparatively little research of this type has been carried out(18).

Consumer protection in contract relations also exemplifies another feature of modern development, the integration of the rules in wider systems. The general problem of "access to justice" illustrates this trend(19). It is not sufficient to introduce a number of rules , or possibilities of administrative measures, for the purpose of protecting consumer interests if the machinery of enforcement is inefficient. Special institutions have therefore been created, and the substantive rules must be seen in their connection with these institutions. In this way the organisation and working of such institutions as the Consumer Ombudsman and the Public Complaints Board in Sweden are connected with the criteria for setting aside unfair contract terms and the contents of the Consumer Sales Act(20).

Consumer protection can in this way be seen as part of the problem of safeguarding "diffused interests". It will then be compared with other aspects of protecting "diffused interests", such as those relating to the protection of the environment(21).

Integration of one set of rules in a wider system will often take the aspect of harmonizing various sets of rules. We then face the question how far such harmonization is possible or desirable. Some of the most important problems of modern contract law are found here, and it is impossible for legal science to ignore them(22).

5. THE LAW OF TORTS

Most of what has been exposed now refers to the law of contracts,and it illustrates how legal science within this field becomes less and less "traditional", even without crossing the frontiers to sociology of law or to law and economics. The same trend can be found in the law of torts. Here the decisive factors have been two. The importance of the relationship between tort liability and insurance has been recognized and new legislation and new kinds of insurance have drawn attention to the aims and purposes that should be pursued by tort liability and other systems of compensation.

In this field we also find a process of integration, which characterized modern conditions and which influences legal science. Tort liability is traditionally seen from the point of view of the courts. The interrelation between tort and insurance is then on the whole a minor matter. However, if we look at tort liability from the point of view of a party suffering injury or damage or loss, or of society as a whole, tort damages prove to be just one source of compensation besides those of social and private insurance. Tort damages are mostly paid by insurers because of liability insurance, and the rules and conditions of such insurance are therefore of vital importance to both the person suffering injury or damage and to the tortfeasor. Further, the differences between personal injury and damage to property are found to be considerable, in spite of the fact that according to the Tort Damages Act, the principles that apply are mainly the same. These features cannot be ignored by legal science(23).

Here we also find a feature which was mentioned *à propos* contracts, i.e., that legislation is just one means of reform which is an alternative to other means. There are striking illustrations in the development of the Swedish "security insurance for work-connected injuries", "patient insurance" and "pharmaceutical insurance". Each of these are institutions set up by private insurers in collaboration with authorities (including representatives of the Ministry of Justice) and private organizations, because they were considered to offer advantages in comparison with legislation for the same purpose(24).

Regardless of whether reform is to be carried out by legislation or by other means, problems of a new type emerge.How does compensation work? What arguments are relevant when preparing a reform? A great amount of new writing is concerned with questions of this type.

Empirical research on the working of tort liability has principally been carried out in the USA(25). It has proved, for one thing, that the study of legislation and case law will not give us a reliable picture of how tort law works as a means of compensation or of deterrence. No similar research has yet been carried out in Scandinavia, but the American experience teaches us at least that we cannot form any reliable idea of the working of tort law without empirical investigations.

Empirical research will rarely provide, or even purport to provide, any result relating to the goals and purposes that should be pursued when reforming tort liability and other means of compensation. Law and economics will, on the other hand, provide it, at least if we accept economic efficiency as a goal. Tort liability is one of the favourite topics of law and economy. Unfortunately, the opinions differ strongly among the protagonists, and the relevance of their findings is also unclear, especially for those who will not acknowledge economic efficiency as the supreme goal of a system of compensation(26).

6. VALIDITY OF ARGUMENTS IN TORT LAW

The current discussion, which aims at reforming the law of compensation for personal injury, has revealed a deep--going uncertainty regarding the relevance and value of the arguments that are commonly advanced. Some of them, particularly the one of estimating the deterrent effect of tort liability, can possibly be answered by empirical research. The same may be true of economic efficiency. But there is a more fundamental problem: since we cannot provide full compensation for all losses that occur as the result of unforeseen circumstances, how should we decide who is to be entitled to such compensation? Should fault on the side of a tortfeasor be a requirement, should strict liability be imposed regarding motor traffic, etc.? This problem is often stated in the form of the so-called "bathtub argument". When there is a proposal for improving the compensation to the victims of motor-traffic accidents, it is asked why they should have better compensation than those who are injured when they slip in their bathtubs. It is argued that the need for compensation is the same, regardless of how the injury arose. But is this a genuine argument or is it fallacious? Are there other arguments that are more pertinent? Should we discard the centuries-old notion that hazardous activity should be subject to strict liability?

It is remarkable that recently a work of a thousand pages has been published in Denmark devoted wholly to questions of legal policy relating to the compensation for personal injuries(27). Even if one can reproach the author for dwelling too long upon a number of details, there is no doubt that the questions which he discusses are of the greatest importance. Personally I am not always convinced that the author is right, but I am equally not convinced that he is wrong. From other countries, one can particularly point to the report of the Pearson Commission as an illustration of the complexity of the problems which face us when the future of tort liability is at issue(28).

CONCLUSIONS

These are but a few hints of the modern development of legal science dealing with contracts and torts. Another person, writing about the same field, would probably have made another choice in characterizing the present trends. Someone dealing with other fields of law and legal science would surely have found other tendencies. These would probably not have much in common with what has now been exposed, except that they would demonstrate the wide range of modern legal science and the difficulties facing those who want to pursue it.

Those who write about legal theory might contend that the development I have sketched has no bearing on the problems with which they are concerned.This is possible, but I have difficulty in seeing how they can then deny that they live in an ivory tower, out of contact with the real problems of legal science. They might also add that they have nothing to offer which might be of avail to legal scientists facing the problems that have been mentioned here. This would be sad, because who can then help the scientist who feels deeply uncertain before the questions regarding method, strategy and values which he meets in his work?

NOTES

(1) I hope that I shall not be considered unduly immodest by referring largely to my own works. They are the ones that I know best. As the object is no more than to provide examples, they should suffice as well as any others. In many fields, the international literature is immense, and it is almost impossible to point out the leading works. As a further excuse I may mention that, even when writing a paper of this character, it is difficult not to be influenced by other work that the writer is doing at present.

(2) Within the field of contracts, examples are the commentary by P.Westerlind on Chapter 4 of the Code of Real Property (forming part of the commentary _Jordabalken 1-5 kap.,_ Stockholm 1971), K.Grönfors, _Sjölagens bestämmelser om godsbefordran_ (Stockholm 1982), and P.Westerlind et al., _Kommentar till kon-_

sumentkreditlagen (Stockholm 1980), all of which are concentrated on the interpretation and construction of the statutes, although they often get beyond it. A classic Swedish commentary on the law of sales, T.Almén, Om köp och byte av lös egendom (4th ed., Stockholm 1960), has the form of a commentary but in fact covers the whole law of sales, including a great number of questions that are not regulated by the statute.

(3) Apart from the commentaries, "practitioners' books", like the ones found e.g. in England and the Federal Republic of Germany, are rare in Sweden.

(4) K.Rodhe, Obligationsrätt (Stockholm 1956) answers this description. So will, I hope, my own forthcoming work "Speciell avtalsrätt II.Kontraktsrätt". These two works differ largely in other respects. Whether the similarities or the differences are more important must be left to the judgment of others.

(5) Two different types of approach, the one focused on statutory rules, the other dealing systematically with a number of situations, are illustrated by two works on somewhat similar topics, B.Bengtsson, Om jämkning av skådestand (Stockholm 1982) and K.Grönfors, Skadelidandes medverkan (Stockholm 1954).

(6) A comparison may be made between the UN Convention on Contracts for the International Sale of Goods (of 1980) and the Scandinavian Sale of Goods Acts. I have commented on some of the differences in "the UN Convention on International Sales of Goods - an Outsider's View", in Ius inter nationes, Festschrift für Stefan Riesenfeld (1983) pp.71 ff. and in "UN--konfentionen om internationella köp och den nordiska köprätten", Tidsskrift for rettsvitenskap 1983, pp. 449 ff.

(7) The principal example is Chap. 2, Sec.1, of the Tort Damages Act, which prescribes that anyone who intentionally or negligently causes personal injury or damage to property is liable to pay damages.

(8) As an example it may be mentioned that according to Chap. 3, Sec.1, an employer is liable for faults of his employee committed "i tjänsten" (in the scope of his employment). This requirement is well known from all or most legal systems, but the two words cannot give any guidance for solving the complex problems that arise.

(9) As an example, Chap. 3, Sec. 9 may be mentioned. It states that the state and municipalities are not liable according to the rules of vicarious liability for damage or injury caused by the fault of a ship's pilot.

(10) A comprehensive work which treats this subject from an international point of view is E.H.Hondius, *Standaardvoorwaarden* (Den Haag 1978).

(11) Cf. J.Hellner, "Rechtsreform durch Gesetzgebung im Bereich des Verbraucherschutzes", in *Festschrift für Konrad Zweigert* (Tübingen 1981), pp. 827 ff.

(12) I have published a booklet on *Information om livförsäkring till konsumenter* (Information to Consumers About Life Insurance) (Stockholm 1981).

(13) Cf.F.Reynolds, "The applicability of general rules of private law to consumer disputes" and J.Hellner, "A comment", in *Law and the Weaker Party, An Anglo--Swedish Comparative Study* Vol.III (Abingdon 1982), pp.91 ff.

(14) Cf. *supra*, n.6.

(15) It is almost impossible to know what works to cite, so I have chosen, more or less at random, D.A.Rice, *Consumer Transactions* (Boston 1975), and E.von Hippel, *Verbraucherschutz* (Tübingen 1979).

(16) See e.g. R.A.Posner, *Economic Analysis of Law* (Boston 1977).

(17) See for Sweden e.g. L.Werin, *Ekonomi och rättsystem* (Stockholm 1982), pp.230 ff.

(18) An early and well-known investigation is S.Macaulay, "Non-contractual Relations in Business: A Preliminary Study",*American Sociological Review* 1963, pp.55 ff.

(19) M.Cappelletti (ed.), *Access to Justice*, Vol.1-4 (Milano 1978).

(20) Cf.J.Hellner, "The Consumer's Access to Justice in Sweden", *Rabels Zeitschrift 1976*, pp.727 ff.

(21) Cf.M.Cappelletti (ed.), *Access to Justice and the Welfare State* (Firenze 1981).

(22) This is main theme in a number of contributions to the collection *Gutachten und Vorschläge zur Überarbeitung des Schuldrechts*, hrsg. vom Bundesminister der Justiz (Köln 1981).

(23) I have surveyed the development in "Ersättningsrätt", *Svensk rätt i omvandling, Studier tillägnade Hilding Eek et al.* (Stockholm 1976), pp.169 ff.

(24) Cf.J.Hellner, "Haftungsersetzung durch

Versicherungsschutz in Schweden", Fleming/Hellner/von Hippel, _Haftungsersetzung durch Versicherungsschutz_ (1980), pp.24 ff.

(25) See e.g. _Economic consequences of automobile accident injuries_ (Department of Transportation, Automobile Insurance and Compensation Study, Washington D.C. April 1970).

(26) Cf. I.Englard, "The system builders: a critical appraisal of modern American tort theory", _Journal of Legal Studies_ 1980, pp.27 ff.

(27) Bo von Eyben, _Kompensation for personskade I. Reformering av ulykkeskompensationen_ (Copenhagen 1983). Von Eyben deals extensively with the "bathtub argument".

(28) _Royal Commission on Civil Liability and Compensation for Personal Injury: Report_ (1973: Cmnd 7054).

ABSTRACT

Work relating to contracts and torts is chosen here as an example of modern legal science that neither is wholly traditional nor belongs to such fields as sociology of law or "law and economics". The new trends depend less on new insights into permanent problems than on the emergence of new legal institutions.

The control of standard form contracts is mentioned as an example of how legal science must integrate the rules applied by the courts with other measures, such as advance control by administrative agencies. The distinction between arguments _de lege lata_ and _de lege ferenda_ is blurred when legislation is percieved to be just one of several means of control. Various other aspects of integration of rules and measures in their treatment by legal science are mentioned.

The role of legal science in laying down the foundations for the evaluation of legal rules and institutions, and in carrying out these evaluations, is stressed.

Within the law of torts, new institutions and new legislation lead to new questions regarding the validity of both traditional and new arguments advanced in the debate.

Finally it is asked whether legal theory takes account of the development that has been sketched, and whether it can offer any help to the solution of problems regarding evaluations that have appeared.

THE ECONOMICS OF TRADE LAWS

Göran Skogh

Economic models of legal rules often focus on the question of the optimality of such rules in maximizing wealth or the monetary value of trade. Surely, the question of optimality is of fundamental importance in the economic analysis of legal systems. A shortcoming of such an approach as it appears in the Law and Economics literature, however, is that a clear-cut distinction is not always made between "positive" (i.e. descriptive) theories of legal rules, the empirical question on how existing legal systems work, and the normative issue of how the system ought to function. Indeed, one pioneer in the field (Richard Posner, 1977), has claimed that most of the American Common Law doctrines both are, and ought to be wealth-maximizing(1). Such a view tends to largely wipe out any meaningful distinction between a positive and normative theory.

The importance of Posner's contributions to the economic analysis of legal rules is undeniable. His normative theory has been criticized, however, by both philosophers and economists. One of their objections is that the goal of the legislators (or judges in Common Law systems) need not be to support wealth-maximization. Another is that his theories, as presented, can usually not be verified.

Criticisms such as these of Posner's work have sometimes been seen as a critique of the entire field of law and economics. One should note, however, that many economic studies of law are based on testable theories. In the present paper I present a theory of this sort regarding trader's reactions to changes in the legal framework.Trade laws are here defined as rules pertaining to trade, e.g. contracts, remedies, breach, entry and the like enforced by the public legal system.

An idea basic to this theory is that traders adjust to existing legal rules in a manner making the outcome of trade as large as possible. The focus of the theory is the influence which legal changes can have upon relations between traders. The economic analysis of relations between traders here differs from a traditional economic

A. Peczenik et al. (eds.), Theory of Legal Science, 469–482.

analysis of market regulation - which is usually limited to questions of the impact of laws on competition, prices and quantities of goods(2) - in that emphasis is placed on the informal control systems which traders develop when they feel they cannot rely on trade laws or on public justice.

In section 1, the problem of establishing credibility in markets without public enforcement is analysed. In section 2 the economic impact of freedom of contract and of public enforcement of contracts and trade practices is examined. In section 3 the objectives of different trade laws are discussed. Finally, in section 4 the treatment of those various matters by the theory is summarized.

1. CREDIBLE TRADE WITHOUT PUBLIC ENFORCEMENT

A basic idea in economic theory is that exchange takes place when the parties assume it to be in their self-interest, each regarding the exchange as a gain. The increase in utility or wealth for the parties concerned is due to the redistribution of goods to the party that values it most, that is, is willing to pay the most for it (in terms of money or of other goods)(3). Possible gains from trade do not suffice, however, to make trade take place. The parties may, for example, be geografically far apart, the goods may be hard to store or impossible to transport, or negotiations may seriously complicate trade.

To circumvent such difficulties trade may be organized in institutionalized forms. The evolution of market-places is an example. Trade at a common place during a given period of time is beneficial to traders. It reduces transaction costs such as those of search and of transportation. Moreover, on competitive markets established prices develop, reducing costs of negotiation. With predetermined prices, there is no room for negotiation. Both buyers and sellers have to accept the competitive price if they want to trade.

A major problem in trade is to establish credibility, especially if there is no reliable public enforcement of contracts. Assume, for example, that two traders meet far away from home for the first time, both have goods to exchange. Assume as well that there is no authority that controls the parties and nothing that hinders robbery. If both know this they may not even attempt an exchange. This example has the characteristics of both parties being potentially able to gain from trade, but of their being tempted to cheat by failing to compensate the counterparty. Since both realize this possibility, both decline to participate in trade. In game-theoretical literature this problem is formalized in a matrix called the "prisoners' dilemma".

The prisoners' dilemma may end up in a practical "solution" if sanctions against cheating, robbery or fraud exist. For such rules to be enforced, some sort of justice system is needed. Public justice is just one form of justice, however(4).

Expectations regarding future exchange, for example, may reduce the favorability of cheating. If there is an indefinite stream of possible future transactions the costs of cheating may be large. Trade may be simplified, therefore, if goods are transacted regularly. This provides one explanation of why traders tend to specialize in trade in certain goods to a limited number of customers; the parties know that both will come back for future exchanges and that they thus can trust the other. The continuing of trade by families or clans (where control can be passed from father to son, for example) or by particular trade organizations is also conductive to reliable trade. Investment in good-will and brand-names creates credibility in a similar way. When cheating is detected, the offender can be punished by exclusion. Exclusion is an especially harsh punishment if the offender is unable to find new trading partners. Hence, an organization that monopolizes trade has the power to establish and enforce the rules of the market(5).

Contracts that involves obligations over time either in terms of credit or of insurance, may be much more complex than the momentary exchange situations discussed above. The need for credit is based on the fact that production, delivery and payment take place at differing times. The risk for both parties in credit and in insurance is connected with uncertainty regarding the future. Contracts attempting to cover all the many contingencies possible would tend to be very extensive and to be complicated to interpret and enforce. Trade involving long-term obligations does exist, and has existed in most cultures(6). In Scandinavia, for instance, relatively advanced trade had developed already during the early

Viking period. More complicated credit and risk-sharing contracts became important there during the Middle Ages. After enforceable trade customs had developed in the towns, those towns around the Baltic gradually submitted to the rules of Lybeck, which, through being a superior power, protected trade and provided a state-like justice system. Enforceable contracts during that period encouraged trade and a further division of labour. Many historical monuments in the area remind one of the rich cultural life which flourished during the Hanse period. In a similar way in many other parts of the world as well, trade developed and long-term contractual obligations evolved without the existence of formal trade laws being necessary.

Trade without trade laws can also occur on black markets. If the quality of the good traded there is hard to identify, the buyer may easily be cheated. Private, often violent and arbitrary, enforcement develops on such markets, though long term agreements between traders may reduce violence and caprice. Such agreements are often made where they appear favourable, in spite of police endeavoring to prohibit trade. Illegal international trade in drugs, for example, takes place within multi-national organizations with highly developed internal control. The "maffia" is such an organization.

2. THE ECONOMICS OF NON-MANDATORY TRADE LAWS

What will be examined next is the situation where a public justice system exists which approves and protects private contracts, and where, when conflicts between parties occur on points regarding which no explicit agreement has been made, the parties have the right to a public trial in which the question of liability is settled by a judge. In the analysis that follows, the costs of maintaining the legal system will be neglected. The non-mandatory trade laws in effect under such conditions have the advantage of reducing transactions costs. To clarify this, let us for a moment assume there to be no transaction costs. Under such conditions costs for search, negotiation and contracting would be nil, enforcement being carried out by the public justice system at zero costs. Such a lack of transaction costs could well lead the parties to make contracts complete in all respects in order to eliminate all risk of future conflict. All doubts then would disappear regarding whether mutual agreement existed or not and regarding the respective parties' obligations in case of breach, bankrupcy, impossibility, etc. The contract would therefore clarify all duties of both parties in all possible situations. Obviously, such a complete contingent-claim contract would need to be very extensive, but this would not matter as long as negotiation and contracting costs were zero.

If such contracts existed, public enforcement would be simple. If parties turned to the court the judge would need only to read the contract and enforce the remedies stipulated in it. In a world of this sort the trading partners would not have to rely on private sanction systems. Complicated contracts could be made between strangers. Anyone could conclude a contract with anyone else, ensured that the contract would be fullfilled and damages paid under all conditions where this was called for.

It is of interest to observe that the public enforcement of contracts in such a manner would support competition since complete freedom would exist in the choice of contract partners from among those partners potentially available.

In reality, of cause, complete contingent-claim contracts do not exist. Everything cannot be foreseen and included in contracts, and even if such contracts were possible, they would be very time consuming and costly to implement. The complete contract is thus only a hypothetical construction. In practice, a simple handshake may suffice to complete a contract. Similarly, parties may be explicit in agreeing on price, quantity and the date of delivery but leave everything else unspecified. The reasons for excluding most contingencies from a contract are obvious. Doing otherwise is costly; instead of ex ante agreements on all possible contingencies, the parties rely on ex post judgment and enforcement in case a conflict arises. In light of what has just been said, a number of hypotheses can be presented regarding how contracts are formed. It will be assumed here that traders are rational in calculating negotation costs, benefits of trade and the benefits explicit publically enforceable contracts.
is were zero.

Hypothesis 1: When the values involved in potential conflicts are large relative to the negotiation and contracting costs, a contract will be relatively extensive and explicit. For instance, in buying hand tools one often accepts the receipt in the shop as the only "contract", whereas in purchasing a large and expensive machine one should tend to demand a more clearly specified contract. This hypothesis is an obvious one and easy to verify.

Hypothesis 2: When parties feel they can rely in private enforcement, contracts tend to be less extensive and often to be implicit in character. Relatives, neighbours, employees, etc. make many important agreements without any explicit contract at all. Similarly, when parties trade repeatedly they need not rely on extensive contracts, a phenomena which has been studied by Macauley (1963). The same should be the case when the seller has good-will or a brand name to protect.

Hypothesis 3: Public justice functions as a substitute and/or complement to private control. Private control may take the form of informal sanctions or of agreements concerning arbitration. Hence, when public justice is costly or arbitrary, the parties tend to rely more on private control. For example, if the public trail is time-consuming and the outcome hard to foresee the traders will tend to more readily agree on arbitration.

Hypothesis 4: When the quality of a product is hard for traders to evaluate, they tend to rely on long-term relations. For example, the services of medical doctors, lawyers and dentists may vary considerably in quality without consumers knowing much about these differences. To counteract uncertainty here consumers tend to choose the same agent repeatedly. In the long run, failures can then more readily be detected, and that tends to make the seller of such services more reliable(7).

From Hypothesis 3 it follows that legal changes will (at least in the long run) result in changes in transaction costs and in relations between traders. To evaluate a legal system correctly it is thus necessary to know the costs of public justice, as well as contracting costs and the costs of private alternatives such as investments in good-will and reliance on long-term relations among traders.

The same hypothesis leads to an interesting question: What manner of solving conflicts should a public justice system employ in order that it can be successful in competing with private control? The answer is that a public justice system will have to judge in a manner that keeps negotiation and enforcement costs for traders as low as possible. In attempting to maximize utility or profit, traders will tend to rely on public conflict solving when they expect it to be the cheapest alternative. The existence of trade laws which are widely accepted among traders thus contributes to keeping transaction costs low, and this in turn increases potential gains from trade.

In analysing how trade laws can best be fitted to the needs of those who trade, consider first the hypothetical case where the outcome is completely arbitrary (or follows some procedure that yields results stochastically). Such a system may somewhat be better than no public justice at all, since private justice in the forms of duells, vendettas etc. may be equally arbitrary and more cruel(8). Nevertheless, if court judgments take no heed of the intentions and interests of the parties, the parties will then tend to draw up contracts that are relatively explicit, and also to rely rather much on internal private control.

Compare now the case of completely stochastic judgment just described with that of the judge being completely informed about what the parties would have formally agreed to by contract - if they had been able to formulate such a contact with no thought to the question of transaction costs. Assume also that the judgment arrived at always does reflect the ex ante intentions of the parties. Here parties will have no reason to invest resources in drawing up a contract. A handshake or a phone call should suffice for reaching an agreement. Later, if a conflict develops, one party or both can go to court knowing that the outcome will in principle be an acceptable one. An all-knowing judge, therefore, who correctly grasped the will and the interest of the parties would tend in the long run to keep negotiation costs and private enforcement costs at a minimum. Reality is, of course, somewhere between these two extremes. Judges cannot be fully informed about the ex ante will of the parties and court judgments are not completely stochastic. In most legal systems the court relies on trade customs when no explicit agreement between the parties was made. This practice accords fairly well with the principle that a judge should endeavor to take account of the will of the parties. Since if both parties are familiar with usual trade customs when they make an agreement, there is reason to believe them to have accepted such trade customs - provided there was no explicit agreement to the contrary. Even if the parties are not familiar with trade customs as such, it can be assumed they would have agreed with them if the matter had been raised.

At the same time, there are many conflicts in which the judge has no simple trade law or trade custom to follow. One way out of this problem is to compare the conflict in question with roughly parallel cases where the parties have agreed to some explicit form of risksharing. Such analogies are common in legal practice. The judge may also directly, or indirectly, try to imagine what the parties would have agreed to by contract - if they could have done so without transaction costs. Here economic models of efficient legal rules can be useful, models involving optimal contingency contracts in a world without transaction costs. A basic justification for such an approach is that, if the court shares risks in ex post conflicts in a manner that would have been accepted in ex ante negotiations, this will tend to minimize transaction costs in trade.

3. OBJECTIVES OF TRADE LAWS

A priori there is no obvious reason why trade laws or any other existing laws should be efficient in some general sense. Legislators may have targets other than that of efficiency. There is some evidence nevertheless that legislators have aimed to wealth-maximize trade through bringing the law into line with the will of the parties. For example, the Swedish Trade Act is founded on older trade customs and on the liberal principle of freedom of contract. Rules on duress, unconscionability, assent, etc. express the principle that a contract, in order to be valid, must have come about through the will of independent and conscious parties. When there is a valid contract, it is to be honored. The principle that damages are to be based on opportunity costs is considered as being in line with ex ante agreements as these would have been made in competitive markets. Doctrines of a similar sort exist in most countries.

Legislators and courts may sometimes want to counteract certain outcome which would normally came about in trade, often with the objective of redistributing income so as to favour a specific group of traders, for example consumers. Here as in the cases considered before, it is important to know the reactions to this which traders might tend to show.

Mandatory laws which conflict with agreements made on the market are not automatically followed by those who trade. If both parties believe that they can gain from exchange there are usually ways around the law: parties may even trade in direct opposition to the law. Trade will usually be reduced under such circumstances, but trade will continue even if transaction costs are relatively high. Suppose, for example, that the court consistently judges in favour of those consumers who are poor. This will increase the risk in selling to poor people. In order to obtain the same goods as other buyers poor consumers may therefore have to accept a higher price, or be forced to sign a contract in which they give up the rights otherwise provided them. If their legal rights cannot be cancelled by contract, service (or other qualitive aspects of goods) may be lower for poor customers paying the ordinary price. The reason for this is that sellers tend not to sell on competitive markets if there is no compensation for the risks taken. If higher prices or inferior purchase conditions for poor people are then outlawed, the poor may have trouble in finding trading partners. The poor consumer may then have to enlist the aid of another buyer in order to obtain the goods. Alternatively, if the seller can rely on private enforcement, he may sell to the poor consumer knowing that the latter will not utilize his legal rights.

One reason for interventions in consumers markets has been the assumption that the seller has monopolistic dominance there. However, mandatory consumer legislation only to dissolve a cartel, for example, does not necessarily end the problem of monopoly power. Similarly, mandatory terms in a contract may be counteracted by higher prices or by lowered product quality. A more efficient way to protect customers may be to encourage competition, e.g. by the allowance of free entry to the market.

Some trade laws have the explicit purpose of eliminating trade. The aim here may be paternalistic. Thus, authorities may refuse to accept agreements made among traders, as in the case of the drug markets, or of prostitution and gambling, which are regulated or outlawed in many countries. Limited entry to markets may also be introduced in order to give privileges to specific traders. Domestic traders may, for example, be protected through duties and customs. Professionals are often protected by laws regulating entry to their field.

The existence of black markets shows that outlawed trade nevertheless take place. When the trade itself is illegal, public protection of it is practically eliminated. Transaction costs and private enforcement costs in such markets are therefore relatively high. The social consequences of the private control may be as serious as the negative effects of the trade itself. Prices may be high as on the illegal drug market, the quality of goods (i.e. drugs) may be poor, cheating may be frequent and private enforcement may be arbitrary and brutal.

4. SUMMARY

According to the theory of exchange trade take place when both parties expect to gain. Transaction costs, however, such as costs of transportation, storage, search of counter-parties, negotiations, contracting, and enforcement of agreements, make trade complicated. On markets with limited public protection of its rules credibility has to be created. To overcome distrust, repetitive trade with a given party may be established; the common interest of future trade here motivates the parties to loyalty. Trade can also be organized within groups which possess strong internal control, such as families or organizations like the maffia. Trade organizations like the Hanse in medieval Scandinavia represent still another way in which trade is sometimes disciplined. If a publicly enforced trade law is based a) on freedom of contract and b) on the principle that the conflict solving in the court should be in accordance with the intentions of the parties, then it will reduce transaction costs and simplify trade. Existing trade laws have in part these aims and in part such aims as redistributing wealth e.g. so as to favour certain consumers, or hindering trade, e.g. in drugs. Measures of this sort often have negative side effects which are important to analyze.

A shortcoming in some of the Law and Economics literature on contract laws is that theories are formulated in a way which makes them more or less impossible to verify or falsify. In contrast, a theory as that presented here concerning transaction costs and the reactions of traders to different legal frameworks involves hypotheses that are possible to test, in the case of the present theory the following:
1) The greater the importance of a transaction is relative to the contracting costs the more extensive the explicit contract will tend to become. 2) The greater the chances of a party to sanction the counter-party, the less extensive the contract will tend to be. 3) The more arbitrary or costly public justice is, the more traders will tend to rely on private control. 4) The more difficult it is for the trading parties to evaluate the quality of the product and/or the more costly and complicated explicit contracts are to enforce, the more parties will tend to rely on long-term relationships.

These hypotheses are such that they can be tested empirically. Research here, casting light as it would on the impact of legislation and legal practice upon relations between traders, would be of practical relevance, regardless of whether law-makers' objectives are to encourage trade or to reduce it.

NOTES

*) Associate professor, Department of Economics, University of Lund. The support which the Swedish Council for Research in Humanities and Social Sciences provided for the present research is gratefully acknowledged. Thanks for their comments are also due to Robert Goldsmith, to Rolf Johansson and to the participants of a workshop in political economy at the Department of Economics in Lund.

(1) The research field "Law and Economics" is well established in the U.S. There is an increasing interest in the subject in Europe as well. For a survey of recent developments, see Veljanovski (1982). An introduction to the economic analysis of contract law is found in Kronman and Posner (1979). In the essay "Utilitaria-

nism, Economics, and Legal Theory" Posner (1979) develops the view that wealth-maximization ought to be a legal norm. This view has been criticized, among others by both Kronman (1980) and Veljanovski (1981).

(2) Relational economics is a new field developed in the main within the framework of labour economics and in the theory of the firm: see e.g. Williamson (1975) and Klein, Crawford and Alchain (1978). For theories of private enforcement, see Telser (1980) and Williamson (1982).

(3) There is a well established and formalized economic theory on the gains from trade, see e.g. Hirshleifer (1980), ch.7.

(4) The prisoners' dilemma and the possibilities of "solving" it through repeated exchange have been thoroughly studied in the game theoretical literature. For an introduction, see Luce & Raiffa (1957). The prisoners dilemma is also used in formalized economic theories of social institutions: see Schotter (1982) and Hirchleifer (1982).

"Solutions" to the prisoner's dilemma may rest on norms or ethics concerning cheating and fraudulent behaviour. There is no simple theory, however, of how "moral" attitudes come above here. One possibility is, of course, that certain norms are already in the genetic code of the human mind. Another possibility is that all norms are a result of learning. The feeling of what is right and wrong in business may be connected as well with philosophical or religious views.

(5) Trade organizations which control the market usually motivate their existence with the argument that they limit "unjust" competition. To the extent that they limit fraudulent behavior and the spread of misleading information such organizations can reduce transaction costs on the market. Among economists it is commonly argued, on the other hand, that cartelization results in monopoly profits which harm the market. It is not unreasonable to assume that the good and the bad effects are both present.

(6) Malinowski (1926) made a famous study of the Melanesian society in northeastern New Guinea. His main conclusion was that considerable exchange took place among the savages, the trade following strict norms.

The same families or groups exchanged with each other repeatedly. The obligations of traders were far-reaching.

(7) There is an extensive literature in economics on the principal-agent relation concerning optimal contracts between traders who cannot easily identify the performance of the other.

(8) Private enforcement may also be insufficient. Exclusion from further trade is not always enough to deter fraud or robbery, for example. That is perhaps a reason why such serious violations of trade rules are usually included in the criminal code.

REFERENCES

Hirshleifer, J. (1980) Price Theory and Applications. Second Edition. Prentice-Hall, New Jersey.

Hirshleifer, J. (1982) "Evolutionary Models in Economics and Law: Cooperation versus Conflict Strategies". Research in Law and Economics. Vol.4, pp.1-60.

Klein, B., Crawford, R.G. and Alchian, A. (1978) "Vertical Integration, Approbriable Rents, and the Competitive Contracting Process". The Journal of Law and Economics, no.21,pp.297-326.

Kronman, A. and Posner R.(1979). The Economics of Contract Law. Little Brown. Boston

Kronman, A.(1980). "Wealth Maximization as a Normative Principle". The Journal of Legal Studies. Vol.9, no.2, pp.227-242.

Luce, D. & Raiffa, H. Games and Decisions. John Wiley & Sons. New York. Macaulay, S.(1963) "Non-contractual Relations in Business; A Preliminary study". The American Sociological Review. Vol.28 p.145 ff.

Malinowski, B.(1926) Crime and Custom in Savage Society. Routledge & Kegan Paul. London.

Posner, R.(1977). Economic Analysis of Law. Little Brown. Boston.

Posner, R.(1979) "Utiliatarianism, Economics, and Legal Theory". The Journal of Legal Studies. Vol.8, pp.103-140.

Schotter, A.(1981). The Economic Theory of Social Institutions. Cambridge University Press.

London.

Telser, L.(1980). "A theory of Self-enforcing Agreements". The Journal of Business. Vol. 53, no.1, pp.27-44.

Veljanovski, C.(1981). "Wealth Maximization, Law and Ethics - On the Limits of Economic Efficiency". The International Review of Law and Economics. Vol.1, no.1. pp.2-28.

Veljanovski, C.(1982). The New Law-and-Economics. A Research Review. Centre for Socio Legal Studies. Oxford.

Williamson, O.(1975). Markets and Hierarchies: Analysis and Implications. Macmillan. London.

Williamson, O.(1982) "Mitigating Contractual Hazards using Hostages to support Exchange". Center for Study of Organizational Innovation. University of Pennsylvania. Discussion Paper 126.

CHAPTER 7

INTERDISCIPLINARY BRIDGES BETWEEN LEGAL RESEARCH AND OTHER SCIENCES

SYNOPSIS

Aleksander Peczenik

This chapter starts with some optimistic programmes and ends with a caveat.

Jörgen Dalberg-Larsen gives some reasons why legal research should take up new approaches, especially concerning changes of both the legal system and the society in which it works. First, newer statutes often authorize the administration to make new rules of law, not merely to apply pre-existing rules. Second, new statutes often specify some purposes and leave to the administration the finding of the best means. Third, new types of administrative agencies are often created, adapted to these new functions. Fourth, the state interventionism creates a more and more complex legal system. In Dalberg-Larsen's opinion, all this demands that the lawyers take up new research methods, more integrated with social sciences. For example, a jurist employed in the Department of Environmental Policy must not only interpret the law but also gather empirical data about pollution etc. A divorce lawyer must not only inform his client about the divorce laws but also predict the client's living conditions after the divorce. The author concludes, that the legal dogmatics should also assimilate new social-scientific methods, despite the normative character of legal research. After all, even social sciences include some normative research.

François Ost and Michel van de Kerchove state more precisely the character of an interdisciplinary theory of law. It should adopt an explanatory external point of view and answer such questions as, "Why does a legal system give priority to such and such a concept of validity?", etc. Such a theory must be interdisciplinary, not merely

A. Peczenik et al. (eds.), Theory of Legal Science, 483–485.

multidisciplinary. It must find problems whose solution demands a synthesis of several disciplines, not merely a collection of unrelated pieces of information. When developing its external explanatory point of view, a legal scientist must thus take into consideration the internal point of view of the legal system. He must avoid two dangers, that is, the risk of reducing the law to facts and the risk of pseudo-scientific speculation. The authors conclude that the interdisciplinary theory of law search for legal means better suited to social results and perhaps even encourage criticism of the socially dominant legal rationality.

The idea of an external point of view that respects the internal point of view is also one of the suggestions made by Vittorio Villa. He points out the connection of this idea with hermeneutics and its role in the proposed unity of science. The hermeneutical approach is unavoidable, since the legal researcher had to do with a "double hermeneutic". In other words, he meets two social practices interacting with each other, the social practice of the scientific community, and the social practice which is the object of investigation. In spite of this, one may also profit from some ideas developed in the philosophy of natural science. Villa's broad vision of integration is possible because he assumes a broad definition of science, as any activity aiming at representation, reduction to unity, and explanation of phenomena. This definition fits the traditional legal dogmatics and it also fits the projected integrative science of law.

Hubert Rottleuthner is, however, less optimistic. He gives an extensive historical background of legal philosophy, legal theory and legal sociology and concludes as follows. Legal philosophy is rather practical and normative. Legal theory is primarily conceptual and logical. Legal sociology is more oriented towards empirical reality and more critical. Yet, legal sociology rarely satisfies the standards of empirical discipline. The empirically proceeding sociology of law could be established in the United States because of the fact that the conceptual foundation-questions were "innocently" excluded. The conclusion is that a further rapprochement of legal philosophy, legal theory and sociology of law is possible, but a consolidation is not, since the distinction of practical-normative, analytical-conceptual and empirical approaches is too significant.

Kazimierz Opalek shares this view and adds the following. In recent times, there is an integrational "overload", manifested in changing superficial fashions. In the author's direct experience, he met the succeeding waves of fascination with psychological theories by Petrazycki, Hägerström and others; Viennese positivism; sociology; informatics; deontic logic; argumentation theory; Oxford linguistics; modern empirical linguistics; and the revival of axiology. No one can master all this at once. Moreover, a concentration on an extra-legal discipline may convert a jurist into a logician, philosopher, sociologist, etc. and lead to a loss of the juridical point of view. This leads to unsatisfactory solutions of legal problems. The main reason for this risk is the danger of neglecting the normative dimension of the law. The law consists primarily of norms, conceived not as purely linguistic phenomena but as unions of a decisional act and its linguistic expression. Psychological experience, legal conduct and legal values are only indirectly legal, due to their relation to the norms. In consequence, the main field of integration of legal science with other disciplines should not consist in the social sciences but rather in a general theory of norms, metaethics, philosophy of language, argumentation theory and decision theory. The starting point of the integration should consist in the paradigmatically legal problems, dealt with by the study of law. Integration is meant to give help in solving them.

ON BRIDGING THE SO-CALLED GAP BETWEEN NORMATIVE LEGAL DOGMATICS AND EMPIRICAL-THEORETICAL SOCIAL SCIENCE

Jørgen Dalberg-Larsen

1. In the following I'm going to deal with a theme which has been frequently debated during recent years, but which I still think important though to discuss nationally as well as internationally. Briefly, it is about the practical requirement for and the theoretical problems involved in studying modern legal systems within new categories of 'legal science' which cannot be distinctly defined either as legal-dogmatic in the current sense of the word or as purely sociological, economic or the like(1).

I'm going to lead off with a brief treatment of some previous attempts at establishing a connection between legal science and social science in different ways. Next, I'll touch on some features of modern legal systems hampering a traditional legal-dogmatic approach to essential legal branches and essential legal problems. Finally, I'm going to debate what kinds of possibilities are open and relevant for a legal-scientific treatment of these problems and legal branches, with special attention to the use of legal-sociological research within the framework of a practically oriented legal science.

2. The debate on the need for a shift of legal science in the direction of the other social sciences has as its decisive condition the establishment of a number of new empirico-theoretical social sciences which took place in the last century. From the point of view of classical natural law there is of course no fundamental dividing line between the principles to be followed in order to obtain valid cognition within law and within other fields of social life. However, especially at the end of the last century, first in Germany and later on in many other countries, some ideas were formulated indicating that traditional legal-scientific research should be inspired or even replaced by social science, which means a new form of science strongly influenced not by natural law but by natural science, expecially concerning the conception of

A. Peczenik et al. (eds.), Theory of Legal Science, 487–496.

theory and the importance of quantitative methods.

Very roughly, the most important new projects of legal social science can be divided into three categories:(2)

The first comprises the projects trying to 'make scientific' the basic elements of legal dogmatics, especially the statements describing the contents of law in a specific legal system. It is regarded as necessary from a scientific point of view that exact criteria are pointed out enabling a verification or falsification of such statements. Different 'realistic' theories can be placed in this category, e.g. the theory of Alf Ross, but also the theory of the legal sociologist Theodor Geiger, according to whom the efficiency quota of a rule had to be measured, on a result of which conclusions could be drawn as to whether specific rules should be considered to be in force or not.

The second category comprises theories which for practical or theoretical reasons aim at a changeover of the main task of legal science from being a description and interpretation of the law in force to something else. This might be a truly empirico-theoretical science, i.e. with the aim of establishing causal relations between legal and non-legal phenomena, or it might be a science defining legal politics as the most important field of research within which you have to build up a more or less exact knowledge of how to realize specific aims in society by means of legal rules.

Finally, there are some schools of thought having as their aim the founding of a new form of practically orientated legal science, where traditional views and new social-scientific views are combined within legal science. An example of this is 'die Interessenjurisprudenz', picturing among other things empirical studies of the actual strength of various interests in different fields combined with studies of how to promote specific interests in society, through judicial decisions, as important extensions of the current legal-dogmatic research.

Of course, it will always be a combination of many factors which determines the emergence of new legal--theoretical points of view, and other combinations which decide the fate of these new thoughts. This also applies to the above-mentioned ideas. Looking at the reasons adduced by supporters of the respective schools for encouraging lawyers to follow their thoughts, you will, however, find a clear tendency showing that supporters of the first category and of the first variant of the second category are arguing in favour of their ideas on the basis of internal scientific criteria (i.e. with a view to general scientific requirements), whereas the supporters of the legal-political variant within the second category and supporters of the last category are, to a larger extent, using arguments based on practical and social considerations(3).

In what follows I will deal primarily with some practical reasons why legal science ought to take up new kinds of research, especially those concerning changes of the legal system and changes in the society in which it functions. However, I want to emphasize that in my opinion no hard and fast line can be drawn between practical, social, political, and scientific reasons. The very idea of law as a phenomenon in continuous change and the idea of law as one among several instruments for social engineering have turned up in a certain period of the legal development and are based on a certain conception of society. This conception is connected with specific political points of view and is also a decisive basis for some social--scientific schools, whereas others have a tendency to take up other attitudes. It can thus be maintained that, ultimately, the personal attitude to fundamental legal-philosophical and political problems is decisive for what legal views you think ought to be prevailing within legal science - whether, for example, you primarily ought to argue backwards in the direction of given legal rules or forwards, thus taking an interest in the social impact of law, including the impact of the judicial decision.

3. Now I shall outline some of the alterations that have characterized the Nordic legal systems, especially during the period after World War II(4). These alterations have not taken place to the same degree within all legal branches. Within some traditional fields which up till now have received much attention from legal science the legal material is even almost unchanged. Considered from the special angle of these legal branches, the result of this will, of course, be that traditional legal dogmatics is found functional and sufficient even to-day.

These alterations are among other things the following:

1. An increasing use of framework laws containing no precise rules of law, but, instead, authorizing the

administration to make concrete rules of law.

2. A large quantity of laws (including framework laws), the most essential concrete contents of which is specifications of which purpose or purposes you want to have implemented within the field in question.
3. A large number of new types of organs both working within the law-making process and in the process of law application. These organs often have a very mixed composition of a rather impenetrable character, consisting of representatives of different interests, laymen, political representatives, experts in different branches of knowledge (including social science), and finally jurists(5).
4. Because of the growing extent of legal control of all conditions of life it will more and more often be necessary to use rules of law and legal principles from several legal branches for good advice in or the solution of conflicts in one particular case.

Altogether, it means that concurrently with the increasing complexity of society and increasing state intervention, law itself has become more complex and to a still larger extent influenced by changing socio-political views(6).

4. Now, what do these alterations of the legal systems mean to a legal-scientific treatment of the rules of law? In my opinion they imply among other things that it becomes a most problematic thing to maintain, without further proof, that the modern legal system naturally ought to be treated strictly within the already established categories of legal science. Stig Strömholm calls these categories legal dogmatics, sociological legal science, and logical legal science and thinks that they should be kept apart because of basic differences in their goals and methods. According to Strömholm the first category, i.e. legal dogmatics, views the legal system from within and particularly deals with the solution of especially complicated legal problems of interpretation, basing both on the ordinarily accepted legal methodology and hermeneutics and on the specific dogmatic tenets within the legal discipline which covers the subject in question(7). As a contrast to a legal dogmatic lawyer a legal sociologist never answers the lawyers' questions as he studies the legal system from outside(8). A sociologist can address himself to anybody, whereas legal dogmatics is

defined not only by its method but also by the groups to which it addresses itself, that is primarily jurists, but also other practicians.

In this place I shall not deal more generally with this distinction nor discuss logical legal science. Instead, I shall give special attention to the question of what a recommendation of a maintenance of this distinction with the contents given to it by Strömholm means to the study of new legal branches, especially as to some new disciplines within public law(9).

A very decisive consequence of such a recommendation is that you are easily led into transferring generally accepted principles of method and interpretation to new legal branches where the character of the legal material requires a break with these principles. Another consequence of following these instructions is that you can hardly start your legal studies at all if you have to build on the dogmatic tenets of a field which has not yet been dogmatically explored. In practice, this last difficulty will often be coped with simply by applying the tenets of a previously studied apparently related field. By this procedure, however, you will end up with the same problems that the use of generally accepted methods presents, to wit: the fields of application of particular methods and principles are expanded beyond those they derive from and are adapted to. In the history of legal science there are numerous examples of legal scholars engaging in the study of a new legal branch taking as their starting point the principles and methodology known from the study of other well-established parts of law. A most characteristic example is the application of the principles and methods of civil law to the study of public law in Germany at the end of the last century(10).

As far as I can see, there are essential practical and theoretical reasons why you should not straight away plunge into traditional legal-dogmatic investigations in new legal branches. Instead, you have to pause a little trying to consider, both from internal and external points of view, what is particularly characteristic of the subject in question.

A characteristic feature of many new legal branches is the fact that the importance of the traditional rôle of the judge is rapidly declining. Whether this tendency is regretted or welcomed, it is anyhow a fact which determines that the law-court point of view considered as the only practically relevant one is most unrealistic and consequently unpractical(11). The characteristic feature of many new disciplines of legal science is, however, that, in spite of all practical requirements, the legal problems have still been considered primarily from the previously current points of view, i.e. just from the judge's seat. To give an example: within administrative law this has had the consequence that the question on what grounds an administrative act can be declared invalid during a court trial is set up as the most central question or even the only legal question within this branch of inquiry.

If, however, you accept that the view of the judge is not the only possible one within a practically orientated legal science, whether it is due to reflections like this one or to the fact that the judge only seldom or perhaps never gets into the picture, or perhaps that you think that different actors in the same case might make different claims to legal science, or finally that lawyers deal with other things than concrete solution of conflicts - then the field of possible forms of legal science will anyhow be drastically extended(12). If you think that also non-jurists ought to be considered as practicians, i.e. addressees of a practically orientated legal science, then the field will be extended even more.

5. Let us illustrate this by a few examples. A jurist employed in the Department of Environment Policy and dealing with environmental cases must, of course, as an actor in the system, adopt an internal attitude to the relevant rules of law and consider these as binding. He also has to solve difficult specific legal problems. But does this prevent him from using socio-scientific views and relying decisively on empirically collected data by solving these problems? Of course not, no more than the judge is prevented from taking these data into consideration. The difference between the rôles of the judge and the administrator is rather that the person administering such a legal branch must be considered obliged to make himself acquainted with the specific problems and knowledge of the branch in question, whereas the judge is assumed to have no more relevant knowledge than that of the man in the street at the start of the case. The administrator is especially bound by the rules regulating his own field, and if these rules comprise some precepts about environmental protection, he can hardly be considered to be taking his task

seriously if he doesn't try to find the accessible knowledge of the subject in question. Therefore, he has to master approved methods exceeding the current legal ones(13).

A lawyer who has been given the task of drawing up a standard contract within a special field must of course elaborate it in such a way that it isn't in conflict with the indispensable rules of law. You might say that this is a consequence of his acting as a lawyer and of the resulting internal view of the existing law. In order to safeguard the interests of his client as well as possible, however, he must also elaborate the contract in a way that takes into account the greatest possible number of the conflicts which can be expected to arise. It might also be relevant to include other possible consequences in the considerations before the contract is elaborated(14). For a lawyer in such a situation it is not at all inconceivable that a broad empirico-sociological investigation of the problems which others have been confronted with in similar situations, combined with an investigation of how legal contracts in general have a conflict-preventing effect, would be of rather considerable practical interest.

It is not sufficient for a lawyer advising a client in a divorce case only to give information about the existing law if the task has to be carried out satisfactorily. He must also try to give an estimate of the client's conditions of life and possibilities of making a living after a possible divorce as these questions are often the client's greatest problems. It is not a question of solving a single complicated legal case or of predicting the solution, but of setting forth an estimate of the joint consequences of many different codes of civil law and public law in a specific type of situation. Such a qualified estimate would probably presuppose knowledge which can be derived from an empirico-sociological investigation of the social-legal consequences of a number of divorces covering a longer period after the divorce. By so doing, you would also come to a better understanding of the general mechanisms of the mode of operation of law in a society like ours(15).

6. But after all, it might be argued, aren't there some quite fundamental differences between legal science in a dogmatic sense and other social sciences which exclude all

mixed forms? It is often said that dogmatics is normative compared with other sciences, that it takes an internal view on its object, that it operates with qualitative methods, and that it has quite special relations with practice (legal practice). I do not think that the differences are so clear and fundamental as is sometimes presumed, but that, partly, they are due to historical tradition rather than to any logical or practical necessity.

In my opinion it is very misleading to set legal dogmatics against all other social sciences. The object of many other sciences is to put forward specific recommendations to specific groups. This holds for economics or part of it, the theory of organization, the science of administration, and so on. Legal dogmatics instructs how to solve legal conflicts according to the existing law, and many other sciences give guidance on how you ought to act if you want to obtain a maximum margin of profit, to establish an efficient organization etcetera, etcetera.

Other branches of science, too, are strongly attached to one specific profession, e.g. theology and medicine. As for the internal point of view demanding that, during your studies, you identify yourself with a rôle within the legal system, there are influential schools with social sciences holding that such identification will always be necessary in order to obtain relevant knowledge of society, and there are also schools advocating the use of qualitative analyses instead of quantitative ones(16).

The above has not been written in order to deny very substantial differences in many respects between typical forms of legal dogmatics and other branches of social science, including the ones just mentioned. In legal dogmatics you take great pains to solve quite specific questions of legal interpretation, where, in other sciences, you usually have a more generalizing aim. But once again: This feature is not uniquely characteristic of legal dogmatics.

Even if law as an object of study can be regarded as a very special one in many ways, I see no basic theoretical reasons why the study of law shouldn't emancipate from the traditional forms, but only practical problems which are partly due to the great weight of historical tradition. Criteria for what is considered valuable legal research are widely transmitted from generation to generation and like other ingrained habits can only be changed over a long period of time. As I have argued above, I think that some features of the modern legal systems seem strongly to favour the introduction of new points of view in legal research. And it is also obvious that nowadays much more legal-scientific literature of widely different character with a socio-scientific strain is being written than previously.

NOTES

(1) Cf.e.g. Jørgen Dalberg-Larsen, Retsvidenskaben som samfundsvidenskab, Copenhagen 1977, espec. part 4, incl.ref., Roland Dubischar, Theorie und Praxis in der Rechtswissenschaft, Freiburg, München 1978, Werner Krawietz, Juristische Entscheidung und wissenschaftliche Erkenntnis, Wien-New York 1978, and Hubert Rottleutner, Rechtstheorie und Rechtssoziologie, München 1981, espec. p.190-215.

(2) Cf. Dalberg-Larsen, opus cit. p.84-91 and part 3, and idem 'Four Phases in the Development of Modern Legal Science', in Scandinavian Studies in Law 1979, p.77-105.

(3) For such practical arguments cf. Dubischar, opus cit., espec. p.91 seqq.

(4) In this context I particularly refer to the Danish legal system, and I cannot preclude the possibility that a few of the points in question apply rather badly to the legal systems in other countries.

(5) Cf.e.g. Vilhelm Aubert, Rettens sosiale funksjon, Oslo 1976,p.277 seqq.

(6) Of course, this has also taken place in many traditional fields of civil law, as will appear from what is said below in section 5.

(7) Cf. Stig Strömholm, 'Hur många rättsvetenskapar finns det?',Tidsskrift for rettsvitenskap 1978, p.621-54, espec. p.640-41.

(8) Cf. Stig Strömholm, opus cit., p.644.

(9) I want to emphasize that I find Stig Strömholm's characterization of various types of legal science very telling in most respects.What I want to contest is particularly the distinct boundary between internal and external points of view on law and between the question which the legal profession wants to get a reply to and other types of questions.

(10) For the political motives for and consequences of

this see Dalberg-Larsen, opus cit. (note 1), p.266 seqq., and p.323-24 incl.ref.

(11) Cf. Vilhelm Aubert, opus cit., a.o.p.211-15, and Jan Hellner, 'Syften och uppgifter för rättsvetenskaplig forskning', in Festskrift til Alf Ross, Copenhagen, 1969, 5 seqq.

(12) Cf.Jørgen Dalberg-Larsen, 'Er der grundlag for en juridisk metodelaere', T.f.R. 1969, p.568 seqq., espec.p.596 seqq., and Stig Jørgensen, Law and Society, Aarhus 1970, chapter I.

(13) Thus, the internal viewpoint here implies that external viewpoints are also taken, i.e. that you try to register and analyze the actual impact of law in society. Whether the actor in question is a lawyer or e.g. a sociologist won't change that; however, the implication is that the task becomes easier or more difficult depending i.a. on the concrete education of the person to perform it.

(14) For contracts as a means to prevent conflicts see e.g. Jørgen Dalberg-Larsen, Retssociologi, problemstillinger og teorier, 3. ed., Copenhagen 1977, a.o. p.124 seqq.

(15) Cf. H.H.Andrup a.o., Dansk underholdsret, Esbjerg 1978. The research concerning legal aid has also contributed to the elucidation of such connections.

(16) Cf. e.g. Dalberg-Larsen, opus cit. (note 1), chapter 3.

TOWARDS AN INTERDISCIPLINARY THEORY OF LAW

François Ost and Michel van de Kerchove

The formulation of an interdisciplinary theory of law presupposes that several conditions, which we plan to examine in turn, are met. First, we have to conceive a discourse on law which, whilst not adopting the internal point of view of private citizens and officials, does not reduce the specific nature of this point of view but takes it into account. Chapter I will try, therefore, to define and justify an external discourse which takes account of this internal point of view. Chapter II will envisage an interdisciplinary theory of law which, while respecting the preceding conditions, aims at supplying a real explanation of the legal phenomenon, and not only a description and a systematisation of this phenomenon. In this chapter, we will also focus on the problems inherent in such an approach.

1. INTERNAL AND EXTERNAL POINTS OF VIEW ON LAW

Our starting point is Hart's very judiciously formulated idea that "one of the difficulties facing any legal theory anxious to do justice to the complexity of the facts is to remember the presence of both these points of view and not to define one of them out of existence"(1).

We feel, however, that a few observations are required on the nature of these two points of view. Faced with the problem of the rules of conduct of a social group, Hart argues that the external point of view is that of "an observer who does not himself accept the rules"(2), whilst the internal point of view is that of "a member of the group which accepts and uses them as guides to conduct"(2).

Furthermore, Hart distinguishes between two kinds of external points of view. The first, which might be qualified as "simply" external, consists of an observer who, without accepting the rules himself, "asserts that the

A. Peczenik et al. (eds.), Theory of Legal Science, 497–508.

group accepts the rules, and thus may from outside refer to the way in which they are concerned with them from the internal point of view"(2).

The second, which might be qualified as "radically" external, consists of an observer "who does not even refer, in this way, to the internal point of view of the group"(2).

We believe, as Hart himself argues, that legal theory may attain a scientific status and reflect the specific nature of its object only by adopting a "simply" external point of view, i.e., an external point of view which takes account of the internal point of view. We feel, however, that a few further comments are required even if, overall, we support Hart's position on this point. We believe that Hart's distinction between internal and external points of view needs clarifying in three spheres, first, with respect to the nature of the act (acceptance) which the adoption of an internal point of view involves; secondly, with respect to the object (the law) of such an acceptance; and, lastly, with respect to the persons likely to adopt such a point of view.

1.1 The Act of Acceptance

As to the nature of the act of acceptance which the adoption of an internal point of view involves, steps should be taken, first of all, to see the wide range of its parts, and consequently, the different degrees of intensity which may be attributed to such an attitude. Our view will therefore, have to be pluralist and relativistic.

In this respect, the supporters of legal positivism and natural-law theories appear to accept the common presupposition that a norm is necessarily valid or invalid, on the basis of a single criterion which seems to be, for some of them, legality (formal validity) of the efficacy of the norm, and for others its legitimacy.

In contrast, it appears that the acceptance of a norm depends on these three criteria at one and the same time, but to sharply varying degrees.

1.2 The Object of Acceptance

Certain distinctions must also be drawn about the object of acceptance. Although this acceptance may relate individually to each norm belonging to a legal system, it is also capable of being viewed in relation to a given legal system, considered in its entirety, or even in relation to the existence of a legal system in general. Now, one of these forms of acceptance does not necessarily involve the other, even if they tend to converge most of the time. Furthermore, as Hart underlined it, a legal system does not consist of the simple addition of a set of standards of conduct but presumes the existence of "secondary" rules which "specify the ways in which the primary rules may be conclusively ascertained, introduced, eliminated, varied, and the fact of their violation conclusively determined"(3). While some of these secondary rules are formulated explicitly, others remain implicit.

It is advisable to note that even these largely implicit rules, which could be called the system's "rules of the game", may or may not be adopted by private persons and officials. Their acceptance therefore contributes, as much as the system's other norms, to defining the internal point of view which interests us here.

1.3 The persons

Turning to the persons likely to adopt an internal or external point of view, a distinction should be drawn between different categories of individuals, or, rather, different types of "roles" that such individuals may adopt. Whilst the points of view adopted by private citizens and officials do not pose a basic problem in this study, a few remarks should be made about those adopted by legal doctrine (including legal dogmatics) and the different conceivable types of legal sciences (cf. figure 1).

1.3.1 Legal doctrine We are faced with the problem if we try to define the position of legal doctrine in terms of a theatrical metaphor on account of the doctrine's dual role. Although this role may be considered as a quest for knowledge and may claim, in this respect, to be scientific, it is also undeniable that legal doctrine may be considered to be an action inasmuch as it exercises a practical function as a complementary source of law. We may therefore consider that the author of doctrine, provided that he remains within the limits of his theoretical function, is a spectator of the performance going on before his eyes and plays no active role in this performance. As soon as he collaborates in the show, even indirectly, by means of practical suggestions, the author of doctrine leaves the spectator's ranks, although not coming visibly on stage. We might say that he plays the role of prompter, both collaborating with the actors, should one of their number forget his lines, and at the same time situated off stage, nearly at the level of the ordinary spectators.

Using a more orthodox terminology, we will qualify the scientific side of legal doctrine as "legal dogmatics" and its practical side as "doctrine" (in its strict sense).

Legal dogmatics consists of the systematic study of a specific legal system(4). The kind of question that it examines is: "in such legal system, does such law apply to such situation?", or "in such legal system, what is the meaning (the meanings) of such concept?" "What is the function (the functions) of such an institution?" So as to reply to these questions, legal dogmatics defines, classifies and systematizes its object(5).

Seen from this angle, we will say that legal dogmatics develops a descriptive external point of view which takes account of the internal point of view of private citizens and officials.

Generally speaking, however, this attempt at systematisation by means of legal dogmatics leads to a form of collaboration in the collective task of creating law. As legal concepts are not univocal, as laws cannot provide for all the eventualities of everyday life, and as there may not necessarily be any coherence between the laws, it is inevitable that attempts at definition, classification and systematisation lead to people "taking sides" and that the author of doctrine is forced to indicate preferences and to argue consequently. Provided that an accurate distinction can be drawn between this moment and the previous theoretical moment, we are facing doctrine in the strict sense of the term.

The following type of question is examined: "has the judge to enforce such law in such case?", or "in such situation, has the citizen to comply with such law?" Such questions show that the author of doctrine somehow adopts the normative point of view inherent in the legal systems; there is an imperceptible passage from external and descriptive analysis to the plea calling for the argument which is considered as having to be imposed.

We will consequently qualify this approach as the scholarly internal point of view. Putting into relation the term "internal", which presumes an acceptance of the rule, and the term "scholarly", which presumes on the contrary a theoretically detached view, clearly illustrates the hybrid - but not less operational - nature of such an approach and calls to mind the approach of the judge whose "authoritative internal point of view" might also, in certain respects, be qualified as "scholarly". It is well-known how frequently case-law and legal doctrine interact and add weight to each other.

1.3.2 Legal science. The application of the term "science" to such or such discourse obviously depends on the view held about scientific work(6). Legal dogmatics may therefore claim, in its own view, to be scientific. In this study, we will only try to specify the position, with respect to the distinction between internal and external point of view, of several discourses which set out to produce legal science.

Generally speaking, and making use once again of our theatrical metaphor, scientists have the position of spectators; in their role of theoreticians, they are outside the intrigue and observe it with a critical eye. Their point of view is therefore "external". It is clear, however, that a wide variety of external points of view emerge; furthermore, the extent to which these views reflect the points of view of legal subjects also varies. New distinctions must therefore be introduced.

1.3.2.1 An analytical external point of view.
This category covers the usual position of general theory of law which sets out to analyse, at an extremely generalised level (even more generalised than the dogmatics approach), the properties of legal systems. The following sort of questions are analysed: "what is meant when it is said that a law is valid, that a law is clear, that an institution is defending the general interest?". In each of these instances, legal theory uses series of concepts which are supposed to clarify the positions adopted by the various participants in legal life. A critical question however, which we must pose in the case of legal dogmatics is to what extent this analysis does not merely reproduce, in scholarly language, representations conveyed through legal practice, and to what extent therefore this external point of view does not itself tend to endorse these internal representations.

1.3.2.2. A reducing external point of view
This category covers the position of legal theories which, by adopting a radically external point of view and disregarding the representations of the various participants in legal life, result in reducing the very object of which they claim to give an account. This is the mistake underlying the "economistic" Marxist theories of law which, by failing to take the legal structure and the supporting representations seriously, reduce legal data to the rank of epiphenomena of economic modes of production. This is equivalent to a theatre critic who gives an account of the actors' performances solely in terms of the employment situation amongst professional actors or who explains the audience's reaction to the play in terms of the audience's socio-professional composition.

1.3.2.3. An explanatory external point of view. This category qualifies the kind of legal science which we propose. It has both an explanatory and inter-disciplinary character. It tries to reply to the following sorts of questions: "why does a legal system give priority to such or such concept of validity?" or "why is a legal system based on such concept or such institution?". Its position is simply external inasmuch as the essential epistemological distance from its object does not result in the disintegration of the representations conveyed by the legal system. This sets it apart from the reducing theories which we have just discussed. On the other hand, as it is not restricted to describing or analysing "internal" representations, but tries to explain them by means of hypotheses borrowed from other theoretical fields, it should be in a better position than dogmatics and legal theory to escape from the implicit rules which govern the working of a legal system. For instance, it will strive to restore, behind the rationalising discourse of the agents, the deep-rooted practical logic which determines their conduct, thereby even revealing the efficacy of such or such an implicit assumption which is generally left unexamined. We are all too well aware, however, of the many problems inherent in such a project. One must permanently take care not to get too involved in its object, and at the same time not become too detached from it. These problems will be discussed in chapter II.

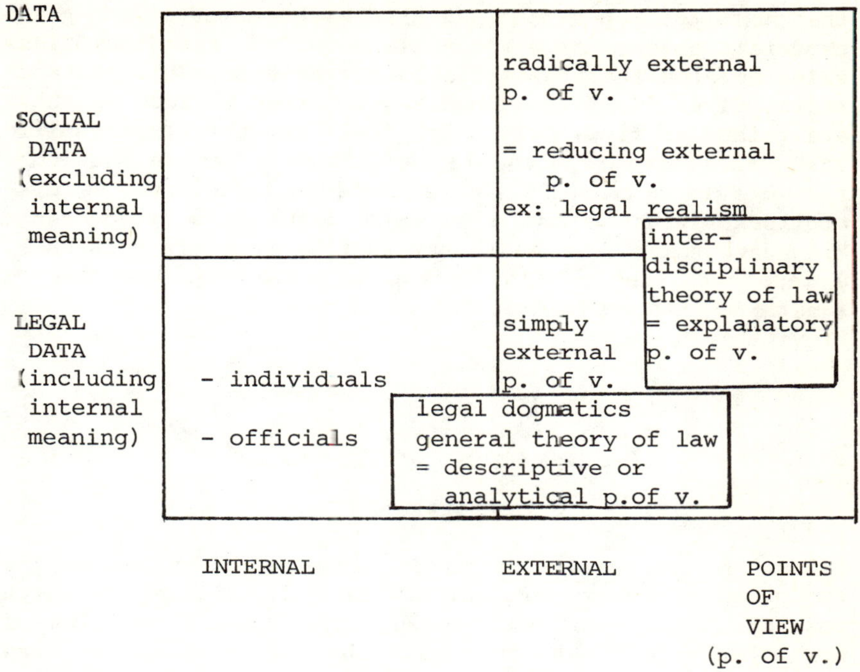

figure 1

2. CONDITIONS FOR AN INTERDISCIPLINARY THEORY OF LAW

Three of the specific problems raised by the constitution of an interdisciplinary legal science will be discussed here: 1. the status of interdisciplinary research; 2. the articulation of internal and external points of view in this research; 3. the variety of "interests" (as defined by Habermas) underlying this research.

2.1. The status of interdisciplinary research

It is obviously not enough merely to multiply the number of theoretical approaches to a question in order to create the right conditions for interdisciplinary scientific work. A distinction might be made, therefore, between multidisciplinarity, transdisciplinarity and interdisciplinarity.

2.1.1. Multidisciplinarity. This type of work consists of the juxtaposition (cf.figure 2), with respect to a given empirical object, of a collection of different disciplines which develop their specific points of view. This juxtaposition gives rise, of course, to as many objects as there are points of view. Only a "magical" concept of scientific work would make it possible to imagine a common problematic emerging from the simple combination of various disciplines(7). If we use a metaphor based on Wittgenstein's "language games", we could say that we face, in this case, a mere coexistence of different languages, giving rise to something like scientific Babel.

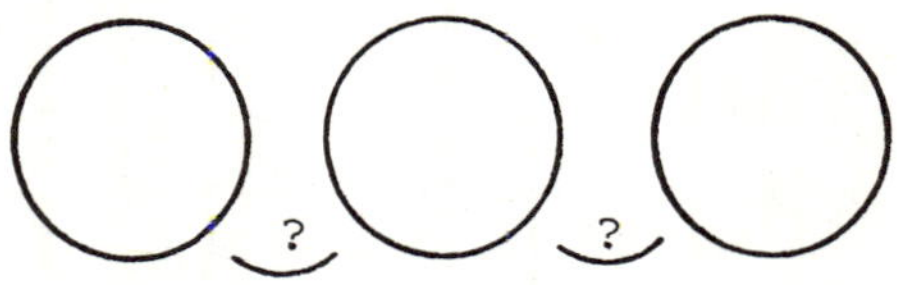

figure 2

2.1.2. Transdisciplinarity. In this case, by abandoning the point of view peculiar to each discipline, research results in an autonomous science through the creation of new theoretical objects and the application of new methods. Transdisciplinarity involves the strict integration of various disciplines(8) (cf.figure 3). Returning to the metaphor of language, it could be said that the aim is the creation of a new, common language, like a scientific esperanto.

figure 3

2.1.3. Interdisciplinarity. In this case, research is carried out on the basis of the theoretical field of one of the disciplines concerned which develops problematics that partly overlap those developed by the other disciplines (cf. figure 4). It therefore involves the articulation of sciences, eventually leading to their partial reorganisation. It could be said that one "linguistic game" is confronted with the other, and that each language tries, through a dialogue, to benefit from the genius or the "Weltanschauung" peculiar to the other language(8)

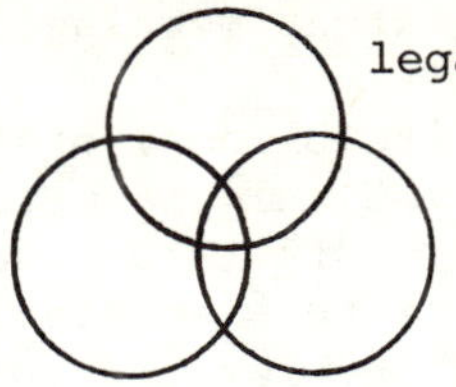

figure 4

In the light of this typology, it becomes obvious that only interdisciplinarity permits truly "focused" research. Multidisciplinarity and transdisciplinarity, on the other hand, function rather as "scientific utopias": the first because it fails to construct a common theoretical field, the second because it goes beyond all known scientific fields. However, the interdisciplinarity emerges as a relatively unstable intermediary position: it can at any time veer into the simple juxtaposition of approaches (i.e. multidisciplinarity), just as it can result, at certain stages of research, in a process of enquiry that is transdisciplinary in nature. In this case, the risk is for one discipline to overdetermine or dominate the other. Thereby, it raises an epistemological danger of distorting concepts and causing the dissolution of problematics in one discipline, and an axiological danger of strengthening the forms of legitimacy involved in the other discipline. This leads us directly to the difficulty of articulating the internal and external points of view in interdisciplinary legal science.

2.2. Articulation of internal and external points of view in interdisciplinary research

In the first chapter, it was noted that the interdisciplinary theory of law aims at developing an explanatory external point of view which takes into consideration the internal point of view of the legal system. Schematically, it is possible to say that this approach is caught between two opposite dangers. Either the internal point of view and its scholarly rationalisation are inadequately taken into account, thereby reducing law to fact, or legal science does take these into account and risks ideological connivence and pseudo-scientificity.

However, we can notice that these difficulties are shared by all disciplines which, while remaining scientific, nevertheless try to study meaningful or symbolic phenomena.

Moreover, the interdisciplinary process can be conducted at an even more radical level. Rather than just submitting the object (law, in this case) to the criticism of external disciplines, a further step should be taken to rid these disciplines of any form of connivence with legal evaluations. This would take the form of a "permutation of positions" (cf. figure 5). Each discipline concerned could propose to the other disciplines its point of view on the object of their common consideration (the problem child, land development, consumer's protection, etc.) and then submit its methods, its paradigms, its view of the world to the other's criticism. In this case, the interdisciplinary process would consist of reorganising the various theoretical fields and allowing each discipline to arbitrate in turn the various issues at stake(10). Even if, at the present stage of legal research, this schema is still largely prospective, it is the only guarantee that one of the different participants in the interdisciplinary dialogue does not formulate both the questions and the answers.

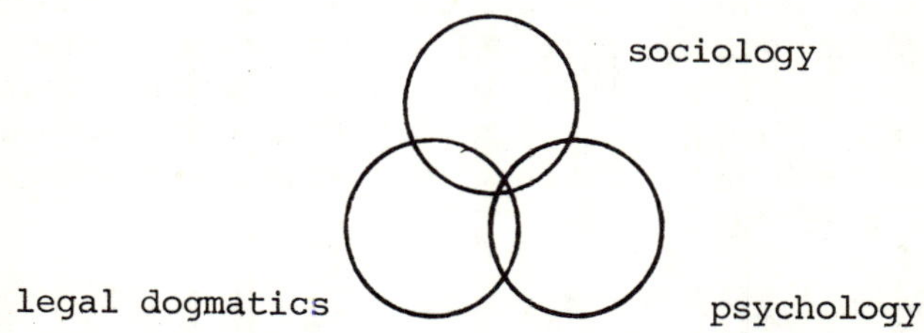

figure 5

2.3. Variety of interests underlying interdisciplinary research

The dialogue between legal specialists and other scientists may be viewed in at least two different perspectives. On the one hand, interdisciplinarity can be considered at a <u>practical</u> level: the dialogue takes place, for instance, between the legislator and the sociologist or the judge and the psychiatrist, tending to improve legal practices. On the other hand, it can be conceived at a theoretical level and geared to improving the understanding of legal phenomena.

Although we are concentrating on the second case, it should nonetheless be recalled that both of these two views of interdisciplinarity may be rooted in two distinct types of interest which, using Habermas's terms, could be described as "technical" and "emancipatory"(11).

Inasmuch as interdisciplinarity is based on a technical interest, the aim is to ensure that the methods employed are better suited to the results sought, without questioning the results themselves. In our particular case, the consultation of experts by jurists would ensure that legal means are better suited to social results. Since this type of dialogue does not discuss the social results, it probably tends to strengthen the socially dominant rationality. For instance,"legislative sociology" as conceived by contemporary legal sociology reveals a willingness to rationalise "legal production" in order to ensure greater harmony between the social norm and the legal norm(12). Similarly, it appears that the reasons that prompt many legal theorists to recommend interdisciplinary research are also based on the view that the legal system needs to be rationalised(13)

Conversely, the hypothesis that an "emancipatory interest" is at the root of interdisciplinary activity implies a reexamination of the social and individual factors that influence legal theory and practice. This in turn encourages criticism of the socially dominant rationality and therefore leads to the quest of alternative results.

The interdisciplinary theory of law we recommend should take this second path even if, in the same way as in any science, there is always a possibility of involuntary diversion.

NOTES

(1) Hart, H.L.A.: 1961, The concept of law, Oxford, p.88.
(2) Ibidem, p.87
(3) Hart, H.L.A.: 1961, op.cit., p.92
(4) In this respect, cf. Kalinowski, G.: 1969, Querelle de la science normative, Paris, p.1. Cf. also Kelsen,

H.: 1962, _Théorie pure du droit_, translation of the 2nd edition by Ch.Eisenmann, Paris, p.99

(5) On this question, cf. in particular four papers presented at the 11th World Congress on Legal Philosophy (Helsinki, August 14-20, 1983): Van Hoecke, M.: "La systématisation dans la dogmatique juridique"; Alchourron, C.: "Systematization and Change in the Science of Law"; Bulygin,E.: "Legal Dogmatics and the Systematization of Law"; Aarnio, A.:"On the Change of a Legal System".

(6) On this question, cf. van de Kerchove, M. and Ost, F.: 1978, "Possibilité et limites d'une science du droit", _Revue interdisciplinaire d'études juridiques_, 1,1-39.

(7) In this respect, cf. Gusdorf, G.: 1968, _Encyclopédie Universalis_, T. 8, Paris, _V° Interdisciplinaire_, p.1089.

(8) With respect to illusions inherent in transdisciplinary work, cf. Jochimsen, R.: 1974, "Zur gesellschaftspolitischen Relevanz interdisziplinärer Zusammenarbeit", in Holzhey, H.: 1974, _Interdisziplinar_, Basel-Stuttgart, pp.12-13; cf. also Palmade, G.: 1977, _Interdisciplinarité et idéologies_, Paris, p.21.

(9) For an analysis of interdisciplinarity as an articulation of different "language games", cf. Malherbe, J.F.: 1980, "Eléments pour une analyse du travail interdisciplinaire", _Revue philosophique de Louvain_, pp.91-97.

(10) Cf. Resweber, J.P.: 1981, _La méthode interdisciplinaire_, Paris, pp.57 and 159, on this point.

(11) Habermas, J.,1973, _La technique et la science comme "idéologie"_, translated by J.R.Ladmiral, Paris, p.145.

(12) Commaille, J.: 1982, "Esquisse d'analyse des rapports entre droit et sociologie. Les sociologies juridiques", _Revue interdisciplinaire d'études juridiques_, 8, p.16.

(13) Van Hoecke, M.: 1983, "La systématisation dans la dogmatique juridique", paper presented at the 11th World Congress of social and legal philosophy at Helsinki, August 14-20, (to be published); Aarnio, A.: 1983, "Some conceptual foundations of legal policy research", _Philosophical perspectives in jurisprudence_, Helsinki, p.222.

LEGAL SCIENCE AND HERMENEUTIC POINT OF VIEW

Vittorio Villa

1. A DEFINITION OF 'LEGAL SCIENCE'

In this paper I will offer some methodological reflections on legal science and its relations with the natural sciences and above all with the 'human sciences'. The basic assumption throughout is that the intellectual activity which I call 'legal science' is truly a science on the grounds of its substantial epistemological similarity to the other scientific disciplines. This presupposes a unitary image of science, such as is made possible by post-positivist conceptions. Such a unitary image embraces all varieties of science while allowing of a partial methodological autonomy within each of the particular sciences(1). Here there is a useful analogy to be drawn between science and a game which comprises within itself many sub-games connected by 'family resemblances'(2). Putting it shortly, there are methodological analogies and similarities among all the particular scientific 'sub-games', and this justifies us in referring them all to a common image of science.

If this point of view is accepted, then the contention that the discipline of legal study can be reconstructed as a science must be justified by showing that there are some real points of contact between it and the other scientific disciplines. But of course legal science will differ from other sciences in many important aspects, mainly on account of the distinctive features of the 'legal' as a piece of experience, and of the consequential necessity for appropriately distinctive criteria.

In this paper, however, I assume, without further investigations(3), that we can truly speak of 'legal science' in the sense of an activity which is cognitive in character, aiming (as in other fields) at the representation, reduction to unity, and explanation of phenomena(4). In this as in other sectors of experience the theoretical goal is to construct hypotheses which are explanatory of

A. Peczenik et al. (eds.), Theory of Legal Science, 509–520.

the relevant 'slice of reality' (in this case the 'legal reality' composed of rules, actions, institutions,etc.), that is, which give an account of (legal) phenomena in a 'law-like way', not by way of scientific laws in the strict sense, but by way of general theoretical principles and categories (for instance, 'rights' or 'duties'). In every discipline, in natural sciences as well as in legal science, the same general epistemological principle is applicable: that is, the principle of seeking a simplification of experience(s), making a unity out of diversity, order out of disorder, regularity out of anomalies. For example, what other purpose do theories of legal order serve if not to attempt to construct general explanatory hypotheses which are able to reduce into unitary form the ostensible disorder and diversity of a complex bundle of legal rules and principles?

It will be seen that this conception of legal science is a very broad one and that for the moment I make no assumptions either about specific issues of methodology or (*a fortiori*) about the general theory of law. I would submit that this broad conception of mine is preferable to one which treats 'legal dogmatics' as exhaustive of legal science. This 'legal dogmatics' thesis, at least in some recent formulations(5), ascribes to legal science certain specific methodological and theoretical parameters which apply only to some cultural contexts, but not to all. To treat legal dogmatics as the specific activity of interpreting and systematizing the content of a legal order, is already to presuppose that legal science has a specific kind of subject matter (a body of codified law) and must therefore have a specific and determinate methodology.

My definition is pitched at a high level of abstraction. There is here no pretention to mere portrayal of some specific and independently existing legal praxis. In the case of the real 'day to day activity' of legal exposition or legal scholarship, there is nothing which can be seen as scientific activity in its pure form. Here what we have is an all but inextricable mixture of discourses diverse in nature, belonging indeed in part to the universe of scientific discourse, but in part to those of politics, ethics and 'technology' as well. My definition provides an 'optical' instrument, which enables one, however, to discern in this mixture the genuinely scientific elements and to separate them from the rest.

2. LEGAL SCIENCE AND NATURAL SCIENCES: MODELS AND ANALOGIES

The sense in which I speak here of sciences as games and of similarities between legal science (as 'human science') and natural sciences is completely different from that advocated by legal theorists influenced by neopositivism (for example by Ross(6)). These theorists maintain a methodological monism according to which scientific method is essentially and by definition unitary, and is the method of the natural sciences. The natural sciences' criteria are the sole available criteria. Such criteria, by contrast with the particular content of scientific theories, constitute the unalterable and historically invariant part of science. They must therefore (like Ross's verification principle) remain in force with unchanging content regardless of the ontological and theoretical character of the 'domain' to which they are applied.

My position is, on the contrary, pluralist. It follows the new post-positivist conceptions of science, according to which methodological criteria are (at least partially) context-dependent. By this I mean that the specific methodological criteria, even if they descend from the same general methodological principle, are subject to change in relation to the context-situation, that is to the empirical, theoretical, ontological features of the domain to which the general principle is applied. In this view, it is in the last analysis the task of scientific communities to choose, in their 'normal' activity, the methodological criteria which must be put at work. But it is not an arbitrary affair; on the contrary, scientific communities' choices are subjected to considerable constraints. These constraints arise not only from the concrete and material organization of scientific institutions in a certain social context and from their relations with the wider society. There are also epistemological and metaphysical constraints which depend on the image of science and on the world view which is predominant in the historico-cultural context of the scientific community in question.

This conception implies a quite radical change of attitude in the traditional way of discussing the problem of the relations between natural sciences and legal science. Rather than treating natural sciences as a direct source of methodological imperatives for the scientific study of law (the methodological imperialism adopted by Ross), we should use the natural sciences, in the post-positivist reconstruction of them, as analogue models(7) against which to appreciate more clearly and distinctly the methods "in fact" at work in legal science. This can lead to a more lucid discrimination among the 'materials' before us - the whole complex of the work carried out by jurists - enabling us to pick out the operations which are the most significant from a methodological point of view. This cannot but lead to a better understanding - understanding through reconstruction - of the nature and the meaning of legal methods in their triple relationship with epistemology, with scientific communities and their paradigms, and with legal phenomena. And that is not all. Using the analogue model as a 'tool of observation' may well lead to the discovery of 'new facts', albeit of a sort not perceptible to the 'naked eye', and make it possible to 'see' aspects and implications of legal method in a new and deeper way. We could discover, for example, the empirical quality of some activities of jurists, as that of interpreting rules ('quasi-observation') and controlling legal theories ('quasi empirical control')(8).

It is worth noting that it would be a great mistake to treat the kind of interdisciplinary approach I take to the problem of legal science as though it allowed only of one-way traffic between natural sciences and legal science (and the other human sciences), the former always playing the role of model for the latter. There is no single discipline, not even physics, which can today arrogate to itself the right to prescribe criteria of scientificity to other disciplines. In truth all the science-games are connected to each other by a set of mutual and reciprocal family resemblances. So, depending on circumstances, any particular science can legitimately use any particular other science as a model, the respective roles of any pair being reversible in different circumstances.

3. LEGAL SCIENCE AND HUMAN SCIENCES

That there are analogies between legal science and natural sciences should certainly not lead us to ignore the important differences between them. As a human science, legal science deals with different phenomena and postulates entities of a different kind from those of the natural sciences, and these differences are reflected at the methodological level as well.

Which, we may then ask, are the most significant differences between legal phenomena (rules, institutions, actions performed by judges, officials and citizens, arguments produced by judges, speeches and accounts presented by the agents as to the intentions, motives, reasons, values, commitments which they advance as explaining or justifying their behaviour or attitude in relation to rules, etc.) and natural phenomena?

One of the most important and significant differences is that the former, and human phenomena in general, are unlike natural phenomena in that they cannot exist except as an aspect of human activity. They are constituted by standing, and sometimes deliberate, practices of individuals and social communities; furthermore, they have a meaning and a significance for science, only in so far as they are in themselves meaningful and important in the lives of individuals and communities to which they belong. These 'cultural facts', therefore, count as 'phenomena' only to the extent that they have some meaning intrinsic to themselves, apart from any interpretation which may be put upon them by science. Phenomena such as 'legal rules', for example, genuinely constitute social phenomena, amenable to scientific investigation, only in so far as they possess a meaning and an importance for the social life of agents; this meaning is already ascribed by them, prior to any intervention of interpretative categories applied by scientists. This pre-existing meaning is, therefore, constitutive for the very existence of the phenomena.

This feature of legal rules, and of the other legal phenomena, gives rise to a quite complex problematic, common to all the human sciences, to which the name 'double hermeneutic' has been well given(9). Here there are two social practices which meet and interact: the social practice constituted by the organized scientific work of legal communities of scholars, and the object of investigation, which is, in its turn, broadly speaking, a social practice of the community, at least in the sense that the objects investigated are the products of a social practice and are embodied in it. This situation has important consequences for legal method, for the very reason that in this domain (and in all the domains of human sciences) the frames of interpretation and of theoretical qualification adopted by scientific communities to some extent interact with the frames of interpretation and the common sense knowledge used by the agents.

It is just here that methodology meets hermeneutics. The peculiarity of the hermeneutic relationship between the observer and the 'human reality', in which the objects investigated by the scientist-observer are 'materials' of the same kind as those out of which his own social life is built, gives rise to many complex problems of methodology. The problems are perhaps most acute in the cases in which the scientist's task is (like in anthropology) to understand an alien culture. But many other problems are truly common to all these disciplines.

Such issues cannot, however, be clearly investigated here. Anyway, I would add that there is no contradiction here with the ideas advanced in the first part of the present paper. That for some purposes the natural sciences are useful as 'analogue models' for legal science by no means excludes making reference for other purposes to other human sciences as models for the study of legal science. Nor is there any objection to seeking inspiration from hermeneutics. There is no inconsistency in the 'game theory of science' (as we may call it) between empirical method and hermeneutic method (in the moderate approach of which I will speak later).

4. 'EXTERNAL' OR 'INTERNAL' POINT OF VIEW?

Against that background I will deal now with one only of the problems connected with the peculiar subject-object relationship which we find in human sciences. The problem is as to the kind of posture and the kind of methodological attitude which human scientists should adopt towards human objects. My particular concern will, of course, be with the posture and the attitude of the jurist towards the legal phenomena, and in particular towards the legal rules. Coming back to the 'games analogy' and considering social practices as if they were games, we could ask this kind of question: must the jurist be a 'player' like the others, that is a whole-hearted participant in the game, or does he have to limit himself to watching the game as a concerned spectator?

This is a very important problem in which all human sciences, such as sociology, ethnomethodology (as a domain of sociology), anthropology, linguistics, etc., are involved. In legal science the issue has recently been put in a clear light by Hart's work, and particularly by his conceptual distinction between the 'internal' and the 'external' point of view(10). I cannot here dwell on the details of this well-known distinction, but will simply adopt it as a conceptual tool for framing in a suitable way some typical patterns of solutions adopted in legal theory to the problem of the correct posture and the right attitude of the jurist towards a legal system.

The first, 'external', solution, connected, in a quite radical approach, to American and Scandinavian legal realism (Ross), and, in a more moderate approach, to some forms of legal positivism (for example, in Bobbio's position(11)), is strictly dependent on an empiricist epistemology. This position denies that there could be any difference as to proper attitudes between natural sciences and human sciences. Legal study can be 'scientific' only if it adopts the method of the natural sciences, or, at least, follows it as a model as far as it is possible. This means that the jurist must try to reduce normative phenomena to 'hard facts', or, at least, must treat them as being analogous to hard facts. From this kind of epistemological perspective it follows, of course, that jurists must adopt an external point of view and must refuse, not only to hold a personal position on the value choices and the interpretative frames belonging to the agents, but also to take any account of these attitudes of the agents, unless they could be considered as strictly empirical phenomena.

In none of all these conceptions is there to be found any place for dialogue or communication between the legal science game and the law game. The scientists must conduct themselves as 'neutral spectators', who must avoid being or seeming to be 'fans' of any of the players of the game.

The second, 'internal', solution is suggested by all those (for example by Aarnio, in my opinion(12)) who are persuaded that the jurist must adopt an internal point of view. According to this opinion it is impossible to approach the rules correctly or to understand their meaning adequately, without standing inside the legal system, without actually or potentially participating in the law game. The jurist is a participant with the special task of observing how the other players go about playing the game, that is, for example, of reproducing the meanings and the commitments towards the rules adopted by the agents.

In this second position, in contrast with the first one, there is a sort of communication, the beginning of a dialogue between the observer and the legal agents whose activities are under investigation. But it is a communication which ends up as a soliloquy on the part of the legal agents, because the scientist limits himself to listening, or at best to making more clearly audible whatever the agents have to say.

I am persuaded that both these positions are unsatisfying and must be given up. About the first position, to tell the truth, there is nothing more to say after the conclusive criticism to which behaviourism has been recently subjected, both on the methodological and on the substantive level. If we leave out from the "legal field of experience" legal agents' attitudes, value choices, common sense knowledge and shared meanings, which constitute legal practice as a social practice within a form of life, very little of scientific value remains to be described.

The second position goes wrong when it states that a necessary condition for the understanding of legal (and human) phenomena is participating, becoming full members of that specific community, in short playing the same game. 'Participating for understanding' (as we may call this methodological recipe) means that the scientist, in order to understand some particular aspect of the social life of the agents, must adopt the same rules of actions, the same common sense knowledge, the same attitudes and meanings as the agents themselves. In all these positions the criteria (rules, principles, frames of meanings, and at least a conception of rationality) adopted by the participants in that practice are not only the privileged data which we must take into account in giving descriptions and explanations, but constitute also the privileged criteria for evaluating the trustworthiness and the conformity to 'reality' of the descriptions given by the scientists.

5. LEGAL SCIENCE AND THE HERMENEUTIC POINT OF VIEW

We should however note, at this point, that theses of this kind conflict with the new post-positivist epistemological perspectives. According to these new images of science, the methodological criteria are - to some extent - produced by the praxis of scientific communities, of course not in an arbitrary way, but respecting many kinds of constraints. Therefore, in the last analysis, it must be for the scientific communities themselves to have the responsibility of deciding, even in human sciences, what counts as soundness in explanation, as trustworthiness in findings of fact, etc.. We cannot make to the scientist the suggestion that he look simply at the way the game is actually played, leaving the situation entirely unchanged. In the legal field, for example, we cannot, in describing the various kinds of legal arguments adduced by judges, leave these modes of reasoning exactly as they are (as Aarnio does(13)), with all the inconsistencies, contradictions, mistakes, which they always carry with them. I think, on the contrary, that the jurist's task is just that of selecting the most significant modes of legal reasoning, of formulating the criteria determining what counts, in that context, as good or bad argument.

At this point we may feel to be at a dead end. If the legal scientist is neither a spectator of the game nor a participant, what can his role be?

Fortunately, however, the situation is not so desperate as it may seem at first sight. There are already, in legal science, and in human sciences in general, some ideas and suggestions which try to mediate between the full detachment and the indifference of the 'external observers' and the 'full membership' of the participants. An example of this new attitude in the human sciences, which comes from sociology, is supplied by certain theses of Anthony Giddens.According to Giddens, it is absolutely necessary, as hermeneutics stresses, that in studying social phenomena we take into account what human beings say and do, their attitudes and the meanings they assign to their conduct. This means "showing a respect for the authenticity of people's thinkings and doings"(14). But, continues Giddens, "a respect for the authenticity on the level of meaning has to be distinguished from the validity of propositions about the world that are expressed as beliefs within a particular meaning frame". Even if scientific descriptions and theories cannot pretend to arrive at the truth, this does not mean that they have to be "at the mercy" of the descriptions or of the theories implicitly accepted by the participants in a given social practice or in a given community(15). To sum up, what Giddens proposes is a moderate hermeneutic approach, in his words "a hermeneutically informed social theory"(16).

A second example comes from the legal field and is constituted by certain theses of Neil MacCormick, drawn from the Hartian conceptual distinction between the 'external' and 'internal' point of view. MacCormick realizes that from this distinction a simple opposition between the external observer and the fully committed participant does not result at all. The 'external point of view' contains another standard situation, its 'non-extreme variant', where, in Hart's words, "the observer may, without accepting the rules himself, assert that the group accepts the rules, and thus may from the outside refer to the way in which they are concerned with them from the internal point of view"(17). This is the methodological standpoint, called 'hermeneutic point of view', which is advocated by MacCormick. According to MacCormick, the jurist must, indeed, stand on the external side of the legal system; but this does not mean that his task is that of describing simple empirical connections among psychic and social facts. On the contrary, his specific theoretical task is that of understanding and of seeking "to describe legal rules as they are held from the internal point of view regardless of any commitment he himself has for or against these rules in their internal aspect"(18).

These tentative proposals to which I have referred, even if in rather too vague terms, are basically correct and take a middle course between a behaviourism which is empiricist in character and the radical relativism of the more extreme hermeneutic tendencies. They are, therefore, suggestive of a methodological approach which is worth developing further, if only because it does not postulate frontal oppositions between human and natural sciences. I think that adopting a moderate hermeneutic approach in human sciences is consistent with postulating (in a post-positivist perspective) the thesis of the unity of science. The obvious differences between the hermeneutic and the empirical method could on this basis,be explained as internal articulations of the same general methodological principles.

NOTES

(1) My conception uses the following distinction between 'epistemology' and 'methodology': by 'epistemology' I mean a general theory of scientific knowledge regarding the goals, the structure and the referents of scientific enterprises; by 'methodology' I mean the discipline which, on the basis of a certain epistemology, offers a - partial and selective - reconstruction of the most significant and relevant criteria "currently used" (from the point of view of that accepted methodological perspective, of course) by scientists to judge the soundness of their methods of pursuing scientific inquiries; e.g., criteria stipulating what is the correct way to conduct observations, to approach the task of theory building, to establish empirical controls, and to choose among rival theories.But methodology, in my opinion also has a 'normative task', because it can in its turn criticize, on the basis of the presupposed image of science, the criteria it reconstructs, suggesting improvements in them or improved versions of them. For this conception of 'methodology', see, for example, G.RADNITZKY, "Popperian Philosophy of Science as an Antidote Against Relativism", in Essays in Memory of Imre Lakatos, ed. by R.S.Cohen, P.K.Feyerabend, and M.W.Wartofsky, Dordrecht, 1976, pp.507-509.

(2) The analogy between sciences and games, obviously based on the later works of Wittgenstein, is developed by D.L.PHILLIPS, Wittgenstein and Scientific Knowledge. A Sociological Perspective, London, 1977, pp.154-168.

(3) For a more detailed discussion of the problem of 'legal science', see my book Teorie della scienza giuridica e teorie delle scienze naturali. Modelli e analogie, Giuffre', Milano, 1983.

(4) Here I follow the definition proposed by E.di

ROBILANT, "Prospettive sul ruolo del giurista nella società tecnologica della nuova Europa, in _Materiali per una storia della cultura giuridica_, 2, 1980, p.510.

(5) Cf., for example, A.AARNIO, _Philosophical Perspectives in Jurisprudence_ Helsinki, 1983, p.163, and A.PECZENIK, _The Basis of Legal Justification_, Lund, 1983, p.118.

(6) A.ROSS, _On Law and Justice_, London, 1958,p.67. A reconstruction of Ross's methodology is given in my essay "Implicazioni meta-teoriche nella teoria del diritto di Alf Ross", in _Rivista internazionale di Filosofia del diritto_, 2, 1979, pp.258-288.

(7) By 'analogue model' I mean a set of representations or assumptions on a determinate object (or artefact), which is chosen (or built) by scientists in order to shed light on certain relevant affinities between it and a scientific (or meta-scientific) object which is, at the moment, under investigation. Cf.P.ACHINSTEIN, _Concepts of Science. A Philosophical Analysis_, Baltimore, 1968,p.210.

(8) For this, see again my book (supra,n.(3)), ch.IV.

(9) I use the expression 'double hermeneutic' in the sense given by A.Giddens, _New Rules of Sociological Method: a Positive Critique of Interpretative Sociology_, London, 1976, p.79.

(10) H.L.A.HART, _The Concept of Law_, Oxford, 1961, pp.55-56, 86-88.

(11) See N.BOBBIO, _Il positivismo giuridico_, Torino, 1961, pp.160 ff., and _Giusnaturalismo e positivismo giuridico_, Milano, 1965, pp.105 ff.

(12) Cf.A.AARNIO, _On Legal Reasoning_, Turku, 1977, pp.148, 297-298.

(13) A.AARNIO, _On Legal Reasoning_, cit., p.148.

(14) A.GIDDENS, _New Rules of Sociological Method_, cit.,p.145.

(15) _Ibid._,p.145.

(16) A.GIDDENS, _Profiles and Critiques in Social Theory_, London, 1982, p.5.

(17) H.L.A.HART, _The Concept of Law_, cit.,p.87.

(18) Cf.N.MACCORMICK, _H.L.A.Hart_, London, 1981,pp.33-43.

LEGAL THEORY AND SOCIAL SCIENCE

Hubert Rottleuthner

To tackle the relations of neighboring scientific disciplines historically is no foregone conclusion. A view of their present relations - a systematic view - hammering out distinctions and locating points of contact where one can fertilize the other or detect blind spots should be rather instructive. However, a historical retrospective of the evolution of related sciences can also be made with a systematic purpose: to point out reciprocal relationships in an exemplary field of discussion.

1. THE ORIGIN OF SOCIOLOGY IN THEORIES OF NATURAL LAW

From a historical viewpoint an asymmetrical relation exists between legal philosophy, legal theory and the social sciences to the extent that a specifically social scientific outlook -on law too- only appears relatively late. No doubt, the social theory of the ancients was not free from references to reality, but nevertheless, these were embedded in a natural-law quest for correct social arrangements. An independent social science developed out of the tradition of natural law only in proportion as the primacy given to a practical orientation was overlaid by an experiential-scientific orientation. I see the beginnings of this evolution in the theoretical efforts during the 18th century to digest the pressure of experience which had built up primarily from accounts of the newly discovered lands.

Travel accounts were the favorite reading matter of the house-bound bourgeoisie. What to the bourgeois reader represented merely an accumulation of curiosities and exotic ticklers -hence, extremely remote possibilities- confronted natural law social theory, of course, with ticklish problems. Assumptions about human nature, about the truly correct social arrangements, were challenged as mere speculation and projection, and,thus, threatened with

A. Peczenik et al. (eds.), Theory of Legal Science, 521–529.

invalidations. The Eurocentric, anthropologically founded viewpoint was exposed to the danger of a fundamental relativization.

Putting aside the daily repression of these problems by "exoticizing" and furnishing of curio galleries, on the theoretical level at least four reactions can be distinguished clearly:

1. The relativization of one's own perspective; European self-conceptions are consciously adopted by fictively assuming the perspective of the other. In the role of an outsider -whether an imported American or a member of a faraway high culture traveling on a diplomatic mission- European civilization is described, often for amusement, but frequently with a socio-critically alienated intent.

2. The abundance of information accumulated in the travel accounts of distant lands -their customs, morals, legal and political arrangements- could not be merely compiled or described systematically (encyclopaedically). As soon as an attempt was made to explain different social phenomena, a theoretical assimilation began. The foreground appeal of the unknown dissolved in a detached perspective in which the causes of social phenomena were called into question.

We find both reactions -the assumption of an external, alienated viewpoint and a detached explanatory perspective- in an author of the 18th century who is frequently regarded as a progenitor of sociology: Montesquieu. In his "Persian Letters" he permits the French society of his time to see itself in a mirror held by a stranger. In his "Spirit of Laws" he assimilates the most important travel accounts of his time (along with the accounts of ancient authors) in the framework of a theory which seeks to explain the multifariousness of social arrangements by a series of factors. Montesquieu works with biological, geographical, climatic, religious and ethico-moral factors which today appear rather vague to us, but, at any rate, an attempt is made to establish relationships ("rapports") between explained conditions and phenomena requiring explanation.

3. The huge expansion of illustrative material for a philosophical anthropology and of the spectrum of societal forms weakened the edifice of natural law. A radical reaction against such a relativizing pressure of experience can also take the form of avoiding all experience in questions of social praxis, correct conduct, and opti-

mal social arrangements. As is well known, Kant chose this path by apriorizing and formalizing practical philosophy. What is interesting biographically in this instance is that Kant himself liked to read travel accounts, and in table talk with guests he gave the impression that he himself had seen all those things in distant lands which he knew how to describe graphically. How he, in this manner, experienced the world in Konigsberg is found in his work. Except for a few allusions, he kept his practical philosophy radically free of empirical references. Practical reason is pure, autonomous, that is to say, gives itself its own laws, without the necessity of borrowing discoverable "empirical" laws. Kant shrinked from the pressure of experience into the sealed shell of the metaphysics of morals and then chatted at table about the the empirical facts.

4. While Montesquieu himself took literary advantage of the relativization and scrutinized it explanatorily -but accepted it as fact- a concept subsequently spread which in the long run appeared to canalize the pressure of experience: the concept of progress. Prodded by the accumulation of technico-scientific knowledge, the idea of progress now was transferred to moral-social evolution. It was well suited to the systematization of information about strange societies, because it allowed them to be localized on different steps of a rank ordering. This kind of focus was still lacking in "Spirit of Laws", where Greeks, Romans, Chinese, Germans, Moluccans, Indians, etc. were treated evenhandedly and synchronously. The idea of progress found its full unfolding, then, less than a century later, in Hegel's philosophy of history. Hegel, like Kant, distinguished nature and mind; however, autonomous mind had objectified itself in the world (as "second nature"), and these objectifications in the historical world can come to be known and be "rationally" understood as being "rational" sequences themselves, namely, as "progress in consciousness of freedom". This principle of world history turns out to be a unique bridging structure: in his philosophy of world history Hegel absorbs historical experience totalistically, but experiences which contradict an optimistic progress glance off the world spirit as merely subjective "moralizing". The principle of progress contains a judgement on the movement of world history; at the same time it struggles against application as a maxim of practical orientation, as a postulate or value

standard for future conduct and a rationale for established institutions. The world spirit strides forward, but only at twilight does it allow its extent to be surveyed.

The great social theorists of the 19th century up to Max Weber wrote variations on the historico-philosophical theme of progress (or, somewhat more neutrally, evolution) and connected it to the natural law question of the conditions of social arrangements. The consequence frequently is theories which emphasize a single factor in stages of evolution and social integration -entirely opposite to Montesquieu: whether it is the dialectic of productive forces and relations of production in Marx and Engels, the process of differentiation borrowed from biology in Spencer, forms of social solidarity in Durkheim, the rationalization of world view in Max Weber, etc. Thereby, the normative implications of the concept of evolution are immense, whether existing conditions are criticized by means of optimistic projections of the future or the present and perhaps also the near future is viewed as an epoch of decline.

Since the end of the natural law tradition in the 18th century placed state and law at the center of the theories, they can be characterized as social scientific. And yet Hegel unfolded his sociology in the framework of the "Philosophy of Right". The social theories of the 19th century grew out of the legal philosophical tradition in two senses: the legal-philosophical inheritance is still recognizable at their roots, but the weighting is, as it were, overturned. Law becomes a phenomenon which is "derivable", socially explicable, has no unique history (as Marx said), "expresses" basic underlying social facts, is one of as well as the central index for occidental rationalization, etc.

Within sociology, after Max Weber, interest in dealing empirically with law almost entirely disappeared. And with the increasing extension of empirical social research the great theories of progress and evolution resign themselves to speaking of "social change". Sociology proceeds à la Montesquieu, but, thanks to elaborate empirical and statistical methods, on a higher level. Monocausal explanations are superseded by multi-variable approaches; normative implications, such as of progress, are eliminated if possible, and, moreover, sociologically distancing means are worked with a language which makes one see things differently; a view which cuts across the foregone conclusions of everyday life, putting apparently disparate items into functional relations; and the analysis of latent or objective functions.

2. THE DIFFERENTIATION OF LEGAL PHILOSOPHY, LEGAL THEORY, AND LEGAL SOCIOLOGY

Recent legal sociology as a special discipline, did not evolve out of sociology. What might appear as a split-off of an autonomous legal sociology from general sociology in the 19th century, perhaps in the work of Eugen Ehrlich (Fundamentals of a Sociology of Law, 1913) and in the writings of the members of the "Pragmatic Instrumentalism"--movement (as Robert Summers calls it) is much more a reaction within the juridical world against the formalism of conceptual jurisprudence. It was jurists, Ehrlich and others of the "free law" group as well as the "legal realists", who articulated the insufficiency of a purely conceptual treatment of the allegedly autonomous normative ether of legal arrangements. This legal-sociological reaction led the way to developments which in the second half of the 19th century produced a specific form of jurisprudence. While in the 18th century an experiential-scientific area already had been carved out of practical philosophy, the leftover "remnant" of practical philosophy experienced a radical alteration through two developments: through the growing positivization and institutionalization of law - a legal "objectification" of the world spirit, as it were - and through the scientification of its operation. Thereby, in the creation of a quasi-closed sphere of valid law, the subject area of a legal-scientific analysis was constituted. Jurisprudence is sedimented practical philosophy to the extent that it finds a subject area in which practical questions already appear to be addressed through governmental activities. It restricted itself to detail work on problems which remained open and to the systematization of solutions which had been found and decisions which had been made. This could develop -in a science sui generis- into something independent from historical, social and political references.

Neo-Kantianism reflected the tendencies towards scientification of the 19th century, in that it was understood as philosophy of science (today one would more readily say "theory of science"). The task is no longer one of a critique of pure theoretical or practical reason but of scientific reason as it is practiced in the natural and cultural (or value) sciences. Legal science was raised to a meta-level, as is well known, in Kelsens's "Pure Theory of Law", whose fundamentals he developed at the beginning of this century -based on Neo-Kantianism and enriched phenomenologically. Thereby, in the foundations of the legal science in which the former tradition was administered, the tradition of practical philosophy was divided into the disciplines of legal philosophy, legal theory and sociology of law.

Legal philosophy occupied itself in a practical and normative way with questions of correct dealing and good social arrangements. Indeed, it descended more and more into a juristic philosophy which served the procurement of legitimation of jurists (with the period of National Socialism as absolute low-point). Legal theory in the narrower sense (pure theory of law and analytical legal theory) primarily treated conceptual and logical problems. It is "normative" in the sense that it deals with "normative facts" as its subject, but no longer in the sense that it formulates norms itself. In contrast to this conceptual purity, sociology of law laid stress upon the necessity of an analysis of the social reality of law- its formation, evolution, administration and effects.

The relations of these basic legal-scientific disciplines continue to be not without their own problems. An attractive as well as revealing scientific-historical constellation arises from the fact that Kelsen subjected Ehrlich's "Fundamentals of a Sociology of Law" to a sharp critique and that from this there developed a discussion ending in a mutual lack of understanding. Kelsen pressed the point that a sociological view of law is guided by an external, namely, juristic view of law, without which the demarcation of the subject area would be entirely impossible. Ehrlich, for his part, evaded all efforts towards a conceptual clarification of basic questions of legal sociology and insisted on the empirical - in any case non-juristic - grasping of law as a social phenomenon.

A strained relationship developed between sociology of law and legal philosophy to the extent that sociology of law understood itself to be critical of ideology, thus also critical of the ideology of a juristic philosophy with its legitimation of existing relations. Then, besides research into the facts of the rule of law (A.Nussbaum), which rather put at issue the removal of the information deficit about the social use of legal norms, during the twenties a critical trend in sociology of law was able to unfold, in Germany, carried on primarily by labor law specialists (Sinzheimer, Fraenkel, F.Neumann, Kahn-Freund). These critical approaches were ended radically in 1933, whereas legal philosophy in the period of National Socialism could zealously weave its veil of legitimation. The critical tradition of the sociology of law could be taken up again in the Federal Republic of Germany only at the end of the sixties.

Work in sociology of law was characterized, until the sixties, by the fact that - separated from the rest of sociological development - it rarely satisfied the standards of an empirical discipline. This was true of merely descriptive research on the facts of the rule of law and the sociology of law based on critical social theory, as well as of American legal realism. It was also characterized by a certain predilection for basic questions, the search for a "sociological concept of law", thus by a vacillation between legal theory and the sociology of law.

That it was possible for an empirically proceeding sociology of law to be established in the last two decades, primarily in the USA in the framework of "judicial research", is attributable to the fact that the conceptual foundation questions - one is almost inclined to say, in American innocence - were excluded. This kind of innocently empirical legal sociology, nevertheless, could not be spared analytical purification. I myself have attempted, as it were, to substantially move the discussion between Kelsen and Ehrlich to a new level of understanding, in order to make the results of analytical legal theory fruitful for the work of the sociologists of law and to provide a clarification of more or less unproblematic conceptual presuppositions in sociological analyses of law. On the other hand, the sociology of law could point out factual propositions in the theories of everyday life in legal theoretical and legal philosophical analyses which were unproven but empirically testable (e.g., on the function of law, the use of rules, etc.). Thus, if the sociology of law requires analytical cleansing, legal philosophy and legal theory are dependent upon empirical enrichment.

3. TENDENCIES TOWARDS INTEGRATION

In recent years, I presume, the sociology of law has found company in other sociological currents in so far as these employ empirical social research. However, on the level of a theoretical sociology which borrows the findings of empirical sociology merely illustratively, a rediscovery of law appeared in the work of Niklas Luhmann.

Just as social theory attached an ever smaller importance to the role of state and law, practical philosophy lost contact with governmental-legal problems. Like analytical meta-ethics, it purified itself of practical questions, comparable to legal theory. Only with the renaissance of practical philosophy in the form of theory of argumentation which followed its analytical purification was the ground prepared for a new integration of practical philosophy, legal philosophy and legal theory. Alexy's thesis that juristic argument is a "special case" of general practical discourse is the clearest proof of that.

4. FURTHER POSSIBILITIES OF INTEGRATION

A further rapprochement of legal philosophy, legal theory, and sociology of law in a general theory of law appears possible - but not a consolidation, because the distinction of practical-normative, analytico-conceptual and empirical approaches is too significant. But the new approaches to theory of argumentation mentioned above already demonstrate how legal philosophy and legal theory can at least be mutually enriched. Legal sociologists can contribute to that relation the results of research about actual processes of consensus formation, of conflict control generally, content analysis of real argumentation, observation of procedural lapses, etc.

An important implication of this possible development should be pointed out. The scientific character of legal science in recent times has been challenged primarily from the social scientific side. Through a critique of method based in theory of science, jurisprudence has been given a new scientific thrust. If it is to satisfy methodological standards, legal science ought to push forward as social science. Most reactions from the juristic side to these demands aim at the reservation of a "proper" field, say that of "juristic rationality", however, whereby the achieved scientifically theoretical level would slip regressively or resignedly. In contrast, a science of law founded in language and argumentation theory actually presents itself as a counterprogram moving on a comparable level towards an understanding of legal science as social science. Not entirely a counterprogram - I have pointed to possible points of contact - but nevertheless a different weighting, which perhaps can be more suitable to the practical character of jurisprudence.

A further field of rapprochement may be presented by theories of legal evolution. Work is conceivable in which proposed approaches to evolutionary theories of law would be subjected to an analytical critique, proposals and projections of future legal evolution would be formulated and justified - what functions should law still have, which should it leave to be otherwise fulfilled, which legal alternatives or alternatives to law are possible - against a background of assumptions informed by sociology of law about actual functions and practical alternatives. A contribution such as the present one - a historical reconstruction of the launching of theory in systematic perspective - must also be integrated in such theories of evolution.

INTEGRATION BETWEEN LEGAL RESEARCH AND SOCIAL SCIENCE

Kazimierz Opałek

I. INTRODUCTION

The idea of integration of the study of law with social sciences was ardently proclaimed in the period immediately after World War II. The programming activities were followed by manifold attempts at carrying such integration into effect. Integration, one can say, came into fashion, and was considered a panacea for various shortcomings of the study of law. In the literature an enthusiastic tone dominated, accompanied by complaints that the jurists of the older generation do not properly estimate the profits of integration, and that the representatives of social sciences frequently do not reciprocate the promptness in the study of law to cooperate with them(1).

Although the period after World War II is a striking one from the point of view of the concentration on conscious and programmatic integrational attempts, it can hardly be considered as the beginning of the era of integration. It can be observed that the integrational trend started already at the beginning of our century, in the time of the crisis of the traditional study of law and of the rise of realistic legal theories. Although the latter did not explicitly speak of integration, they factually attempted at it. As legal theories were elaborated, based on the views that the law is a psychological phenomenon (e.g. L.Petrazycki), or a social one (e.g. E.Ehrlich), or else that the law has psychological aspects together with social ones (e.g. school of Uppsala), then necessarily the study of law began to be linked with such disciplines as psychology and sociology. In the thirties even conscious integrational programs started; the leading role in this respect has to be ascribed to K.N.Llewellyn and his methodological contributions(2).

Since these initial undertakings some 80 years have already passed, and since the climax of integrative tendencies a period of about 40 years; it seems, then, pos-

A. Peczenik et al. (eds.), Theory of Legal Science, 531–549.

sible and required now to consider the experiences of the study of law in the field of integration without one-sided enthusiasm and without a priori condemning those, who, among jurists and beyond, were not inclined - or still are not - to submit to integration. One has to try and show not only the attainments due to integration, but also the shortcomings of the latter, while attempting at explaining their causes. Last but not least, our aim will be also that of considering the further perspectives of the integration of the study of law with other sciences, and - as far as it will turn out to be required - of suggesting ways of modifying the integrational "strategies". Such a conception of the paper seems to make the latter apt to stimulate discussion, in which various observations and integrational experiences could be exchanged.

II. INTEGRATION IN VARIOUS SCIENCES. THE CASE OF THE STUDY OF LAW

The case of the study of law.The processes mentioned above could be observed not only in the study of law, as the needs of and attempts at integration are not something exclusive and peculiar of this branch of science. They are known to have penetrated the area of natural sciences, the problems of which, however, will not be dealt with her for obvious reasons. When concentrating on social sciences it is easy to see that they have their common and boundary-problems, which leads up to mutually making use of the results obtained, and to cooperation in research. As it was pointed out, the processes of specialization have caused leaving such problems out of account, and later reflection on this fact has in turn reasserted their importance and paved the way to integration(3).

Although the problem of integration is one we are dealing with in all social sciences, it has in each of them a somewhat different character. And so, e.g., sociology (together with social psychology) is a commonly desirable partner for integration, but - at least as far as general sociology is concerned - is not so much dependent on integration as other social sciences, possibly because it is theoretically and methodologically more advanced. On the other hand, for political science the integration is of crucial importance, since it used to be not an independent discipline, but a conglomerate of various investigations, carried out in a marginal way within such disciplines as history, study of law, sociology, and political economy. Hence even the term "political sciences", is also in use nowadays in some circles. In order to become a separate "political science" (in singular), "political sciences" had to accomplish the task (still by no means wholly fulfilled) of integrating the dispersed "political" fragments of the disciplines mentioned above(4).

As to the study of law it is nowadays stressed (the point being elaborated in detail especially in Polish legal theory) that its subject matter is composed, or else rather consisting of a set of objects such as linguistic ones, psychological experiences, human conduct, and the values (the so-called planes, or levels of law in the ontological sense). This again leads up to the need of employing differentiated research-methods (the so-called planes, or levels in the methodological sense)(5). Debatable points of this conception will be touched upon later (VI); now it will suffice to say that the lines of legal research are diversified and have connections with investigations of linguistics and logic, of psychology, sociology, but also political science and political economy, of ethics, general axiology and philosophy; hence the need of making use of the data, theories, and methods, elaborated by these disciplines. The multi-plane character of law makes the task of integration in the study of law specific and important.

III. RESTRICTION OF THE FIELD OF INTEGRATION TO SOCIAL SCIENCES AND LATER EXTENDED VERSION OF INTEGRATION

Among the potential partners of integration with the study of law we have, then, social sciences on the one side, and on the other disciplines included in humanities, philosophy, and logic as a formal science. However, not only in programs and in research practice of the period after the II. World War, but also in the earlier decades of our century, the integration of the study of law was distinctly restricted to social sciences. The current formulas of the recent period were: "Law and integration of social sciences", or "Law and behavioural sciences"; the title of the programmatic paper of K.N.Llewellyn, "Law and social sciences - especially sociology"(6) is also significant in this respect. In the beginnings of our century such formulas were not coined yet, but the research practice of the reformers of the study of law went in just the same direction (orientation of psychology and/or sociology). In order to explain this, one has to take into account two factors as the most important ones.

Firstly, from the beginning of the XX. century we are dealing with the fight of the realistic trends against what was called "legal formalism". Recently, a greater moderation in this respect can be observed, but certain prejudices are still there, and manifest themselves also in the neighboring social sciences. Hence such disciplines as linguistics and logic could not be popular as partners in integration, for they would only be apt to reinforce the alleged formalism.

Secondly, positivism, and later on neopositivism, proclaimed the fight against the humanities and the maximalistic trends in philosophy (the latter being "unscientific"), excluded valuations from science, as well as the possibility of applying logic to normative statements. To this camp belonged realistic theories of law, adopting the patterns of natural sciences and tending towards shaping up the study of law as an empirical social science. In this way the study of law would join the most widely conceived integrational program of the "unity of science" proclaimed by neopositivism(7). Such tendencies became reinforced again in the fifties, in the period of so-called behavioral revolution in social sciences, when the plan of the unification of social sciences as empirical behavioral sciences was elaborated and its realization started. The study of law was influenced by this current, the effect of this influence being in particular the rise of empirical sociology of law as a separate discipline(8). The attitudes of the researchers within the study of law were, of course, differentiated: their division into the "traditionalists", mostly of the older generation, and "innovators" of the younger one as well as that into moderate and radical reformers became apparent. The latter were even ready to transform the "scientifically conceived" study of law into a branch of sociology(9).

The integrational activities adopting this course established connections of the study of law with social sciences, with sociology in particular. These connections can be considered as lasting: their main field is the sociology of law. In the last decades, however, wider conceptions of integration of the study of law also with other disciplines began to prevail, as is testified by recent publications and scientific meetings on this subject(10). Here the view of the multi-plane character of law and of the methods of its research finds its expression. The impact can also be noted from the return to "great problems" of philosophy - in opposition to neopositivist philosophical minimalism - , especially to axiological ones, as being particularly important in the study of law. Seeing norms merely as "pure emotives" becomes obsolete, and hence the attempts at constructing systems of logic of norms, and deontic logic, partly inspired also by practical needs in collecting, transforming, and retrieval of legal data (legal informatics). To the widening of the field of integration has contributed also the development of linguistics and philosophy of language. And so linguistics and logic, ethics, general axiology and philosophy, became included in the field of the integrational activities of the study of law, which amounts to a retreat from the initially narrower program of integration.

IV. PROFITS OF INTEGRATION

The account will be but a brief one here, as these profits are rather obvious, and as the difficulties and shortcomings of integration seem to be more important, the latter will be discussed below in a more detailed way.

The integration made it possible, firstly, to investigate problems wholly neglected before (apart from some quite primitive and superficial notions), that is to say, problems of the social (or psycho-social) operation of law, but also those of the social factors shaping up the law. These problems come under the common heading: Law and Society. The main contribution here is that of sociology, but political economy and political science have also their share in these investigations.

Secondly, integration made it possible to deal in a more competent way with problems already under investigation in the study of law; the ways of approaching them were hitherto mostly those elaborated in the course of the long-standing legal practice. Here belong problems nowadays called "linguistic-logical", which were investigated in studies concerning legal integratation, the campus of juristic reasonings, the structure of the norm and of the system of norms. Normative theories of interpretation, those of the "legal syllogism", and of the nature of basic legal and juristic concepts are the well-known results of these studies. In the light of modern logic, linguistics, and philosophy of language quite a number of the traditional views became checked, developed, corrected, or rejected. Some problems already revealed in the extra-legal disciplines mentioned above, were identified as present also in the law.

Thirdly, the study of law, and especially the theory or philosophy of law, referred in the past to some contributions of other sciences but in a superficial, fragmentary, or belated way. E.g., the psychological problems of law came into fashion in the German study of law at the turn of the XIXth and XXth centuries, being treated in the jurists' own manner with but a scarce share of the contemporary findings of psychology. It is worth mentioning that the psychological theory of law of L.Petrazycki, influenced by this trend and outstanding in its perspicacity, dealt, admittedly, with the works of psychologists of the nineteenth century in a critical way, but it seems to be hardly reliable in confrontation with successive contributions in psychology(11). The second instance is that of philosophical problems. The ties of the study of law with philosophy were close in the past; they weakened in the XIXth century, partly under the influence of the positivistic proclivities, the result being the deficiencies of legal philosophical findings. The contacts were revived in our century and enabled making use of the contributions of various philosophical schools in the axiological, but also ontological and epistemological problems of law. One has to observe, however, that, as a rule, only narrow groups of philosophers are engaged in these contacts, adhering to different trends in philosophy, and so these contacts are very much dispersed. One can hardly speak of a wide - and deep - impact of modern philosophy on the study of law. Anyway, the ties of the latter with psychology and philosophy became reinforced.

This account of integrational contracts and profits is not - it hardly could be - exhaustive. It shows, however, that the study of law in the course of integration has widened the scope of problems under research, has attained more competent and strict devices of dealing with problems investigated for centuries and has reinforced and actualized contacts with disciplines, from which it was previously drawing profits in an insufficient way.

One can distinguish two levels of this integration. The first is the assimilating by the legal scholars themselves of the results and methods of other disciplines, and applying them in their work; the second the teamwork on some problems of legal scholars together with those from other disciplines. As to logic, linguistics, ethics, and philosophy one can speak mainly of integration of the first level, while in the case of social sciences, and mainly sociology, not only the first, but also the second level of integration was attained. Integration of the first level has contributed to the widening and modernizing of the knowledge of new generations of legal scholars. Integration of the second level has produced many valuable interdisciplinary works. Mentioning even the most important ones is not possible, and would be also not expedient, since they are widely known(12).

V. DIFFICULTIES AND SHORTCOMINGS OF INTEGRATION

1. Integrational "overload" and its negative consequences

On the one hand, it seems justified to criticize the one-sidedness of the narrow formula of "the integration of the study of law with social sciences"; and to advocate the "harmonious" integration of it both with the latter and with other disciplines, the problems of which are related to those of the study of law. On the other hand one has to ask, how and who in the study of law is able to carry out such comprehensive tasks? One has to observe, that the integrational "duties" are not equally divided in the study of law. The majority of legal scholars are legal dogmatists, either indifferent toward the matter of integration, or expecting it to be accomplished by somebody else. The much smaller group of historians of law is partly inclined to restrict themselves to descriptive "micrography"; only some of them make use of the conceptions and methods of other disciplines - mainly sociology and political economy. Practically, the whole burden of integration is carried by legal theorists who are not very numerous and whose possibilities have to be realistically evaluated.

In the course of my scholarly activities I have experienced, or observed, the succeeding "waves" of integration, beginning with the "psychological fascination" (in connection with L.Petrazycki's and A.Hägerström's theories); it was followed by the admiration for the Circle of Vienna and conceptions influenced by it. In turn, in the fifties, one witnessed the strong impact of sociology and social psychology, but soon began the era of legal cybernetics and informatics. Then the interest became focussed on the logic of norms and deontic logic, on topica and argumentation theory, on language philosophy of Oxford, and on modern empirical linguistics. The revival of axiological considerations also began to be felt on a much larger scale than in the first half of the century, when they were cultivated only by a small number of legal philosophical schools.

It is beyond discussion that these successive integrational interests have contributed to the progress in legal theoretical thinking, which in turn has to some extent influenced also other legal scholars, mostly those of younger generation. The objection has to be made, however, that there is a limit in assimilating novelties which can hardly be transgressed. Apart from that, the main objective of integration is not improving "the general education" of legal scholars, but the application of results and methods of extralegal disciplines in concrete research, and here we meet with the dangers of integration.

The integrational "overload" has the effect that the conceptions of other disciplines are frequently adopted but superficially, without efforts towards their deeper adaptation to legal problems. Many a time they are transferred mechanically, pretentiously introducing new concepts and terminology, into the study of law, not inquiring if their operational value is greater than that of the conceptual-terminological apparatus hitherto applied, and if they are really adequate to legal problems. The accumulating integrational trends lead up to the chaotic use of incoherent conceptions, produced by various schools of thinking. Integration becomes a matter of changing fashions, giving rise to dilettantism(13). The legal theorist runs this danger in the first place, since he is not able to make competently use of the contributions of numerous disciplines, nowadays highly advanced, and requiring - each of them - a great and long effort, if one seriously wants to penetrate their "arcana". As matters stand, the case in which the legal theorist knows all, but nothing well, is unfortunately not a rare one, just as in that of legal theorist "the best jurist" in an extra-legal discipline, and the best specialist in the latter among the jurists.

2. The phenomenon of the "secondary specialization"

In this situation is the retreat from integration so conceived into the direction of concentrating on one discipline, or even some part of it, quite natural. It amounts to specialization of individual scholars, or their groups, into integration with particular disciplines. Such specialization is either being completed by means of additional university studies - alongside those in the law - in the chosen branch (e.g. philosophy, logic, sociology), or by means of individual preparatory work in the given discipline. The qualification for integrative research, attained in such way, promises much more, as far as the niveau of this research and its results are concerned, than the integration of the sort characterized in the preceding section. Admittedly, according to some opinions, the legal scholar - in spite of such additional specialization - will always remain an outsider in his "second discipline", unable to hold a fully competent dialogue with it(14). The matter is disputable: one has certainly to discern among the potential additional disciplines those which are apt to be more easily mastered (e.g., sociology, political science) from those of specially hard demands (e.g., logic and semantics). As some practical experiences show, however, another danger is also involved here. Namely, the concentration of efforts on certain extra-legal discipline may lead up to the change of the focus of interest, and in effect to the transformation of the jurist into a logician, a philosopher, or a sociologist. This happens the more easily as each discipline is governed by its "own laws" concerning its points of emphasis, and so, imperceptibly, a reorientation on the research interests and ways of putting problems in a given extra-legal branch of science may occur. In such case the scholar is lost for the study of law and its integrative work.

If the latter does not obtain, the integration, contrary to its initial designs, leads up to a new specialization (e.g., sociology of law, legal cybernetics and informatics, logic of norms and deontic logic as applied to the law). It has some positive features, because in this way deeper connections between the study of law and given extra-legal disciplines can be established. Also here, however, some dangers can be observed. Firstly, it comes easily to a new isolation of such borderline-disciplines from the wider context of relevant legal problems (e.g., by virtue of "cult of technique", applied in empirical research, in informatics, or in logical calculi). Secondly, the borderline-disciplines frequently tend to attain solutions of non-legal problems, someway "on the example of the law" (e.g., in the sociology of small groups, in building up informatical, or logical systems)(16).

3. Lack of awareness of the peculiarities of law and its study

This problem was already mentioned partly above (V 2.), but the following has to be added. The integrational experiences show, that the integration is quite frequently marked by disbelief in the study of law, by underestimating its hitherto attainments, or simply by considering it as "unscientific". For many promoters of integration everything seems to begin now, from this moment, anew. Hence the propensity to find correct what other disciplines say, and to reject what the "traditional" study of law ventured to claim; hence at once other theories, methods, concepts and terms. Instead of arriving, by way of a critical comparative reflection, at some synthesis of the conceptions of the study of law and the given extra-legal discipline, or disciplines, one is rather inclined to force the study of law to adopt some extra-legal standpoint. It leads up, on the one hand, to unsatisfactory solutions of legal problems, and on the other hand, to introducing, under the pretence of collaborating with the study of law, problems, hypotheses and theorems, relevant in an extra-legal domain, but not in the law. In other words, what we have in mind is the inadequacy of many integrational conceptions in relation to the law, as they not account for the peculiarities of law and its study. The question of the "normative dimension" is of crucial importance here: on the one hand, the sociological conceptions, oriented on the naturalistic interpretation, are guilty of improper reductions of 'ought' to facts, and on the other hand, the logical conceptions, suited to problems of the statements with cognitive meaning, also get rid of normativity someway, when they operate with the prediction theory, or introduce the axiom of "ought entails can", or else favour too much the analogies between the logic of norms and deontic logic on one side, and the modal logic on the other, which leads up to well-known paradoxes(16).

One has to mention at last that as a result of integration some new areas of research (to which the status of separate disciplines can not always be ascribed) come into existence, but their connections with law are somehow distant and secondary. Here belong, e.g., legal-(psycho-) sociological, legal-politicological, legal-informatical investigations.

VI."PLANES" OF LAW AND THE DIRECTLY AND INDIRECTLY LEGAL OBJECTS

From the above (V,1-3) could be inferred that, firstly, the field of integration has to be reduced to some reasonable and realistic dimensions, secondly, in integration some degree of specialization is needed, but not tending towards getting rid of legal problems, and that, thirdly, in integration some synthesis of the critically examined contributions (in theses and methods) of the study of law, has to be elaborated, with those of given extra-legal disciplines. One has to observe, however, that these postulates have a limited force, since the course of integrational processes cannot be "decreed", but perhaps only regulated to some extent.

In order to make the above three points precise we shall try to distinguish the main from the secondary fields of integration. This may enable us to suggest some integrational preferences, and to differentiate the ways of approach to problems of the respective two orders.

The convenient starting point in dealing with these questions will be the conception of the "planes" of law, mainly in the ontological sense, as presented above (II). The legal objects, distinguished in it, are (a) norms, conceived as linguistic - logical creations, (b) psychological experiences (those of understanding, evaluating law, and of motivating conduct), (c) legal conduct, and (d) legal values. Are the objects of all these kinds "legal" in the same sense and to the same extent? This question has to be answered in the negative. The norms are directly "legal", but the objects of the further categories not. The psychological experiences under discussion are the experiences of understanding and of evaluating law as norms, as well as those of making choices of conduct under the influence of law as norms; they are, then, "legal" but indirectly, by virtue of being related to the directly legal objects, which are the norms. The so-called legal conduct is in fact a compound of heterogeneous objects: we deal here, firstly, with conduct influenced by law (which has to be treated not apart from the motivational experiences, mentioned under (b), but as constituting together with the latter homogeneous psycho-social phenomena)(17). To the so-called legal conduct belongs, however also the conduct legally qualified. While "the conduct influenced by law" is the category of indirectly legal objects (as related factually to the law as norms), in "the conduct legally qualified" the legal object is in reality the legal qualification (so the norm as directly legal object) of the conduct. In other words, "the conduct legally qualified" is not an object from the sphere of factuality, an empirical one, but a normative one(18). And in the end, as to values, not entering into disputable philosophical problems connected with them, one has to adopt the view that they are but indirectly legal objects as values of law, values related to law (as norms).

We have, then, on the one side, norms as directly legal objects and, on the other side, psychological legal experiences, conduct influenced by law, and legal values, as indirectly legal objects. One has to consider now separately the concept of the norm, which is a controversial one.

The conception of the ontological "planes" of the law is adopting the linguistic (or linguistic-logical) concept of the norm. This conception is open to doubt mainly because of the fact that it concentrates as a rule on the norm as semantic creation, leaving aside the pragmatic problems of the norm as actual fact (norm-utterance), while in this second character only the norm can be included into factual dependencies, as a fact influencing psychological experience and conduct. As to the values the matter is more complex: if they are reduced to value-judgments, they are included into the sphere of factual dependencies; if, however, they are treated as extra-empirical objects sui generis, then they stand in some ideal relation to the norm, either conceived also as an ideal object, or as an empirical one (the conception of the "manifestation" of values in reality).

Linguistic pragmatics, however, is not able to supply all (and full) answers to the questions of norms as facts. In accordance with some suggestions (which are still in need of being further elaborated and developed), especially those contained in J.L.Austin's theory of performatives, we adopt the view that the norm is essentially a non-linguistic phenomenon, an act of deciding as to the way of conduct of the addressee, a phenomenon, in which the making use of linguistic formulations is but of an instrumental character (the use of words prescribed for the decisional act to be "happy", and necessary for its communication)(19). These problems have some complicated aspects which cannot be discussed here. We limit ourselves to stating that the norm as a directly legal object can be considered in a two-fold character: as the decisional act, and as the linguistic expression of this act (the latter undergoing analysis on the syntactic, semantic, and pragmatic levels, just as linguistic expressions in general).

In this section we have considerably revised the conception of "planes" in the ontological sense, which was, however, not our main objective. The latter was drawing conclusions as to focal points of integration, and ways of its realization in various fields.

VII. THE MAIN AND THE SECONDARY SCOPES OF INTEGRATION

As the norm is the "directly legal object", which is tantamount to saying that law is norms, then the study of law in the strict sense has as its subject-matter norms - conceived (comp.VI above) as decisional acts concerning addressees' conduct, and as linguistic expressions of these acts. Such statement seems to be rather commonplace, but here it is a relevant since it contradicts both the view that the law consists of four categories of equally "legal" objects (norms, psychological experiences, conduct, and values), and the view that the sole directly legal object is psychological experience (L.Petrazycki), or conduct (American legal realism), or else values as the "material ought" (natural law doctrines)(20). According to our view, the norm is the directly legal object, while the psychological experience, conduct and value are but indirectly legal objects and as such do not constitute the subject-matter of the study of law in the strict sense.

Such differentiation of "legal objects" offers some suggestions for the problem of integration of the study of law and other disciplines. The main and direct field of integration will be that of the problems of the legal norm, and so the partners of integration in this field will be the branches of science able to contribute to solving these problems. Here belong in particular: general theory of norms and meta-ethics, philosophy - and mainly philosophy of language, linguistics and logic, topica and argumentation theory, and last but not least - the theory of decisions. It is striking that contrary to the formula of "integration of the study of law with social sciences" the latter are not placed in the main field of integration. Social sciences, however, have also a place in integration - a secondary, but important one.

The starting-point of the integration in the main field as characterized above are the paradigmatically legal problems, that is to say, those typically dealt with by the study of law. Integration is meant to help in solving them. It corresponds to the above (VII) postulates as to limiting the scope of integration, as to the concentrating in integration - in the first place - on problems the study of law is peculiarly occupied with, and - which is closely connected with the latter - as to attempting at a synthesis of the contributions (theoretical and methodological) of the study of law and such contributions of the disciplines mentioned above which are relevantly related to the problems and their solutions, offered by the study of law. This integration has to take place within the scope of the study of law itself (as delimited above), and to be carried out mainly by legal scholars - which seems to be within the limits of their possibilities, as the main field of integration is narrowed. It does not exclude specialization in different directions (e.g., logical, linguistic-philosophical, etc.).

Such problems as legal-psychological, legal-sociological (or jointly: legal-psycho-sociological), legal-axiological, legal-historical, the border-line problems of law and politics and law and economics, do not belong to the scope of the study of law in the strict sense, and as integrational areas are of a different character. These problems are not specifically legal, but essentially belong to the extra-legal spheres of psychology, sociology, political science, political economy, history, and axiology. They are partly dealt with the already separated disciplines, such as (psycho-) sociology of law, axiological philosophy of law, and history of law. The main role in the integration of these territories has to be ascribed to the representatives of the respective extra-legal disciplines, in collaboration with the legal scholars, although the inspirations to research-initiatives, and also the institutional divisions, are not essential as far as the character of the research itself is concerned. E.g.,the legal history and sociology of law owes much to the inspirations of legal scholars, such as F.K.von Savigny in the first instance, and R.Pound in the second; the disciplines in question also mostly belong institutionally to law-faculties, but they are essentially history and sociology, and their research historians and sociologists (if they were even initially jurists). We want to stress once more that - in our opinion - the border-line field mentioned above, already specialized as separate disciplines, or not, are parts of psychology, sociology, political science, political economy, history, and general axiology, and not of the study of law as defined above. The role of the jurists in these scopes of integration consists in making the non-jurists aware of the peculiarities of law and in introducing the "legal point of view" into the jointly undertaken research.

It is striking that in these theses on the integration one can find some resound of the frequently critized view of H.Kelsen, who located the sphere of the so-called legal factuality outside the study of law, and offered the research in this sphere to the psychologists and/or sociologists(21). Moreover, there is here even some resound of the view of the radical protagonists of the integration of the study of law with social sciences, that the study of law (specifically conceived) would form a part of sociology. One has to observe that Kelsen, in spite of some exaggerations, had the right intuition as to the difference between the problems 'specifically legal', and legal but accessorily; and in what pronounced the radical reformers of the study of law there is also a "rational core", although not with reference to the whole study of law, but to some secondarily legal field.

In sum, after the above considerations the view can be adopted, that the hitherto results of integration in many important respects correspond to expectations, and that the integration is still an undertaking worth attempting. One has only to consider various levels of integration connected with law, and differences between integrational work in the field of the study of law in the strict sense and beyond.

NOTES

(1) Comp.L.Nowak-S.Wronkowska, "Zagadnienie integracji nauk prawnych w polskiej literaturze teoretycznoprawnej" ("Problem of integration of the study of law in the Polish legal theoretical literature"), Studia Metodologiczne, v.V, 1968.

(2) Comp., a.g., K.N.Llewellyn, "A realistic jurisprudence: The next step", Columbia Law Review, v.XX, 431,1930; Legal tradition and social science method, Brooking Institution, Essays on research in the social sciences, New York 1931; "On reading and using the newer jurisprudence" Columbia Law Review, v.XL,581, 1940.

(3) Comp. A.Malewski, O zastosowaniach teorii zachowania ("On the applications of the theory of behaviour"), Warszawa 1964, Part I: "Dezintagracje nauk spolecznych i droga prowadząca do jej przezwyciezenia" ("Desintegration of social sciences and the way towards its surmounting")p.3-29.

(4) K.Opalek, Zagadnienia teorii prawa i teorii polityki ("Problems of legal and political theory"), Warszawa 1983, Ch.VII: "Dzisiejszy stan nauk politycznych" ("The present status of political science"), p.131-148.

(5) Comp.J.Wróblewski, "Zagadnienie przedmiotu i metody teorii państw. i prawa" ("Problem of the subject-matter and method of the theory of State and law"), Państow i Prawo 11, 1961; A.Peczenik, Plaszczyzny badania prawa" ("The planes of legal research"), Państwo i Prawo 2, 1968; K.Opalek, The complexity of law and of the methods of its study, Scientia, v.CIV,1969.

(6) K.N.Llewellyn, "Law and social sciences-especially sociology", Harvard Law Review, v.LXII,1949.

(7) Comp.esp.O.Neurath, "Soziologie im Physikalismus", Erkenntnis, v.II,1931; Empirische Soziologie: Der wissenschaftliche Gehalt der Geschichte und Nationalökonomie, Wien 1931; and later, Foundations of the social sciences, Chicago 1944.

(8) Comp., e.g., L.M.Friedman-S.Macaulay(eds.), Law and the behavioral sciences, I and II ed., Indianapolis 1969, 1977; M.Rehbinder, Rechtssoziologie, Berlin-New York 1977; H.Rottleuthner, Rechtstheorie und Rechtssoziologie, München 1981, and the literature cited there.

(9) Comp.A.Podgórecki, Socjologia prawa ("Sociology of law"), Warszawa 1962; "Socjologia a nauki prawne" ("Sociology and the study of law"),Panstwo i Prawo 5-6, 1962; "Problem klasyfikacji nauk prawnych" ("Problem of classification of legal disciplines") Studia Socjologiczne 1, 1967, and the following discussion in Poland.

(10) Comp.D.Grimm (ed.), Rechtswissenschaft und Nachbarwissenschaften, v.I, Frankfurt a.M.1973, v.II, München 1976; Internationales Symposium Philosophie und Wissenschaften als Basis der Jurisprudenz, Retzhof bei Graz, May 1979 (Materials edited in: Rechtstheorie, Beiheft 1, 1979, and Archiv für Rechts- und Sozialphilosophie, Beiheft 7, 1980).

(11) Comp. J.Kowalski, Psychologiczna teoria prawa i państwa Petrazyckiego ("The psychological theory of law and State of Leon Petrazycki), Waszawa 1963, Ch.VII 3, p.180 f: K.Opalek (ed.) "Z zagadinien teorii nauki i teorii prawa Leona Petrazyckiego" ("Problems of the theory of science and legal theory of Leon Petrazycki"), Warszawa 1969 (passim).

(12) N.Luhmann, Ausdifferenzierung des Rechts. Beitrage zur Rechtssoziologie und Rechtstheorie, Frankfurt a.M. 1981.Ch.9: "Rechtstheorie im interdisziplinären Zusammenhang", gives a very different evaluation of integrative efforts. The study of law is, according to him, still in the state of "interdisziplinäre Isolierung", its interdisciplinary contacts and hopes were reasonable in the early decades of the XX century, when the production of social sciences" (...) war nach Volumen und Inhalt auch für Juristen zugänglich" - what is not the case now (p.191). Luhmann

stresses the importance of integration for legal theory but in his opinion: "Die Suche in interdisciplinären Grenzgebieten nach begrifflichen Anhaltspunkten für den Aufbau einer allgemeinen Rechtstheorie hat bisher bestenfalls Präliminarien eingetragen" (p.222).

(13) K.Opałek, Peculiarities of legal science and the problem of integration, Archivum Juridicum Cracoviense, v.I,1968 on similar phenomena in social sciences see (e.g. P.A.Sorokin, Fads and foibles in modern sociology and related sciences, Chicago 1956; E.Gellner, The alchemists of sociology, Inquiry, v.II 2,1959,p.128 f.

(14) Such views are advanced especially with regard to logic, analytical philosophy, and linguistics.

(15) As to sociology of small groups see the remarks of P.Selznick on "Jury Project", "The sociology of law" in Sociology today. Problems and prospects, New York 1960, p.125. Such approach can also frequently be observed in studies belonging to the theory of norms, and deontic logic (see, e.g. C.E.Alchourrón - E.Bulygin, Normative systems, Wien-New York 1971 (Introduction).

(16) Prediction theory in: A.R.Anderson, "A reduction of deontic logic to alethic modal logic", Mind 67, 1958, p.100 f; comp. also G.H.von Wright, "On the logic and ontology of norms", Philosophical logic, ed. by J.W.Davis, D.J.Hockney, W.K.Wilson, Dordrecht 1970, p.70 f.(see criticism in: A.Aarnio, R.Alexy, A.Peczenik, "The foundation of legal reasoning", Rechtstheorie, v.XII, 1981, p.430 f.) "ought entails can", in: G.H.von Wright, "Norm and action.A logical enquiry", London 1963, Ch.VII; comp. also R.M.Hare, Freedom and reason, Oxford 1963, p.51 f; there is a vast literature on paradoxa of deontic logic, comp. on this subject G.Kalinowski, La logique des norms, Paris 1972, Ch.III and IV.

(17) Comp. A.Delorme, "Sociologiczna a psychologiczna plaszczyzna badańia prawa" ("The sociological and the psychological plane of legal research"), Panstwo i Prawo 10, 1968.

(18) K.Opałek, "Law as social phenomenon", Archiv für Rechts- und Sozialphilosophie, V.IVII, 1971, p.47 f.

(19) J.L.Austin, How to do things with words, ed. J.O.Urmson, Oxford 1962, p.8,14 f; J.Woleński, Z

zagadnień analitycznej filozofii prawa", ("Problems of the analytic philosophy of law"), Kraków 1980, p.69 f., 89, 91 f., 93 f.

(20) Comp. K.Opalek, "Problemy metodologiczne nauki prawa") ("Methodological problems of the study of law"), Warszawa 1962, Ch.II.

(21) H.Kelsen,"Hauptprobleme der Staatsrechtslehre. Entwickelt aus der Lehre vom Rechtssatz, Tübingen 1911, p.V, 42, 343; Reine Rechtslehre, II ed. Wien 1960, Part III.

CHAPTER 8

ANALYSIS OF LEGAL NORMS AND JURISTIC PROPOSITIONS

SYNOPSIS

Lars Lindahl

This chapter deals with the detailed analysis of different kinds of legal norms and juristic propositions. The paper by Jan Evers gives analytical exposition of Karl Olivecrona's theory of legal rules and of Olivecrona's criticism of legal voluntarism. Causal or "weak" voluntarism, merely holding that legal rules are intentional and can be explained as means to ends within a social context, is distinguished from semantical or "strong" voluntarism, maintaining that the correctly interpreted content of a legal rule is identical with what is willed by the law-giver. As Evers points out, the target of Olivecrona's criticism is strong voluntarism rather than weak voluntarism, his arguments against strong voluntarism being mainly of four kinds, viz. that the theory is circular, that some versions of it are metaphysical, that the theory is in conflict with empirical facts and that it is based on an inadequate semantical analysis of imperative expressions. The central core of Olivecrona's own theory of legal rules as so-called "independent imperatives" is held to be that the content of a legal rule is fixed in a complex process of legal interpretation, not presupposing any unitary commanding subject behind the law.

While (as Evers points out) Olivecrona divides legal rules into prescriptive rules and performative rules, a more wellknown distinction is that between rules of conduct and power-conferring rules. Rules of the lastmentioned kind are the subject of Tore Strömberg's paper though the author prefers the expression "norm of competence" to "power-conferring rule". Arguing against tendencies of reductionism in legal classification, the

A. Peczenik et al. (eds.), Theory of Legal Science, 551–552.

author maintains that norms can usefully be classified by their linguistic formulation and that logical equivalence between two norms does not guarantee that the norms should be classified as belonging to the same group. Norms of competence, in particular, can be recognized by being formulated in terms saying (approximately) that a person fulfilling certain requirements "is able" or "has the competence" to bring about such and such legal effects by some specified physical act. The author agrees that a norm of this kind may be logically equivalent to another norm formulated by saying that if a person, fulfilling certain requirements, executes some specified physical act, such and such legal effects will ensue; it is maintained, however, thaty by virtue of its formulation this latter norm should be classified as a "norm of qualification" rather than as a norm of competence. The author surveys the Scandinavian literature on norms of competence and, in a final comment, maintaining that Kelsen's basic norm and Hart's rule of recognition are norms of competence in the appropriate sense, raises some doubts as to whether such "ultimate" norms could even in principle be reduced to norms of conduct.

While Evers and Strömberg deal primarily with norms and norm formulations issued in a legislative process, the paper by Riccardo Guastini is concerned rather with the analysis of some juristic sentences, in particular sentences of the two forms "the rule _r_ is valid" and "it is obligatory that _p_" as they are used in juristic discourse. The author maintains that both kinds of sentences are ambiguous. Since "rule" can refer either to a legislative sentence or to a meaning-content the assertion of a rule's being _valid_ can either mean that some specific sentence has been produced by the legislator in accordance with given secondary rules of production or it can express the acceptance of a specific meaning-content within the legal system, such acceptance being the outcome of a complex juristic process of refining and systematizing the law. A juristic assertion that some action is _obligatory_ can also be understood in various ways. When used "directly" it will express a norm of conduct, although, as the author argues, being a juristic assertion it would express what the author calls an "iterated norm". When used "obliquely" it can be understood as a descriptive report on a legislative act of commanding or as an elliptical interpretative statement on the meaning content of some legislative sentence.

KARL OLIVECRONA'S THEORY OF LEGAL RULES AS INDEPENDENT IMPERATIVES

Jan Evers

Karl Olivecrona's non-voluntaristic theory of legal rules can, of course, be described and discussed without any reference to the voluntaristic theories it was intended to replace. In his own exposition, however, criticism and construction go hand in hand, and it is easy to misunderstand parts of his theory if one does not clearly see the difference between them and other views. I will, therefore, first describe the kind of theory Olivecrona did not accept, briefly state the main reasons for and against it and then try to summarize Olivecrona's own standpoint.

The classical - and still widely accepted -(1) will-theory of norms interprets imperatives, and legal rules as a species of these, as in some sense "expressions of will". Both individual commands and general rules are thought to emanate from an <u>imperator</u> and to serve as a means for him (or them) to motivate the addressee to comply with his will (or wishes). This thesis can, however, be understood in two different ways:as a <u>causal</u> theory about the creation of rules or as a <u>semantical</u> theory about the interpretation of them. This point has too often been overlooked and is in need of some explanation.

As a <u>causal</u> theory the will-theory contains statements of the following type:

1) Rules of law are intentional, caused by voluntary acts.
2) In a social context rules of law can be explained as means to ends, the latter being human ideals, interests etc.

The falsity of the causal theory would imply that legal rules created themselves or fell down from heaven or were instituted blindly. Since all these consequences are false, so is their antecedent, which means that the causal theory is true. It is, moreover, uncontroversial - at least in the broad sense here given it - and it is not the

A. Peczenik et al. (eds.), Theory of Legal Science, 553–558.

target of Olivecrona's criticism of voluntarism. On the other hand the causal theory is without interest for the theory of interpretation.

As a semantical theory the will-theory goes much further and makes assumptions of the following kind about the meaning of rules:

3) The content (=correct interpretation) of a legal rule is identical with the content of the will of the law-giver.

This thesis is thought to hold not only at the moment when the rule is promulgated but as long as it is valid. Or,as Windscheid puts it, in interpreting a law it is necessary "sich in die Seele des Gesetzgebers hineinzudenken".

I will call adherents to the causal theory "weak voluntarists" and to the semantical theory "strong voluntarists". In traditional legal theory weak voluntarism is more or less taken for granted. Strong voluntarism is either explicitly defended or tacitly presupposed. The notion of the will of the law-giver is sometimes considered to refer to reality, sometimes to be a "necessary fiction" (Jellinek:"Der Staatswille is eine juristische Denknotwendigkeit").

There are various forms of strong voluntarism but they have - as far as I can see - two main premisses in common, which makes it possible te reduce the theory to the following syllogism:

p1: All legal rules are imperatives
p2: All imperatives are expressions of will
C: Therefore, all legal rules are expressions of will.

If the premisses are true, the consequence holds, and if it does one is forced to ask: Who is the will behind the law?

The weak part of the argument is the second premiss:- imperatives are classified as expressive(2). Very clearly one finds this view in Austin: "Every rule is a command and every command is a signification of desire". A great number of equivalent statements from other authors could be cited.

Olivecrona's objections against strong voluntarism(3) can be divided in four groups: a) logical, b) philosophical, c) empirical an d) semantical.

ad a) The main logical counter-argument is that strong voluntarism is circular. In his commentary on Bentham Olivecrona writes:

"Bentham fully realized that the lawgiving autho-

> rity must exist and possess power independently of the legal order if the law is to be conceived as the expression of its will. But did the sovereign as represented by Bentham really have such a position? Was it independent of the existing legal order?
> The sovereign was described as the sovereign <u>in a state</u>. But a state presupposes the existence of a <u>legal</u> order. For how could we speak of a 'state' in any conceivable sense of the word if there is not an organization of power? There can be no state in an unorganized mass of people. Organization implies, however, that there is a set of rules which the members of the organization feel bound to follow more or less consistently. But the rules regulating the functioning of the state organization are precisely such rules as are called rules of law. When the sovereign is conceived as being sovereign in a state, a legal order is supposed to exist. The explanation is circular." (<u>Law as fact</u>, p. 67).

The accusation of circularity is directed towards <u>all</u> forms of strong voluntarism.

ad b) Some forms of strong voluntarism (e.g. the German doctrine of the "Staatswille" or the French of "la volonté générale) are classified as <u>metaphysical</u>, and hence as neither true nor false. They can be suggestive or inspiring (as, I take it, some kind of legal poetry) but they are not theories properly so called.

ad c) In order to be able to defend some forms of strong voluntarism, their advocates are driven to make improbable assumptions about social facts. One example is Austin's (and Bentham's) view that legal rules are commands from a group of superiors towards which the population shows "habits of obedience". The theory forgets that the superiors in question hardly would be obeyed if they were not considered to be <u>legitimate</u>. It is basically people's habit of obedience towards the law that enables the rulers to rule, and not conversely.

ad d) As strong voluntarism is a theory of interpretation semantical objections are particularly important. In the syllogism given above the second premiss stated that "all imperatives are expressions of will"; that is, <u>expressive</u> sentences. It is probably true that if someone gives an order, a command, he normally <u>wants</u> that the commanded action be performed or that the state of affairs which would then result is brought about. This is, however, not at all necessary and it is not what the imperative _means_:

> "--- a wish is a common motive for issuing a command, but by no means the only possible one. In many situations commands are given for other reasons. People in a hierarchical organization, for instance, often have to give commands to their subordinates because they feel obliged to do so. Giving a certain command can even be utterly distasteful to him who has to give it. We may think of a company commander in the First World War who had to order his men to go over the top in face of the machineguns, though he personally held this particular offensive action to be senseless from a military point of view." _(Law as fact_, p.122)

It could be retorted that the commander expressed the will of a higher commander, that is _indirectly_ expressed the latters wish. Even if this _could_ be the case, it does not _have to_ be the case. To make this point more precise: If I say "Oh, if I were in London" and then go on to add: "but I have no wish to be in that city", then I am clearly paradoxical. But if I order someone: "Go to London" and then add: "but I would personally prefer that you stayed in Lund", I am by no means linguistically misleading. The first sentence ("Oh, if I were in London") is exclamatory, _expresses_ a wish, whereas the second ("Go to London") is _prescriptive_ and is only accidentally caused to be uttered by a wish. Or, to put it in yet another way: The reasoning that if A orders B to do X, then A wants B to do X, is an _inductive_ and not a _deductive_ inference.

Why, and in what sense, did Olivecrona characterize legal rules as _independent_ imperatives?

Firstly, strong voluntarism can be said to treat legal rules as _dependent_ imperatives because, according to this theory, to give a correct interpretation of a legal rule is identical with a correct understanding of the _imperators_ interpretation of the rule.

Since, secondly, Olivecrona cannot accept this view he concludes that even if - as weak voluntarism states - legal rules are intentionally created by lawgivers, - the understanding of these rules does not imply investigations into the minds of members of parliament of ministers. Parliament issues texts not commands, and these texts are treated by legal professionals according to traditional methods of contextual interpretation, where other statutes, precedents, doctrines etc are taken into account. Legal rules are semantically independent of any unitary, commanding "subject" behind the law. Their content is fixed in an interpreting process, which is very complex, and the understanding of these complicated procedures is hindered, not helped by oversimplifying models of the strong voluntarist kind.

Olivecrona further divides legal imperatives into two classes: prescriptive imperatives and performative imperatives, the former being action-imperatives (e.g. rules of penal law), the latter state-imperatives (e.g.rules conferring rights). It seems possible to reduce Olivecrona's standpoint to a counter-syllogism as follows:

p1: All legal rules are imperatives

p2: All imperatives are either prescriptive or performative.

C : All legal rules are either prescriptive or performative.

Formulated in this way the syllogism does not allow the consequential problem of strong voluntarism ("Who is the will behind the law?") to arise at all. The problem is no problem because it has no foundation in facts.

A concluding remark: Everyone who is familiar with the work of Karl Olivecrona will see that I have taken certain liberties in interpreting it. I have rearranged his historical disposition into a systematic one and introduced concepts - weak and strong voluntarism - which he did not explicitly use. I have also omitted some arguments which he might have considered to be of great importance. My excuse is that I interpret his work not only by reading his texts (although I think I know them fairly well) but also by remembering many long and detailed discussions during three decades.

NOTES

(1) A detailed account of various forms of voluntarism is given in Karl Olivecrona: Law as fact, second ed. 1971

(2) I do not find it necessary to discuss the highly implausible view - which has been held - that imperatives are true of false descriptions of wishes.

(3) Certain kinds of specified weak voluntarism are also criticized but space does not allow any discussion of these questions.

NORMS OF COMPETENCE IN SCANDINAVIAN JURISPRUDENCE

Tore Strömberg

It may perhaps be convenient to explain the use of the word competence in the above title. In Wesley Newcomb Hohfeld's scheme of fundamental legal conceptions the term competence does not appear(1). There we find instead the word power. It is very likely that in this connection the word competence sounds somewhat strange to English ears. In his work The Concept of Law the Englishman H.L.A.Hart uses the term power in the same sense as Hohfeld did(2). There he disputes John Austin's legal theory as consisting only of the commands of the Sovereign connected with a threat of something disagreeable in case of disobedience(3). Unlike Austin, Hart objects that law does not contain only rules of action but also, as an essential element, rules of a kind which he calls power-conferring rules.

Just as there could be no crimes or offences and so no murders or thefts if there were no criminal laws of the mandatory kind which do resemble orders backed by threats, so there could be no buying, selling, gifts, wills, or marriages if there were no power-conferring rules; for these latter things, like the orders of courts and the enactments of law-making bodies, just consist in the valid exercise of legal powers(4).

We observe that the power created by power-conferring rules is termed a legal power by Hart. He lays stress upon the fact that the power is founded not on fear but on a respect for authority. It is something other than the factual power that, according to Austin, makes the commands of the Sovereign into law. However, the term power is ambiguous. By a legal power I can mean among other things the legal freedom to do something that is not forbidden, e.g. to take up a bushel of water from the open sea. Both Hohfeld and Hart mean something else, namely a power to cause by an utterance of will a legal alteration, e.g. the power of the English Parliament to legislate or the power of two adult persons to enter into a contract. For this kind of legal power it may be rational to reserve the term

A. Peczenik et al. (eds.), Theory of Legal Science, 559–570.

(legal) competence.

In Scandinavian legal theory the term competence (in Danish _kompetence_, in Norwegian _kompetanse_, in Swedish _kompetens_) has become naturalized. The authors use it not only in their own language but also when they write in English. On the other hand, the Scandinavian statutes have not adopted the term.

The idea of legal competence is a fundamental ingredient in the ideology of the law and it surely has its roots far in the past. The idea of a special type of norm as the basis of legal competence does not seem to be as old and in any case it has not found a clear expression in juridical science until the modern period. Indeed, not until our own time, the norms of competence have become the object of profound logical analysis.

In the Scandinavian theory of law the norms of competence were brought to the fore in the middle of this century and they have since then enjoyed a secure a firm position in reasoning about the structure of legal norms and their logical classification. Different opinions have appeared not only on the contents and the ideal form of norms of competence but also on the need for them as an independent category of norms. In these respects a Danish, a Norwegian and perhaps also a Swedish line can be distinguished. One of the aims of this short essay is to make plain the dissimilarities in the views of the different authors. I think that the best way to do this is to describe, firstly, a certain conception as a sort of model and then to compare the opinions of each author with this model.

Law in the juridical sense determines the relations among men. That is a commonplace. Law is not possible without communication, without an exchange of thoughts, among men. Communication occurs almost exclusively with the aid of the spoken or the written language. With the help of words the content of legal norms is transmitted among the members of a society. Therefore it must be possible to express the contents of a norm in words. Yet the possibilities of the language are numerous. What from a linguistic point of view appears as two separate norms or as two separate kinds of norms is in fact sometimes the very same norm or else norms of the same kind. Thus it has become possible to construct on linguistic grounds more types of norms than is logically necessary, and this has actually occurred. So, for instance, both rules of rights and rules of duties have been created, although rules of one of the two types would have been logically sufficient (cf. Hohfeld's scheme on the fundamental legal conceptions and their mutual relations). In consequence of this the question has emerged in the theory of law as to whether and to what extent the number of existent types of norms may be "reduced". Even if we can reduce them it is not yet certain that we will do so. There may be a practical need for a distinction of types of norms which are characterized not only by their content but also by their linguistic form.

I have pointed out these circumstances because the model which I intend to use as a standard in my description of the norms of competence in Scandinavian jurisprudence is a concept partially determined by the linguistic form of the norm. Then, according to the model, whether a norm is one of competence or of another type depends partly on the form given to the norm. Is this manner of defining a type of legal norm appropriate? I think it is, and I will try to explain why.

A proposition (in the logical sense), either a statement or a norm, must be expressed by a sentence (in the linguistic sense) which is called a principal sentence or a principal clause. To the principal clause may be joined one or several subordinate clauses, but this is not necessary. Since legal norms are always conditional it is always possible to construct such a norm with a principal clause together with one or more conditional clauses. The normative, the determinative, in the norm is then always the content of the principal clause, i.e. the legal consequence. The conditional clause represents the facts which satisfy the conditions for the affirmation of the legal consequence. For a classification of norms on the basis of logical criteria the legal consequence is more relevant than the legal facts. A division based on the nature of legal facts is also to be found, namely the division of the legal system into separate parts, such as the law of contracts, the law of torts, and so on. This division is in fact very important, and it has its counterpart in the division of legal science into separate disciplines. However, a competence is a legal consequence of a certain situation and a distinction between norms of competence and other kinds of norms ought therefore to be based on the nature of the legal consequences.

From this starting point I choose as a model for the

concept "norm of competence" a norm determining that <u>a competence</u> arises or exists on certain conditions. As a linguistic phenomenon the norm is then recognized inasmuch as the principal sentence contains one or more words denoting the idea of competence. If it does not do so, then, according to the model, the norm is not one of competence.

Let us select an example of a relatively simple norm of competence: If a person is ordained as a minister in the Swedish Church he acquires a competence - in keeping with the marriage ritual of the Church - to marry a man and a woman. In this norm three essential criteria can be distinguished:

1) The <u>conditional</u> sentence contains the claim that the competent person shall belong to a certain class of person (clergymen in the Swedish Chuch).
2) The <u>principal</u> sentence allots to the person a special kind of competence (the capacity of creating a matrimonial bond between a man and a woman).
3) The <u>principal</u> sentence makes the competence dependent on the execution of a defined physical act (the pronunciation of certain words in a certain situation).

In short, the norm of competence says that a certain person is able to bring about a certain legal effect by executing a certain physical act. Even more briefly it can be said that the norm decides <u>who</u> can produce <u>what</u> and <u>how</u> he can do it.

In most statutory rules nothing is said about the manner in which the competence ought to be exercised. It is then understood that the competence is exercised by an utterance that the legal effect takes place, a linguistic act that in modern logical terminology is called a performative(5). A Swedish statutory rule, for instance, that gives to a judge the competence to appoint an administrator of a bankrupt party's estate, does not tell how that shall be done. Everybody finds it self-evident that the judge shall do this by saying or writing that he appoints Mr X as an administrator of the estate of the bankrupt Mr. Y. The requisite procedure may also be a complicated and minutely regulated one in other respects, as it is in the enactment of legislation by a parliament or in the election of its members. In these cases the procedure is determined not by the rule which established the competence but by other rules. The above mentioned marriage ritual is also a complex procedure. In this ritual some steps are obligatory for the validity of the act, whereas others are irrelevant from this point of view.

It is necessary that a competence is exercised only by the execution of a defined act which can easily be distinguished from other kinds of acts. Otherwise it would be impossible to determine, for instance, whether a man and a woman are spouses or not. Matrimony itself cannot be observed but the facts by which matrimony is considered to have arisen can be observed. Consequently the requirement that a defined procedure be observed is an obligatory element in a complete norm of competence. The fact that prescriptions as to competence in codes and statutes are seldom complete is a different matter. It may even happen that such a prescription is absent altogether or, in other words, is tacitly presupposed.

In the example of a norm of competence such as the one chosen above the requisite procedure is placed in the principal sentence as an element of the legal consequence. Yet there are authors who conceive of the procedure as a condition for exercising the competence, and for this reason it is natural to ask whether the correct place for the procedure is not perhaps in a conditional sentence rather than in the principal sentence.

Let us put the procedure in a subordinate sentence and see what the result is. The norm may then have the following form: If a person is ordained as a minister in the Swedish Church and if - in keeping with the marriage ritual of the Church - he declares a man and a woman to be husband and wife, - - - - - ? What will the rest be like? Apparently the legal consequence will be that the man and the woman become husband and wife.

If we compare the norm expressed in this way with the supposed model for a norm of competence, we find that the legal consequence is not the same. It does not contain the idea of competence and it is according to the model therefore no norm of competence. Of course, for all practical purposes it functions in the same way as a norm of competence and it is thus equivalent to such a norm. But what are we to call the norm in its changed linguistic form? I have called imagined qualities such as matrimony or spouse, legal qualities, and in consequence I call the norm <u>a norm of qualification</u>. The legal consequence in the norm is the emergence, the coming into being of a legal quality.

Without changing the import of the norm as a guide to human behaviour a norm of competence can evidently be transformed into a norm of qualification, provided that the two kinds of norms are defined in the manner sketched above.

Apparently this applies to all norms of competence. The transformation can no doubt also be carried out in the opposite direction. In substance, the norm is the same independently of the transformation. This is no more peculiar than that a statement can be expressed in different terms. It is true that from a logical point of view there is no hindrance to calling a norm of competence whether it appears in the one form or the other. That I nevertheless prefer not to do so is owing to the fact that there are many norms of qualification which cannot be transformed into norms of competence. The definition of these would be unnecessarily circumstantial if it were to exclude those norms which substantially coincide with norms of competence. Norms of qualification which cannot be transformed into norms of competence are, e.g., the following: If a child is in the common custody of its parents, both of them are its trustees (the Swedish Code of Parenthood, ch.11, sec. 1). - A building is an accessory to the landed property (and is consequently real estate; the Swedish Code of Land, ch.2, sec.1). - A purchase made between two merchants in their profession is a purchase in trade (the Swedish Sale of Goods Act, sec.4).

It is true that these other norms of qualification could in principle be rendered superfluous by incorporating their contents with other norms which are connected with them, but this would be extremely unpractical, and no legislator would entertain such an absurd idea. By sacrificing the term norm of qualification we would lose a valuable instrument for the classification of legal rules.

But now _in medias res!_ In Scandinavian jurisprudence the first fairly complete theory about norms of competence was shaped by the Danish scholar Alf Ross (1899-1979). The main sources of his views on these norms are his now classical works _On Law and Justice_(6) and _Directives and Norms_(7). In the latter he described the norms in question as follows:

Competence is the legally established ability to create legal norms (or legal effects) through and in accordance with enunciations to this effect. Competence is a special case of power. Power exists when a person is able to bring about, through his acts, desired legal effects.

The norm which establishes this ability is called a norm of competence. It states the conditions necessary for the exercise of this ability. These conditions usually fall into three groups: (1) those which prescribe what person (or persons) is qualified to perform the act which creates the norm (personal competence); (2) those which prescribe the procedure to be followed (procedural competence); and (3) conditions which prescribe the possible scope of the created norm with regard to its subject, situation and theme (substantial competence)(8).

Ross's statement that a norm of competence establishes a competence seems to agree with the supposition in the above model that the norm is characterized by its legal consequence, which is the origin of a legal competence. Even so, the later part of the quotation arouses uncertainty as to Ross's conception of the logical structure of the norm. It prescribes conditions,he declares, for the exercise of the competence (i.e. not for its origin), and the conditions fall into three groups. Only the first of these groups ("personal competence") is to be found as a condition in my model. The two others have been relegated to the legal consequence.

If we go back to Ross's analysis of the norms of competence in On Law and Justice, we can understand more clearly why it is difficult, even impossible, to link up his conceptions of these with the scheme of conditions-legal consequences. According to Ross the norm of competence is properly understood as an indirectly expressed norm of conduct. It is directive to the effect that such directives as have been created in conformity with a declared mode of procedure shall be regarded as norms of conduct(9). This means, for instance, that the norm on the legislative power tells me that I am to obey the statutes promulgated in virtue of this power. Thus, as a norm of conduct the norm of competence has no content of its own, but rather borrows its content from one or more other norms; it becomes what we might call a supernorm. Ross also meant that the norms of competence can in this way be reduced to norms of conduct(10). This opinion is in harmony with his extreme realism and his identifying of the legal order with a social reality served by the ideology of law only as a "scheme of interpretation"(11).

However, a norm of competence according to Ross contains a pattern of conduct in another sense, and he therefore called it also a norm of procedure(12). "Secondly, it must be observed that any well-developed legal system, being institutional and dynamic, contains not only _norms of conduct_, which prescribe how to act, but also _norms of competence_, which provide how new valid and binding norms may be created through the performance of _actes juridiques_"(13)

Ross' views on the norms of competence have had some influence on Danish as well as Norwegian authors. Like Ross, the Norwegian scholar Torstein Eckhoff makes a trisection though not, as Ross did, of competence as such, but of the questions concerning competence, which he "after the pattern of Ross" terms personal, material and procedural questions(14). He conceives of the norm of competence as a combination of claims, but here he is more lucid than Ross in that he regards the claims not as conditions _on the exercise_ of the competence but _on the origin of the validity of a legal act_(15). Yet he offers some reservation on his thesis. As to substance, Ross and Eckhoff probably meant the same thing, but by his choice of words Eckhoff makes clear that a person can have a legal competence without being entitled to exercise it either at all or beyond certain limits. In this case the validity of a legal act is not excluded by the fact that the competent person was not allowed to make use of his competence to bring about the legal effect.

Among Scandinavian authors, the Norwegian Nils Kristian Sundby (1942-1978) is the one who has treated the norms of competence most exhaustively. In his great treatise _Om normer_ (On Norms)(16) he devoted about 100 pages to the concept of competence and to the norms of competence. The pattern from Ross can be discerned also in Sundby. The latter made use of a trisection, not of the conditions for the exercise of a competence nor of the questions concerning competence, but of the norms themselves, which, employing a somewhat divergent terminology, he divided into personal, material and formal norms of competence(17). Like Eckhoff, he conceived these norms as prescriptions of conditions for the validity of a legal act(18). He added that the norms are in a certain sense thereby provided with a sanction, namely nullity. This sanction enters when someone does not "obey" a norm of competence and thus "breaks" it. Sundby also differed from Ross in that he set up norms of qualification as a special category of norms. He even devoted a chapter of 40 pages to this category. He included in this chapter the norms of competence, which constituted, as he saw it, norms of qualification of a special type: "They are a sort of norm of qualification, namely norms determining the linguistic actions which are to be accepted as valid acts of creating norms"(19).

Sundby's charactization of the norms of competence as conditions for validity seems to suggest that he regarded the linguistic form as insignificant for the concept of norm of competence, and that irrespective of its linguistic form, such a norm was thus at the same time a norm of competence and a kind of norm of qualification.

The coupling together of competence and legal validity by Eckhoff and Sundby - which seems to me to be quite correct - has caused some difficulty in cases where a pretended transgression of the limits of a competence does not result in an absolute nullity but only in the possibility for a party to get an act nullified on due demand, and also in some exceptional cases where not even this possibility exists. In my opinion these problems are due to the fact that the texts of codes or statutes are not always logically satisfactory.

Not surprisingly Ross' influence has been greater in Denmark than in Norway. The Danish scholars Preben Stuer Lauridsen and Sven Gram Jensen both share Ross's opinion that the legal system can be exhaustively described in terms of norms of conduct and norms of competence, and they find no use for the norm of qualification concept. Lauridsen criticises Sundby's and my own classifications of norms and declares the norm of qualification concept to be superfluous. Concurring with Ross he claims that norms of competence can be reduced to norms of conduct (norms of duty)(20). Since Gram Jensen, like Lauridsen, does not use the concept of qualification, he does not consider the form to be significant for the designation of a norm as a norm of competence. Transformed into a norm of qualification (according to my terminology) the norm is in the eyes of Gram Jensen still a norm of competence. Indeed, he takes the matter one step further, considering this form of the norm to be its most correct expression(21). From a logical point of view there is, as far as I can see, no objection to that.

The question just touched upon - whether norms of competence may be reduced to norms of conduct - is in Scandinavian jurisprudence somewhat unclear. The question is properly seen as two questions, which are not always kept apart. One question is whether a single norm of competence can be made superfluous by replacing it by a norm of conduct. In many cases the answer must be in the affirmative. Assume that a rule gives to all ordinary policemen the right (=the competence) to order drivers of motorcars to stop for certain checks. The same normative and practical result can be attained by a rule obliging drivers to stop on an order from a policeman. Yet the latter rule presupposes a rule of the legislators' competence to give prescriptions to drivers. Can this other rule of competence in its turn be replaced by a rule of conduct? And so on. Finally we are thus confronted with the other question, i.e. whether all norms of competence in a legal system can be eliminated.

A declaration by Alf Ross indicates that he answered this question in the affirmative: "Furthermore, any norm of competence may be transcribed as a norm of conduct, whereas the converse does not hold"(22). The Swedish scholar Karl Olivecrona (1897-1980) touched upon the same question, but he was not so categorical and did not take a definite position: "It is perhaps thinkable to give to all legal rules an equally simple form and thus express them as rules of action and nothing else"(23). Among other authors, Sundby shares the opinion of Ross(24), whereas Eckhoff seems to be a little doubtful(25).

The problem of a total reduction of norms of competence leads us to Hans Kelsen's theory of a basic norm and to H.L.A.Hart's theory of a rule of recognition(26). Is it necessary to presuppose such an ultimate norm and - if it is - is it necessary that this norm be a norm of competence? Is it otherwise possible to conceive of a legal system as being valid and its rules as being binding in the usual sense among lawyers and among people in general? Is it not necessary that the ability to perform the action logically and temporally precede the execution itself of the action? If God created the universe, are we not inclined to imagine that he possessed the power to do so beforehand? Can we deny this question without abolishing the idea of power in our thinking? And are we willing to do that?

To sum up: There are various views on norms of competence in Scandinavian jurisprudence(27). I think, however, that all authors agree with the opinion that this type of rule is an important instrument in the hands of the legislator and an indispensable element in the organisation of society. In legal science and in legal education it is also indispensable. To examine this instrument and to improve it are tasks worthy of the general theory of law.

NCTES

(1) Wesley Newcomb Hohfeld, Fundamental Legal Conceptions, New Haven 1923.
(2) H.L.A.Hart, The Concept of Law, Oxford 1961.
(3) See John Austin The Province of Jurisprudence Determined (1832), edited by H.L.A.Hart, London 1954.
(4) Hart, op.cit., p.32.
(5) On the notion of performative, see J.L.Austin, How to Do Things with Words, Cambridge, Mass. 1962. See also Karl Olivecrona, Law as Fact, 2nd ed. London 1971, pp. 217 ff. (Legal Performatives).
(6) Alf Ross,Om ret og retfaerdighed, Copenhagen 1953 and 1966. (Also published in English: On Law and Justice, London 1958).
(7) Alf Ross, Directives and Norms, London 1968.
(8) Ross, Directives and Norms, p.130.
(9) Ross, Om ret og retfaerdighed, p.45 (On Law and Justice, p.32).
(10) Ross, op.cit., p.197 (p.162), Directives and Norms, p.120.
(11) Ross, Om retog retfaerdighed, pp.26 f., 41 (On Law and Justice, pp.17 f., 29).
(12) Ross, op.cit., pp.45, 251 (pp.32, 207).
(13) Ross, Directives and Norms, p.118.
(14) Torstein Eckhoff, Rettskildelaere (The Doctrine on Sources of Law), 3rd ed. Oslo 1980, p.39.
(15) Op.cit., pp.37-39, 45 f. See also Torstein Eckhoff and Nils Kristian Sundby, Rettsystemer (Systems of Law), Oslo 1976, pp.99, 103.
(16) Nils Kristian Sundby, Om normer, Oslo 1972.
(17) Op.cit., pp.358-62.
(18) Op.cit., pp.357, 363.
(19) Translation from op.cit., p.151.
(20) Preben Stuer Lauridsen, "Om jus og normer" (On Ius and Norms), in TfR 1978, pp.123 f.
(21) Svend Gram Jensen, Hvad er retfaerdighed? (What is Justice?), Copenhagen 1983, pp.21, 28 f.

(22) Ross, *Directives and Norms,* p.120.

(23) Translation from Karl Olivecrona, *Rättsordningen, Idéer och fakta* (The Legal Order. Ideas and Facts), Lund 1966, p.165.

(24) Sundby, *Om normer*, pp.393-96.

(25) Eckhoff and Sundby, *Rettssystemer*, p.106.

(26) Cf.Eckhoff, *Rettskildelaere*, p.39.

(27) The supposed "model" in this paper is in accordance with the representation of norms of competence in the following two works of the present author:
Tore Strömberg, "Om kompetensnormens definition" (On the Definition of the Norm of Competence), in *Festskrift tillägnad Ivar Agge* (Writings in Honour of Ivar Agge), Stockholm 1970.
Tore Strömberg, *Inledning till den allmänna rättsläran* (An Introduction to the General Theory of Law), 8th ed. Lund 1981.

A TENTATIVE ANALYSIS OF TWO JURISTIC SENTENCES

Riccardo Guastini

0. No doubt, one of the most important tasks for a descriptive theory of legal science is the analysis of actual juristic sentences. However, in juristic discourse (legal doctrine, or legal dogmatics) sentences of different kinds are met. To which among them is it proper to pay special attention? In the existing literature reductionist tendencies sometimes prevail. On the one hand, juristic discourse is often reduced to a set of sentences of just one kind only. On the other hand, each author usually ascribes to these sentences a univocal meaning by means of a stipulation.

For instance, Kelsen states that most juristic sentences (Rechtssätze) are deontic sentences (Sollsätze). However, it is not entirely clear whether a juristic deontic sentence is to be understood either as a validity-judgment, or as an interpretation of human behaviour through norms(1). Hart seems to suggest that juristic sentences are but internal statements of law, understood as propositions(?) about the existence or validity of rules(2). Alchourrōn and Bulygin, just like Von Wright, picture juristic sentences (although not all of them) as norm-propositions, viz. puzzling propositions "about" norms(3). On the other hand, according to von Wright norm-propositions are norm-formulations, i.e. deontic sentences(4). On the contrary, to Niiniluoto norm-propositions are statements about some social state of affairs(5). According to Ross actual juristic sentences are to be understood as a mixture of propositions (predictions) about the law in force and directives de sententia ferenda(6). Finally, according to Aarnio most juristic sentences are interpretative statements, namely accounts of the "contents of legal norms". However, he suitably distinguishes among "norm-propositions", "norm-contentions", and "norm-recommendations"(7).

My point is that no reductionism is justified. As a matter of fact, jurists utter (write) sentences of several

A. Peczenik et al. (eds.), Theory of Legal Science, 571–581.

kinds. And, above all, each of these sentences is not provided with one, univocal, meaning, but can be used (and indeed it is used) in a lot of different ways.

In the present paper, I shall confine myself to propose a tentative analysis of two distinctive juristic sentences as used by continental jurists. Their standard form is the following:

(1) The rule _r_ is valid

(2) It is obligatory that _p_

However, I dare not hope my analysis to be exhaustive. As a matter of course these two sentences are but a small part of actual legal dogmatics.

1. THE SENTENCE

(1) The rule _r_ is valid

obviously is a sentence of the second-order language (meta-language). It bears upon an entity of legislative language: i.e., a rule, which amounts to its object-language. However, the meaning of (1) is equivocal. One could think that the ambiguity of (1) essentially depends on the unclear meaning of the word 'valid', and it partly does. Nevertheless, I think it is more fruitful to begin the analysis from the word 'rule', the meaning of which is not less unclear. Indeed, in my opinion, the meaning of 'valid' is but a dependent variable of the meaning of 'rule'(8).

The word 'rule' has a lot of senses, but two of them seem to be especially relevant in this context, and it is important to keep them distinct.

(a) Sometimes jurists use the word 'rule' to denote the sentences of legislative discourse. In this connection, however, I shall substitute the word 'rule' by the phrase 'normative sentence'.

(b) Sometimes, on the contrary, jurists use the word 'rule' to denote not just the legislative sentences, but their meaning contents. In such a connection, I shall substitute 'rule' by the term 'norm'.

Now, in the context of (1), the word 'rule' is used sometimes in the first sense, and sometimes in the second one. As a consequence, the sense of 'valid' changes as well.

1.1. When 'rule' is used in the first sense (as normative sentence), then (1) could properly be re-written in the following form:

(3) The normative sentence S is valid.

This clearly means that the normative sentence S has been produced by the proper subject according to the proper procedure(9). In other words: S has been produced in accordance with given secondary norms of production (i.e., norms about norm-creating acts). Therefore, in the context of (3), 'valid' means: produced in accordance with norms of a superior degree. With regard to this, we could speak of 'formal validity'. Formal validity is a property not of norms, but of normative sentences(10).

The sentence (3), if understood this way, expresses a statement on the relationship between a given behaviour (a human act) and a norm (or set of norms) of production(11). The behaviour at issue is a speech act: one of those speech acts which are collectively designated by means of the word 'legislating'. You also could say that (3) is a "constative" about the "happy" performance of a speech act(12).

Maybe one would say that (3) is a proposition in the logical sense of the term, namely an entity of which you could say that it is true or false. But I am not quite sure of this. The production of the normative sentence S is a matter of fact. But the norms which regulate such a normative act (hence its "happiness", too) are a matter of interpretation. Thus, the utterance of (3) presupposes both the establishing of certain facts, and the interpretation of certain normative sentences: viz., those normative sentences which regulate norm-creating acts.

1.2. When the word 'rule' is used in the second sense (as norm), the analysis of (1) becomes harder. First of all, two possible situations have to be distinguished. Jurists sometimes refer to a norm which is expressed by a legislative sentence: I shall name this 'explicit norm'. Sometimes, on the contrary, they refer to a norm which is not expressed by any legislative sentence, but is held to be implied in (and/or entailed by) either a combination of legislative sentences, or a combination of explicit norms: I shall name this 'implicit norm'.

In the first instance, (1) could be re-written this way:

(4) The explicit norm $\underline{N}_e$ is valid

Such a sentence means that $\underline{N}_e$ is consistent with a given set of already accepted norms.

In the second instance, (1) could be re-written in the following way:

(5) The implicit norm $\underline{N}_i$ is valid

Such a sentence means that $\underline{N}_i$ can be derived from a given set of normative sentences or norms.

In these contexts, the word 'valid' (as a predicate of norms, not of normative sentences) is a tool to express the acceptance of a norm as an element of a normative system(13). In this connection, one could speak of 'material validity'. In accordance with the political doctrine which states that the law ought to be consistent, an explicit norm is accepted inside a system if it is not in conflict with other already accepted norms. In accordance with the philosophical thesis which states that normative (legislative) discourse is provided with logical properties analogous to cognitive discourses, an implicit norm is accepted inside a system if it can be derived (by means of some logical or quasi-logical procedure) from the already accepted norms(14).

As used this way, (1) is no factual sentence, the truth of which can be controlled with any observational procedure. On the contrary, (1) is a part of a complex juristic discourse which, as a whole, amounts to a work of refining and systematizing the law. This paper is not the proper place to attempt a deep analysis of such juristic activities. Anyway, I shall suggest a possible direction for inquiry. 'Refining the law' denotes the activity of handling a given set of explicit norms in order to draw out of them (other) implicit norms. 'Systematizing the law' denotes the activity of accepting or refusing some explicit norm as an element of a system, which is claimed to be consistent.

Refining and systematizing the law are something different from the plain truth-functional description (on a meta-linguistic level) of legislative discourse. They rather are direct interventions of jurists inside the legislative discourse itself(15). In other words, by refining and systematizing the law, jurists depart from the descriptive meta-language of science, and act as legislators(16). They do not describe, but transform legislator's language(17).

2. THE AMBIGUOUSNESS OF THE SENTENCE

(2) It is obligatory that p

mainly depends on the fact that it is susceptible of two uses: a direct use, and an indirect, or "oblique", use (inverted commas use)(18). The importance of such a distinction is a matter of course: the direct use of (2) has a prescriptive function, while the indirect one, as a general rule, has a descriptive function.

2.1. When used directly, (2) is no meta-linguistic sentence about a norm. On the contrary, it is a sentence which directly bears upon behaviour (the behaviour which is described by p). When used directly, (2) expresses a norm of conduct: the norm to the effect that it is obligatory that p. The juristic use of (2), however, is clearly different from the legislative one, which is prescriptive too. Such a difference, perhaps, can be clarified by saying that, in the direct juristic use, (2) repeats or iterates a legislative norm. It expresses an iterated norm(19).

Some phrases equivalent to 'iterated norm' are present in literature with at least two different meanings. In a first sense, 'iterated norm' can be understood as second-order norm: a norm which deals with (other) norms or norm-creating acts(20). In a second sense, 'iterated norm' can be understood as repeated norm. In the present paper, I do use 'iterated norm' in this second sense, as referring to the phenomenon of repeating a pre-existing norm. I also could say: an iterated norm is a new utterance of a normative sentence which was already uttered by the legislator. Hence, an iterated norm is no second-order norm: it is a norm of conduct.

But why should a jurist iterate a pre-existing legislative norm? In my opinion, the iteration of a norm expresses both a normative attitude(21) of approval towards that norm, and a piece of advice to the effect that such a norm ought to be obeyed or fulfilled. After all, you also could say that (2), if understood this way, is a hiddenly meta-linguistic sentence, the explicit form of which should be the following:

> (6) You ought to obey the norm to the effect that it is obligatory that *p*

However, (6) expresses a meta-norm ('You ought to obey'), the normative contents of which are but the contents of the object-norm itself ('It is obligatory that *p*').

2.2. The sentence

> (2) It is obligatory that *p*

is susceptible not of just one oblique use, but of (at least) three oblique uses.

2.2.1. Sometimes jurists use (2) as a synonym of

> (7) The legislator has commanded that *p*

Indeed, legislative discourse can be easily represented as a set of utterances which perform directive speech acts. Some of such acts are: commanding, prohibiting, authorizing, permitting, derogating(22).

Well then the sentence (7) is a meta-linguistic sentence provided with a descriptive meaning. It just describes the performance of a speech act by the legislator. (Here is no explicit reference to the validity or "happiness" of such an act.) Hence, (7) is a factual sentence, the meaning of which is a proposition in the logical sense of the term.

2.2.2. Sometimes jurists use (2) as an elliptical way of saying

> (8) The normative sentence *S* means *M* (*M* = It is obligatory that *p*)

However, on its turn, the sentence (8), which is the standard form of an interpretative sentence, can be used in two different ways.

First of all, the utterance of (8) can be the performance of an interpretative speech act(23). The act of interpreting a sentence amounts to ascribing a definite meaning to that sentence. The explicit performative form of (8) is the following:

> (9) I ascribe to S the meaning M (M = It is obligatory that p)

It is a stipulation, which presupposes a set of evaluations and choices, since no sentence (especially no normative sentence) is provided with just one univocal meaning, so that people cannot simply take knowledge of it(24).

Therefore, the utterance of (9) does not belong to the descriptive use of language. In this connection, one could better speak of an 'ascriptive' use of language. Indeed, a clear-cut definition of 'ascriptive language' seems to be quite hard. Anyway, a tentative definition could be the following(25).

First: whereas descriptive and prescriptive languages can refer either to linguistic or non-linguistic entities (states of affairs), ascriptive language refers only to linguistic entities, such as 'meanings', 'rights', 'obligations', and so on(26).

Second: whereas descriptive and prescriptive languages cannot affect or change the state of affairs to which they refer, ascriptive language "puts into being" the entities to which it refers. Describing is to say that a state of affairs actually is the case. Prescribing is to command that a state of affairs be put into effect. Whereas ascribing is to "create" states of affairs of a special (linguistic) kind(27).

To ascriptive language, in this sense, belong stipulative definitions, interpretative sentences, some legal rules (which can hardly be reduced to prescriptive sentences), and so on.

2.2.3. Sometimes jurists use (2) or (8) not to ascribe some meaning to a normative sentence, but to report that, as a matter of fact, a meaning so-and-so has been ascribed by someone to a normative sentence. In this sense, (8) is a constative, and could be re-written as follows:

> (10) The normative sentence S has been interpreted in the sense M (M = It is obligatory

that p)

The sentence (10) is a descriptive one. It does not perform, but describes an interpretative speech act. Whereas interpretation (ascription of meaning to a sentence) is something analogous to stipulative definitions, the description of other people's interpretation is quite similar to lexical definitions. Therefore, (10) is a proposition in the logical sense of the term.

3. The preceding analysis was designed to draw out some features of the peculiar intricacy of actual juristic discourse.Summing up the results of such an analysis, no less than seven juristic sentences can be singled out.
(a) Sentences which describe the actual performance of a legislative speech act (commanding, etc.).
(b) Sentences which state the happiness of such a performance (i.e., which state the formal validity of the uttered normative sentence).
(c) Sentences which ascribe a meaning to a normative sentence (meaning-ascribing sentences).
(d) Sentences which describe the meaning actually ascribed to a normative sentence (meaning-describing sentences).
(e) Sentences by which an explicit norm is accepted as a member of a normative system.
(f) Sentences by which an implicit norm is accepted as a member of a normative system.
(g) Sentences which iterate a pre-existing norm.

NOTES

(1) H.Kelsen, Reine Rechtslehre, Deuticke, Wien, 1960, ch.III, sect.16; H.Kelsen, "Value Judgments in the Science of Law", 1942, in H.Kelsen, What Is Justice? Justice, Law, and Politics in the Mirror of Science, University of California Press, Berkeley and Los Angeles, 1957, ch.VIII.
(2) H.L.A.Hart, The Concept of Law, Clarendon, Oxford, 1961, ch. VI, sect. 1, and ch.4, sect. 2.
(3) C.E.Alchourrón, E.Bulygin, Normative Systems, Springer, Wien and New York, 1971, ch.VII; C.E.Alchourrón, E.Bulygin, Introducción a la metodología de las ciencias jurídicas y sociales, Astrea, Buenos Aires, 1974, ch.VII.

(4) G.H.von Wright, Norm and Action, Routledge & Kegan Paul, London, 1963, esp.ch.VII.

(5) I.Niiniluoto, "On Truth and Argumentation in Legal Dogmatics", in Rechtstheorie, Beiheft 2, 1981; I.Niiniluoto, "On the Truth of Norm Propositions", in Rechtstheorie, Beiheft 3, 1981.

(6) A.Ross, On Law and Justice, Stevens, London, 1958, pp.44 ff.

(7) A.Aarnio, Philosophical Perspectives in Jurisprudence, Acta Philosophica Fennica 36, Helsinki, 1983, ch.VIII. A lot of important (non-reductionist) suggestions can be found in A.Wedberg, "Some Problems in the Logical Analysis of Legal Science", in Theoria, 17, 1951.

(8) Cf.A.G.Conte, "Studio per una teoria della validità, 1970, in R.Guastini (ed.), Problemi di teoria del diritto, Mulino, Bologna, 1980.

(9) Cf.J.Wróblewski, "Verification and Justification in the Legal Science", in Rechtstheorie, Beiheft 1, 1979; J.Wróblewski, "Tre Concetti di validità", in Rivista trimestrale di diritto e procedura civile, 36, 1982; A.Aarnio, Philosophical Perspectives in Jurisprudence, cit., ch.VII.

(10) Cf.R.Guastini, "I giuristi alla ricerca della scienza. Rileggendo Bobbio", to be published in Revista de ciencias sociales (Valparaíso, Chile), 1984.

(11) Cf. J.Wróblewski, "Statements on the Relation of Conduct and Norm", in Études de logique juridique, IV, 1970; K.Opałek, "Norm and Conduct. The Problem of the 'Fulfillment' of the Norm", in H.Hubien (ed.), Le raisonnement juridique, Bruylant, Bruxelles, 1971; K.Opałek, "Directives, Optatives, and Value Statements", in Études de logique juridique, VII, 1978.

(12) Cf.K.Opałek, "Les normes, les énoncés sur les normes, et les propositions déontiques", in Archives de philosophie du droit, 17, 1972; K.Opałek, "'Doing Things with Words' and the Law", in Anuario de filosofía del derecho, 17, 1974.

(13) Cf.A.Aarnio, Philosophical Perspectives in Jurisprudence, cit., ch.VII.

(14) Cf.C.E.Alchourrón, E.Bulygin, Normative Systems, cit., esp. pt.II; A.Aarnio, On Legal Reasoning, Turun Yliopisto, Turku, 1977, ch.III, sect.4; A.Aarnio, Denkweisen der Rechtswissenschaft, Springer, Wien und

New York, 1979, ch.4.

(15) R.Guastini, "I giuristi alla ricerca della scienza", cit.

(16) Cf.Z.Ziembiński, "Le langage du droit et le langage juridique. Les critères de leur discernement", in _Archives de philosophie du droit_, 19, 1974.

(17) Cf.A.Peczenik, "Non-Equivalent Transformations and the Law", in _Rechtstheorie_, Beiheft 1, 1979; A.Aarnio, R.Alexy, A.Peczenik, "The Foundation of Legal Reasoning", in _Rechtstheorie_, 12, 1981; A.Peczenik, _The Basis of Legal Justification_, Lund, 1983; R.Guastini, "I giuristi alla ricerca della scienza", cit.

(18) R.M.Hare, _The Language of Morals_, Clarendon, Oxford, 1952, _passim_.

(19) U.Scarpelli, "Le 'proposizioni giuridiche' come precetti reiterati", in _Rivista internazionale di filosofia del diritto_, 44, 1967.

(20) T.Mazzarese, "Metanorme e linguaggio deontico: un'analisi logico", in _Materiali per una storia della cultura giuridica_, 12, 1982.

(21) Cf. R.Guastini, "Metateoria degli atteggiamenti normativi", in _Materiali per una storia della cultura giuridica_, 12, 1982; R.Guastini, _Lezioni di teoria analitica del diritto_, Giappichelli, Torino, 1982, pt.I, ch.2.

(22) H.Kelsen, _Allgemeine Theorie der Normen_, Manz, Wien, 1979, p.1. On directive speech acts cf.J.R.Searle, "A Taxonomy of Illocutionary Acts", 1975, and "Indirect Speech Acts", 1975, both in J.R. Searle, _Expression and Meaning_, Cambridge U.P., Cambridge, 1979.

(23) Cf.A.Aarnio, _On Legal Reasoning_, cit., sect. 4.2; A.Aarnio, _Philosophical Perspectives in Jurisprudence_, cit., ch.VIII.

(24) Cf.J.Wróblewski, _Meaning and Truth in Judicial Decision_, Juridica, Helsinki, 1983 2, esp.ch.s. II and IV; J.Wróblewski, "L'interprétation en droit: théorie et idéologie", in _Archives de philosophie du droit_, 17, 1972.

(25) Cf. H.L.A.Hart, "The Ascription of Responsibility and Rights", 1949, in H.L.A.Hart, _Essays in Jurisprudence and Philosophy_, Clarendon, Oxford, 1983.

(26) Such a suggestion can be read in K.Opałek, "The Problem of 'Directive Meaning'", in _Festskrift till Alf Ross_, Juristenbundets forlag, København, 1969;

K.Opałek, "On the Logical-Semantic Structure of Directives", in *Études de logique juridique*, IV, 1970.

(27) Cf.A.G.Conte, "Aspekte der Semantik der deontischen Sprache", in A.G.Conte, R.Hilpinen, G.H. von Wright (eds.), *Deontische Logik und Semantik*, Athenaion, Wiesbaden, 1977; A.G.Conte, "Konstitutive Regeln und Deontik", in E.Morscher, R.Stranzinger (eds.), *Ethik. Akten des Fünften Internationalen Wittgenstein-Symposiums*, Hölder-Pichler-Tempsky, Wien, 1981; G.Carcaterra, *Le norme costitutive*, Giuffrè, Milano, 1974; G.Carcaterra, *La forza costitutiva delle norme*, Bulzoni, Roma, 1979; M.-E.Conte, "La pragmatica linguistica", in C.Segre (ed.), *Intorno alla linguistica*, Feltrinelli, Milano, 1983; R.Guastini, "Teorie delle regole costitutive", to be published in *Rivista internazionale di filosofia del diritto*, 1984.

CHAPTER 9

LOGICAL AND PREFERENCE-THEORETICAL STRUCTURES IN THE LAW

SYNOPSIS

Lars Lindahl

Though all the papers of this section presuppose some acquaintance with logic, the topics discussed are widely different. The paper by Antonio Martino et al. aims at a formal model for computerized analysis of some part of a legal code. The model sought for is to meet the requirement of being operational in computer application while not (as some earlier models) presupposing consistency and completeness of the system represented. In the formal model proposed by the authors, the various legally relevant conditions and their negations, conjunctions etc. are represented by finite sequences of units and zeros, the representation obtained being a Boolean algebra isomorphic to the free Boolean algebra generated by the conditions. Since some conditions may be incompatible (this meaning that their conjunction is an empty case), the disjunction of empty cases is identified and discarded while those sequences representing non-empty cases are associated to the particular solutions provided for in the code. Taking advantage of the notions of elementary case, maximal solution and minimal solution, well known from the work of Alchourrón and Bulygin, the authors show that, within the model, situations of incompleteness and inconsistency can be identified and sometimes avoided by reformulation. The paper ends with a detailed discussion where the model is applied to those articles of the Italian Civil Code relating to the legal capacity of emancipated minors.

Ghita Holmström's paper forms part of that development where deontic logic is supplemented by a logic of action and a logic of possibility, such supplementation providing

A. Peczenik et al. (eds.), Theory of Legal Science, 583–585.

the means for combining or nesting concepts or operators of various kinds. Mrs. Holmström's paper deals in particular with the notion of practical possibility and its combination with action concepts and deontic concepts. Different types of capacity are distinguished and it is suggested how the various types can be defined within a formal language with modal operators and quantifiers. The second half of the paper deals with deontic modalizations of statements relating to a persons's capacity and it is indicated that different types of rights may emerge from various such modalizations. In the final section there is a discussion of the wellknown question whether (and in what sense) Shall implies Can. The author suggests that if it shall be that a person performs a certain action, what is implied is not that the person _has_ the capacity of performing this action but rather that it _shall be_ that he has such capacity (this meaning that the consequent is itself a deontic sentence).The author ends by proposing that the implication could preferable be seen as a "deontic implication" in the sense that we have a general principle according to which it _shall_ be that the consequent just referred to holds if the antecedent does.

Lars Lindahl's paper develops a theory for _requirements_, _urgency_ and _worth_, mainly within a preference-theoretical framework. The first half of the paper contains an informal discussion of the central notions and their relevance to moral and legal theory. It is argued that a "grading" of performances (for instance as acceptable, good, etc.) occurs in law as well as in morals, and that such grading is associated with the idea of a plurality of "levels of accomplishment" and with the idea that some features of a performance are more "urgent" or have more "worth" than other such features. The author indicates that grading according to "fixed levels" (signified by words such as "acceptable", "good" etc.) is a special case of comparative grading in general where a preference ordering among alternatives is assumed, and also suggests how a preference-theoretical framework suitable for requirements, urgency and worth and incorporating classical _deontic_ logic may be provided. In the formal part of the paper, the concepts of level requirements, urgency and worth are defined in terms of preference and for each of the concepts defined, its proper "logic" is stated. In a final section the author discusses the interrelation between arguments from urgency and arguments from worth, suggesting that in politics, law and morals, argument from "aggregate worth" is often opposed by argument from "individual urgency". It is maintained that such disputes can sometimes be resolved by taking into account that (as shown in an appendix) the notions of preference, urgency, and worth are in fact interdefinable.

AUTOMATED ANALYSIS OF LEGISLATION

Antonio A. Martino
Carlo Biagioli
Paola Mariani Biagini
Gilberto Dini
Fiorenza Socci Natali
Daniela Tiscornia

1. INTRODUCTION

This paper will illustrate the most recent developments in a research project which has been in progress over the last four years at the Istituto per la Documentazione giuridica of Florence of the Italian National Research Council. The research group, coordinated by Professor Martino, is made up of four legal researchers, three of whom also have training in linguistics and one in informatics, and by a mathematician.

Our aim is to create and evaluate an automated system capable of analysing a positive legal corpus. It is of now well known how difficult it is to "navigate" in the legal world. The proliferation of legislation has created a disorderly superimposition of norms requiring legal operators to do long and complex research, as the solution to a case requires them to identify exactly which and how many are the norms applicable, which of these are in force and how they should be placed in relation to each other.

This paper analyses the concepts of consistency and completeness applied to a simple model of the legal system.

The legislator has received proposals from many sectors which would result in a better formulation of legislation but little has yet been done to put order into the already existing legislation, that is, to create an instrument capable of "taking stock", of analysing and restructuring the law in such a way that the legislator's efforts are not frustrated by the insertion of new laws in an already "polluted" legal system.

The achievements made in the study of the logical analysis of the law provide valid methodological assistance in this direction. The identification and analysis of the logi-

A Peczenik et al. (eds.), Theory of Legal Science, 587–605.

cal relations between norms could lead to a systematic revision and a rational reorganisation of the law.

It is, therefore, necessary to create a "model" which reproduces the logical relations between norms. In other words, a "model" which carries out the logical operations implied by legal operators in their daily work of identifying norms and establishing the ties between them.

2. THE FIRST ARITHMETICAL MODEL

At the beginning of our research project, we singled out the various phases which had to be dealt with and these were then kept strictly autonomous. This subsequently allowed us to go back over some of the phases for modifying or improving them without affecting the work already done.

These phases are, as follows:

1) definiton or adaptation of a "model" for the rational reconstruction of a normative system;
2) identification of a legal corpus on which to experiment the model;
3) passing from the technical legal language of legal propositions to formal language;
4) identification of a model isomorphic to the rational structure of the normative system;
5) creation of methodologies for rendering the model operational and, in particular, for the creation of computer programs;
6) simulation of the system's data input and output.

Results of the first phases(1) have been analysed and the parts seeming the least satisfactory have come to our notice. In the first place, the operational model implied the consistency and completeness of the system i.e. the capacity for identifying contradictions and gaps was missing.

Secondly, the analysis was always carried out by referring to a single objective (consequence) and to conditions which are important for it. It was not possible to compare two actions which had some relationship between them. These two features are common to all the extant literature on this subject(2).

3. THE NEW MODEL

We will now describe the work we have been doing which enabled us to provide a new solution to the problems discussed above. Later in this paper we will present a section on the results obtained from the application of the analysis of the norms of the Italian Civil Code relating to minors emancipated by marriage.

3.1 Theoretical Assumptions

We will always make reference here to the six phases of our work in order to give clarity to our exposition.

This first phase will concentrate on choosing and adapting a model which is not only capable of providing deontic solutions to a certain number of situations or cases but which is also capable of providing information about the consistency and/or completeness of a system of norms. For these concepts we have referred to the theoretical assumptions of Alchourron and Bulygin set out in their book, Normative Systems(3). We have previously referred to the concept of "case" as a circumstance or situation in which it is possible to ask if a certain action is permitted, obligatory or forbidden in a particular normative system.

Such cases may be generic or individual (concrete situations placed in time and space).

We will take a universe of elementary generic cases into consideration. They will, in other words, be characterized by a conjunction which contains for each property (or condition) either the property itself or its negation. This universe has two fundamental features:

1) it is conjunctively exhaustive of all the individual cases;
2) the cases are mutually exclusive.

Given the general norm of correlating a generic case with a solution, this norm, as a consequence, will resolve all the individual cases which, in reality, exemplify that generic case.

When a normative system does not provide a solution for a generic case a normative gap occurs and the system proves to be imcomplete. The incompleteness of the system arising in this way at the level of generic cases(4) consequently effects the whole range of individual cases referring to them.

If, vice versa, a normative system provides two contradictory deontic solutions for the same case, it will be defined as an inconsistent system.

It should be underlined at this point that when we are discussing incompleteness and inconsistency in relation to the lack or the abundance of deontic solutions, we are referring to maximal deontic solutions or, to put it another way, solutions which determine all the deontic contents of a universe of possible actions. In our case this means obligation, prohibition and bilateral permission (permitted to do or not to do something).

The incompleteness of the system, therefore, appears in the lack of maximal solutions to one or more of the elementary cases provided for in the universe under consideration. The inconsistency of the system consists, on the other hand, in having at least two different maximal solutions for the same case (in some cases even between a maximal solution, for example Ph and a minimal solution p). This does not detract from the fact that an incomplete system may have partial solutions when minimal deontic solutions are given for it (permission).

3.2 Representation of Legal Sentences

It is necessary, to bring the model "to life", to isolate the data to be inserted in the structures we have defined above. Therefore, once several actions regulated by the law, have been identified, the "corpus" can be defined. It should be noted that we have not paid attention to assembling all the norms referring to the selected actions because the evaluation of our results is concerned with the accuracy of the model rather than its overall legal validity.

The subsequent phases produce representations of data formalized in the model. We will now no longer be discussing norms but rather "normative sentences", and for two reasons: in the first place because the norms are significant sentences; in the second place because in dealing with sentences we eliminate the problem of knowing whether the logic used is that of merely sentences or of norms. Thus, it is necessary to pass from the natural language of the normative sentences to formal language (see note 3, at p. 615).

Whoever has faced this problem knows how very complex it is to find homogeneous criteria for linguistic analysis limiting, as far as possible, the arbitrariness of their interpretation.

An adequate solution seemed to us to limit the analysis to the syntactical structure of the normative sentences, translating the syntactical connectors between the sentences into logical operators of conjunction and disjunction(5). In this way, once the "constituting" normative sentences (one or more sentences expressing either a condition or an action) were isolated, we limited ourselves to representing the logical relations between them.

Each normative sentence was substituted by a symbol identifying the condition or the action, giving us, in this way, a list of conditions and a list of actions which have been modalized deontically.

The finite sequences were written in the subsequent phase. The calculation methodology will vary, as we shall see, according to the new objectives of the analysis: completeness and consistency.

3.3 Calculation of Completeness and Consistency

3.3.1 Some Prelinimary Notes. We begin by defining a model isomorphic to the rational structure of a normative system.

We will refer to a set ($\mathbb{N}$) whose elements are finite sequences formed by unit elements (1) and zero elements (0). Let

$$k \in \mathbb{N} \quad \text{and} \quad X \in \{0,1\}^k;$$
$$N_j(X), \quad j=1,\dots,k$$

will indicate the projection of X on the j-th factor, for example

$$\text{if } X = (1,0,1,1) \text{ then } N_3(X) = 1.$$

Moreover we consider, as usual, the following operations:

$$\wedge, \rceil, \vee$$

on $\{0,1\}^k$ defined by

$$N_j(X_1 \wedge X_2) = \begin{cases} 1 \text{ if } N_j(X_1) = N_j(X_2) = 1 \\ 0 \text{ otherwise} \end{cases}$$

$$N_j(\rceil X) = \begin{cases} 1 \text{ if } N_j(X) = 0 \\ 0 \text{ if } N_j(X) = 1 \end{cases}$$

$X_1 \vee X_2$ is defined using the de Morgan laws.
$1\!\!/_k$, $\emptyset_k$ will be the finite sequences of $\{0,1\}^k$ formed respectively of all the unit elements or all zero elements.

$\{\{0,1\}^k, \wedge, \vee, \rceil, 1\!\!/_k, \emptyset_k\}$ is a Boolean algebra which we will specify with B_k (eventually omitting the index if there is no possibility of ambiguity).

In what follows O, P, F, V will indicate the modal operators of obligation, permission, bilateral option, prohibition and, on rarer occasions, we will consider the operator U defined by $\rceil$F.

If $c_1,\ldots,c_n$ are conditions, the free Boolean algebra generated by $c_1,\ldots,c_n$ will be denoted by $A(c_1,\ldots,c_n)$ (A when there is no possibility of ambiguity). If $c_1,\ldots c_n$ are independent, $A(c_1,\ldots c_n)$ is isomorphic to $B_k k = 2^n$, and the following isomorphism is set

$$\psi : A \to B$$
$$C_i = \psi(c_i)$$

is the finite sequence obtained by writing i-times the finite sequence formed from 2^{n-i} unit and 2^{n-i} zero.
It is obvious that a different ordering of the conditions results only in a change of the isomorphism (which moreover has been chosen arbitrarily,

For example, if c_1 and c_2 are conditions

$$C_1 = 1100 \qquad \rceil C_1 = 0011$$
$$C_2 = 1010 \qquad \rceil C_2 = 0101$$

Each elementary case in this isomorphism is represented by a finite sequence which has all the elements zero except one.

3.3.2 Universe of Consistant Cases. As we often meet non-independent conditions when analysing a normative corpus, some of the elementary cases, which have formally been generated by them, are empty and can, therefore, be ignored in the above-mentioned analysis. Thus we will proceed directly to quotient the algebra which has been taken into consideration with respect to the ideal of the empty cases. Once, for an objective h of the analysis, the relevant conditions, $c_1(h),\ldots,c_n(h)$ have been given (that is, the universe or properties according to the terminology used by Alchourron and Bulygin) possible incompatibilities between these conditions are listed. Therefore, the propositional form I which identifies all and only the incompatible cases is determined.

A finite sequence I of 2^n elements remains associated to this form in the isomorphism defined in the preceding sections.

The characteristic finite sequences of relevant conditions relative to consistent cases are obtained from those previously taken into consideration, suppressing all the elements corresponding to units in the finite sequence I.

It is obvious that the resulting finite sequences are shorter. Furthermore, it should be noted that:

a) possible repetitions of incompatibilities in I do not (due to the law of idempower) alter I;
b) possible incompatibilities which are derived from those already considered are taken into consideration automatically. (For example: if $c_1 \wedge c_2$ and $\rceil c_2 \wedge c_3$ are empty cases, also $c_1 \wedge c_3$ results this way).

We will set out here, by way of example, the cases of incompatibility between conditions relevant for the analysis of art. 84 of the Italian Civil Code relating tot he legal capacity to marry. This and all the articles relative to the subject have been the object of analysis in one of our previous papers(6).

Art. 84: Minors cannot contract marriage.

The Court, on the request of the interested party (...) may (...) allow the marriage of those who have reached sixteen years of age.

r18	= the interested party has turned 18 years of age
r16	= the interested party has turned 16 years of age
tbar16	= the interested party has the permission of the court to contract marriage based on age.

R18 = 11110000
R16 = 11001100
TBAR16 = 10101010

We selected the following incompatiblities or empty cases:

R18 ∧ TBAR16
R18 ∧ ⌉R16
⌉R16 ∧ TBAR16

from which:

(R18 ∧ TBAR16) ∨ (R18 ∧ ⌉R16) ∨ (⌉R16 ∧ TBAR16) = I

I = (1010 0000) ∨ (00 11 0000) ∨ (00100010) = (10110010)

from which:

R18 = (1 000); R16 = (1110); TBAR16 = (0100)

3.3.3 Determining Conjunctions. The computer, from the determining conjunctions of the conditions relevant for every objective h dealt with, deduces, in a manner analogous to I, a disjunction of the finite sequences associated with Oh, Ph, Vh, Fh, P⌉h which will be denoted as OH, PH, VH, FH, P⌉H respectively (OH, for example = ∅ if h is obligatory in no case).

We assume in general that an action h is permitted in every case in which it is modalized by strong obligation (Alchourron and Bulygin should be referred to for the definition of strong and weak operators)7)). This assumption is not necessary.

The characteristic finite sequences of the weak operators $O_w h$, $P_w h$ etc. will be denoted as $O_w H$, $P_w H$ etc., and will be determined by negation of the finite sequences of strong operators. More precisely:

$$N_j\,(P_w H) = \begin{cases} 1 \text{ if } N_j\,(VH)) = 0 \\ 0 \text{ if } N_j\,(VH) = 1 \end{cases}$$

$$N_j\,(O_w H) = \begin{cases} 1 \text{ if } N_j\,(P(\rceil H)) = 0 \\ 0 \text{ if } N_j\,(P(\rceil H)) = 1 \end{cases}$$

etc.

Analogous to the remarks made in the preceding section:

a) possible repetitions in the list of the determining conditions do not alter the finite sequences of the consequent (however, the reporting of possible redundancies is lost);
b) if c_1 and c_2 are conditions such that "c_1 implies c_2" (alias $c_1 \wedge \neg c_2$ is an empty case) and if "c_2 then, for example, Oh" the algorithm is able to dermine automatically that "if c_1 then Oh".

3.3.4 Analysis of Maximal and Minimal Completeness. After having determined the finite sequences OH, VH, PH, P¬H, FH and those relative to the weak operators for every objective h present in a normative corpus, the two following finite sequences are determined:

$$X(\alpha,h) = OH \vee FH \vee VH$$
$$Y(\alpha,h) = X(\alpha,h) \vee PH \vee P\neg H.$$

The system is maximally (minimally) complete with respect to the objective h if and only if

$$X(\alpha,h) = 1 \quad (Y(\alpha,h)\ 1)$$

It is immediately clear that if α is complete maximally, it is also minimally:

$$X(\alpha,h) = 1 \rightarrow Y(\alpha,h) = 1,$$

Where $X \neq 1$ $(Y \neq 1)$, the system presents some cases of maximal (minimal) incompleteness for the objective h and the computer points out the elementary cases of incompleteness in the following way: for each j such that $N_j(X) = 0$ there is an elementary case X_e of incompleteness and X_e^j is determined by:

$$\text{if } N_j(C_i) = 1 \quad X_e \wedge C_i = X_e \quad (X_e \wedge \neg C_i = 0)$$
$$\text{if } N_j(C_i) = 0 \quad X_e \wedge C_i = X_e \quad (X_e \wedge C_i = 0)$$

These indications may reveal structural gaps or may suggest that the legal operator modifies either the determining conjunctions on the basis of different interpretations or principles based on common experience (or corrections to the incompatibilities when any of them may have been neglected). In the former example the determining conjunction, coming from the first two sections (α), can be given a superficial interpretation.

$\neg$r18 V_m (m = to contract marriage)

r16 $\wedge$ tbar16 P_m

$X(\alpha, m) = (0111)$ is obtained. Therefore the elementary case r18 $\wedge$ r16 $\wedge \neg$tbar16 represents a case of maximal incompleteness. As Y=X such a case is also a case of minimal incompleteness.

3.3.5 Analysis of the Inconsistencies

$INC_1\ (\alpha, h) = OH \wedge VH$ $INC_4\ (\alpha, h) = OH \wedge P\neg H$

$INC_2\ (\alpha, h) = OH \wedge FH$ $INC_5\ (\alpha, h) = VH \wedge PH$

$INC_3\ (\alpha, h) = VH \wedge FH$ $INC_6\ (\alpha, h) = FH \wedge (OH \vee VH)$

The system is consistent with respect to the objective h if:

$$INC\ (\alpha, h) = INC_4\ (\alpha, h) \vee INC_5\ (\alpha, h) \vee INC_6(\alpha, h) = \emptyset$$

If $INC\ (\alpha, h) \neq \emptyset$ the computer, by proceeding in the same way as it did for incompleteness, will point out the elementary cases in which there is more than one deontic solution and what these are. In the example: $INC_1 = 0 = INC_2 = INC_3 = INC_4 = INC_6$

$$INC_5 = (0111) \wedge (0100) = (0100),$$

hence $\neg$r18 $\wedge$ r16 $\wedge$ tbar16 is an elementary case of inconsistency. In this case it is evident that the inconsistency derives from a superficial analysis resulting in the writing of the determining conjunctions. The legal operator may, therefore, reformulate the determining conjunctions as follows:

$\neg$r18 $\wedge \neg$(r16 $\wedge$ tbar16) V_m

r16 $\wedge$ tbar16 P_m

The new system α_1 is consistent, minimally complete in the case of $\bar{\alpha}$ but maximally incomplete in a subsequent case. In fact

$VM = (0011)$ $OM = (0000) = FM$ $PM = (0100)$
$P\neg M = (0011)$
$X(\alpha_1, m) = (0011)$

and therefore the elementary case corresponding to the subsequent zero ($\neg$r18 $\wedge$ r16 $\wedge$ tbar16) is a case of maximal incompleteness. (Even in this case, given by way of example, the incompleteness derives from the interpretation r16 $\wedge$ tbar16 Pm rather than from the possible r16 $\wedge$ tbar16 Fm). The procedure followed for removing the inconsistency of α may especially be followed each time the inconsistency derives from norms where a hierarchy can be established (temporal or otherwise) and, in particular when a norm is subsequently introduced containing a clause repealing that which was previously in contrast with it.

The computer can be programmed for automatically deriving all the consistent systems obtainable from a given system resolving, in the various formally possible ways, the cases of inconsistency.

It is almost superfluous to note that this possibility has a practical result only if the cases of inconsistency are not too numerous.

In the example the formal alternative α_2 would be

$\neg$r18	V_m
$\neg$($\neg$r18) $\wedge$ r16 $\wedge$ tbar16	P_m

from which PM = (0000) that is α_2 reduced only to the first sentence.

Up until now we have avoided calculating the characteristic finite sequences of the weak operators for the examples because we believe that these are useful mainly in the analysis of single cases.

Moreover, it should be noted that the procedure does not necessarily require the analysis to take into account only one objective even if the properties of cinsistency or completeness deal with each objective, without excluding that the system is, e.g., consistent with objective h_1 and inconsistent (or incomplete) for the other objective h_2.

It is also interesting to note that the assertion "if the system α is consistent and complete maximally (as regards an objective h) the strong and weak operators coincide" has an equivalent in the following proposition "if the system α is consistent and complete to the maximal (as regards an objective h), that is

$$X(\alpha,h) = \not{1} \text{ and}$$
$$INC(\alpha,h) = \emptyset,$$

then the finite sequences associated with the strong and weak modal operators of h coincide". In fact, assume for clarifying our ideas $OH \neq O_w H$, that is there exists j such that

$$N_j(OH) \neq N_j(O_w H)$$

Two cases may be present:

a) $N_j(OH) = 1 \qquad N_j(O_w H) = 0$

b) $N_j(OH) = 0 \qquad N_j(O_w H) = 1$

In the first case

$$N_j(O_w H) = 0 \rightarrow N_j(P\rceil H) = 1 \rightarrow N_j(OH \wedge P\rceil H) = 1;$$

that is

$$N_j(INC_4(\alpha, h)) = 1$$

and α is therefore inconsistent. In the second

$$N_j(O_w H) = 1 \rightarrow N_j(PH) = 0 \rightarrow N_j(VH) = 0 = N_j(FH),$$

from which

$$N_j(X) = 0$$

and α is incomplete.

3.3.6 Introduction of Data in the Subsequent Phases. The procedure proposed for rectifying any possible cases of inconsistency or incompleteness may also be applied for introducing the normative corpus in more than one phase.

It should be kept in mind that new normative propositions are introduced in a subsequent phase and it is not necessary to trace the finite sequences of the preceding conditions. Rather the following procedure may be followed. If a new proposition (for any objective) contains relevant conditions which were not previously present the new finite sequences associated with the conditions already present may be obtained by repeating 2^p times each element of the preceding finite sequence and then proceeding for the new conditions as we have suggested in the first section.

Obviously the quotient operation must then be repeated if the conditions considered show incompatibilities which were not previously introduced. The finite sequence of the modal operators will then be traced.Nothing stops us from specifying to the computer that the new normative propositions are to be considered hierarchically higher than the previous ones in order to solve possible cases of inconsistency (when appropriate).

3.3.7 Analysis of Decision-Making. Decision-making aspects of the normative corpus dealt with here is still the object of research and a discussion on it is postponed for the present. We will limit ourselves to merely pointing out some possibilities.

Given a case C (elementary or otherwise) identified by a proposition form in the relevant conditions, a finite sequence C of bit may be associated with this from the comparison with which the possible elementary cases of C, showing inconsistency, may be extracted with the finite sequence x and t. The eventual deontic solutions can be determined, excluding from C the previous cases from the comparison with the finite sequences charactistic of the various modalities of actions dealt with.

4. PRACTICAL APPLICATION

The Italian Civil Code groups together all the norms relating to minors emancipated by marriage (arts. 390-397) under the heading "On Emancipation". From these we have isolated the norms on the capacity of the emancipated person.

4.1 Analysis of the Articles

Art. 390 (Emancipation by law): A minor is emancipated by law on marriage (s. 82), the article, which is operative with regard to emancipation. Therefore, the symbol E, representing the emancipated minor, contains the two requisites ("age of minority" and "marriage") which never appears among the conditions. The fact that E as a present condition in the finite sequence is the conjunction of two conditions would create ambiguity when it is negated. In fact, the norms do not take into consideration the fact that the conditon E is absent. The legal universe in which the analysis is carried out is, therefore, that of the emancipated minor.

Art. 394 (Capacity of the emancipated minor). Emancipation confers on the minor the capacity to perform acts which do not exceed ordinary administration (OA) (396, 397, 472, 474).

The emancipated minor may, with the assistance of the guardian, draw capital on the condition that it will be used uitably and he may come before the courts either as a plaintiff or as a defendant.

For other acts exceeding ordinary administration, the permission of the tutelary judge is necessary, in addition to the consent of the guardian. For acts set out in art. 375 permission must be given by the court on the advice of the tutelary judge, if the guardian is not a parent.

The article deals with the capacity of E in relation to various legal acts. A legal definition both of acts of ordinary administration (OA) and of facts exceeding OA is missing from the norm.

Several acts exceeding OA are individually named. The norm, moreover, deals jointly with "other" acts, which have been given the symbol AX. They are understood as meaning "all acts of extraordinary administration (SA) excluding those explictly dealt with".

The article, in the construction of the finite sequence, has been dealt with together with art. 392 sub-section 3 ("If the minor subsequently contracts a new marriage the guardian will also assist him in the acts set out in art. 165") and with art. 397 still dealing with the legal capacity of the emancipated minor.

As we have already pointed out, we have not, as yet, considered the norms relating to the relationship between the guardian and the emancipated minor (arts. 392, 394, v.c. and 395).

Art. 396 (Failure to comply with preceding norms). Acts carried out without complying with the norms set down in art. 394 may be annulled on the petition of the minor or his heirs or on showing cause.

Art. 397 (Emancipated minor authorized to carry on a business enterprise). An emancipated minor may carry on a business enterprise (2195) without the assistance of the guardian, if he has been given permission by the court, with the advice of tutelary judge and on hearing the guardian.

The permission may be revoked by the court on the petition of the guardian or by court itself, in both cases with the advice of the tutelary judge and on hearing the emancipated minor.

The emancipated minor who has been given permission to carry on a business enterprise, may carry out acts on his own, which are in excess of ordinary administration, even if they are unrelated to the carrying on of the business.

Always taking into consideration the legal capacity of the emancipated minor, the only pertinent subsection is sub-section 3, although we include even sub-section 1 in our interpretation. Sub-section 2 has been omitted because it is a procedural norm.

It should be noted that the acts of extraordinary administration (SA), in coherence with what has been said above, have been symbolized by the conjunction of the "other" acts with all the acts of SA expressly indicated in the norm.

The list of conditions and consequences is as follows:

COND.

AC = assistance of guardian
ICR = suitable use of capital
M2 = new marriage
COC = consent of guardian
GT = permission of tutelary judge
AX = other acts exceeding ordinary administration
CG = parent guardian
TBA = permission of the court
TBAI = permission to carry on a business enterprise

CONS.

R = drawing capital
GIU = coming before the courts
A165 = acts ex art. 165
A375 = acts ex art. 375

4.2 Construction of the Deontic Scheme

Therefore, art. 394 deals with the legal capacity of the emancipated minor specifying the acts he may carry out, under certain conditions and having fulfilled certain requisites.

The absence of the latter is not expressed in terms of explictly prohibiting the carrying out of various acts but lays down their voidability (on the petition of the minor, of his heirs and on showing cause) (art. 396).

By not modalizing the text of art. 396 as a deontic prohibition and therefore by inserting only art. 394 in the model, knowledge is limited to what and when the emancipated minor may do, but not to what and when he may not do.

An artificial gap is, therefore, produced in this way.

Let us take as the example the art. 394 sub-section 2 which regulates the drawing of capital (R), in compliance with the conditions "assistance of the guardian" (AC) and "suitable use of the capital" (ICP). The + and - signs indicate the presence of a condition or its negation. We obtain, therefore, the following finite sequence:

+AC	+ICP	PR
-AC		?
-ICP		?

The other interpretation, instead, represents art. 396 as an implicit prohibition in the deontic scheme, in as far as it relates to behaviour in contrast with the prescriptive rules laid down in art. 394. In other words, art. 396, which declares the acts carried out in failing to comply with the abovementioned norms as voidable, would be equal to laying down the prohibition on carrying out such acts, each time one of the determining conditions is missing.

+AC	+ICP	PR
-AC		VR
-ICP		VR

It is, however, debatable whether to define them as prohibited, as they are valid if there is no annullment.

All of this gives us reason to assume that the deontic scheme of our model is complete.

If we keep in mind that "the capacity" to carry out "valid acts" is a legal status rather than a deontically modalized action, then the consequent may be represented as a "non deontic goal".

This solution, however, goes beyond our attempts to experiment with the application of the deontic scheme.

The obstacle may be avoided by shifting the consequent out the effects provided for in art. 396, by not reading the two articles as regulating the capacity of the emancipated minor, but as going to the voidability of the acts of the emancipated minor.

We are not dealing, on a logical level, with a distortion of meaning, but rather with the level of the organization of the text, as the legislator entitles art. 394 "Capacity of the emancipated minor", which means "the capacity" becomes the subject-matter of the entire article.

For example, the sub-section regulating the drawing of capital by the emancipated minor, translated according to this criterion, informs us about the voidability, (E) or not of the act of drawing the capital. In the scheme it appears as follows:

+AC	+ICP	VAR
+AC		FAR
-ICP		FAR

The analysis and the formalization of the examined norms, which are necessary for carrying out the calculation of completeness and consistency on them, has demonstrated the difficulty existing in applying the deontic scheme, as we have seen. For example, we are dealing with associating the deontic prohibition modality to behaviour which does not comply with the prescriptive norms but which results in very different legal consequences such as annullment, nullity and prohibition-explicit sanctions.

We will put this complex problem aside for the moment as it will be the topic for further research. We will, instead, concentrate on the evaluation of completeness and consistency which are the subject-matter of this paper.

4.3 Example of the Calculation of Completeness and Consistency

On the basis of the conditions and consequences set out above and their most recent interpretation, the following finite sequences may be obtained:

AC	ICP	VAR
$\neg$(AC $\wedge$ ICP)		FAR
TBAI		VAR
COC	GT	VAAX
$\neg$(COC $\wedge$ GT)		FAAX
TBAI		VAAX

By way of example we have calculated the completeness and inconsistency for the last finite sequence, as follows:

COC = 0000 1111 $\neg$COC = 1111 0000
GT = 0011 0011
TBAI= 0101 0101

Once we have the incompatibility between the condition relating to the permission of the tutelary judge (GT) and the permission of the Court (TBAI), the finite sequence representing the incompatibility will be:

$$I = CT \wedge TBAI = 00010001$$

Having filtered through the incompatibility the finite sequences representing the conditons become:

$$COC = 000\ 111$$
$$GT = 001\ 001$$
$$TBAI = 010010$$

Therefore, the action: "annullment of the acts of extraordinary administration" will be forbidden in the following cases:

$$VAAX = (COC \wedge GT) \vee TBAI$$

for which the finite sequence associated with VAAX will give:

$$VAAX = [(000111) \wedge (001001)] \vee 010010 = 000001 \vee 010010 = 010011$$

analogously, the action will be facultative if:

$$FAAX = \neg(COC \wedge GT) = 111110$$

The calculation of maximal completeness (x= 1) will be given by the disjunction of the finite sequence representing all the maximal modalities ($V \vee F \vee O$):

$$X = (010011 \vee 111110 \vee 000000) = 1$$

The consistency (INC = 0) will be given by the conjunction of the modalities equal to 0 giving us:

$$INC = VAAX \qquad FAAX = 010010$$

the presence of two 1's in the finite sequence indicating 2 elementary cases of inconsistency. By comparison with the finite sequences associated with each condition we can see that the cases of inconsistency are:

1) COC -GT +TBAI
2) -COC -GT +TBAI

NOTES

(1) C. Biagioli et al., 'Experiments on the 'Model of Sanchez-Mazas'", in: Deontic Logic, Computational Linguistics and Legal Information Systems, vol. II (1982), North-Holland, Amsterdam, pp. 215 ff.;
G. Dini, "A Modifief Version of an Arithmetical Model, for Legal Information Systems", in: Deontic Logic, Computational Linguistics, and Legal Information Systems, vol. II (1982), North-Holland, Amsterdam, pp. 205 ff.;
C. Biagioli et al., "Un modello automatico per l'analisi dei sistemi normativi: una proposta sperimentale" in: Proceedings of the III International Congress "Legal Informatics in the National and International Communities", Rome, 9-14 May 1983.

(2) A.A. Martino, "Contributo logico informatico all', analisi della legislazione", in: Informatica e diritto, 2, 1982, p. 63.

(3) C.E. Alchourron and E. Bulygin, "Normative Systems", Springen Verlag, Vienna-New York, 1971. The propositions of Normative Systems refer to the concepts of consistency and gaps and to their theoretical setting. We will not, instead, assume the postulate that legal orders are consistent and complete.

(4) C.E. Alchourron and E. Bulygin, op. cit. p. 27 ff.

(5) During this phase of our work we have borne in mind L.E. Allen, "Una guida per redattori giuridici di testi normalizzati", in: Informatica e diritto, 1, 1979, part. 2, p. 61 ff.

(6) C. Biagioli et al., op.cit. p.

(7) C.E. Alchourron and E. Bulygin, op.cit. p. 122 ff. "Definiton of strong permission: to say that p is strongly permitted in the case of q in a system α means that a norm which gives p permission in q is a consequence of α. Definition of strong prohibition: to say that p is strongly prohibited the case of q from the α system means that a norm which gives p prohibited in q is a consequence of α."

RIGHTS AND PRACTICAL POSSIBILITIES

B. Ghita E. Holmström

1. ACTUAL CAPACITY AND POTENTIAL CAPACITY

In this paper I shall deal with the concepts <u>actual capacity to act</u> and <u>potential capacity to act</u>, the former of which I call <u>practical possibility</u>. In ordinary language people say that

x can do p

when they mean that

x has the practical possibility to see to it that p.

So I then discuss what must be included in this ordinary language concept "can". In discussing <u>types of practical possibility</u> I restrict myself here to one-agent types and make some observations about various connections to positions of rights both in Kanger's model and in Lindahl's.

The underlying legal problem is that of changing the legal position - for oneself, a community or a country. I want to underline the importance of the distinction between potential possibility and actual - that is practical possibility to act, e.g. when we talk about changing our legal status.

Let us start with two examples:

(1) Charles is a bachelor and is going to change his legal position by marrying Mary.

(2) Robert, who is a citizen of Finland (the state y), has the right versus Finland to leave the country.

These two examples are of the formal form

A. Peczenik et al. (eds.), Theory of Legal Science, 607–615.

(1') x sees to it that r by seeing to it that m
(2') x has versus y the right of type T_n with respect to S.

Now we are ready to introduce our concepts.

(1) the act description E(x,m,r) for 'x sees to it that r by seeing to it that m'
(2) the agent-relative necessity N(x,p) for 'it is the case that p independently of what x does'
(3) the agent-relative possibility M(x,p) for 'it is possible for x that p'
(4) the logical symbols, the negation mark ~, the conjunction and disjunction marks '&' and 'v' respectively, and the existential quantifier (Ex) and the universal quantifier (x).

It is reasonable to suppose that N(x,p) is of type S4 in Lewis' system and it seems equally reasonable to define M(x,p) as the dual of N(x,p), i.e.

(DfM) M(x,p) = df ~N(x,~p)

Now, let p be an action of the specified form

x sees to it that r by seeing to it that m

or an action of the open form

x sees to it that r.

The connection between these two concepts is given by the definition

(DfE) E(x,r) = df (Em)E(x,m,r)
(DfE1) 'x sees to it that r' is by definition equal to 'there is some (measure) m such that x sees to it that r by seeing to it that m' (This may be shortened: 'x sees to it that r by (seeing to) some measure m'.)

The basic form is a three-place relation between the agent x, the result r and the measure m. This relation turns out to be a <u>partial ordering</u>.
When p in the model expression is replaced by an act des-

cription (1) we get an expression for some kind of possibility

(5) M(x,E(x,m,r)) for
(5.1) it is possible for x to see to it that r by seeing to it that m

Let us take two examples:

Ex 1. It is possible for Charles to change his legal position by marrying Mary

and

Ex 2. It is possible for Robert to change his legal position by acquiring another citizenship

However, these express only a potential capacity but do not assert that the agent x really and actually has the possibility concerned. Namely, as we notice, it is possible for every bachelor to change his legal status by marrying Mary.
So the specified potential capacity as described (above) does not entail that there is some specified actual capacity. The only thing it really entails is an unspecified potential capacity i.e.

(6) M(x,(Em)E(x,m,r))
(6.1) it is possible for x that there is some (measure) m such that x sees to it that r by seeing to it that m.

But if it happens to be the case that Mary wants to marry Charles and other difficulties can be overcome, then we could say that Charles, as a matter of fact, has the specified actual capacity

(7) (En)(M(x,E(x,n,m)&E(x,m,r,)))
(7.1) there is some measure n such that it is possible for x to see to it that m by seeing to it that n and to see to it that r by seeing to it that m.

An actual capacity to act, then, means that there is a specified measure or at least that there is some measure m for Charles for fulfilling his task. The latter expresses an unspecified actual capacity to act, i.e.

(8) (Em)M(x,E(x,m,r)).

2. "CAN"

Normally we say that

x can see to it that r

or

x can do r

when we mean that he has the actual capacity to act. E.g.

Charles can marry Mary

or

Robert can leave his country and change his legal status by acquiring another citizenship

But is is not very clear in what respect Charles "can" marry Mary or Robert "can" acquire another citizenship. Are these some potential, hypothetical "cans" or what are they? (It is e.g. not a contradiction to say: Robert can leave the country but he can't).
We have got four descriptions of capacity

a. specified potential capacity to act
b. unspecified potential capacity to act
c. specified actual capacity to act
d. unspecified actual capacity to act

When we say that the agent

x has the practical possibility to (see to it that r) do r

this means exactly the same as saying

x has the actual capacity to see to it that.

This includes both the opportunity and the ability.

The four distinctions for possibility with respect to actions show that "can" is a very vague and imprecise expression for the agent's capacities in relation to some y. So if we use the expression "can do" at all, we take it as a shorthand for both forms of the actual capacity to act i.e. for the practical possibility and include in this both the opportunity and the ability. E.g.

Robert can go abroad.

If this is true it means that

1. in relation to y and z (y and z are states) he has an opportunity to go abroad
   - he gets a passport to travel and cross the borderline
   - he gets the permission to leave his own country (y) and to enter another country (z)
   - and, furthermore, he has the available tickets, money etc.
2. he has the ability
   - he can handle the practical details, has the physical health and strength, knows the language and so on.

Some people in some countries and people in general in other countries are not permitted to go abroad. Even if it were possible to go abroad by taking an aeroplane they don't have the practical possibility because they do not have the opportunity e.g. they do not get a passport. From this we conclude that an agent sometimes has the practical possibility with respect to certain results while others lack it. But sometimes the same agent might have the practical possibility whereas in other situations he lacks it.

3. TYPES OF PRACTICAL POSSIBILITIES

Analogously to Lars Lindahl's development of Kanger's types of rights we can construct types of practical possibilities. However, we must keep in mind some logical restrictions which hold here but not in Lindahl's system. (Among others there is the axiom (9).)

(9) $\models (Em)M(x,{\sim}E(x,m,r))$

(9.1) for all agents x there is always some measure m, such that it is possible for x to omit from seeing to it that r

(By omission here I mean exactly the same as not seeing to it that r.)
Then our starting point will be to consider the following statements ((10)(a)-(b)) and their negations.

(10) a. (Em)M(x,E(x,m,r))
b. (Em)M(x,E(x,m,~r)
c. ~(Em)M(x,E(x,m,r))
d. ~(Em)M(x,E(x,m,~r))

(10.1) it is for some m practically possible for x to see to it
a. that r by seeing to it that m
b. that not-r by seeing to it that m
it is not for any m practically possible for x to see to it
c. that r by seeing to it that m
d. that not-r by seeing to it that m

These four types of practical possibility can be restricted in the law and can form the grounds for the Kanger-Lindahl types of right concerning practical possibilities. (See Kanger & Kanger 1966 and Lindahl, 1977).

For each of the alternatives ((10)(a)-(d)) it either May be the case (is permitted) or Shall not be the case (is forbidden). E.g. it Shall not be practically possible for Robert to acquire another citizenship.
According to a generally accepted definition in deontic logic,

(DfShall) Shall p = df ~May~p

May p can be replaced by its dual ~Shall~p.

As it seems reasonable to ascribe permissions, prohibitions and obligations to a lawgiver or some agent, these deontic operators will be considered agent-relative. This has been taken into account in (2') where it is said that

x has versus y the right of type T_n with respect to S

Now, to say that

Robert has the right versus Finland that he can leave the country

is to say exactly

(12.1) according to the laws of Finland it is permitted that Robert has the practical possibility to leave the country

or still in other words

(12) May(y,(Em)M(x,E(x,m,r))

This simple type of right concerning practical possibilities corresponds to the type power in Kanger's system. (See Kanger, S. & Kanger, H., 1966). Similarily we could construct other simple types of right that correspond in the same way to those in Kanger's system.
Take the remark above, which says that in some countries, it shall be the case that the citizen x can leave the country, whereas in others it shall be the case that the citizen x cannot leave the country. This can be expressed by the formulas:

(13) Shall(y,(Em)M(x,E(x,m,r))
(14) Shall(y,~(Em)M(x,E(x,m,r))

or in terms of permission

(13.1) ~May(y,~(Em)M(x,E(x,m,r))
(14.1) ~May(y,(Em)M(x,E(x,m,r)).

These two formulas correspond to x's claim versus y with respect to leaving the country and immunity with respect to not leaving the country, respectively.
Still one branch of this game would be possible but that is connected to a philosophical question. If we share the opinion that "Shall" → "Can" then we get bridges between Kanger's diagrams and the diagrams for practical possibility. But first we have to observe the difference in Kanger's concept of action and that of ours.

4. DOES "SHALL" IMPLY "CAN"?

To demand a practically impossible action would be considered cruel, both legally and morally. In other words, a legal system or a norm system would be considered inconsistent if it allows claims with respect to x's doing r in conjunction with the practical impossibility for this very agent to perform this action. Therefore, it will be a principle of ours that

(15) ~(Shall(y,E(x,m,r)) & (m)~M(x(E(x,m,r))

(15.1) Not: according to y it Shall be the case that x sees to it that r by measure m and it is for all measures m practically impossible for x to see to it that r.

Does this mean that we are suggesting that "Shall" → "Can"? Consider the following situations

(i) It shall be the case that Robert acquires another citizenship
(=Shall(y,E(x,m,r))

(ii) It is practically possible for Robert to acquire another citizenship
(=(Em)M(x,E(x,m,r)))

(iii) It shall be the case that it is practically possible for Robert to acquire another citizenship
(=Shall(y,(Em)M(x,E(x,m,r)))

It seems reasonable, at least in some cases, as in the example above, to restrict the scope of (i) to imply that it shall be the case that (ii). So what we are saying is not in the first place, that "Shall" implies "Can", but that "Shall do" implies "shall be practically possible" or in other words

(16) ⊢Shall(y,E(x,m,r)) → Shall(y,(Em)M(x,E(x,m,r))

(16.1) x shall see to it that r, by measure m, implies that x shall have the practical possibility to see to it that r.

For example, if it shall be the case that Robert acquires another citizenship then it shall be the case that it is practically possible for him to acquire another citizen-

ship.
Let us take still another example. Consider the following sentences (1)-(3):

(1) It shall be the case that Charles changes his legal position by marrying Mary
(2) It is practically possible for Charles to change his legal position by marrying Mary
(3) It shall be the case that it is practically possible for Charles to change his legal position by marrying Mary.

As was the case earlier we could also here impose the restriction that (1)→(3). But then we have to notice that the "lawgivers" in (1) and (3) should be the same institutions or persons. Otherwise we have the impression of staying in two different normsystems. And Charles could feel a little bit uncomfortable if he on the one hand has the obligation versus Mary to marry her and on the other hand does not have the practical possibility to do so. Then he would say that according to him it shall be the case that he has the practical possibility to marry Mary.
Accordingly, we find it reasonable to state that according to logical rules

> it shall be the case that, if it, according to some legal system S shall be the case that x sees to it that r (by seeing to it that m), then according to the legal system S, it shall be the case that it is practically possible for x to do r.

This gives us the concluding axiom:

(17) $\vdash$Shall(Shall(y,E(x,m,r)) →
Shall(y,(Em)M(x,E(x,m,r)))).

REFERENCES

Kanger, S. & Kanger, H. (1966). "Rights and Parliamentarism", Theoria, vol. 32, pp. 85-115.

Lindahl, L. (1977), Position and Change: a study in law and logic. D. Reidel, Dordrecht.

REQUIREMENTS, URGENCY, AND WORTH*

Lars Lindahl

I

In moral theory, a distinction is often made between meeting the moral ideal and just measuring up to the standards of minimal moral decency. Failing to attain the ideal is not considered blameworthy to the same extent as not complying with requirements of minimal acceptability(1).

Such a distinction implies differentiating between various levels of accomplishment, or "levels of perfections", as we might also say. The standards of the ideal pertain to a higher level of perfection than the standards of mere acceptability.

A differentiation between levels of perfection can also be made in the field of law. Thus, when evaluating the performance of a judge in a law suit, we may hold that it is an ideal that the judge shall reason as carefully as Ronald Dworkin's famous judge Hercules, whose decisions are based on an elaborate theory where particular data and general principles are in reflective equilibrium(2). But having a conception of the ideal is consistent with insisting that there are a number of minimal requirements with which any acceptable judge must comply. For instance, it is minimally required of any judge that he shall pay heed to the very basic principles of due process (such as "audiatur et altera pars"). Naturally, a judge whose performance does not comply with the requirements of minimal acceptability will be more severely censured than a judge who merely fails to meet the ideal of judicial excellence and perfection.

The aim of this paper is to give the outlines of a conceptual framework for a theory of levels in law and morals; in particular, the ideas of legal or moral requirements and the notions of urgency and worth will be central. In section II, the relevant notions will be informally explained and in sections III-IV, they are related to some problems of legal and moral theory and of

A. Peczenik et al. (eds.), Theory of Legal Science, 617–638.

preference theory. In section V (the most substantial part), there follows a semi-formal exposition with explicit definitions and a number of logical rules. In the concluding section VI, there is a brief discussion of the interrelations between urgency and worth.

II

When we differentiate between levels, the number of levels assumed will depend on how fine a differentiation we need in the situation at hand. In the examples of this section, I shall be content to assume just two levels, viz. an ideal level and a level of minimal acceptability.

For each level, it is important to distinguish between what is necessary and what is sufficient for attaining this level. The conditions necessary for the attainment of a level will be called the requirements of this level.

If the level of minimal acceptability is attained, things are not so bad as they would have been if not even this level had been attained. This indicates that also the level of minimal acceptability (not only the ideal level) is what I shall call a "positive" level. In what follows, when I speak of levels, I shall mean positive levels in this sense.

If a performance attains the ideal level it also, a fortiori, attains the level of minimal acceptability (but not vice versa). Therefore, if P is a requirement of minimal acceptability, P is also, a fortiori, required at the ideal level (but again, not vice versa). Thus, in the scheme,

Level	Requirements
Ideal	P, Q, P & Q
Minimal	P

P is required at the minimal level, and therefore, also at the ideal one, while each of Q, P & Q, is required only at the ideal level.

In the case illustrated by the scheme, we may say that P, which is required already at the minimal level, is a <u>more urgent</u> requirement than Q, P & Q (which are required only at the ideal level). Thus, for instance, the requirement that a judge shall respect fundamental principles of due process, i.e., P, is more urgent than the more sophisticated requirements that hold only for the ideal judge Hercules, i.e., Q, P & Q.

So far, we have been concerned with level <u>requirements.</u> But that P is required (i.e., necessary) for the attainment of some level does not, of course, imply that P is sufficient for the attainment of this level, i.e., that P <u>guarantees</u> the attainment of the level. And so, a natural counterpart of the notion of a level requirement is that of a <u>level guarantee</u>. Thus, starting from the scheme of level requirements just indicated, we may turn to the corresponding scheme of level guarantees, i.e.,

Level	Guarantees
Ideal	P & Q
Minimal	P, P & Q.

Since the attainment of the ideal level implies attainment of the minimal level, all ideal guarantees are also minimal guarantees, but not vice versa.

The main reason for introducing the notion of level guarantees in the present essay is that is serves to improve our understanding of urgency by elucidating the distinction between urgency on the one hand and what might be called <u>worth</u> on the other. In the scheme of guarantees, since P & Q guarantees the ideal level (and, hence, also the minimal level) while P guarantees only the minimal level, P & Q will be said to have <u>more worth</u> than P.

From our schemes, it can be seen that while, on the one hand, P & Q has more worth than P, on the other hand it holds that P is more urgent than P & Q. Therefore, it will be clear that urgency and worth are two different notions.

In some cases, the relations of being more urgent than and having more worth than will coincide. Thus, in our example, P is both more urgent than and has more worth than Q. But this coincidence of the two relations in some cases is rather one more reason for sharply distinguishing between the notions. Urgency (in law and morals) is a notion closely associated with the idea of requirements and obligations in the sense of deontic logic. Worth, on the other hand, is associated rather with the idea of merit or "goodmaking power". As we shall see later, the urgency of a whole combination of good things never exceeds the urgency of the least urgent component, whereas the worth of the whole combination may well exceed the worth of even those of its components having most worth.

III

With respect to actions of private individuals and officials, i.e., what is required of agents, many legal theorists will hold that in the field of law there is only one positive level. That is, for any P, either P is required (at this one level) or non-P is required, or neither P nor non-P is required, there being no need to distinguish between requirements at different levels. This seems to be the view of legal positivists as well as of opponents of positivism like Ronald Dworkin. Thus, Dworkin and the legal positivists might agree that _if_ both P and non-P are permitted for an agent (i.e., neither P nor non-P is required of him), then both P and non-P are within the agent's "leeway" or _Spielraum_ for action, as far as the law is concerned. The controversy between Dworkin and the legal positivists rather concerns two other questions. One is the question of the appropriate method for establishing whether something _is_ legally required or not, the other is the question whether a judge deciding a so-called "hard case" _has_ a _Spielraum_ or not. As is wellknown, according to legal positivism, in a hard case the judge will have a _Spielraum_ with some options where he can exercise his "discretion" and, if he acts within his _Spielraum_, his decision cannot be legally criticized at all, although, from a moral or political point of view, it might be. According to Dworkin, on the other hand, there is no judicial _Spielraum_ at all. The reasoning of a judge in a hard case is always susceptible to evaluation and criticism from other lawyers as to whether it was "the right answer" and such evaluation and criticism is not (as the legal positivists will have it) to be called moral or political as distinguished from legal(3).

Concerning our distinction between levels, it may first be noted that in the linguistic expressions of enacted law, there is, in fact, a legislative distinction between "Shall" and "Ought" (Swedish "Skall", "Bör"), which, for either of the two schools, makes it difficult to maintain the doctrine that what is required is just simply required and what is non-required is just simply so. In legislative language, if something _Shall_ be, it is more strongly obligatory than if it only _Ought_ to be, and this seems to presuppose that there is more than just one level of accomplishment.

Secondly, concerning the difference between Dworkin and

legal positivism, Dworkin's view, though he emphasizes "ideal theory" rather than "minimal theory", can easily be combined with a differentiation between levels. Looking at legal positivism, on the other hand, some of its adherents might hold that the "ideal" requirements incumbent on Dworkin's judge Hercules belong to a different normative system (moral or political) as compared with such legal requirements accepted by the positivist school. I.e., there would not be a distinction between requirements at two levels (ideal and minimal), but rather a distinction between the requirements of two different systems. Against this, it may be argued, however, that there is at least some normative system, concerned with decisions of judges in hard cases and relevant for investigation and elaboration by lawyers, where the distinction between ideal and minimal level can be made. For our present purposes, the question whether this system is to be called strictly legal or not may be disregarded.

_If_ we _do_ distinguish between levels of accomplishment, a particular problem will be how such a distinction is related to the notion of _obligation_ or _duty_. One line of argument might be that the notion of obligation (or duty) is appropriate only for the requirements at one single level, namely the minimal one. This would mean that not all of the requirements at the ideal level would be obligations in the proper sense. In other words: Attaining the ideal is something that "goes beyond" fulfilling obligations. (In moral theory, such a view can be associated, in particular, with the philosopher Leonard Nelson and his distinction between "Pflichtlehre" and "Ideallehre".(4))

In view of the affinity between "P is required" and "P is obligatory"(5), an alternative way of dealing with obligations, however, would be to make the notion of obligation relative to the particular levels in view. For instance, distinguishing between an ideal level and a minimal level, a corresponding distinction would be made between "P is obligatory at the ideal level" and "P is obligatory at the minimal level". And so also, in law and morals, the statement "P is a more urgent requirement than Q" would be translated into "P is more strongly obligatory than Q".

In the present paper, I shall not take any stand on the issue of the two alternatives now referred to and I shall accordingly prefer to speak of level requirements rather than of obligations at various levels. But it is worth emphasizing that, for each level L, the basic logical characteristics of "P is required at level L" are the same as those assumed, in the classical version of deontic logic, for "P is obligatory". What is said about requirements at various levels in law and morals will, therefore, also refer to obligations in the sense of classical deontic logic.

IV

In a theory of levels, it is important to distinguish between grading according to fixed levels, on one hand, and merely comparative grading on the other. The first kind of grading means subsuming some particular performance under a level concept (for instance, "excellent", "good", "acceptable") according to the meaning criteria for this concept in the context under consideration(6). Merely comparative grading, on the other hand, means only establishing whether, according to comparative criteria, a performance is better or worse than other alternatives, i.e., it means merely establishing a preference order among alternatives(7).

The examples of section I-III have mainly suggested grading according to fixed levels and such grading is, indeed, a principal legal concern. If an action is deemed to be legally acceptable (in the sense that none of its features is prohibited), the level of acceptability referred to is a fixed level with special meaning criteria. And if, in some cases, we do distinguish between an ideal level and a level of minimal acceptability, each of these levels will have meaning criteria of its own (albeit vague, context-dependent and contested).

In the law there are, however, also frequent examples of merely comparative grading, i.e., cases where it is sufficient merely to establish a preference ordering among alternatives. Thus, there is the case where two judges in a court disagree on the decision of a hard case. Then, the decision advocated by the two judges are two alternatives and the disagreement between the judges will concern the (legal) preference ordering of these alternatives.

Also in merely comparative grading, we may use the notion of levels, since of two alternatives, *a* will attain a *higher* level than *b* if *a* is *preferred* to *b*. But in such a case, the notion of levels will be used merely for comparing alternatives in point of preference and the levels are not "fixed" in the sense referred to earlier.

When we speak of merely comparative grading as distinguished from grading according to fixed levels, the reason is that the latter kind of grading is a special case of comparative grading in general. In other words: Grading according to fixed levels is also comparative grading but the converse does not hold. If a is acceptable whereas b is not, then a is better than b. But from the statement that a is better than b we cannot, for instance, draw any conclusions as to whether b is acceptable or not. This is so, since grading according to fixed levels depends upon assumptions, which go beyond what is assumed in merely comparative grading(8).

In any comparative grading, the preference ordering of the alternatives will depend on some properties of these alternatives, viz. those properties that are relevant to the preference ordering in view. Thus, to each of the two dissenting judges in our earlier example, the very reason for preferring his own decision to that of his opponent will be that, in his opinion, it has certain properties which the alternative decision lacks. As is wellknown to any lawyer, the relevent properties may be of many different kinds: one of the decisions may be in accordance with the wording of a statutory rule, whereas the other decision is not, one of them may be supported by precedent whereas the other is not, one of them may be deemed to have some favorable causal consequences which the other will not have, and so on. It will emerge from our exposition in section V that if the judges do agree on which properties the alternatives actually have, their disagreement on the preference ordering will reflect a disagreement on the legal urgency, and the legal worth, of the properties in view.

The conceptual framework for requirements, urgency and worth, to be developed in section V, will only be concerned with comparative gradings in general and with comparative notions of levels, urgency and worth. The special additional features needed for a theory of fixed levels (and fixed degrees of urgency and worth, like "very urgent" etc.) will not be dealt with.

Since the idea of preference orderings among alternatives is a standard feature in preference theory, it is appropriate to indicate here that the approach to be adopted in this essay, focusing on the notions of requirements, urgency and worth, is different from the most common approach within economic and philosophical preference theory. In fact, it has more affinity with the ideas of the nineteenth century economist Carl Menger and some of his modern followers.

The value theory of Carl Menger and the early Austrian school of economists was based on ideas of the various "urgency" of different want satisfactions and on the various "importance" of disposing over different quantities of goods(9). In the mainstream of economic thought, however, these theories were supplanted, first by the theory of marginal utility and then by the theory of indifference curves(10).

The indifference curve technique was designed for comparisons between so-called commodity bundles such that a marginal loss in some component commodity can be compensated by a gain in another component, this resulting in the consumer's being indifferent between two bundles. The legal analogy of this would be to handle combinations of legal requirements as bundles of legal "goods", where a marginal "loss" in the fulfilment of one requirement can always be compensated by a sufficient "gain" in the fulfilment of some other requirement, this resulting in strict legal indifference between alternatives(11).

The approach adopted in the present essay, however, is rather to look at the requirements in each area of the law as a finite number of "constraints", ordered in point of their legal urgency. Such a view of legal requirements is more akin to the original ideas of Carl Menger and the early Austrian school. In modern preference theory, it is most closely connected with some branches of so-called lexicographic preference theory(12).

The theoretical framework developed in section V is not, however, lexicographic in the usual sense of the word.Rather, it may be seen as the first and simplest part(13) of a theory aimed at filling the gap between lexicographic preference theory, on the one hand, and legal theory and deontic logic, on the other.

V

Though it can be shown (see the appendix) that in the kind of situations in view the notions of preference, urgency and worth are, in fact, interdefinable, in this section I choose to start from the preference relation showing how the notions of level requirement, level guarantee, urgency and worth can be defined and what logical rules will hold for these notions. Thus, I shall start by assuming that a number of alternatives a, b, c, ... are ranked by a preference relation $\gtrsim$. The expression $\underline{a} \gtrsim \underline{b}$ is to be read "$\underline{a}$ is at least as good as $\underline{b}$". (Italics $\underline{a}$, $\underline{b}$, $\underline{c}$ ... are used as variables for alternatives among a, b, c,...) The relation of strong preference > ("better than") is defined, as usual, by saying that $\underline{a} > \underline{b}$ iff, i.e., if and only if, $\underline{a} \gtrsim \underline{b}$ but not $\underline{b} \gtrsim \underline{a}$. The indifference relation $\sim$ ("equally good as") is defined by saying that $\underline{a} \sim \underline{b}$ iff $\underline{a} \gtrsim \underline{b}$ and $\underline{b} \gtrsim \underline{a}$.

The alternatives may have many different properties but I assume that there is a finite number of properties F,G,H,.... relevant for the preference ordering in view. I shall use the expression F$\underline{a}$ for expressing that alternative $\underline{a}$ has property F.

A triple consisting of a set of alternatives a,b,c,..., a set of properties F,G,H,... and a preference ordering, will be called a preference structure, where the set of alternatives together with the set of properties is called the <u>situation</u> component in the structure.

The properties F,G,H,... primarily in view in a situation can be combined into <u>compound properties</u> by Boolean operators $\sim$, $\wedge$, $\vee$, $\supset$, $\equiv$, analogous to the usual connectives in sentential logic. Thus, for instance, $[F \wedge G]$, $[F \vee G]$, $[F \supset G]$, are compound properties which the alternatives will either have or lack. The definitions of compound properties are provided by the following scheme (where P, Q, are either primary properties or compounds of such):

$[\sim P]\underline{a}$	iff	not: $P\underline{a}$,
$[P \wedge Q]$	iff	$P\underline{a}$ and $Q\underline{a}$,
$[P \vee Q]\underline{a}$	iff	$P\underline{a}$ or $Q\underline{a}$,
$[P \supset Q]\underline{a}$	iff	$[\sim P]\underline{a}$ or $Q\underline{a}$,
$[P \equiv Q]\underline{a}$	iff	$P\underline{a}$ if and only if $Q\underline{a}$.

(For convenience, inner brackets will be dropped when there is no risk for a misunderstanding. Thus, for instance, [[~P] ∧ Q] will be written [~P ∧ Q].)

While negation, conjunction, disjunction etc. are operations forming new properties out of old ones, I shall also define two relations between properties. Thus (if P and Q are properties) I shall say that, in a situation, P _comprises_ Q iff for any alternative _a_ in this situation it holds that if P_a_ then Q_a_. Furthermore, I shall say that, in a situation, P _coincides_ with Q iff for any alternative _a_ in this situation it holds that P_a_ if and only if Q_a_.

By a property P being _considered_ in a situation (of some preference structure) I shall mean that the following holds for P in this situation:

a) P is either one among the properties F, G, H, ... primarily in view in the situation, or P is a compound of such properties; and
b) Among the alternatives a, b, c,... in the situation there is some alternative _a_ such that P_a_ and also some alternative _a_ such that not P_a_.

All the primary properties F, G, H,... in a situation are, naturally, considered in this situation. Some _compounds_ of these properties need not, however, be considered in the situation. For instance, if, with respect to some situation, it holds that P comprises Q, then the compound [P ⊃ Q] will not be a property considered in this situation and neither will the compound [P ∧ ~Q]. (This result follows, since any alternative in the situation will have property [P ⊃ Q] and no alternative in the situation will have property [P ∧ ~Q].)(14)

Among the complex properties considered in a situation, an important group consists of those properties that I shall call _basic compound_ properties. A basic compound formed from F,G,H,... is a compound formed by the "conjunction" operator ∧ by taking exactly one member from each of the pairs (F,[~F]), (G,[~G]), (H,[~H]), ..., as a conjunct. For instance, if F and G are the only properties considered, the basic compound properties will be [F ∧ G], [F ∧ ~G], [~F ∧ G], [~F ∧ ~ G]. If there are n primary properties, there will be 2^n basic compound properties, though it may be that not all of these are considered in the situation at hand. Each alternative in a situation will have exactly one basic compound property.

In the present paper, I shall only have in view such situations and preference orderings where a number of basic assumptions are fulfilled. These assumptions are as follows:

I The preference relation $\gtrsim$ between the alternatives is transitive and connected, i.e., it is a so-called weak ordering(15).
II The number of primarily relevant properties F,G,H,... is finite.
III If alternatives a and b have exactly the same primarily relevant properties, then $a \sim b$.
IV There is an alternative a and an alternative b such that $a > b$.

For any situation and preference ordering fulfilling assumptions I-IV, it will hold that the alternatives can be arranged in preference classes such that:

1. every alternative belongs to exactly one such class;
2. $a \sim b$ iff both of a and b belong to the same preference class;
3. if a and b have the same basic compound property, they belong to the same preference class;
4. the number of preference classes is finite;
5. the preference classes can be numbered and referred to as classes $c_0, c_1, \ldots, c_n$ in such a way that any member of a class with a higher index is better than any member of a class with a lower index.

As an example, starting from just two primary properties F,G, we may, for instance, have the following ordering (where { a:....a } is an abbreviation for "the set of all a such that ....a"):

$$C_3 = \{ a : [F \wedge G]a \}$$
$$C_2 = \{ a : [F \wedge \sim G]a \}$$
$$C_1 = \{ a : [\sim F \wedge G]a \}$$
$$C_0 = \{ a : [\sim F \wedge \sim G]a \}$$

This would mean that any alternative with property $[F \wedge G]$ is better than any alternative with property $[F \wedge \sim G]$, which, in turn, is better than any alternative with property $[\sim F \wedge G]$, which, in turn, is better than any alternative with property $[\sim F \wedge \sim G]$.

Of the preference classes $c_0, c_1, \ldots, c_n$ in a situation I shall say that c_0 is non-positive, whereas all of $c_1, \ldots, c_n$ are positive preference classes.

The positive preference classes are used for defining what I shall call the level properties $L_1,\ldots,L_n$ of the alternatives. These properties are defined by the following scheme:

$$\begin{array}{lll} L_n\underline{a} & \text{iff } \underline{a} \text{ is at least as good as the member of} & c_n \\ \cdot & \text{-"-} & \cdot \\ \cdot & \text{-"-} & \cdot \\ \cdot & \text{-"-} & \cdot \\ L_1\underline{a} & \text{-"-} & c_1 \end{array}$$

Thus, the number of level properties will be finite and the level properties will be numbered from 1 to n, as the corresponding preference classes in the scheme. Property L_n is the highest level property while L_1 is the lowest one. It also follows that if L_j is higher than L_i ($1 \leq i \leq j \leq n$), then $L_j\underline{a}$ implies $L_i\underline{a}$, but not conversely.

In the examples just referred to with four preference classes c_0, c_1, c_2, c_3, we shall obtain the following result for the three level properties L_3, L_2 and L_1, corresponding to the positive preference classes c_3, c_2, and c_1:

$L_3\underline{a}$ iff $[F \wedge G]\underline{a}$;
$L_2\underline{a}$ iff $[F \wedge G]\underline{a}$ or $[F \wedge {\sim}G]\underline{a}$; i.e., iff $F\underline{a}$;
$L_1\underline{a}$ iff $[F \wedge G]\underline{a}$ or $[F \wedge {\sim}G]\underline{a}$ or $[{\sim}F \wedge G]\underline{a}$; i.e. iff $[F \vee G]\underline{a}$.

The next move concerns the definition (relative to a situation and a preference ordering) of the important notions of level characterization, level requirement and level guarantee. Using the abbreviations,

Char_i P for: P is an L_i-characterization,
Req_i P for: P is an L_i-requirement,
Guar_i P for: P is an L_i-guarantee,

the general definitions will run as follows ($1 \leq i \leq n$):

Char_i P iff: For any $\underline{a}$, $L_i\underline{a}$ if and only if $P\underline{a}$,
Req_i P iff: For any $\underline{a}$, if $L_i\underline{a}$ then $P\underline{a}$,
Guar_i P iff: For any $\underline{a}$, if $P\underline{a}$ then $L_i\underline{a}$.

(Thus, Char_i P means that L_i coincides with P, Req_i P that L_i comprises P and Guar_i P that P comprises L_i.)

Any situation fulfilling our basic assumptions will be such that for each L_i of this situation there is a primary or compound property P such that $Char_i$ P. Thus, in the example referred to, we have the following level characterizations:

$Char_3$ [F ∧ G],
$Char_2$ F,
$Char_1$ [F ∨ G].

Since, by definition, $Char_i$ P iff both Req_i P and $Guar_i$ P, for each L_i there will always be at least some P such that Req_i P and $Guar_i$ P. On the other hand, it does not hold generally that Req_i P implies $Guar_i$ P or, conversely, that $Guar_i$ P implies Req_i P. In fact, in the example, among the level requirements and level guarantees are the following ones (the lists are not exhaustive):

<u>Level requirements</u>
Req_3 [F ∧ G], Req_3 F, Req_3 G, Req_3 [F ∨ G],
Req_2 F, Req_2[F ∨ G],
Req_1 [F ∨ G] .

<u>Level guarantees</u>
$Guar_3$ [F ∧ G] ;
$Guar_2$ F, $Guar_2$ [F ∧ G],
$Guar_1$ [F ∨ G] , $Guar_1$ F, $Guar_1$ G, $Guar_1$ [F ∧ G].

And so, while, in the example, F is both a level 2 requirement and a level 3 requirement, G is only a level 3 requirement. And while F is both a level 2 guarantee and a level 1 guarantee, G is only a level 1 guarantee.

With respect to situations fulfilling our basic assumptions, the following rules will hold for level requirements and level guarantees:

<u>Rules for level requirements</u>
1. If P comprises Q, then: if Req_i P, then Req_i Q.
2. If Req_i P and Req_i Q, then Req_i [P ∧ Q].
3. If Req_i P, then not Req_i [~P].
4. If Req_i P, then Req_j ($1 \leq i \leq j \leq n$).

(It is easily seen that rules 1-3 for level requirements are basically the rules for "Obligatory" in deontic logic).

Rules for level guarantees

1. If Q comprises P, then: if $Guar_i$ P, the $Guar_i$ Q.
2. If $Guar_i$ P and $Guar_i$ Q, then $Guar_i$ [P ∨ Q].
3. If $Guar_i$ P, then not $Guar_i$ [~P].
4. If $Guar_j$ P, then $Guar_i$ P ($1 \leq i \leq j \leq n$).

Rules relating level requirements and level guarantees

1. If Req_i P, then not $Guar_j$ P ($1 \leq i < j \leq n$).
2. If Req_i P and $Guar_j$ Q, then Q comprises P ($1 \leq i \leq j \leq n$).

A further development of the theory will now be made by introducing the notions of the urgency and the worth of a property with respect to a situation and a preference ordering. We start from the following definitions (where P, Q, are primary or compound properties considered in the situation in view):

> P is at least as urgent as Q
> iff: For any i ($1 \leq i \leq n$), if Req_i Q, then Req_i P.
> P has at least as much worth as Q
> iff: For any i ($1 \leq i \leq n$), if $Guar_i$ Q, then $Guar_i$ P.

From the assumptions made concerning the preference relation $\gtrsim$ and its connection with the properties, it follows that each of the two relations now defined is a weak ordering of the properties P,Q,R,... considered in the situation at hand.

Since each of the two relations is a weak ordering we may simplify our notation by introducing two (ordinal) measures u and w, where u(P) indicates the urgency of P and w(P) indicates the worth of P, with respect to a situation and a preference ordering. The definitions are obvious:

> $u(P) \geq u(Q)$ iff: P is at least as urgent as Q; and,
> $w(P) \geq w(Q)$ iff: P has at least as much worth as Q.

(Here, $\geq$ is the relation "greater than or equal to" between non-negative integers, not to be confounded with $\gtrsim$ as the relation "as least as good as" between alternatives.) We get the following results:

"Greater than":

$u(P) > u(Q)$ iff: For some i ($1 \leq i \leq n$), Req_i and not Req_i Q.

$w(P) > w(Q)$ iff: For some i ($1 \leq i \leq n$), $Guar_i$ P and not $Guar_i$ Q.

"Equal to":

$u(P) = u(Q)$ iff: For any i ($1 \leq i \leq n$), Req_i P if and only if Req_i Q.

$w(P) = w(Q)$ iff: For any i ($1 \leq i \leq n$), $Guar_i$ P if and only if $Guar_i$ Q.

With respect to those situations and preference orderings fulfilling our basic assumptions, it obviously holds that if P coincides with Q, then $u(P) = u(Q)$ and $w(P)=w(Q)$. More interestingly, in such situations, each of the two notions of urgency and worth will obey a different basic principle governing its use. In order to state the principles in a concise way, if i, j, are two numbers (not necessarily different), we shall use the expression min(i,j) for denoting the smallest number among i,j ($\min(i,j)= i$ iff $i \leq j$). The basic principle for urgency, following from our assumptions, will then be the following one (where P, Q, are properties considered in the situation):

$$u([P \wedge Q]) = \min(u(P),u(Q)).$$

Informally: the urgency of a conjunction equals the urgency of its least urgent conjunct. From the basic principle, a number of other principles for urgency can be derived as corollaries. For instance, it follows that:

If $u(P) > u(Q)$, then $u(P) > u([P \wedge Q])$ and $u(Q) = u([P \wedge Q])$,

i.e., if P is more urgent than Q, then P is more urgent than $[P \wedge Q]$ and the urgency of $[P \wedge Q]$ equals the urgency of Q.

If P comprises Q, it follows that P coincides with $[P \wedge Q]$, and, hence, in this case, it holds that

$$u(P) = u([P \wedge Q]) = \min(u(P),u(Q)).$$

That is, if P comprises Q, then $u(P) = \min(u(P),u(Q))$.

From this, since obviously, $u(Q) \geq \min(u(P),u(Q))$, we obtain the following important principle:

If P comprises Q, then $u(Q) \geq u(P)$.

And it is a trivial matter to derive a number of corollaries such as, for instance:

$$u(P) \geq u([P \wedge Q]),$$
$$u([P \vee Q]) \geq u(P).$$

Here, the last statement is equivalent to:

$$u([P \vee Q]) \geq \max(u(P),u(Q)),$$

(where max (i,j) denotes the greatest number among i,j).

The basic principle for worth, different from the one stated for urgency, is as follows (where, again, P,Q, are properties considered in the situation):

$$w([P \vee Q]) = \min(w(P),w(Q)).$$

Informally: the worth of a disjunction of properties equals the worth of that disjunct which has least worth. From this principle, by procedures analogous to the ones just followed, we can derive a number of principles for worth as counterparts of the principles previously derived for urgency. Thus, we get:

If $w(P) > w(Q)$, then $w(P) > w([P \vee Q])$ and $w([P \vee Q])=w(Q)$,

(i.e., if P has more worth than Q, then P has more worth than [P ∨ Q] and the worth of [P ∨ Q] equals the worth of Q. Furthermore, we get:

If P comprises Q, the $w(P) \geq w(Q)$,

from which, in turn, it follows,

$$w(P) \geq w([P \vee Q]),$$
$$w([P \wedge Q]) \geq w(P).$$

And here, again, the last statement is equivalent to:

$$w([P \wedge Q]) \geq \max(w(P),w(Q)).$$

From the foregoing, it is apparent how the notions of urgency and worth will differ with respect to conjunction ($\wedge$) and disjunction ($\vee$). For conjunction, we have:

$$u([P \wedge Q]) = \min(u(P),u(Q)),$$
$$w([P \wedge Q]) \geq \max(w(P),w(Q)),$$

whereas for disjunction we have:

$$u([P \vee Q]) \geq \max\,(u(P),u(Q)),$$
$$w([P \vee Q]) = \min\,(w(P),w(Q)).$$

VI

The basic difference between urgency and worth, reflected in our definitions and basic principles for the two notions, may be briefly summarized in a pair of slogans: The urgency of a property depends on how bad its absence is bound to be; the worth of a property depends on how good its presence is sure to be.

In the absence of further argument it cannot be held that the notion of worth is more basic than the notion of urgency, or vice versa. In legal and political history, however, some reformers have concentrated on the <u>worth</u> of proposals, advocating the goodmaking characteristics of some reform to the benefit of present and future generations. Others, like Karl Popper with his program of so-called "piecemeal engineering", have rather advocated concentration on what is <u>urgent</u> such as rendering help to those who suffer and trying to prevent suffering(16).

Likewise, in moral and legal decision making, emphasis may be put on the legal or moral <u>worth</u> or on the moral or legal <u>urgency</u> of the features of a decision. Thus, some may argue that the moral or legal worth of benefiting many people is generally greater than the worth of benefiting one person. That is, if p_1, $p_2 \ldots$, p_n and q are persons it may be argued that even if the moral or legal worth of benefiting q somewhat exceeds the worth of benefiting any of the other persons $p_1, p_2, \ldots, p_n$, taken singly, still the worth of benefiting <u>all of</u> $p_1, p_2 \ldots, p_n$ is, in the case at hand, greater than the worth of benefiting person q.

This might be called an argument from "aggregate worth". But against this, the advocate of urgency may argue that if the moral or legal urgency of benefiting q is greater than the urgency of benefiting any of the other persons p_1, $p_2 \ldots p_n$, taken singly, then, surely, the urgency of benefiting q is greater than the urgency of benefiting <u>all of</u> p_1, $p_2 \ldots, p_n$. This argument (which accords with our principle that the urgency of a conjunction equals the urgency of its least urgent conjunct) might be called an argument from "individual urgency". And so it seems that argumentation starting from worth can be opposed by argumentation starting from urgency(17).

As indicated in the appendix, however, starting from a complete order of urgency, we can derive a corresponding order of worth, and vice versa. And also, starting from a complete ordering of either kind we may derive a preference ordering of alternatives corresponding to the worth ordering or urgency ordering we assumed at the beginning.

In the example just mentioned, once the opponents agree on a preference ordering among alternatives (i.e., decisions), their dispute is resolved. Therefore, it appears that a sufficient condition for obtaining a solution is that the complete urgency ordering of the advocate of urgency will generate the same complete worth ordering as the one held by his opponent advocate of worth. Then, the problem is solved because the opponents will have the same urgency ordering, the same worth ordering and the same preference ordering among alternatives. As to the question whether the urgency ordering of one party actually does correspond to the worth ordering of the opponent, it can be answered by applying the definitions indicated in the appendix.

APPENDIX: <u>Interdefinability between urgency, worth and preference</u>

A. Urgency and worth defined in terms of preference.

For preference structures fulfilling assumptions I-IV, the following definitions of urgency and worth are equivalent to those given in the text and the urgency or worth ordering will be a weak ordering of the properties P,Q,R,... considered in the situation:

i. $u(P) \geq u(Q)$ iff: For any $\underline{a}$ such that not $\underline{Pa}$ there is some $\underline{b}$ such that not $\underline{Qb}$ and $\underline{b} \gtrsim \underline{a}$.(18)

ii. $w(P) \geq w(Q)$ iff: For any $\underline{a}$ such that $\underline{Pa}$ there is some $\underline{b}$ such that $\underline{Qb}$ and $\underline{a} \gtrsim \underline{b}$.

B. Preference defined in terms or urgency.

Assuming a weak urgency ordering of properties P,Q,R,...considered in a situation (letters Φ , ψ , being used as variables for such properties) and assuming the logical principles for urgency stated in the text, the following two definitions are equivalent and the preference ordering obtained will be a weak ordering of the alternatives:

i. $\underline{a} \gtrsim \underline{b}$ iff: For any Φ such that not $\Phi\underline{a}$ there is some ψ such that not $\psi\underline{b}$ and $u(\psi) \geq u(\Phi)$.

ii. $\underline{a} \gtrsim \underline{b}$ iff: $u([\sim B(\underline{b})]) \geq u([\sim B(\underline{a})])$.

(The expressions $[B(\underline{a})]$, $[B(\underline{b})]$ denote the basic compound properties fulfilled by $\underline{a}$ and $\underline{b}$, respectively, and, hence $[\sim B(\underline{a})]$, $[\sim B(\underline{b})]$ denote the negations of those properties.)

C. Preference defined in terms of worth

Assuming a weak worth ordering of properties P,Q,R,... considered in a situation and assuming the logical principles for worth stated in the text, the following two definitions are equivalent and the preference order obtained will be a weak ordering of the alternatives:

i. $\underline{a} \gtrsim \underline{b}$ iff: For any Φ such that $\Phi\underline{b}$ there is some ψ such that $\psi\underline{a}$ and $w(\psi) \geq w(\Phi)$.

ii. $\underline{a} \gtrsim \underline{b}$ iff: $w([B\underline{a}]) \geq w([B(\underline{b})])$.

D. Worth defined in terms of urgency

Assuming a weak urgency ordering of properties P,Q,R,...considered in a situation and assuming the logical principles for urgency stated in the text, the following definition will yield a weak worth ordering of the same properties P,Q,R,...:

$w(P) \geq w(Q)$ iff: For any $\underline{a}$ such that $\underline{Pa}$ there is some *b* such that Q*b* and $u([\sim B(b)]) \geq u([\sim B(a)])$.

E. Urgency defined in terms of worth

Assuming a weak worth ordering of properties P,Q,R,... considered in a situation and assuming the logical principles for worth stated in the text, the following definition will yield a weak urgency ordering of the same properties P,Q,R,...:

> $u(P) \geq u(Q)$ iff: For any *a* such that not P*a* there is some *b* such that not Q*b* and $w([B(b)]) \geq w([B(a)])$.

NOTES

* I want to thank Mr.Jan Odelstad for helpful comments. I also gratefully acknowledge financial support from the Tercentenary Foundation of the Bank of Sweden.

(1) See, for instance, T.G.Roupas, 'The value of life', 7 *Philosophy and Public Affairs* (1978),p.154.

(2) R.Dworkin, *Taking Rights Seriously* (1977), pp.105 ff.

(3) Dworkin, op.cit., in particular Chapters 4 and 13.

(4) L.Nelson, *Vorlesungen über die Grundlagen der Ethik. Zweiter Band. System der philosophischen Ethik und Pädagogik* (1932). Cf. K.E.Tranøy, *On the Logic of Normative Systems,* vol. I-II (1953; diss. Cambridge, mimeographed) and 'Asymmetries in ethics', 10 *Inquiry* (1967), pp.351-372. On Nelson and Tranøy, see also R.Ohlsson, *The Moral Import of Evil* (1979).

(5) On this affinity and the so-called Kanger and Anderson constants, see B. Hansson, 'An analysis of some deontic logics', in R.Hilpinen (ed.), *Deontic Logic: Introductory and Systematic Readings* (1971) pp.128 ff.

(6) Cf. G.H.von Wright, *The Varieties of Goodness* (1963), Chapters I-VI.

(7) Cf.C.G.Hempel,*Fundamentals of Concept Formation in Empirical Science* (1952).

(8) In some contexts, a notion is used which might be called "the Ought of right choice". Thus, it is sometimes said than an agent ought to prefer *a* to *b* (or choose rather *a* than *b*) iff *a* is better than *b*. It might also be said that choosing *a* is acceptable iff

there is no better alternative. Such statements may be regarded as statements on second order grading. In fact, they will refer to a fixed level (acceptability) for grading choices, which, themselves are comparative gradings. Cf. S.Halldén, 'Jämförbarhet inom preferenslogiken' (Comparability within preference logic), in T.Pauli(ed.) Logic and Value: Essays Dedicated to Thorild Dahlquist (1970). Halldén uses principles of deontic logic as second order principles for deriving first order principles on comparability between alternatives.

(9) C.Menger, Grundsätze der Volkswirtschaftslehre (1871), Cap.3.

(10) See, for instance, T.W.Hutchison, A Review of Economic Doctrines 1870-1929 (1953) and J.Hicks, Value and Capital, 2nd ed. (1946) Ch.1.

(11) Cf.B.Schlink, Abwägung im Verfassungsrecht (1976), Kap.6.

(12) See, in particular, N.Georgescu-Roegen, 'Choice, expectations and measurability', 68 Quarterly Journal of Economics (1954), pp. 503-534; J.Encarnación, 'A note on lexicographic preferences', 32 Econometrica (1964), pp.215-217; P.Fishburn, 'Lexicographic orders, utilities, and decision rules', 20 Management Science (1974), pp.1442-1471.

(13) One of the limitations of the present framework is that it deals only with properties which the alternatives either have or lack, whereas it is not indicated how to handle such attributes which apply rather in a more-or-less fashion. These matters will be dealt with in a forthcoming paper where a more comprehensive framework is introduced.

(14) Both the notion of being considered and the notions of comprising and coinciding with are defined relative to a particular situation and, therefore, relative to a particular set of alternatives in view. The set of alternatives in view might be called the "horizon" in the situation. If this horizon is made wider or more restricted, the set of properties considered may change and there may be changes in what properties comprise or coincide with other properties. Firstly, as to what properties will be considered, suppose that in situation S there are alternatives with F as well as alternatives with ~F. Then, if S' differs from S by F being taken as given in S',

the result will be that in S' all alternatives have F while none has ~F, which means that F and ~F are not properties considered in S'. Secondly, as to what properties will comprise or coincide with other properties, suppose that in S there are alternatives with the compound property [F ∧ ~G] as well as alternatives lacking this property. Then, if S' differs from S by the fact that in S' there are no alternatives with property [F ∧ ~G), this will mean that in S' it holds (since any alternative having property F will also have property G) that F comprises G, while the same did not hold in situation S, where the horizon was wider.

(15) Transitivity means that for any a, b, c, if a $\gtrsim$ b and b $\gtrsim$ c then a $\gtrsim$ c. Connectedness means that for any a, b, a $\gtrsim$ b or b $\gtrsim$ a.

(16) K.Popper, The Open Society and Its Enemies, vol.I (4th ed.1962) p.235, note 6. Cf. p.284 f., note 2 and vol II (same ed.) p.304, note 62.

(17) The notion of urgency has recently been emphasized in moral philosophy by T.Scanlon, 'Preference and urgency', 72 Journal of Philosophy (1975) pp.655-669. Cf. A.Sen, 'Equality of What', in The Tanner Lectures on Human Values (1979), pp. 197-220; D.Parfit, 'Innumerate Ethics', 7 Philosophy and Public Affairs (1978) pp.285-301 (see, in particular, p.299). It should also be mentioned that many of the problems discussed in A.Gewirth, Reason and Morality (1978), are related to problems of urgency and worth in the sense of the present essay.

(18) Definition A.i. (as a definition of "P is at least as important as Q") together with the basic principle in the logic of urgency (formulated as: "the importance of a conjunction equals the importance of its least important conjunct") was presented by me at a seminar in Uppsala in March 1980. In a written communication to me after the seminar, Dr.Sven Danielsson made some objections to this earlier version of the theory. The present version, making use of the notion of levels and distinguishing between urgency and worth, was developed at a later stage.

THE PROPERTY RIGHT OF SWEDEN TODAY - OR A REQUIEM OVER AN OUTDATED WAY OF ARGUEING*

Carl Martin Roos

1. THE REPORT AIMS AT AN ANALYSIS OF ARGUMENTATION

The original purpose of this article was to show how the order of property is built up in Sweden today. After quite a lot of work on the subject I realized that if the results were not to become totally trivial I would have to choose a new point of view. A simple positive law analysis or purely sociological survey would mean no more than a review of what is already known and no new fuel would be brought to the debate concerning the right of property.

Gradually it turned out to be possible to obtain a few new points of view by tackling the subject from the angle of argumentation analysis. The purpose of my report then became to illuminate how the conception of property right and related expressions are used in different situations.

Especially interesting is the usage of the term in argumentative political language. Here I restrict myself to a context of legal politics i.e. when the discussion - at least in the end - concerns changes in the current legal system. (The study has created some problems, as suitable models and methods are missing. However, I want to thank professor Alexander Peczenik and professor Lars Heuman for their help and advice).

An authority in social science has questioned my way

* Due to circumstances beyond both the author's and the editors' control this paper could not be inserted into chapter 8, for which it was planned originally (AP/LL/BvR)

A. Peczenik et al. (eds.), Theory of Legal Science, 639–648.

of tackling the subject. It concerned, according to him, only a debate on how different phenomena should be termed. I cannot see that this criticism is accurate. On the contrary, argumentation analysis of political language has since long been appreciated as an important branch of research. A Swedish pioneer in the field writes as follows:

"Finally political language can be investigated from a logical or semantic viewpoint. This ought to be a politically neutral task. On the other hand it is doubtful whether it will be neutral in the sense of refraining from taking a stand on the quality of the language in question. The purpose of such an investigation is to discover the logical relations between the terms, the linguistic functions of certain interesting expressions, the logical pattern of the types of argumentation, the different kinds of ambiguities and vaguenesses and other characteristics". (Gunnar Fredriksson, _Det politiska språket_, Stockholm 1962 s. 12).

2. THE CONCEPT OF PROPERTY RIGHT HAS DIFFERENT FUNCTIONS IN DIFFERENT CONNECTIONS

From now on I will distinguish between three different ways of using the concept of property right.

First, property right has a purely _technical function_ in the legal system of rules. This legal and technical concept of property right is fundamental and requires a detailed treatment.

In the extension of the legal concept of property right we find a _descriptive function_, where property right indicates a relation between persons and things. In this way different social and economic relations can be explained.

The most interesting application of the word property right occurs as mentioned when the expression has an _argueing_ function. This implies that the person who uses it, refers to values that are connected with the established order of property. It is in this sense that property right occurs in different manifestations of civil rights. The property right is also referred to in the debate of legal policy as an argument for desired actions and decisions. It cannot be excluded that the argueing function is sometimes decreasing, so that property right is used as a slogan with a purely _persuading or demagogic_ purpose without the ambition of rational argumentation.

3. THE PROPERTY RIGHT IS AN INSTRUMENT OF TECHNIQUE

In legal philosophy much energy has been devoted to explaining which function property right and other concepts of rights have in the judicial language. There is no great agreement in this field of research, but still I dare, with an amateur's right, conclude what seems to be established as universally valid knowledge. When the concept of property right is included in legal rules of property it does not refer to facts in the visual world. For concepts like metre and kilo there are substantial measures and prototypes. He who says that he has access to the prototype of property right will not be trusted in the first instance. This is what is meant by property right lacking semantical reference. (compare K.Olivecrona, Law as fact, 2 ed. London 1971).

If you regard property right as a component in the legal property language system, the concept might be seen as a bridge ("connecting words", Olivecrona op. cit. p.199) between legal facts and legal consequences, as a way of linking certain assumptions to certain consequences. If, for instance, I buy a car, inherit it or receive it as a gift etc. I obtain a property right to the car. This involves, in its turn, that I can count on the protection of the legal system if the car is stolen or damaged, that the car must be registered as property for tax purposes, that the car is at my disposal etc.

The concept of property right in that way becomes a language-simplifier: I do not have to tell how I got the car and to line up all consequences of my acquisition. The information that I own or have the property right to the car is sufficient.

This theory is since long common knowledge. (See G.Lantz, Eigentumsrecht - ein Recht oder ein Unrecht? 1977 p. 27 with further references to works by A.Ross, T.Eckhoff, P.O.Ekelöf and A. Wedberg). It might be interesting to examine its consequences.

The theory of property right as a link between complexes of rules means, in the first place, that property right in itself has no given meaning, since it is not certain that it is always the same kind of rules that are tied together. There are for instance entirely different rules about how to sell, acquire or have at one's disposal, on the one hand, real estates, and, on the other hand, e.g. cars. The property right to real estate is subject to great restrictions in the owner's disposal. Acquisition demands a written form, registration and sometimes permission from an authority. Cars can be acquired without forms, and the owner can make use of his car rather freely, even if the necessity of a driver's licence, duty of registration and the compulsory tax and insurance involve certain formalities and certain restrictions. In spite of those far-reaching differences it is property right in both cases. Even if the meaning, to a certain extent, is different, there is a common core, which includes, according to Hedenius, a certain amount of property protection and a certain amount of right of disposition ('Analysen av äganderättsbegreppet' in _Filosofi och rättsvetenskap_, ed. B.Belfrage and L.Stille, Lund 1975 p.48 f).

Thus, the concept of property right does not get a precise meaning until it is clear which _object_ it concerns.

The property right can, according to the legal rules, only involve legal objects: personal property and real property. Persons cannot be owned in a legal sense. To make a link between persons a _contract_ like marriage, employment, joint-venture or something similar is required.

The property right also varies between different _subjects_. Thus the possibilities of disposition are fewer for those who are under age or are lacking special permissions or qualities needed to have entire disposal of the property concerned.

It is also important to point out that acquisition of property right is not the only way to enjoy the property. Leasing of a real estate is a form that lies near at hand. It differs from property right because of a certain limit in time. Another important difference is that the price for using property is paid in a different way than in the case of property right.

On the other hand, the tenant is protected legally as if the leased property was his own. In a cooperative building society the parallel to property right is more obvious. Here a right of disposition is connected to the membership in the cooperative association.

Property right cannot be used as a protection against results of personal creativity being used by others. Here special forms are assigned, for instance in patent and copyright law. Sometimes the terminology of property has been used here as well, but that has no positive support in the Swedish legal system. An interesting special case are the rules on property right to the shares of a company. While in other cases property right involves the disposal of physical objects, it here also involves - through the voting rules on a company's general meeting - not only power over enterprise property but also power over employed labour. The property right involves legally - except the right to make use of the share itself - a possibility to exercise power over those who have been connected to the company by employment contract. In spite of this difference there is a clear resemblance to other instances of property right. You could perhaps say that property right to shares belongs to a larger "family concept" of property right (compare Lantz op. cit. p.30).

In conclusion, property right is only used in connection with certain objects, that is goods and real property. The legal status of the owner may vary considerably depending on the object owned. There are alternative means to attain practically the same position to respective types of property as the property right may give.

4. ACCORDING TO CERTAIN LEGAL RULES PROPERTY RIGHT IS THE OBJECT OF PROTECTION

Evidently the concept of property right has another function than that of a purely legal technique in rules on protection of property right. A typical rule is that of the European Convention which states that the right of physical and legal persons to their property should not be violated. (Art.1).

It is evident that the Convention refers to something concrete when using the expression property right. The concept is more than a simple link between legal rules. As the rule exists in an international convention it can hardly refer to rules of property in any single country.

According to natural law there should be a clear and concise ideological content in the concept, which should be valid for every national system of law. The idea that legal rights can be filled with contents in this way has been denied in legal philosophy.(See Lantz op.cit. p. 24 f with references, Olivecrone op.cit.).

Theories on civil rights are, as Gunnar Fredriksson

points out, queerly expressed recommendations for the social system that the inventor of the theory likes (op.cit.p.12). As different authors like different things there is no homogeneous ideological content in the concept of property right. A Danish scholar - O.A.Borum - has referred to a magician's hat. You cannot take out anything else from the concept that you have put in before.

The theory of Hedenius for instance gives a better understanding of the concept. According to him property right has at least two dimensions, protection of possession and right of disposition. Within these dimensions there is a sliding scale with full possession and full disposition as ideals on one end of the scale. How far you can slide along the scale without the right ceasing to be property right is not decided by this theory.

According to this theory the rule in the European Convention on protection of the property right would be meaningless or at least too loose to be a useful tool of protection. In my view you can only get a meaningful content of the rule by constructing it as a reference to an internationally acceptable standard of the rules of property. You should consequently stop on the sliding scale somewhere where the rules of property of most countries end up. In this way the rule will not be especially precise but at least have some meaning. This reasoning seems to be applicable to the case Sporrong - Lönnrot, where the majority of the European Court considered application of certain Swedish building law contrary to Art. 1. (See T.Opsahl and E.Fribergh in Svensk Juristtidning 1983, especially p. 411 f).

Because of these difficulties in determining the concept of property in protection rules of conventions and constitutions, there have been attempts to find other formulations. The Swedish principle on protection of property right in Ch. 2 Art. 18 in the Swedish Constitution does not contain any general rule on protection of property right but only a formulation that compensation will take place if property is expropriated or subjected to a similar procedure.

In the UN Convention on human rights it seems that the difficulties have been circumvented by the probably meaningless sentence that anybody has the right to own property. This rule does not say anything about the desired standard of the property right.

5. THE CONCEPT OF PROPERTY RIGHT IS USED IN ARGUMENTATION ON LEGAL POLICY

The concept of property right is most often used in a rationally argumenting or persuasive function when used outside the legal system of rules. This use of the concept however is far from consistent. It may refer to other value systems than that mentioned in connection with the European Convention, that is good international standard.

As mentioned the concept of property right is often used in persuasive functions. In the Swedish Conservative Party these was about 1960 argumentation for so-called "owners' democracy". This message on property right seemed to be directed to those who owned property and not to those who managed to live on their salary, on loans and renting their homes, thus avoiding to invest in property. Therefore it is in a way contradictory to use the term democracy, especially as there was no intention to distribute the property to more persons. The aim of this promotion of property right was to strengthen the status of property owners without any corresponding change for those who own no or minimal property. This promotion of property right has now practically completely disappeared from Swedish political debate.

As mentioned above the property right to shares, however, has partly another dimension than the right to other property, in so far as there is a possibility for power over people directly connected to shareholder ownership through the rules of voting right at a company's general meeting. A field where property right is still in the centre of the debate and argumentation is the stock market.

Many authors from Renner onwards have pointed out the extension of the concept of property right in connection with the industrialization in the 19th century. This innovation consisted in a gain of legitimacy for industry and contributed to its growth and at the same time the employees were excluded from the property right to the result of their respective work (see Abrahamsson - Broström, _Om arbetets rätt_, 1979, with references). Considering what I just mentioned, it is in a way misleading to use the concept of property right in argumentation of legal policy if you at the same time suppress the power aspect. The property right to shares includes so much more than acquisition and disposition of goods and real property - finally it is a question of the power over industry that is involved.

One example of how inadequate, dangerous argumentation concerning property right to shares may occur. A very common phrase in Sweden is that a "broader ownership" to shares is to be aimed at. Phrases like that give the impression of a struggle to get a more even distribution of shareholding to bigger groups of citizens. In reality the increased spreading of shares only means that industry gets more capital without any distribution at all of the actual power. The control of a company lies as a rule very secure with professional groups of companies and shareholders with well established and even legally secured position (see Roos, _Avtal och rösträtt,_ 1969, especially p. 35 ff. with references).

6. THE PROPERTY RIGHT IS A BAD FOUNDATION FOR ARGUMENTATION OF LEGAL POLICY

I will try to collect the scattered views above in a comprehensive picture of the concept of property right in argumentation of legal policy. What is lacking is a background consisting of an analysis of material where such argumentation takes place. However, I presume that most sentences are of the type "means an intrusion on the property right to B that is attributed to C. (1).

a. It is evident that the concept of property right in sentences like (1) has some meaning and does not consist only in a joining link or language simplifier. This latter function is attributed to the concept of property right in argumentation of legal technique, e.g.: "Because of C's acquisition of the ownership to B... the sanction A will follow (2)". Sentence (2) is in contrast to sentence (1) a statement considering the contents of valid law.

b. Sentence (1) means instead that valid rules must not be changed because of the contents of property right. Statement (1) is consequently founded on an ideal of the contents of property right.

c. As the contents of property right changes between different objects, different subjects and different rulings of law, you cannot find any precise content in this idea from the contents of valid law.

d. Nor can you find a definite content of the idea of property right outside the legal rulings, provided that you

accept the criticism of theories of national law.
e. The concept of property right in sentence (1) has no more substance than follows from the theory of Hedenius, i.e. two dimensions where you cannot find an agreement on the extension of those dimensions. The concept of property right in this theory has no normative meaning (Lantz op.cit.).
f. From what is said above follows that argumentation with the concept of property right as instrument is characterized by a vagueness that increases with the difference in ideological views that may exist between the sender and the receiver of the message. Another way of putting it: the concept of property right only functions as a distinct bearer of meaning in argumentation between persons who have the same ideological presuppositions i.e. the same opinion on what property right should be.
g. It is presumed here that rational methods of argumentation are those where clearness in the formation of concepts is helpful to an exchange between persons with different values. If this is right, the method of property right will not be amenable to fit to rational argumentation, but could rather be described as a way of "flagging values". Similar criticism may also be directed towards concepts of the type "workers' right" if they form parts of sentences like (1) instead of property right.

A better precision and higher rationality in argumentation of legal policy may be achieved if you use concepts where there is greater agreement between persons with different values on their meaning. A discussion on for instance whether <u>the consequences</u> of a certain measure are adequate from different points of view will therefore appear more meaningful than a debate considering whether measure A. conforms to "property right" or "workers' right" respectively. It is presumed that it is an advantage in rational argumentation to be able to convince the opposite party and reach solutions satisfying to both parties. If the aim of a certain argument is mainly to "flag values", it is on the other hand rational to use concepts of rights.

7. THE PROPERTY RIGHT IS PARTICULARLY BAD AS A FOUNDATION FOR ARGUMENTATION CONCERNING THE PROPERTY OF SHARES.

Ownership of shares includes not only power over the shares themselves but potentially also power over the company involved. As mentioned above this is a slide of meaning that is linked with growing industrialism. The difficulties in using the concept of property in rational argumentation should reasonably increase with this complexity. Here the presence of different categories of shareholders with different factual control over the company can be mentioned (on investing shareholders and shareholders in control see i.a. Roos op.cit. especially p. 36).

Thus my conclusion is that the concept of property right is especially inappropriate for rational argumentation concerning the property of shares. The number of vague and unclear dimensions and components in the concept will in this case become greater than in case of property right to objects. The reasoning implies, as mentioned, that rational argumentation means argumentation that may lead to agreement on practical measures between persons with different values. However, my impression is that rational argumentation in this sens is not always aimed at when it concerns the question of the power over companies.

LIST OF PARTICIPANTS

* Indicates contributors to this volume.

* Aarnio, Aulis. University of Helsinki, Faculty of Law, 36270 Huutijärvi, Finland.
* Balekjian, W.H.. University of Glasgow, Glasgow G12 800, Scotland.
  Bankowski, Z.. Faculty of Law, Old College, South Bridge, Edinburgh EH8 9YL Scotland.
* Ben-Menahem, Hanina. Department of Law, The Hebrew University of Jerusalem, Mount Scopus, Jerusalem, Israel.
* Ben-Menahem, Yemina. Department of Philosophy, The Hebrew University of Jerusalem, Mount Scopus, Jerusalem. Israel.
* Bergholtz, Gunnar. Lund University, Faculty of Law, Fersens väg 18, S 211 42 Malmö, Sweden.
* Biagioli, C. Instituto per la Documentazione giuridica, Via Panciatichi 56-16, 50127, Firenze, Italy.
  Bjarup, J.. Faculty of Law, Århus University, Universitetsparken bygn. 340, DK-8000 Århus C, Denmark.
  Blegvad, B.M.. Soldalen 7, DK-2100 Copenhagen, Denmark.
  Blegvad, M.. Soldalen 7, DK-2100 Copenhagen, Denmark.
  Bogdan, M.. Faculty of Law, University of Lund, Box 1165, S-221 05 Lund, Sweden.
  Bondesson, U.. University of Copenhagen, Denmark.
  Bramklev, L.. Faculty of Law, University of Lund, Box 1165, S-221 05 Lund, Sweden.
* Broekman, J.M. Katholieke Universiteit Leuven, Centrum voor Grondslagenonderzoek van het Recht, Tiensestraat 41, B-3000 Leuven, Belgium. Free University of Amsterdam, Van de Boechorststraat 1, 1081 b. Amsterdam, The Netherlands.
* Castañeda, H.N., Indiana University, Bloomington, Indiana 47 405, USA.
  Cederberg, K.. Pettersbergsgården, Industrigatan 16 V, S-112 46, Stockholm, Sweden.
* Cervera, Alejo de. University of Puerto Rico, Faculty of Law, Rio Piedras, Puerto Rico 00931, USA.
* Dalberg-Larsen, Jørgen. Faculty of Law, Århus University, Universitetsparken 340, 8000 Århus C, Denmark.
* Dini, G. Insentituto per la Documtazione giuridica, Via Panciatichi 56-16, 50127, Firenze, Italy.
  Egaña, J.L. Cea. Universidad Catolica de Chile, Casilla 114, D Santiago, Chile.
* Ekelöf, Per Olof. Uppsala University, Faculty of Law, Nybrogatan 40, 11440 Stockholm, Sweden.

* Evers, Jan. Lund University, Faculty of Law, S.Esplanaden 11, 223 54 Lund, Sweden.

Falk, Per. Hälsogränd 26, S-902 62 UMEÅ, Sweden.

Falkemark, G.. University of Gothenburg, Vasagatan 3, S-411 24, Göteborg, Sweden.

* Flodin, Mats. University of Lund, Faculty of Law, Artillerigatan 99, 11530 Stockholm, Sweden.

Flower, R.J.. Le Moyne College, Syracuse, New York 12214 USA.

Fogelklou, Anders. Faculty of Law, University of Uppsala, Box 512, S-751 20, Uppsala, Sweden.

Forell, U.. University of Copenhagen, Vagtelvej 25, DK-2000 Copenhagen F, Denmark.

Forte, David F.. Cleveland State University, Cleveland-Marshall, College of Law, Cleveland, Ohio 44115, USA.

Frändberg, Åke. Säves väg 11, S-752 63, Uppsala, Sweden.

Garzón Valdés, E.. Hohenzollernstrasse 19, D-5300 Bonn, Federal Republic of Germany.

* Golding, Martin, P.. Law School and Philosophy Department, Duke University, Durham, North Carolina 27708, USA.

Grzegorczyk, C.. 103, rue Brancas, 92310 Sèvres, France.

* Guastini, Riccardo. Institute of Philosophy of Law, University of Genova, Via L. Montaldo, 22/161 37 Genova, Italy.

Gunnarsson, B.L.. (FUMS), Hamnesplanaden 5, S-753 23, Uppsala, Sweden.

* Hanen, Marsha, P.. Faculty of General Studies, University of Calgary, 2500 University Drive N.W., Calgary, Alberta, Canada T2N 1N4.

* Hansson, Bengt. Department of Philosophy, Lund University, Kungshuset, Lundagård, S-22350 Lund, Sweden.

Hansson, Bo. Sandgatan 8, S-223 50 Lund, Sweden.

* Hellner, Jan. Faculty of Law, University of Stockholm, Engelbrektsgatan 18, 114 32 Stockholm, Sweden.

* Hermerén, Göran. Department of Philosophy, Lund University, Kungshuset, Lundagård, 223 50 Lund, Sweden.

Hetzler, A. Faculty of Sociology of Law, University of Lund, Sweden.

* Holmström, G. University of Helsinki, Department of Philosophy, Unionsgatan 40 B, 00170 Helsinki, Finland.

Huffman, James L.. The Lewis & Clark Law School, 10015 S.W.Terwilliger Blvd., Portland Oregon 97219 USA.

Höglund, B.. Department of Economics, University of Lund, Sweden.

Jacobsen, E.. Bogø Hovedgade 20, DK-4793 Bogø by, Denmark.

Jacobsson, U.. Faculty of Law, University of Lund, Box 1165, S-221 05 Lund, Sweden.

Jareborg, N.. Faculty of Law, University of Uppsala, Box 512, S-751 20, Uppsala, Sweden.

* Jørgensen, Stig. Århus University, Faculty of Law, Fyrrevaenget 5, 8541 Skødstrup, Denmark.

* Kamstra, Otto W.M.. University of Amsterdam, Faculty of Law, O.Z.Achterburgwal 217-219, 1012 DL Amsterdam, The Netherlands.

* Kerchove, Michel van de, Facultés Universitaires Saint-Louis, Faculté de droit, Boulevard du Jardin Botanique 43, 1000 Bruxelles, Belgium, Lecturer at the Centre universitaire de Luxembourg.

Klami, H.T., University of Turku, Ratsukonkatu 1 as. 23, 20880 Turku 88, Finland.

Klette, H.. Faculty of Sociology of Law, University of Lund.

Knudsen, Ø.Krøgersvej 9, DK-2930 Klampenborg, Denmark.

* Kunneman, Frank. University of Amsterdam, Faculty of Law, O.Z.Achterburgwal 217-219, 1012 DL Amsterdam, The Netherlands.

Lachmayer, F.. Tigergasse 12/12, A-1080 Vienna, Austria

Larsson, S..University of Stockholm, Faculty of Law, Fiskartorpsvägen 160 C, S-106 91 Stockholm, Sweden.

* Leader, S.. University of Reading, Department of Law, Old Whiteknights House, Whiteknights, Reading RG6 2 AH, England.

* Lenoble, J.. Catholic University of Louvain, Faculty of Law, Collège Thomas More, Place Montesquieu 2, B-1348 Louvain-la-Neuve, Belgium.

* Lindahl, L.. University of Lund, Faculty of Law, Elias Fries Park 14, S-223 60 Lund, Sweden.

Luizzi, V.. South-West Texas State University,San Marcos,Texas 78666 USA.

* MacCormick, Neil. The Faculty of Law, The University of Edinburgh, Department of Jurisprudence, The Old College, Edinburgh EH8 9YL, Scotland.

* Mattsson, Mats. University of Lund, Ängelholmsgatan 6, S-214 22 Malmö, Sweden.

Medel, J.L.. Cea Bermúdez, 68, Madrid - 03 - Spain.

Moberg, K.. Faculty of Law, University of Lund, Box 1165, S-221 05, Lund, Sweden.

* Modéer, K.Å.. Faculty of Law, University of Lund, Box 1165, S-221 05, Lund, Sweden

Nelson, A.. Faculty of Law, University of Uppsala, Box 512, S-751 20, Uppsala, Sweden.

* Opałek, Kazimierz. The Jagiellonian University, Faculty of Law, ul. Mazowiecka 2/3, 30-036 Kraków, Poland.

* Ost, François. Lecturer at the Facultés universitaires Saint-Louis (Brussels). Dean of the Law Faculty, Boulevard du Jardin Botanique, 43, 1000 Bruxelles.

* Pattaro, E.. University of Bologna, Faculty of Law, Instituto Giuridico "Antonio Cicu", Via Zamboni, 27/29, 40126 Bologna, Italy.

* Peczenik, Aleksander. University of Lund, Faculty of Law, Källarekroken 34, S-222 47 Lund, Sweden.

Pieck, M.. Creighton University, School of Law, 2500 California Street, Omaha, Nebraska 68178 USA.

Pöyhönen, J.. Lönebostállsg. 7 E 55, SF-00240 Helsinki 24, Finland.

Rehbinder, M.. Institutet för offentlig rätt, Regeringsgatan 11-13, 00100 Helsinki 10, Finland.

* Roermund, B. van. Faculty of Law, Tilburg University, Postbox 90153, 5000 LE Tilburg, The Netherlands.

* Roos, C.M.. Faculty of Law, University of Lund, Box 1165, S-221 05, Lund, Sweden.

* Rottleuthner, Hubert. Freie Universität Berlin, Fachbereich Rechtswissenschaft, Institut für Grundlagen und Grenzgebiete des Rechts, Boltzmannstrasse 3, D-1000 Berlin 33 (West).

* Schött, Roland. Lund University Law School, Box 1165, S-221 05 Lund, Sweden.

Shekawat, V.. Department of Philosophy, University of Rajastan, Jaipur 302004 India.

* Sintonen, Matti. The Academy of Finland and University of Helsinki, Department of Philosophy, Unioninkatu 40B, 00170 Helsinki 17, Finland.
* Skogh, Göran. University of Lund, Department of Economics, Hagtorsgränd 9, S-223 56 Lund, Sweden.

Socci Natali, F.. Instituto per la Documentazione giuridica, Via Panciatichi, 56-16, 50127 Firenze, Italy.

* Strömberg, Tore. Lund University, St. Jörgens park 1, S-223 60 Lund, Sweden.

Summers, R.S.. Cornell University Law School, Myron Taylor Hall, Ithaca, New York 14853, USA.

Sundberg, J.W.F.. Uggleviksg. 9, S-114 27 Stockholm, Sweden.

* Tiscornia, D.. Instituto per la Documentazione giuridica,Via Panciatichi, 56-16, 50127 Firenze, Italy.

Ullerstam, M.. Magistratsvägen 55 N, S-222 44 Lund, Sweden.

* Varga, Csaba. Institute for Legal and Administrative Sciences of the Hungarian Academy of Sciences, H-1250 Budapest, P.O.Box 25, Hungary.
* Villa,V.. Instituto di Filosofia e Teoria Giuridica e Politica Palermo University, Via Stabile 250, 90141 Palermo, Italy.
* Weinberger, Ota. University of Graz, Department of Philosophy of Law, Universitätsstrasse 27, A-8010 Graz, Austria.
* Wild, A.H. de. Evenkamp 19,. 6041 PC Roermond, The Netherlands.
* Wróblewski, Jerzy. University of Łódź, Faculty of Law, Rectorate, ul. Narutowicza 65, 90-131 Łódź, Poland.
* Zirk-Sadowski, Marek. University of Łódź, Faculty of Law, ul. Narutowicza 65, 90950 Łódź, Polen.
* Zuleta Puceiro, E.. Universidad Nacional de Buenos Aires Universidad, complutense de Madrid, Hermosilla 57, 2°, Madrid 1, Spain.

INDEX OF NAMES

INDEX OF NAMES

A

Aarnio, A. 2, 3, 4, 21, 24, 25, 34, 35, 37, 38, 43, 46, 47, 50, 51, 117, 118, 119, 136, 138, 143, 163, 168, 171, 172, 192, 227, 232, 250, 272, 273, 344, 384, 424, 508, 515, 517, 520, 548, 571, 579, 580
Achinstein, P. 520
Agge, I. 570
Åqvist, L. 379
Albert, H. 340, 341, 343
Alchain, 479
Alchourrón, C.E. 39, 42, 48, 49, 50, 51, 144, 508, 548, 549, 571, 578, 583, 605
Aldisert, R.J., 235, 249
Alexy, R. 117, 118, 119, 160, 161, 163, 171, 218, 227, 233, 270, 271, 272, 323, 332, 528, 548, 580
Allen, L.E. 605
Allen, J.W. 421
Allot, A. 137, 143
Almén, T. 377, 378, 379, 381, 382, 464
Anderson, A.R. 548, 636
Andrup, H.H. 496
Apel, K.O. 333
Atiyah,P. 445, 446, 448, 449, 453
Aubert, V. 310, 314, 495, 496
Audi, R. 313
Aune, B. 95
Austin,J.L. 166, 206, 543, 548, 554, 555, 559, 569

B

Bacon, A. 421
Bacon, F. 416, 422
Balekjian, W.H. 67, 183
Balzer, W. 40, 49, 51
Barnes, B. 36, 37, 453
Barry, B. 293
Bengtsson, B. 465
Ben-Menahem, H. 384, 425
Ben-Menahem, Y. 384, 425
Bentham, 166, 451, 545, 554, 555
Bergh, van den G.C.J.J., 62, 63, 64
Bergholtz, G. 192, 307
Berkowitz, D.S. 421, 422

Berman, H.J XIV, 452,
Biagini, P.M. 587
Biagioli, C. 587, 605
Black, D. 336
Black, M. 453
Blackwood, W. 423
Blake, R. 421
Blumer, H. 62, 65
Bobbio, 515, 520
Boëthius, 148, 159
Böhme, G. 57, 63, 65
Bohr, N. 45
Bolingbroke, 413
Boole, 592
Bowen, C.D. 419, 422
Boyle, 71
Bramstäng, 381, 382
Brand, 161
Brierley, J.E.C. 420, 422
Broad, C.D. 379
Broekman, J.M. 64, 65, 118, 119, 133, 143, 144, 158, 159, 160
Bubner, R. 161
Bulygin, E. 39, 42, 48, 49, 50, 51, 144, 163, 167, 168, 169, 172, 271, 508, 548, 571, 578, 583, 594, 605
Bunge, M. 98, 117, 118
Burgley, 421
Burke, 413, 422
Burtham, 157
Bydlinski, F. 317, 357, 358, 359, 362, 381

C

Campbell, 227
Cappelletti, M. 466
Carcaterra, 581
Carnap, R. 8, 39, 48, 49, 51, 223, 232
Cassirer, E. 371
Castañeda, H.N. 68, 70, 71, 95, 98, 116, 117, 119 147, 156, 157, 158, 159, 160, 161
Castberg, F 128
Cervera, A. de 383, 385
Charles, 71
Christensen, B. 379
Christina, (Queen) XI

Ciampi, C. 270
Cohen, M.R. 188
Cohen, R.S. 519
Coke, E. 384, 413, 414, 415, 416, 417, 418, 419, 420, 423
Coleridge, 422
Commaille, J. 508
Cone, R.A. 188
Conte, A.G. 579, 581
Corbin, A.L. 188
Cossio, C. 98, 117
Cranfield, M. 420
Crawford, 479
Crombag, 65
Croce, B. 363, 371, 372
Cross, R. 289, 290
Cueto-Rua, J.C. 272
Cujacio, 14

D

Dacre, 421
Dalberg-Larsen, J. 36, 37, 483, 487, 495, 496
Danielson, S. 638
David, R. 420, 422
Davis, J.W. 548
Degenkamp, T.H. 63
Delorme, A. 548
Denzin, K. 62, 65
Derham, D.P. 188
Derrida, J. 172
Descartes, R. XI, XII, 122,
Dewey, 270, 355
Dietze, G. 378
Dini, G. 378, 587, 605
Diplock, 292
Disraeli 413, 421, 423
Domat 14
Doneau 14
Douglas-Home, A. 423
Dreier, R. 36, 37
Dubischar, R. 495
Duhem, P. 7, 12, 431
Dummett, M. 7, 12
Durkheim, 524
Dworkin, R. 128, 178, 182, 191, 192, 218 270, 276, 277,

278, 279, 280, 281, 282, 284, 285, 290, 291, 293, 304, 337, 343, 355, 445, 617, 620, 636

E

Eckhoff, T. 118, 376, 378, 381, 566, 568, 569, 570
Ehrlich, E. 525, 526, 527, 531
Einstein, 210, 433
Eisenmann, Ch. 507
Ekelöf, O. 316, 317, 357, 376, 381
Elizabeth, I. 413, 415, 420, 421, 414
Elton, G.R. 420, 423
Encarnación, J. 637
Engels, 524
Englard, I. 466
Evers, 551, 552, 553
Eyben, B. von 466

F

Faralli, C. 370
Fassó, G. 317, 363, 364, 365, 366, 367, 368, 369, 370, 371, 372
Faye, J.P. 142, 144
Feigl, H. 118
Feyerabend, P. 212, 213, 214, 519
Fischer, 208, 214
Fishburn, P. 637
Fish, S. 286, 287
Fleming. 466
Flodin, M. 383, 384, 401, 411
Fløistad, G. 172
Foriers, 272
Foucault, M. 172
Fraenkel, 527
Frändberg, Å. 379
Frank, J. 296, 426
Franken, H. 64, 65
Frege, 49
Freud, S. 210
Fried, C. 442, 443, 444, 447, 449, 452, 453
Friedman, L.M. 452, 547
Friedrichs, R.W. 23
Fuller, L. 128, 422, 453

G

Gadamer, H.G. 7, 12, 161, 172

Galileo, G. 122
Geiger, T. 488
Gentile, 370
Geraets, T.F. 270
Gerber, 16
Gewirth, A. 638
Giddens, A. 517, 520
Gillis-Wetter, J. 272
Gilmore, G. 452
Gilmour, I. 423
Glotz, P. 37
Golding, M. 192, 295
Goldsmith, R. 479
Gooderson, R.N. 433, 435
Goodman, N. 116, 450, 453
Gorla, G. 272
Granger, G.G. 187, 188
Gray, C. 417, 418, 423
Greaves, R.L. 420
Greef, J. de 170, 172
Griffiths, J. 343
Grimm, D. 547
Grönfors, K. 404, 410, 464, 465
Groot, De 60, 62, 65
Grotius, H. XIII, 401, 402, 404, 409,
Guastini, R. 552, 571, 579, 580, 581
Gusdorf, G. 508
Gutting, G. 23, 24, 37
Gijssels, J. 65, 66

H

Habermas, J. 7, 12, 163, 171, 173, 218, 227, 502, 506, 507
Hägerström, A. 98, 116, 374, 375, 376, 379, 380, 485, 538
Hailsham, 422
Halldén, S. 637
Haller, R. 38
Hanen, M. 437, 441
Hanson, N.R. 49, 51, 192, 298, 299
Hansson, B. 1, 4, 5, 536
Hare, R.M. 87, 96, 154, 173, 580
Harris, J.W. 34, 35, 37
Hart, H.L.A. 116, 128, 164, 166, 167, 173, 271, 289, 290, 336, 337, 497, 498, 499, 507, 514, 518, 520, 552, 559, 568, 569, 578, 580

Hayek, F.A. von 373, 378, 380, 422
Heck, P. 357
Hegel, 321, 523
Heidegger, 122
Hellner, J. 438, 455, 465, 466, 496
Hempel, C.G. 636
Henson, R.D. 188
Henry VIII 414
Heraclitus 364
Herlitz, N. 423
Herméren, G. 189, 195, 214
Hesse, M. 454
Hicks, J. 637
Hilbert, 49
Hill, C. 421
Hilpinen, R. 96, 581, 636
Hintikka, J. 50, 51
Hippel, von E. 466
Hirschleifer, 479, 480
Hobbes, T. XIII, 384, 401, 402, 404, 410, 417, 418,
Hobsbaum, E.J. 421
Hockney, D.R. 548
Hoecke, M. van 63, 66, 508
Hohfeld, 157, 559, 561, 569
Holmes, O.W. XIV, XV, 452,
Holmström, G. 583, 607
Holzhey, H.
Hondius, E.H. 465
Horne, R. 423
Hubien, H. 579
Hume, 87, 88, 96, 247, 365
Hutcheson, 295, 296, 637

I

Ihering, R. von, 16, 19, 21, 22, 24

J

Jääskinen, N. 37
James I 413, 417
Jansson, J.M. 119
Jellinek, 554
Jensen, S. 567, 569
Jochimsen, R. 508
Johansson, R. 479
Jørgensen, S. 69, 121, 496

K

Kahn-Freund, 527,
Kalinowski, G. 164, 173, 480, 507, 548
Kamstra, O. 4, 53
Kanger, S. 157, 607, 611, 612, 613, 615
Kant, I. 76, 88, 90, 122, 247, 364, 371, 410, 522, 523
Keat, R. 188
Keenan, E. 161
Kekule, 297
Kelsen, H. 16, 23, 69, 118, 119, 146, 147, 159, 160, 163, 164, 165, 166, 167, 173, 188, 383, 384, 401, 408, 409, 411, 506, 525, 527, 545, 548, 552, 568, 571, 578, 580
Kenyon, J.P. 421
Kerchove, M. van de 483, 497, 508
Kisch 63
Klanderman, J.H.M. 62, 63, 66
Klein, B. 479, 481
Kneale, W.M. 159, 160
Knorr, K.D. 36, 37, 38
Koch, J. 312
Kordig, C.R. 192, 299
Koschaker, P. 24
Kotarbínski, T. 227, 232
Kowalski, J. 547
Krawietz, 28, 35, 135, 143, 159, 160, 270, 495
Krajewski, W. 50, 51
Kratzer, 159, 160
Krohn, R 37, 38
Kronman, 479, 481
Krüger, L. 35, 38
Kuhn, T. 3, 4, 5, 6, 7, 8, 10, 11, 12, 19, 25, 26, 27, 29, 33, 34, 35, 36, 37, 38, 39, 40, 49, 51, 124, 125, 132, 143, 301, 302
Kulcsàr, K. 181
Kunneman, F. 4, 53
Kunz, 1

L

Lacan, J. 171, 172
Ladmiral, J.R 508
Lagerroth, F 423
Lakatos, I. 118, 287, 293, 342, 434

Lamm, V. 181
Lang, W. 117
Langdell, C.C. 441, 452
Langemeijer, G.E. 63
Langer, S.K. 188
Laudan, L. 12, 342, 344
Lauridsen, P. 342, 344
Leader, S. 191, 192, 275
Leatherdale, W.H. 454
Lemon, V.R. 292
Lenin, 12
Lenoble, J. 69, 70, 163
Lessing, G.E. 232
Lewis, D.K. 147, 159, 160
Lindahl, L. 116, 157, 159, 160, 161, 411, 437, 551, 583, 584, 607, 611, 612, 615, 617
Lindblom, C.E. 324, 325, 332
Llewellyn, K.N. 272, 531, 533, 546, 547
Locke, J. 374
Lockyer, 422
Lorenz, K. 227, 232
Lucas, J. 118
Luce, D. 480, 481
Luhmann, N. 528, 547
Lyons, D. 243, 244, 251
Lyotard, J.F. 172

M

Macauly, S. 474, 547
Maccormick, N. 159, 190, 191, 228, 233, 235, 250, 270, 272, 273, 276, 292, 310, 314, 316, 317, 347, 348, 349, 350, 351, 352, 353, 354, 355, 359, 360, 361, 362, 518, 520
Macneil, J. 449, 453
Maistre, de 422
Malewski, A. 546
Malherbe, J.F. 508
Malinowski, B. 480
Marcuse, 208, 214
Marc-Wogan, K. 117
Marshall, T. 293
Martino, A.A. 272, 583, 587, 605
Marx, K. 12, 126, 524
Masaryk, 227
Mattson, M. 318, 373, 382

Maxwell, G. 118
Mazzarese, T. 580
Menger, C. 624, 637
Meuwissen, D.H.M. 65
Mey, M. de 55, 56, 58, 63, 64, 65
Micheli, G.A. 272
Mill, 297
Milsom, S.F.C. 420
Modéer, K.A. XI
Molesworth, W. 423
Montesquieu, 522, 523, 524
Morgan, 592
Morris, 250
Morscher, E. 581
Moulines, C.U. 50, 51
Mulder, 62, 66
Musgrave, A. 118, 293, 343

N
Nagel, E. 312
Nagel, I.H. 293, 311
Nelson, L. 621, 636
Neumann, F. 527
Neurath, O. 547
Newton-Smith, W.H. 453
Newton, 210, 433
Niiniluoto, I. 6, 37, 38, 50, 51, 163, 168, 172, 173, 344, 571, 579
Nilstun, T. 214
Nino, C.S. 165, 173
Nowak, L. 331, 333, 546
Nussbaum, 527

O
Odelstad, J. 636
Ofstad, H. 117
Ohlsson, R. 636
Olbrechts-Tyteca, L. 232
Olivecrona, K. XIII, 117, 118, 119, 376, 380, 381, 551, 553, 554, 556, 557, 569, 570
Opalek, K. 484, 531, 534, 546, 548, 549, 579, 581
Oppenheim, P. 118
Ost, F. 483, 497, 508

P

Palmade, G. 507
Parfit, D. 638
Paton, G.W. 188
Pattaro, E. 317, 363, 371
Pauli, T. 379, 637
Pearce, D. 50, 51, 240
Peczenik, A. 18, 21, 23, 24, 35, 38, 68, 69, 97, 117, 118, 119, 133, 135, 136, 140, 143, 144, 160, 163, 171, 172, 189, 190, 218, 270, 272, 315, 342, 343, 383, 483, 519, 520, 546, 548, 580
Perelman, Ch. 63, 66, 218, 226, 227, 232, 270, 271, 272, 273
Persson, I. 202, 203
Peteri, Z. 181
Petrazycki, L. 485, 531, 536, 538, 544, 547
Phillips, D.L. 519
Piaget, 65
Plato, 212, 321
Plucknett, T.F.T. 420, 421, 423
Pocok, J.G.A. 24
Podgorecki, A. 547
Poincaré, H. 86, 88, 96
Polak, J.M. 63, 66
Popper, K. 6, 7, 12, 17, 117, 212, 213, 214, 300, 364, 365, 384, 425, 426, 428, 429, 430, 431, 434, 435, 432, 433, 633, 638
Posner, R.A. 466, 469, 479, 481
Pound, R. 181, 545
Pöyhönen, J. 35, 37, 38
Pufendorf, S. XII, XIII
Putnam, H. 7, 12, 118

Q

Quaritch, B. 420
Quine, W.V.O. 98, 116, 117

R

Radbruch, G. 128
Radnitzky, G. 55, 63, 66, 519
Raffa, H. 481
Ragnemalm, H. 382
Raiffa, 480
Rantala, V. 50, 51
Rawls, J. 218, 231, 284, 286, 401, 403, 404, 410
Raz, J. 165, 167, 173, 318, 373, 374, 378, 380

Rehbinder, M. 547
Reid, 241
Reiter, B.J. 453
Resweber, J.P. 508
Reynolds, F. 465
Rice, D.A. 466
Riesenfeld, S. 465
Ringhofer, K. 161
Robert, A. 172
Robilant, D.I.E. 520
Rodhe, 464
Roeder, K.D. 232
Roermund, G.C., van, 67, 69, 145, 158, 160, 161
Rombach, J. 63, 66
Ross, A. 163, 165, 166, 173, 310, 376, 488, 511, 515, 520, 564, 565, 566, 567, 568, 569, 570, 571, 579, 580
Ross, W.D. 91, 92, 96, 313
Rottleuthner, H. 484, 495, 521, 547
Roupas, T.G. 636
Rousseau, J.J. 401, 403, 404, 410
Rowse, A.L. 420
Rudden, 349
Russell 176
Rystad, G. 421

S

Salisbury, 417
Savigny, F.K., von 1, 14, 15, 17, 19, 20, 23, 545
Scanlon, T. 638
Scarpelli, U. 580
Schlink, B. 637
Schmidt, A.H.J. 64
Schmidt, F. 379
Schmidt, R. 116
Schnur, R. 421
Schött, R. 384, 413, 421
Schotter, A. 480, 481
Schreier, F. 224, 225, 232
Schubert, G. 309, 313, 314
Schumpeter, 389, 394
Scriven, M. 118
Scruton, R. 422, 423
Searle, J.R. 580
Segre, C. 581

Selznick, Ph. 181, 548
Sen, A. 638
Shapere, D. 49, 51
Shklar, 136
Sills, D.L. 181
Silvers, S. 158
Simon, H. 298, 300, 301, 302
Sintonen, M. 3, 35, 39, 50, 52
Sinzheimer, 527
Skelton, A.N. 423
Skinner, Q. 417, 421
Skogh, G. 438, 469
Smith, A. 326
Sneed, J. 3, 7, 12, 40, 49, 51, 52, 342, 343
Socci Natali, I. 587
Soeteman, 63, 66
Sorokin, P.A. 547
Southgate, H. 422
Spelman, H. 419
Spencer, 524
Spengler, 139
Stachey, L. 420
Stavenow, L. 421
Stegmüller, W. 3, 7, 12, 40, 49, 52,
Stephens, 355
Stranzinger, R. 581
Strömberg, T. 117, 411, 551, 552, 559, 570
Stromhölm, S. 379, 490, 491, 495
Struck, G. 270
Stuart, J. 416
Summers, R.S. 119, 144, 313, 316, 343, 345, 355, 525
Sundberg-Weitman, B. 382
Sundby, 566, 567, 568, 569, 570
Swan, J. 446, 447, 448, 449, 453

T

Tammelo, I. 332
Tarski, M. 270, 271
Taruffo, M. 271, 272
Telser, L. 479, 482
Teubner, G. 143
Thorne, S.E. 420, 423
Thornstedt, H. 382
Tiscornia, D. 587
Tomberlin, J. 95, 96

Toulmin, S. 49, 52, 133, 143, 218
Toynbee, A.J. 138, 144
Tranoy, K.E. 636
Trevor-Roper, H.R. 421
Tuchman, B. 420

U
Urmson, J.O. 548
Urry, J. 188
Uusitalo, J. 34, 37, 38, 172

V
Varga, C. 175, 181, 182
Velden, van der 62, 66
Veljanovski, C. 479, 482
Viehweg, 218
Villa, V. 35, 343, 344, 484, 509

W
Wall, P.M. 433
Walter, R. 161
Walton, D. 161
Wartofsky, M.W. 519
Wasserstrom, 273, 297
Weber, M. 138, 177, 181, 523, 524
Wedberg, A. 343, 579
Weinberger, C. 232, 233
Weinberger, O. 160, 189, 205, 218, 231, 232, 233, 250
Weingart, P. 35, 36, 38
Werin, L. 466
Westerlind, P. 464
White, 415, 420, 422, 423
Whitley, R.D. 35, 37, 38
Wichelhaus, R. van 443, 444, 452
Wieacker, Fr. 158, 160
Wilberforce, 250
Wild, A.H. de 24, 63, 65, 118, 143, 315, 335, 343
Williamson, O. 479, 482
Williston, 452
Wilson, W.K. 548
Windscheid, 554
Wingfield-Stratford, E. 418, 419, 420, 421, 422, 423
Wittgenstein, 9, 118, 145, 158, 227, 320, 350, 519
Wolenski, J. 548
Wolin, S. 24

Wright, G.H. von, 39, 48, 117, 147, 548, 571, 579, 581, 636
Wróblewski, J. 190, 191, 253, 270, 271, 272, 273, 313, 332, 333, 546, 579, 580
Wronkowska, S. 546

Y
Yildirim, 188
Young, K. 423

Z
Zagorin, P. 420
Zaller, R. 420
Ziembinski, Z. 580
Ziokowski, M. 332
Zirk-Sadowski, M. 315, 319
Zitting, S. 46
Zuchini, G. 371
Zuleta Puceiro, E. 2, 13, 23, 35, 36, 38, 140, 144, 272
Zweigert, K. 465